12**

CRM SERIES

Centro
di Ricerca
Matematica
Ennio De Giorgi

Ovidiu Costin
Mathematics Department
The Ohio State University
231 W. 18th Avenue
Columbus, Ohio 43210, USA
costin@math.ohio-state.edu

Frédéric Fauvet
Département de mathématiques - IRMA
Université de Strasbourg
7, rue Descartes
67084 Strasbourg CEDEX, France
fauvet@math.u-strasbg.fr

Frédéric Menous
Département de mathématiques, Bât. 425
Université Paris-Sud
91405 Orsay CEDEX, France
Frederic.MENOUS@math.u-psud.fr

David Sauzin
CNRS Paris
and
Scuola Normale Superiore
Piazza dei Cavalieri 7
56126 Pisa, Italia
sauzin@imcce.fr
david.sauzin@sns.it

Asymptotics in Dynamics, Geometry and PDEs; Generalized Borel Summation vol. II

edited by
O. Costin, F. Fauvet,
F. Menous, D. Sauzin

EDIZIONI
DELLA
NORMALE

© 2011 Scuola Normale Superiore Pisa

ISBN: 978-88-7642-376-2
e-ISBN: 978-88-7642-377-2

Contents

Introduction xi

Authors' affiliations xv

Christian Bogner and Stefan Weinzierl
Feynman graphs in perturbative quantum field theory 1
1 Introduction . . . 1
2 Perturbation theory . . . 2
3 Multi-loop integrals . . . 6
4 Periods . . . 7
5 A theorem on Feynman integrals . . . 8
6 Sector decomposition . . . 10
7 Hironaka's polyhedra game . . . 12
8 Shuffle algebras . . . 13
9 Multiple polylogarithms . . . 17
10 From Feynman integrals to multiple polylogarithms . . . 19
11 Conclusions . . . 22
References . . . 23

Jean Ecalle
with computational assistance from S. Carr
The flexion structure and dimorphy: flexion units, singulators, generators, and the enumeration of multizeta irreducibles 27
1 Introduction and reminders . . . 30
1.1 Multizetas and dimorphy . . . 30
1.2 From scalars to generating series . . . 32
1.3 ARI//GARI and its dimorphic substructures . . . 35
1.4 Flexion units, singulators, double symmetries . . . 36

1.5 Enumeration of multizeta irreducibles 37
1.6 Canonical irreducibles and perinomal algebra . . 38
1.7 Purpose of the present survey 38

2 Basic dimorphic algebras 40
2.1 Basic operations 40
2.2 The algebra ARI and its group $GARI$ 45
2.3 Action of the basic involution *swap* 48
2.4 Straight symmetries and subsymmetries 49
2.5 Main subalgebras 52
2.6 Main subgroups 53
2.7 The dimorphic algebra $ARI^{\underline{al}/\underline{al}} \subset ARI^{al/al}$. . . 53
2.8 The dimorphic group $GARI^{\underline{as}/\underline{as}} \subset GARI^{as/as}$. 54

3 Flexion units and twisted symmetries 54
3.1 The free monogenous flexion algebra $Flex(E)$. 54
3.2 Flexion units 57
3.3 Unit-generated algebras $Flex(E)$ 61
3.4 Twisted symmetries and subsymmetries in universal mode 63
3.5 Twisted symmetries and subsymmetries in polar mode 67

4 Flexion units and dimorphic bimoulds 70
4.1 Remarkable substructures of $Flex(E)$ 70
4.2 The secondary bimoulds $ess^\bullet$ and $esz^\bullet$ 79
4.3 The related primary bimoulds $es^\bullet$ and $ez^\bullet$ 87
4.4 Some basic bimould identities 88
4.5 Trigonometric and bitrigonometric bimoulds . . 89
4.6 Dimorphic isomorphisms in universal mode . . . 94
4.7 Dimorphic isomorphisms in polar mode 95

5 Singulators, singulands, singulates 99
5.1 Some heuristics. Double symmetries and imparity 99
5.2 Universal singulators $senk(ess^\bullet)$ and $seng(es^\bullet)$. 101
5.3 Properties of the universal singulators 102
5.4 Polar singulators: description and properties . . . 104
5.5 Simple polar singulators 105
5.6 Composite polar singulators 105
5.7 From $\underline{al}/\underline{al}$ to $\underline{al}/\underline{il}$. Nature of the singularities . 106

6 A natural basis for $ALIL \subset ARI^{\underline{al}/\underline{il}}$ 107
6.1 Singulation-desingulation: the general scheme . 107
6.2 Singulation-desingulation up to length 2 111
6.3 Singulation-desingulation up to length 4 112
6.4 Singulation-desingulation up to length 6 112
6.5 The basis $lama^\bullet/lami^\bullet$. 116

6.6 The basis $loma^{\bullet}/lomi^{\bullet}$ 116
6.7 The basis $luma^{\bullet}/lumi^{\bullet}$ 117
6.8 Arithmetical vs analytic smoothness 118
6.9 Singulator kernels and "wandering" bialternals . 119
7 A conjectural basis for $ALAL \subset ARI^{\underline{al}/\underline{al}}$.
The three series of bialternals 120
7.1 Basic bialternals: the enumeration problem . . . 120
7.2 The regular bialternals: *ekma, doma* 120
7.3 The irregular bialternals: *carma* 121
7.4 Main differences between regular and irregular bialternals 121
7.5 The *pre-doma* potentials 123
7.6 The *pre-carma* potentials 124
7.7 Construction of the *carma* bialternals 126
7.8 Alternative approach 127
7.9 The global bialternal ideal and the universal 'restoration' mechanism 129
8 The enumeration of bialternals.
Conjectures and computational evidence 130
8.1 Primary, sesquary, secondary algebras 130
8.2 The 'factor' algebra *EKMA* and its subalgebra *DOMA* 132
8.3 The 'factor' algebra *CARMA* 133
8.4 The total algebra of bialternals *ALAL* and the original BK-conjecture 133
8.5 The factor algebras and our sharper conjectures . 133
8.6 Cell dimensions for *ALAL* 135
8.7 Cell dimensions for *EKMA* 135
8.8 Cell dimensions for *DOMA*. 136
8.9 Cell dimensions for *CARMA* 136
8.10 Computational checks (Sarah Carr) 137
9 Canonical irreducibles and perinomal algebra 139
9.1 The general scheme 139
9.2 Arithmetical criteria 144
9.3 Functional criteria 144
9.4 Notions of perinomal algebra 147
9.5 The all-encoding perinomal mould $peri^{\bullet}$ 149
9.6 A glimpse of perinomal splendour 150
10 Provisional conclusion 152
10.1 Arithmetical and functional dimorphy 152
10.2 Moulds and bimoulds. The flexion structure . . . 154

10.3 $ARI/GARI$ and the handling of double symmetries . . . 158
10.4 What has already been achieved . . . 160
10.5 Looking ahead: what is within reach and what beckons from afar . . . 164
11 Complements . . . 165
11.1 Origin of the flexion structure . . . 165
11.2 From simple to double symmetries. The *scramble* transform . . . 167
11.3 The bialternal tesselation bimould . . . 168
11.4 Polar, trigonometric, bitrigonometric symmetries 171
11.5 The separative algebras $Inter(Qi_c)$ and $Exter(Qi_c)$ 175
11.6 Multizeta cleansing: elimination of unit weights 180
11.7 Multizeta cleansing: elimination of odd degrees . 189
11.8 $GARI_{se}$ and the two separation lemmas . . . 192
11.9 Bisymmetrality of $ess^\bullet$: conceptual proof . . . 193
11.10 Bisymmetrality of $ess^\bullet$: combinatorial proof . . 195
12 Tables, index, references . . . 198
12.1 Table 1: basis for $Flex(E)$ . . . 198
12.2 Table 2: basis for $Flexin(E)$ . . . 202
12.3 Table 3: basis for $Flexinn(E)$ . . . 202
12.4 Table 4: the universal bimould $ess^\bullet$ . . . 205
12.5 Table 5: the universal bimould $esz_\sigma^\bullet$ . . . 206
12.6 Table 6: the bitrigonometric bimould $taal^\bullet/tiil^\bullet$ 207
12.7 Index of terms and notations . . . 207
References . . . 210

Augustin Fruchard and Reinhard Schäfke
On the parametric resurgence for a certain singularly perturbed linear differential equation of second order 213
1 Introduction . . . 213
2 The distinguished solutions . . . 216
3 Stokes relations for the functions . . . 221
4 Behavior near the turning points . . . 223
5 The residue . . . 226
6 Stokes relations for the wronskians . . . 228
7 Stokes relations for the factors of the wronskians . . . 229
8 Resurgence of the series $\widehat{r}(\varepsilon)$ and $\widehat{\gamma}^{\pm}$ . . . 232
9 Resurgence of the WKB solution . . . 238
10 Remarks and Perspectives . . . 240
References . . . 241

Shingo Kamimoto, Takahiro Kawai, Tatsuya Koike
and Yoshitsugu Takei
On a Schrödinger equation with a merging pair of a simple pole and a simple turning point — Alien calculus of WKB solutions through microlocal analysis 245
References 253

Yoshitsugu Takei
On the turning point problem for instanton-type solutions of Painlevé equations 255
1 Background and main results 255
2 Transformation near a double turning point 261
2.1 Exact WKB theoretic structure of $(P_{\mathrm{II,deg}})$ 261
2.2 Transformation theory to $(P_{\mathrm{II,deg}})$ near a double turning point 265
3 Transformation near a simple pole 269
References 273

Introduction

In the last three decades or so, important questions in distinct areas of mathematics such as the local analytic dynamics, the study of analytic partial differential equations, the classification of geometric structures (*e.g.* moduli for holomorphic foliations), or the semi-classical analysis of Schrödinger equation have necessitated in a crucial way to handle delicate asymptotics, with the formal series involved being generally divergent, displaying specific growth patterns for their coefficients, namely of Gevrey type. The modern study of Gevrey asymptotics, for questions originating in geometry or analysis, goes together with an investigation of rich underlying algebraic concepts, revealed by the application of Borel resummation techniques.

Specifically, the study of the Stokes phenomenon has had spectacular recent applications in questions of integrability, in dynamics and PDEs. Some generalized form of Borel summation has been developed to handle the relevant structured expansions – named transseries – which mix series, exponentials and logarithms; these formal objects are in fact ubiquitous in special function theory since the 19th century.

Perturbative Quantum Field Theory is also a domain where recent advances have been obtained, for series and transseries which are of a totally different origin from the ones met in local dynamics and yet display the same sort of phenomena with, strikingly, the very same underlying algebraic objects.

Hopf algebras, *e.g.* with occurrences of shuffle and quasishuffle products that are important themes in the algebraic combinatorics community, appear now natural and useful in local dynamics as well as in pQFT. One common thread in many of the important advances for these questions is the concept of resurgence, which has triggered substantial progress in various areas in the near past.

An international conference took place on October 12th – October 16th, 2009, in the Centro di Ricerca Matematica Ennio De Giorgi, in Pisa, to highlight recent achievements along these ideas.

Here is a complete list of the lectures delivered during this event:

Carl Bender, *Complex dynamical systems*

Filippo Bracci, *One resonant biholomorphisms and applications to quasi-parabolic germs*

David Broadhurst, *Multiple zeta values in quantum field theory*

Jean Ecalle, *Four recent advances in resummation and resurgence theory*

Adam Epstein, *Limits of quadratic rational maps with degenerate parabolic fixed points of multiplier* $e^{2\pi i q} \to 1$

Gérard Iooss, *On the existence of quasipattern solutions of the Swift-Hohenberg equation and of the Rayleigh-Benard convection problem*

Shingo Kamimoto, *On a Schrödinger operator with a merging pair of a simple pole and a simple turning point, I: WKB theoretic transformation to the canonical form*

Tatsuya Koike, *On a Schrödinger operator with a merging pair of a simple pole and a simple turning point, II: Computation of Voros coefficients and its consequence*

Dirk Kreimer, *An analysis of Dyson Schwinger equations using Hopf algebras*

Joel Lebowitz, *Time Asymptotic Behavior of Schrödinger Equation of Model Atomic Systems with Periodic Forcings: To Ionize or Not*

Carlos Matheus, *Multilinear estimates for the 2D and 3D Zakharov-Rubenchik systems*

Emmanuel Paul, *Moduli space of foliations and curves defined by a generic function*

Jasmin Raissy *Torus actions in the normalization problem*

Javier Ribon, *Multi-summability of unfoldings of tangent to the identity diffeomorphisms*

Reinhard Schäfke, *An analytic proof of parametric resurgence for some second order linear equations*

Mitsuhiro Shishikura, *Invariant sets for irrationally indifferent fixed points of holomorphic functions*

Harris J. Silverstone, *Kramers-Langer-modified radial JWKB equations and Borel summability*

Yoshitsugu Takei, *On the turning point problem for instanton-type solutions of (higher order) Painlevé equations*

Saleh Tanveer, *Borel Summability methods applied to PDE initial value problems*

Jean-Yves Thibon, *Noncommutative symmetric functions and combinatorial Hopf algebras*

Valerio Toledano Laredo, *Stokes factors and multilogarithms*

Stefan Weinzierl, *Feynman graphs in perturbative quantum field theory*
Sergei Yakovenko, *Deflicity of intersections between trajectories of ODEs and algebraic hypersurfaces*
Michael Yampolsky, *Geometric properties of a parabolic renormalization fixed point*

The present volume, together with a first one already published in the same collection, contains five contributions of invited speakers at this conference, reflecting some of the leading themes outlined above.

We express our deep gratitude to the staffs of the Scuola Normale Superiore di Pisa and of the CRM Ennio de Giorgi, in particular to the Director of the CRM, Professor Mariano Giaquinta, for their dedicated support in the preparation of this meeting; all participants could thus benefit of the wonderful and stimulating atmosphere in these institutions and around Piazza dei Cavalieri. We are also very grateful for the possibility to publish these two volumes in the CRM series. We acknowledge with thankfulness the support of the CRM, of the ANR project "Resonances", of Université Paris 11 and of the Gruppo di Ricerca Europeo Franco-Italiano: Fisica e Matematica, with also many thanks to Professor Jean-Pierre Ramis. All our recognition for the members of the Scientific Board for the conference: Professors Louis Boutet de Monvel (Univ. Paris 6), Dominique Cerveau (Univ. Rennes), Takahiro Kawai (RIMS, Kyoto) and Stefano Marmi (SNS Pisa).

Pisa, June 2011

Ovidiu Costin, Frédéric Fauvet,
Frédéric Menous and David Sauzin

Authors' affiliations

CHRISTIAN BOGNER – Institut für Theoretische Physik E, RWTH Aachen, D - 52056 Aachen, Germany
bogner@physik.rwth-aachen.de

JEAN ECALLE – Département de mathématiques, Bât. 425 Université Paris-Sud, 91405 Orsay Cedex - France
Jean.Ecalle@math.u-psud.fr

AUGUSTIN FRUCHARD – LMIA, EA 3993, Université de haute Alsace, 4, rue des Frères-Lumière, 68093 Mulhouse cedex, France
augustin.fruchard@uha.fr

SHINGO KAMIMOTO – Graduate School of Mathematical Sciences, University of Tokyo, Tokyo, 153-8914, Japan
kamimoto@ms.u-tokyo.ac.jp

TAKAHIRO KAWAI – Research Institute for Mathematical Sciences, Kyoto University, Kyoto, 606-8502, Japan

TATSUYA KOIKE – Department of Mathematics, Graduate School of Science, Kobe University. Kobe, 657-8501, Japan
koike@math.kobe-u.ac.jp

REINHARD SCHÄFKE – IRMA, UMR 7501, Université de Strasbourg et CNRS, 7, rue René-Descartes, 67084 Strasbourg cedex, France
schaefke@math.u-strasbg.fr

YOSHITSUGU TAKEI – Research Institute for Mathematical Sciences, Kyoto University, Kyoto, 606-8502, Japan
takei@kurims.kyoto-u.ac.jp

STEFAN WEINZIERL – Institut für Physik, Universität Mainz, D - 55099 Mainz, Germany
stefanw@thep.physik.uni-mainz.de

Feynman graphs in perturbative quantum field theory

Christian Bogner and Stefan Weinzierl

Abstract. In this talk we discuss mathematical structures associated to Feynman graphs. Feynman graphs are the backbone of calculations in perturbative quantum field theory. The mathematical structures – apart from being of interest in their own right – allow to derive algorithms for the computation of these graphs. Topics covered are the relations of Feynman integrals to periods, shuffle algebras and multiple polylogarithms.

1 Introduction

High-energy particle physics has become a field where precision measurements have become possible. Of course, the increase in experimental precision has to be matched with more accurate calculations from the theoretical side. As theoretical calculations are done within perturbation theory, this implies the calculation of higher order corrections. This in turn relies to a large extent on our abilities to compute Feynman loop integrals. These loop calculations are complicated by the occurrence of ultraviolet and infrared singularities. Ultraviolet divergences are related to the high-energy behaviour of the integrand. Infrared divergences may occur if massless particles are present in the theory and are related to the low-energy or collinear behaviour of the integrand.

Dimensional regularisation [1–3] is usually employed to regularise these singularities. Within dimensional regularisation one considers the loop integral in D space-time dimensions instead of the usual four space-time dimensions. The result is expanded as a Laurent series in the parameter $\varepsilon = (4 - D)/2$, describing the deviation of the D-dimensional space from the usual four-dimensional space. The singularities manifest themselves as poles in $1/\varepsilon$. Each loop can contribute a factor $1/\varepsilon$ from the ultraviolet divergence and a factor $1/\varepsilon^2$ from the infrared divergences. Therefore an integral corresponding to a graph with l loops can have poles up to $1/\varepsilon^{2l}$.

At the end of the day, all poles disappear: The poles related to ultraviolet divergences are absorbed into renormalisation constants. The poles related to infrared divergences cancel in the final result for infrared-safe observables, when summed over all degenerate states or are absorbed into universal parton distribution functions. The sum over all degenerate states involves a sum over contributions with different loop numbers and different numbers of external legs.

However, intermediate results are in general a Laurent series in ε and the task is to determine the coefficients of this Laurent series up to a certain order. At this point mathematics enters. We can use the algebraic structures associated to Feynman integrals to derive algorithms to calculate them. A few examples where the use of algebraic tools has been essential are the calculation of the three-loop Altarelli-Parisi splitting functions [4, 5] or the calculation of the two-loop amplitude for the process $e^+e^- \to$ 3 jets [6–15].

On the other hand is the mathematics encountered in these calculations of interest in its own right and has led in the last years to a fruitful interplay between mathematicians and physicists. Examples are the relation of Feynman integrals to mixed Hodge structures and motives, as well as the occurrence of certain transcendental constants in the result of a calculation [16–33].

This article is organised as follows: After a brief introduction into perturbation theory (Section 2), multi-loop integrals (Section 3) and periods (Section 4), we present in Section 5 a theorem stating that under rather weak assumptions the coefficients of the Laurent series of any multi-loop integral are periods. The proof is sketched in Section 6 and Section 7. Shuffle algebras are discussed in Section 8. Section 9 is devoted to multiple polylogarithms. In Section 10 we discuss how multiple polylogarithms emerge in the calculation of Feynman integrals. Finally, Section 11 contains our conclusions.

2 Perturbation theory

In high-energy physics experiments one is interested in scattering processes with two incoming particles and n outgoing particles. Such a process is described by a scattering amplitude, which can be calculated in perturbation theory. The amplitude has a perturbative expansion in the (small) coupling constant g:

$$\mathcal{A}_n = g^n \left(\mathcal{A}_n^{(0)} + g^2 \mathcal{A}_n^{(1)} + g^4 \mathcal{A}_n^{(2)} + g^6 \mathcal{A}_n^{(3)} + ... \right). \tag{2.1}$$

To the coefficient $\mathcal{A}_n^{(l)}$ contribute Feynman graphs with l loops and $(n+2)$ external legs. The recipe for the computation of $\mathcal{A}_n^{(l)}$ is as follows: Draw first all Feynman diagrams with the given number of external particles and l loops. Then translate each graph into a mathematical formula with the help of the Feynman rules. $\mathcal{A}_n^{(l)}$ is then given as the sum of all these terms.

Feynman rules allow us to translate a Feynman graph into a mathematical formula. These rules are derived from the fundamental Lagrange density of the theory, but for our purposes it is sufficient to accept them as a starting point. The most important ingredients are internal propagators, vertices and external lines. For example, the rules for the propagators of a fermion or a massless gauge boson read

Fermion:  $= i\frac{\not{p}+m}{p^2-m^2+i\delta},$

Gauge boson:  $= \frac{-ig_{\mu\nu}}{k^2+i\delta}.$

Here p and k are the momenta of the fermion and the boson, respectively. m is the mass of the fermion. $\not{p} = p_\mu \gamma^\mu$ is a short-hand notation for the contraction of the momentum with the Dirac matrices. The metric tensor is denoted by $g_{\mu\nu}$ and the convention adopted here is to take the metric tensor as $g_{\mu\nu} = \text{diag}(1, -1, -1, -1)$. The propagator would have a pole for $p^2 = m^2$, or phrased differently for $E = \pm\sqrt{\vec{p}^2 + m^2}$. When integrating over E, the integration contour has to be deformed to avoid these two poles. Causality dictates into which directions the contour has to be deformed. The pole on the negative real axis is avoided by escaping into the lower complex half-plane, the pole at the positive real axis is avoided by a deformation into the upper complex half-plane. Feynman invented the trick to add a small imaginary part $i\delta$ to the denominator, which keeps track of the directions into which the contour has to be deformed. In the following we will usually suppress the $i\delta$-term in order to keep the notation compact.

As a typical example for an interaction vertex let us look at the vertex involving a fermion pair and a gauge boson:

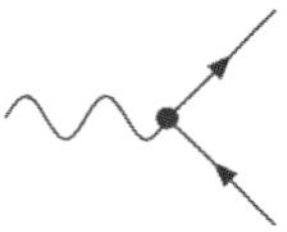 $= ig\gamma^\mu.$ (2.2)

Here, g is the coupling constant and γ^μ denotes the Dirac matrices. At each vertex, we have momentum conservation: The sum of the incoming momenta equals the sum of the outgoing momenta.

To each external line we have to associate a factor, which describes the polarisation of the corresponding particle: There is a polarisation vector

$\varepsilon^\mu(k)$ for each external gauge boson and a spinor $\bar{u}(p)$, $u(p)$, $v(p)$ or $\bar{v}(p)$ for each external fermion.

Furthermore there are a few additional rules: First of all, there is an integration

$$\int \frac{d^4k}{(2\pi)^4} \tag{2.3}$$

for each loop. Secondly, each closed fermion loop gets an extra factor of (-1). Finally, each diagram gets multiplied by a symmetry factor $1/S$, where S is the order of the permutation group of the internal lines and vertices leaving the diagram unchanged when the external lines are fixed.

Having stated the Feynman rules, let us look at two examples: The first example is a scalar two-point one-loop integral with zero external momentum:

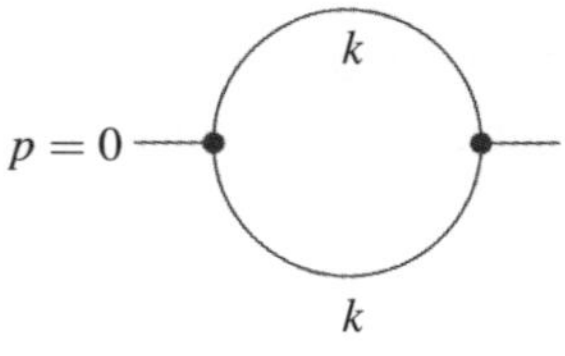

$$= \int \frac{d^4k}{(2\pi)^4} \frac{1}{(k^2)^2} = \frac{1}{(4\pi)^2} \int_0^\infty dk^2 \frac{1}{k^2} = \frac{1}{(4\pi)^2} \int_0^\infty \frac{dx}{x}. \tag{2.4}$$

This integral diverges at $k^2 \to \infty$ as well as at $k^2 \to 0$. The former divergence is called ultraviolet divergence, the later is called infrared divergence. Any quantity, which is given by a divergent integral, is of course an ill-defined quantity. Therefore the first step is to make these integrals well-defined by introducing a regulator. There are several possibilities how this can be done, but the method of dimensional regularisation [1–3] has almost become a standard, as the calculations in this regularisation scheme turn out to be the simplest. Within dimensional regularisation one replaces the four-dimensional integral over the loop momentum by an D-dimensional integral, where D is now an additional parameter, which can be a non-integer or even a complex number. We consider the result of the integration as a function of D and we are interested in the behaviour of this function as D approaches 4. The original divergences will then show up as poles in the Laurent series in $\varepsilon = (4 - D)/2$.

As a second example we consider a Feynman diagram contributing to the one-loop corrections for the process $e^+e^- \to qg\bar{q}$, shown in Figure 2.1.

At high energies we can ignore the masses of the electron and the light quarks. From the Feynman rules one obtains for this diagram (ignoring

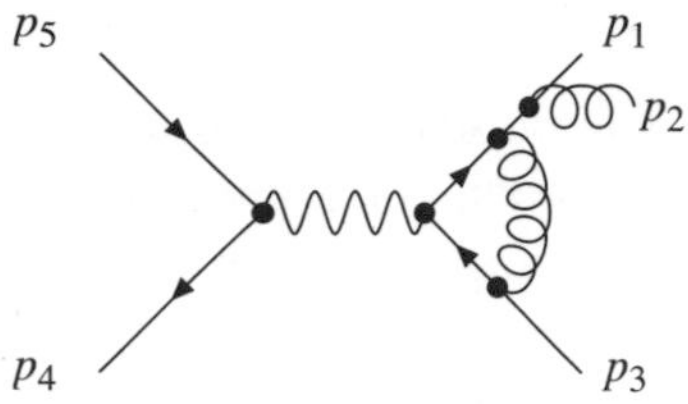

Figure 2.1. A one-loop Feynman diagram contributing to the process $e^+e^- \to q g \bar{q}$.

coupling and colour prefactors):

$$-\bar{v}(p_4)\gamma^\mu u(p_5)\frac{1}{p_{123}^2}\int\frac{d^D k_1}{(2\pi)^4}\frac{1}{k_2^2}\bar{u}(p_1)\not{\varepsilon}(p_2)\frac{\not{p}_{12}}{p_{12}^2}\gamma_\nu\frac{\not{k}_1}{k_1^2}\gamma_\mu\frac{\not{k}_3}{k_3^2}\gamma^\nu v(p_3). \quad (2.5)$$

Here, $p_{12} = p_1 + p_2$, $p_{123} = p_1 + p_2 + p_3$, $k_2 = k_1 - p_{12}$, $k_3 = k_2 - p_3$. Further $\not{\varepsilon}(p_2) = \gamma_\tau \varepsilon^\tau(p_2)$, where $\varepsilon^\tau(p_2)$ is the polarisation vector of the outgoing gluon. All external momenta are assumed to be massless: $p_i^2 = 0$ for $i = 1..5$. We can reorganise this formula into a part, which depends on the loop integration and a part, which does not. The loop integral to be calculated reads:

$$\int\frac{d^D k_1}{(2\pi)^4}\frac{k_1^\rho k_3^\sigma}{k_1^2 k_2^2 k_3^2}, \quad (2.6)$$

while the remainder, which is independent of the loop integration is given by

$$-\bar{v}(p_4)\gamma^\mu u(p_5)\frac{1}{p_{123}^2 p_{12}^2}\bar{u}(p_1)\not{\varepsilon}(p_2)\not{p}_{12}\gamma_\nu\gamma_\rho\gamma_\mu\gamma_\sigma\gamma^\nu v(p_3). \quad (2.7)$$

The loop integral in equation (2.6) contains in the denominator three propagator factors and in the numerator two factors of the loop momentum. We call a loop integral, in which the loop momentum occurs also in the numerator a "tensor integral". A loop integral, in which the numerator is independent of the loop momentum is called a "scalar integral". The scalar integral associated to equation (2.6) reads

$$\int\frac{d^D k_1}{(2\pi)^4}\frac{1}{k_1^2 k_2^2 k_3^2}. \quad (2.8)$$

There is a general method [34, 35] which allows to reduce any tensor integral to a combination of scalar integrals at the expense of introducing higher powers of the propagators and shifted space-time dimensions. Therefore it is sufficient to focus on scalar integrals. Each integral can be specified by its topology, its value for the dimension D and a set of indices, denoting the powers of the propagators.

3 Multi-loop integrals

Let us now consider a generic scalar l-loop integral I_G in $D = 2m - 2\varepsilon$ dimensions with n propagators, corresponding to a graph G. For each internal line j the corresponding propagator in the integrand can be raised to a power ν_j. Therefore the integral will depend also on the numbers $\nu_1,...,\nu_n$. It is sufficient to consider only the case, where all exponents are natural numbers: $\nu_j \in \mathbb{N}$. We define the Feynman integral by

$$I_G = \frac{\prod\limits_{j=1}^{n} \Gamma(\nu_j)}{\Gamma(\nu - lD/2)} \left(\mu^2\right)^{\nu - lD/2} \int \prod_{r=1}^{l} \frac{d^D k_r}{i\pi^{\frac{D}{2}}} \prod_{j=1}^{n} \frac{1}{(-q_j^2 + m_j^2)^{\nu_j}}, \quad (3.1)$$

with $\nu = \nu_1 + ... + \nu_n$. μ is an arbitrary scale, called the renormalisation scale. The momenta q_j of the propagators are linear combinations of the external momenta and the loop momenta. The prefactors are chosen such that after Feynman parametrisation the Feynman integral has a simple form:

$$I_G = \left(\mu^2\right)^{\nu - lD/2} \int\limits_{x_j \geq 0} d^n x \; \delta(1 - \sum_{i=1}^{n} x_i) \left(\prod_{j=1}^{n} x_j^{\nu_j - 1} \right) \frac{\mathcal{U}^{\nu - (l+1)D/2}}{\mathcal{F}^{\nu - lD/2}}. \quad (3.2)$$

The functions $\mathcal{U}$ and $\mathcal{F}$ depend on the Feynman parameters and can be derived from the topology of the corresponding Feynman graph G. Cutting l lines of a given connected l-loop graph such that it becomes a connected tree graph T defines a chord $C(T, G)$ as being the set of lines not belonging to this tree. The Feynman parameters associated with each chord define a monomial of degree l. The set of all such trees (or 1-trees) is denoted by $\mathcal{T}_1$. The 1-trees $T \in \mathcal{T}_1$ define $\mathcal{U}$ as being the sum over all monomials corresponding to the chords $C(T, G)$. Cutting one more line of a 1-tree leads to two disconnected trees (T_1, T_2), or a 2-tree. $\mathcal{T}_2$ is the set of all such pairs. The corresponding chords define monomials of degree $l + 1$. Each 2-tree of a graph corresponds to a cut defined by cutting the lines which connected the two now disconnected trees in the original graph. The square of the sum of momenta through the cut lines of one of the two disconnected trees T_1 or T_2 defines a Lorentz invariant

$$s_T = \left(\sum_{j \in C(T,G)} p_j \right)^2. \quad (3.3)$$

The function $\mathcal{F}_0$ is the sum over all such monomials times minus the corresponding invariant. The function $\mathcal{F}$ is then given by $\mathcal{F}_0$ plus an additional piece involving the internal masses m_j. In summary, the functions

$\mathcal{U}$ and $\mathcal{F}$ are obtained from the graph as follows:

$$\mathcal{U} = \sum_{T \in \mathcal{T}_1} \left[\prod_{j \in \mathcal{C}(T,G)} x_j \right],$$

$$\mathcal{F}_0 = \sum_{(T_1,T_2) \in \mathcal{T}_2} \left[\prod_{j \in \mathcal{C}(T_1,G)} x_j \right] (-s_{T_1}),$$

$$\mathcal{F} = \mathcal{F}_0 + \mathcal{U} \sum_{j=1}^{n} x_j m_j^2 .$$

4 Periods

Periods are special numbers. Before we give the definition, let us start with some sets of numbers: The natural numbers $\mathbb{N}$, the integer numbers $\mathbb{Z}$, the rational numbers $\mathbb{Q}$, the real numbers $\mathbb{R}$ and the complex numbers $\mathbb{C}$ are all well-known. More refined is already the set of algebraic numbers, denoted by $\bar{\mathbb{Q}}$. An algebraic number is a solution of a polynomial equation with rational coefficients:

$$x^n + a_{n-1}x^{n-1} + \cdots + a_0 = 0, \quad a_j \in \mathbb{Q}. \tag{4.1}$$

As all such solutions lie in $\mathbb{C}$, the set of algebraic numbers $\bar{\mathbb{Q}}$ is a sub-set of the complex numbers $\mathbb{C}$. Numbers which are not algebraic are called transcendental. The sets $\mathbb{N}$, $\mathbb{Z}$, $\mathbb{Q}$ and $\bar{\mathbb{Q}}$ are countable, whereas the sets $\mathbb{R}$, $\mathbb{C}$ and the set of transcendental numbers are uncountable.

Periods are a countable set of numbers, lying between $\bar{\mathbb{Q}}$ and $\mathbb{C}$. There are several equivalent definitions for periods. Kontsevich and Zagier gave the following definition [36]: A period is a complex number whose real and imaginary parts are values of absolutely convergent integrals of rational functions with rational coefficients, over domains in $\mathbb{R}^n$ given by polynomial inequalities with rational coefficients. Domains defined by polynomial inequalities with rational coefficients are called semi-algebraic sets.

We denote the set of periods by $\mathbb{P}$. The algebraic numbers are contained in the set of periods: $\bar{\mathbb{Q}} \in \mathbb{P}$. In addition, $\mathbb{P}$ contains transcendental numbers, an example for such a number is π:

$$\pi = \iint\limits_{x^2+y^2 \leq 1} dx\, dy. \tag{4.2}$$

The integral on the right hand side . clearly shows that π is a period. On the other hand, it is conjectured that the basis of the natural logarithm e

and Euler's constant γ_E are not periods. Although there are uncountably many numbers, which are not periods, only very recently an example for a number which is not a period has been found [37].

We need a few basic properties of periods: The set of periods $\mathbb{P}$ is a $\bar{\mathbb{Q}}$-algebra [36, 38]. In particular the sum and the product of two periods are again periods.

The defining integrals of periods have integrands, which are rational functions with rational coefficients. For our purposes this is too restrictive, as we will encounter logarithms as integrands as well. However any logarithm of a rational function with rational coefficients can be written as

$$\ln g(x) = \int_0^1 dt \, \frac{g(x)-1}{(g(x)-1)t+1}. \tag{4.3}$$

5 A theorem on Feynman integrals

Let us consider a general scalar multi-loop integral as in equation (3.2). Let m be an integer and set $D = 2m - 2\varepsilon$. Then this integral has a Laurent series expansion in ε

$$I_G = \sum_{j=-2l}^{\infty} c_j \varepsilon^j. \tag{5.1}$$

Theorem 5.1. *In the case where*

1. *all kinematical invariants s_T are zero or negative,*
2. *all masses m_i and μ are zero or positive ($\mu \neq 0$),*
3. *all ratios of invariants and masses are rational,*

the coefficients c_j of the Laurent expansion are periods.

In the special case were

1. the graph has no external lines or all invariants s_T are zero,
2. all internal masses m_j are equal to μ,
3. all propagators occur with power 1, *i.e.* $\nu_j = 1$ for all j,

the Feynman parameter integral reduces to

$$I_G = \int_{x_j \geq 0} d^n x \, \delta(1 - \sum_{i=1}^{n} x_i) \, \mathcal{U}^{-D/2} \tag{5.2}$$

and only the polynomial $\mathcal{U}$ occurs in the integrand. In this case it has been shown by Belkale and Brosnan [39] that the coefficients of the Laurent expansion are periods.

Using the method of sector decomposition we are able to prove the general case [40]. We will actually prove a stronger version of Theorem 5.1. Consider the following integral

$$J = \int\limits_{x_j \geq 0} d^n x \; \delta(1 - \sum_{i=1}^{n} x_i) \left(\prod_{i=1}^{n} x_i^{a_i + \varepsilon b_i} \right) \prod_{j=1}^{r} \left[P_j(x) \right]^{d_j + \varepsilon f_j} . \qquad (5.3)$$

The integration is over the standard simplex. The a's, b's, d's and f's are integers. The P's are polynomials in the variables x_1, ..., x_n with rational coefficients. The polynomials are required to be non-zero inside the integration region, but may vanish on the boundaries of the integration region. To fix the sign, let us agree that all polynomials are positive inside the integration region. The integral J has a Laurent expansion

$$J = \sum_{j=j_0}^{\infty} c_j \varepsilon^j . \qquad (5.4)$$

Theorem 5.2. *The coefficients c_j of the Laurent expansion of the integral J are periods.*

Theorem 5.1 follows then from Theorem 5.2 as the special case $a_i = \nu_i - 1$, $b_i = 0$, $r = 2$, $P_1 = \mathcal{U}$, $P_2 = \mathcal{F}$, $d_1 + \varepsilon f_1 = \nu - (l+1)D/2$ and $d_2 + \varepsilon f_2 = lD/2 - \nu$.

Proof of Theorem 5.2. To prove the theorem we will give an algorithm which expresses each coefficient c_j as a sum of absolutely convergent integrals over the unit hypercube with integrands, which are linear combinations of products of rational functions with logarithms of rational functions, all of them with rational coefficients. Let us denote this set of functions to which the integrands belong by $\mathcal{M}$. The unit hypercube is clearly a semi-algebraic set. It is clear that absolutely convergent integrals over semi-algebraic sets with integrands from the set $\mathcal{M}$ are periods. In addition, the sum of periods is again a period. Therefore it is sufficient to express each coefficient c_j as a finite sum of absolutely convergent integrals over the unit hypercube with integrands from $\mathcal{M}$. To do so, we use iterated sector decomposition. This is a constructive method. Therefore we obtain as a side-effect a general purpose algorithm for the numerical evaluation of multi-loop integrals. □

6 Sector decomposition

In this section we review the algorithm for iterated sector decomposition [41–47]. The starting point is an integral of the form

$$\int\limits_{x_j \geq 0} d^n x \; \delta(1 - \sum_{i=1}^{n} x_i) \left(\prod_{i=1}^{n} x_i^{\mu_i} \right) \prod_{j=1}^{r} \left[P_j(x) \right]^{\lambda_j}, \quad (6.1)$$

where $\mu_i = a_i + \varepsilon b_i$ and $\lambda_j = c_j + \varepsilon d_j$. The integration is over the standard simplex. The a's, b's, c's and d's are integers. The P's are polynomials in the variables $x_1, ..., x_n$. The polynomials are required to be non-zero inside the integration region, but may vanish on the boundaries of the integration region. The algorithm consists of the following steps:

Step 0: Convert all polynomials to homogeneous polynomials.

Step 1: Decompose the integral into n primary sectors.

Step 2: Decompose the sectors iteratively into sub-sectors until each of the polynomials is of the form

$$P = x_1^{m_1} ... x_n^{m_n} \left(c + P'(x) \right), \quad (6.2)$$

where $c \neq 0$ and $P'(x)$ is a polynomial in the variables x_j without a constant term. In this case the monomial prefactor $x_1^{m_1} ... x_n^{m_n}$ can be factored out and the remainder contains a non-zero constant term. To convert P into the form (6.2) one chooses a subset $S = \{\alpha_1, ..., \alpha_k\} \subseteq \{1, ... n\}$ according to a strategy discussed in the next section. One decomposes the k-dimensional hypercube into k sub-sectors according to

$$\int\limits_0^1 d^n x = \sum_{l=1}^{k} \int\limits_0^1 d^n x \prod_{i=1, i \neq l}^{k} \theta \left(x_{\alpha_l} \geq x_{\alpha_i} \right). \quad (6.3)$$

In the l-th sub-sector one makes for each element of S the substitution

$$x_{\alpha_i} = x_{\alpha_l} x'_{\alpha_i} \quad \text{for } i \neq l. \quad (6.4)$$

This procedure is iterated, until all polynomials are of the form (6.2).

Figure 6.1 illustrates this for the simple example $S = \{1, 2\}$. equation (6.3) gives the decomposition into the two sectors $x_1 > x_2$ and $x_2 > x_1$. Equation (6.4) transforms the triangles into squares. This transformation is one-to-one for all points except the origin. The origin is replaced by the line $x_1 = 0$ in the first sector and by the line $x_2 = 0$ in the second sector. Therefore the name "blow-up".

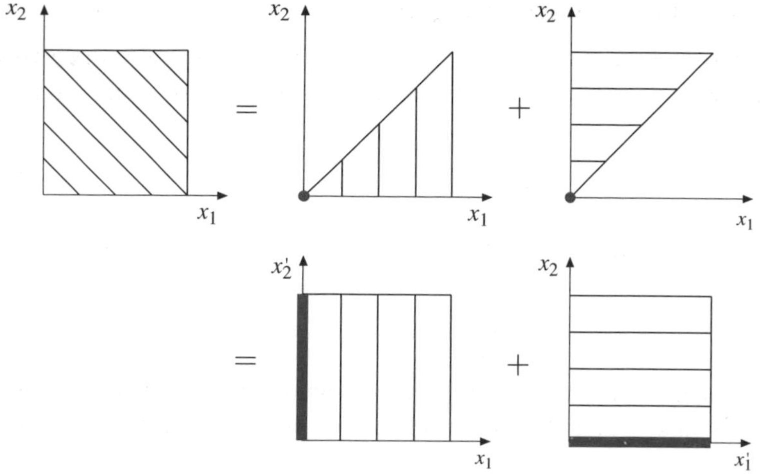

Figure 6.1. Illustration of sector decomposition and blow-up for a simple example.

Step 3: The singular behaviour of the integral depends now only on the factor

$$\prod_{i=1}^{n} x_i^{a_i+\epsilon b_i}. \tag{6.5}$$

We Taylor expand in the integration variables and perform the trivial integrations

$$\int_0^1 dx\; x^{a+b\varepsilon} = \frac{1}{a+1+b\varepsilon}, \tag{6.6}$$

leading to the explicit poles in $1/\varepsilon$.

Step 4: All remaining integrals are now by construction finite. We can now expand all expressions in a Laurent series in ε and truncate to the desired order.

Step 5: It remains to compute the coefficients of the Laurent series. These coefficients contain finite integrals, which can be evaluated numerically by Monte Carlo integration. We implemented[1] the algorithm into a computer program, which computes numerically the coefficients of the Laurent series of any multi-loop integral [45].

[1] The program can be obtained from `http://www.higgs.de/~stefanw/software.html`

7 Hironaka's polyhedra game

In Step 2 of the algorithm we have an iteration. It is important to show that this iteration terminates and does not lead to an infinite loop. There

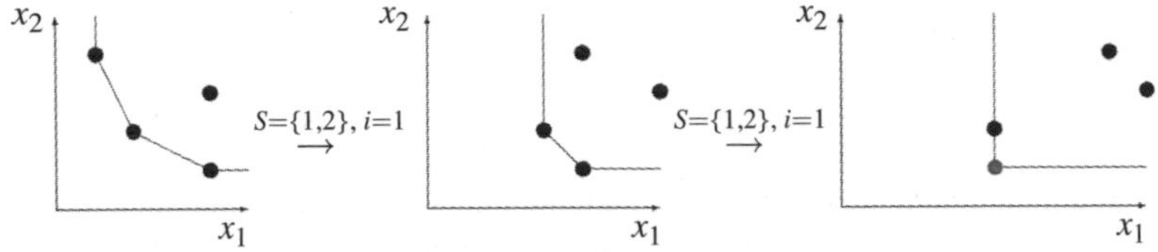

Figure 7.1. Illustration of Hironaka's polyhedra game.

are strategies for choosing the sub-sectors, which guarantee termination. These strategies [48–52] are closely related to Hironaka's polyhedra game.

Hironaka's polyhedra game is played by two players, A and B. They are given a finite set M of points $m = (m_1, ..., m_n)$ in $\mathbb{N}_+^n$, the first quadrant of $\mathbb{N}^n$. We denote by $\Delta \subset \mathbb{R}_+^n$ the positive convex hull of the set M. It is given by the convex hull of the set

$$\bigcup_{m \in M} \left(m + \mathbb{R}_+^n \right). \tag{7.1}$$

The two players compete in the following game:

1. Player A chooses a non-empty subset $S \subseteq \{1, ..., n\}$.
2. Player B chooses one element i out of this subset S.

Then, according to the choices of the players, the components of all $(m_1, ..., m_n) \in M$ are replaced by new points $\left(m_1', ..., m_n'\right)$, given by:

$$\begin{aligned} m_j' &= m_j, \quad \text{if } j \neq i, \\ m_i' &= \sum_{j \in S} m_j - c, \end{aligned}$$

where for the moment we set $c = 1$. This defines the set M'. One then sets $M = M'$ and goes back to Step 1. Player A wins the game if, after a finite number of moves, the polyhedron Δ is of the form

$$\Delta = m + \mathbb{R}_+^n, \tag{7.2}$$

i.e. generated by one point. If this never occurs, player B has won. The challenge of the polyhedra game is to show that player A always has a winning strategy, no matter how player B chooses his moves. A simple illustration of Hironaka's polyhedra game in two dimensions is given in Figure 7.1. Player A always chooses $S = \{1, 2\}$. In [45] we have shown that a winning strategy for Hironaka's polyhedra game translates directly into a strategy for choosing the sub-sectors which guarantees termination.

8 Shuffle algebras

Before we continue the discussion of loop integrals, it is useful to discuss first shuffle algebras and generalisations thereof from an algebraic viewpoint. Consider a set of letters A. The set A is called the alphabet. A word is an ordered sequence of letters:

$$w = l_1 l_2 ... l_k. \tag{8.1}$$

The word of length zero is denoted by e. Let K be a field and consider the vector space of words over K. A shuffle algebra $\mathcal{A}$ on the vector space of words is defined by

$$(l_1 l_2 ... l_k) \cdot (l_{k+1} ... l_r) = \sum_{\text{shuffles } \sigma} l_{\sigma(1)} l_{\sigma(2)} ... l_{\sigma(r)}, \tag{8.2}$$

where the sum runs over all permutations σ, which preserve the relative order of $1, 2, ..., k$ and of $k+1, ..., r$. The name "shuffle algebra" is related to the analogy of shuffling cards: If a deck of cards is split into two parts and then shuffled, the relative order within the two individual parts is conserved. A shuffle algebra is also known under the name "mould symmetral" [53]. The empty word e is the unit in this algebra:

$$e \cdot w = w \cdot e = w. \tag{8.3}$$

A recursive definition of the shuffle product is given by

$$\begin{aligned} (l_1 l_2 ... l_k) \cdot (l_{k+1} ... l_r) = & \; l_1 \left[(l_2 ... l_k) \cdot (l_{k+1} ... l_r) \right] \\ & + l_{k+1} \left[(l_1 l_2 ... l_k) \cdot (l_{k+2} ... l_r) \right]. \end{aligned} \tag{8.4}$$

It is well known fact that the shuffle algebra is actually a (non-cocommutative) Hopf algebra [54]. In this context let us briefly review the definitions of a coalgebra, a bialgebra and a Hopf algebra, which are closely related: First note that the unit in an algebra can be viewed as a map from K to A and that the multiplication can be viewed as a map from the tensor product $A \otimes A$ to A (*e.g.* one takes two elements from A, multiplies them and gets one element out).

A coalgebra has instead of multiplication and unit the dual structures: a comultiplication Δ and a counit $\bar{e}$. The counit is a map from A to K, whereas comultiplication is a map from A to $A \otimes A$. Note that comultiplication and counit go in the reverse direction compared to multiplication and unit. We will always assume that the comultiplication is coassociative. The general form of the coproduct is

$$\Delta(a) = \sum_i a_i^{(1)} \otimes a_i^{(2)}, \tag{8.5}$$

where $a_i^{(1)}$ denotes an element of A appearing in the first slot of $A \otimes A$ and $a_i^{(2)}$ correspondingly denotes an element of A appearing in the second slot. Sweedler's notation [55] consists in dropping the dummy index i and the summation symbol:

$$\Delta(a) = a^{(1)} \otimes a^{(2)} \tag{8.6}$$

The sum is implicitly understood. This is similar to Einstein's summation convention, except that the dummy summation index i is also dropped. The superscripts $^{(1)}$ and $^{(2)}$ indicate that a sum is involved.

A bialgebra is an algebra and a coalgebra at the same time, such that the two structures are compatible with each other. Using Sweedler's notation, the compatibility between the multiplication and comultiplication is expressed as

$$\Delta\left(a \cdot b\right) = \left(a^{(1)} \cdot b^{(1)}\right) \otimes \left(a^{(2)} \cdot b^{(2)}\right). \tag{8.7}$$

A Hopf algebra is a bialgebra with an additional map from A to A, called the antipode $\mathcal{S}$, which fulfils

$$a^{(1)} \cdot \mathcal{S}\left(a^{(2)}\right) = \mathcal{S}\left(a^{(1)}\right) \cdot a^{(2)} = e \cdot \bar{e}(a). \tag{8.8}$$

With this background at hand we can now state the coproduct, the counit and the antipode for the shuffle algebra: The counit $\bar{e}$ is given by:

$$\bar{e}\left(e\right) = 1, \quad \bar{e}\left(l_1 l_2 ... l_n\right) = 0. \tag{8.9}$$

The coproduct Δ is given by:

$$\Delta\left(l_1 l_2 ... l_k\right) = \sum_{j=0}^{k} \left(l_{j+1} ... l_k\right) \otimes \left(l_1 ... l_j\right). \tag{8.10}$$

The antipode $\mathcal{S}$ is given by:

$$\mathcal{S}\left(l_1 l_2 ... l_k\right) = (-1)^k \, l_k l_{k-1} ... l_2 l_1. \tag{8.11}$$

The shuffle algebra is generated by the Lyndon words. If one introduces a lexicographic ordering on the letters of the alphabet A, a Lyndon word is defined by the property

$$w < v \tag{8.12}$$

for any sub-words u and v such that $w = uv$.

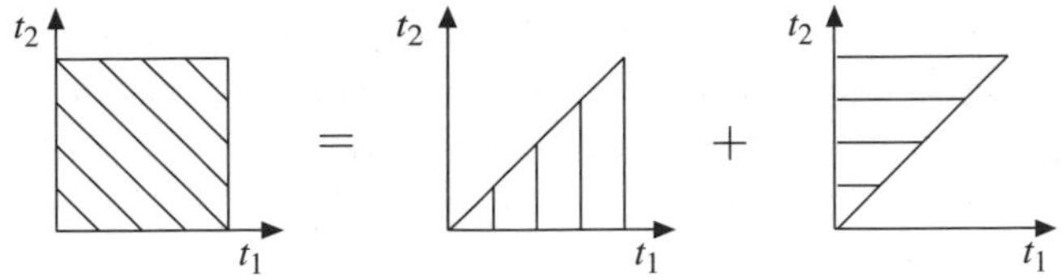

Figure 8.1. Sketch of the proof for the shuffle product of two iterated integrals. The integral over the square is replaced by two integrals over the upper and lower triangle.

An important example for a shuffle algebra are iterated integrals. Let $[a, b]$ be a segment of the real line and $f_1, f_2, ...$ functions on this interval. Let us define the following iterated integrals:

$$I(f_1, f_2, ..., f_k; a, b) = \int_a^b dt_1 f_1(t_1) \int_a^{t_1} dt_2 f_2(t_2) ... \int_a^{t_{k-1}} dt_k f_k(t_k). \quad (8.13)$$

For fixed a and b we have a shuffle algebra:

$$\begin{aligned} & I(f_1, f_2, ..., f_k; a, b) \cdot I(f_{k+1}, ..., f_r; a, b) \\ & = \sum_{\text{shuffles } \sigma} I(f_{\sigma(1)}, f_{\sigma(2)}, ..., f_{\sigma(r)}; a, b), \end{aligned} \quad (8.14)$$

where the sum runs over all permutations σ, which preserve the relative order of $1, 2, ..., k$ and of $k+1, ..., r$. The proof is sketched in Figure 8.1.

The two outermost integrations are recursively replaced by integrations over the upper and lower triangle.

We now consider generalisations of shuffle algebras. Assume that for the set of letters we have an additional operation

$$\begin{aligned} (.,.) : A \otimes A & \to A, \\ l_1 \otimes l_2 & \to (l_1, l_2), \end{aligned} \quad (8.15)$$

which is commutative and associative. Then we can define a new product of words recursively through

$$\begin{aligned} (l_1 l_2 ... l_k) * (l_{k+1} ... l_r) = & \; l_1 [(l_2 ... l_k) * (l_{k+1} ... l_r)] \\ & + l_{k+1} [(l_1 l_2 ... l_k) * (l_{k+2} ... l_r)] \\ & + (l_1, l_{k+1}) [(l_2 ... l_k) * (l_{k+2} ... l_r)] . \end{aligned} \quad (8.16)$$

This product is a generalisation of the shuffle product and differs from the recursive definition of the shuffle product in equation (8.4) through the extra term in the last line. This modified product is known under the names quasi-shuffle product [56], mixable shuffle product [57], stuffle product [58] or mould symmetrel [53]. Quasi-shuffle algebras are Hopf algebras. Comultiplication and counit are defined as for the shuffle algebras. The counit $\bar{e}$ is given by:

$$\bar{e}\,(e) = 1, \quad \bar{e}\,(l_1 l_2 ... l_n) = 0. \tag{8.17}$$

The coproduct Δ is given by:

$$\Delta\,(l_1 l_2 ... l_k) = \sum_{j=0}^{k} \left(l_{j+1} ... l_k\right) \otimes \left(l_1 ... l_j\right). \tag{8.18}$$

The antipode $\mathcal{S}$ is recursively defined through

$$\mathcal{S}\,(l_1 l_2 ... l_k) = -l_1 l_2 ... l_k - \sum_{j=1}^{k-1} \mathcal{S}\left(l_{j+1} ... l_k\right) * \left(l_1 ... l_j\right). \tag{8.19}$$

An example for a quasi-shuffle algebra are nested sums. Let n_a and n_b be integers with $n_a < n_b$ and let f_1, f_2, ... be functions defined on the integers. We consider the following nested sums:

$$S(f_1, f_2, ..., f_k; n_a, n_b) = \sum_{i_1=n_a}^{n_b} f_1(i_1) \sum_{i_2=n_a}^{i_1-1} f_2(i_2) ... \sum_{i_k=n_a}^{i_{k-1}-1} f_k(i_k). \tag{8.20}$$

For fixed n_a and n_b we have a quasi-shuffle algebra:

$$\begin{aligned} & S(f_1, f_2, ..., f_k; n_a, n_b) * S(f_{k+1}, ..., f_r; n_a, n_b) \\ & = \sum_{i_1=n_a}^{n_b} f_1(i_1) S(f_2, ..., f_k; n_a, i_1 - 1) * S(f_{k+1}, ..., f_r; n_a, i_1 - 1) \\ & \quad + \sum_{j_1=n_a}^{n_b} f_k(j_1) S(f_1, f_2, ..., f_k; n_a, j_1 - 1) * S(f_{k+2}, ..., f_r; n_a, j_1 - 1) \\ & \quad + \sum_{i=n_a}^{n_b} f_1(i) f_k(i) S(f_2, ..., f_k; n_a, i - 1) * S(f_{k+2}, ..., f_r; n_a, i - 1). \end{aligned} \tag{8.21}$$

Note that the product of two letters corresponds to the point-wise product of the two functions:

$$(f_i, f_j)\,(n) = f_i(n) f_j(n). \tag{8.22}$$

The proof that nested sums obey the quasi-shuffle algebra is sketched in Figure 8.2. The outermost sums of the nested sums on the l.h.s of (8.21) are split into the three regions indicated in Figure 8.2.

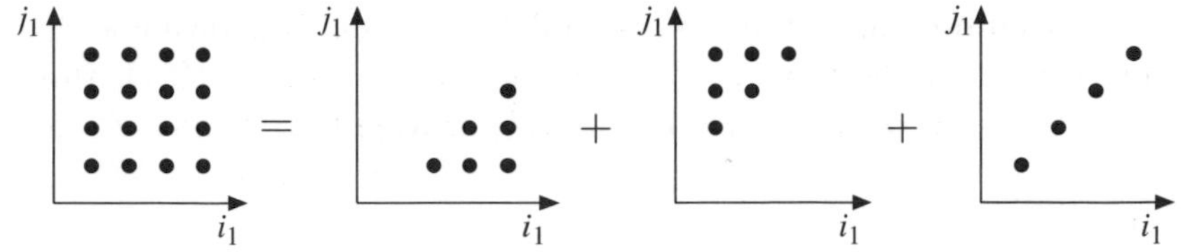

Figure 8.2. Sketch of the proof for the quasi-shuffle product of nested sums. The sum over the square is replaced by the sum over the three regions on the right hand side.

9 Multiple polylogarithms

In the previous section we have seen that iterated integrals form a shuffle algebra, while nested sums form a quasi-shuffle algebra. In this context multiple polylogarithms form an interesting class of functions. They have a representation as iterated integrals as well as nested sums. Therefore multiple polylogarithms form a shuffle algebra as well as a quasi-shuffle algebra. The two algebra structures are independent. Let us start with the representation as nested sums. The multiple polylogarithms are defined by [59–62]

$$\mathrm{Li}_{m_1,\ldots,m_k}(x_1, \ldots, x_k) = \sum_{i_1>i_2>\ldots>i_k>0} \frac{x_1^{i_1}}{{i_1}^{m_1}} \cdots \frac{x_k^{i_k}}{{i_k}^{m_k}}. \tag{9.1}$$

The multiple polylogarithms are generalisations of the classical polylogarithms $\mathrm{Li}_n(x)$, whose most prominent examples are

$$\mathrm{Li}_1(x) = \sum_{i_1=1}^{\infty} \frac{x^{i_1}}{i_1} = -\ln(1-x), \mathrm{Li}_2(x) = \sum_{i_1=1}^{\infty} \frac{x^{i_1}}{i_1^2}, \tag{9.2}$$

as well as Nielsen's generalised polylogarithms [63]

$$S_{n,p}(x) = \mathrm{Li}_{n+1,1,\ldots,1}(x, \underbrace{1, \ldots, 1}_{p-1}), \tag{9.3}$$

and the harmonic polylogarithms [64, 65]

$$H_{m_1,\ldots,m_k}(x) = \mathrm{Li}_{m_1,\ldots,m_k}(x, \underbrace{1, \ldots, 1}_{k-1}). \tag{9.4}$$

In addition, multiple polylogarithms have an integral representation. To discuss the integral representation it is convenient to introduce for $z_k \neq 0$ the following functions

$$G(z_1, \ldots, z_k; y) = \int_0^y \frac{dt_1}{t_1 - z_1} \int_0^{t_1} \frac{dt_2}{t_2 - z_2} \ldots \int_0^{t_{k-1}} \frac{dt_k}{t_k - z_k}. \tag{9.5}$$

In this definition one variable is redundant due to the following scaling relation:

$$G(z_1, ..., z_k; y) = G(xz_1, ..., xz_k; xy). \tag{9.6}$$

If one further defines

$$g(z; y) = \frac{1}{y - z}, \tag{9.7}$$

then one has

$$\frac{d}{dy} G(z_1, ..., z_k; y) = g(z_1; y) G(z_2, ..., z_k; y) \tag{9.8}$$

and

$$G(z_1, z_2, ..., z_k; y) = \int\limits_0^y dt\ g(z_1; t) G(z_2, ..., z_k; t). \tag{9.9}$$

One can slightly enlarge the set and define $G(0, ..., 0; y)$ with k zeros for z_1 to z_k to be

$$G(0, ..., 0; y) = \frac{1}{k!} (\ln y)^k. \tag{9.10}$$

This permits us to allow trailing zeros in the sequence $(z_1, ..., z_k)$ by defining the function G with trailing zeros via (9.9) and (9.10). To relate the multiple polylogarithms to the functions G it is convenient to introduce the following short-hand notation:

$$G_{m_1,...,m_k}(z_1, ..., z_k; y) = G(\underbrace{0, ..., 0}_{m_1-1}, z_1, ..., z_{k-1}, \underbrace{0..., 0}_{m_k-1}, z_k; y). \tag{9.11}$$

Here, all z_j for $j = 1, ..., k$ are assumed to be non-zero. One then finds

$$\text{Li}_{m_1,...,m_k}(x_1, ..., x_k) = (-1)^k G_{m_1,...,m_k}\left(\frac{1}{x_1}, \frac{1}{x_1 x_2}, ..., \frac{1}{x_1 ... x_k}; 1\right). \tag{9.12}$$

The inverse formula reads

$$G_{m_1,...,m_k}(z_1, ..., z_k; y) = (-1)^k\ \text{Li}_{m_1,...,m_k}\left(\frac{y}{z_1}, \frac{z_1}{z_2}, ..., \frac{z_{k-1}}{z_k}\right). \tag{9.13}$$

Equation (9.12) together with (9.11) and (9.5) defines an integral representation for the multiple polylogarithms.

Up to now we treated multiple polylogarithms from an algebraic point of view. Equally important are the analytical properties, which are needed for an efficient numerical evaluation. As an example I first discuss the numerical evaluation of the dilogarithm [66]:

$$\text{Li}_2(x) = -\int_0^x dt \frac{\ln(1-t)}{t} = \sum_{n=1}^{\infty} \frac{x^n}{n^2}. \tag{9.14}$$

The power series expansion can be evaluated numerically, provided $|x| < 1$. Using the functional equations

$$\begin{aligned} \text{Li}_2(x) &= -\text{Li}_2\left(\frac{1}{x}\right) - \frac{\pi^2}{6} - \frac{1}{2}\left(\ln(-x)\right)^2, \\ \text{Li}_2(x) &= -\text{Li}_2(1-x) + \frac{\pi^2}{6} - \ln(x)\ln(1-x). \end{aligned} \tag{9.15}$$

Any argument of the dilogarithm can be mapped into the region $|x| \leq 1$ and $-1 \leq \text{Re}(x) \leq 1/2$. The numerical computation can be accelerated by using an expansion in $[-\ln(1-x)]$ and the Bernoulli numbers B_i:

$$\text{Li}_2(x) = \sum_{i=0}^{\infty} \frac{B_i}{(i+1)!} \left(-\ln(1-x)\right)^{i+1}. \tag{9.16}$$

The generalisation to multiple polylogarithms proceeds along the same lines [67]: Using the integral representation equation (9.5) one transforms all arguments into a region, where one has a converging power series expansion. In this region equation (9.1) may be used. However it is advantageous to speed up the convergence of the power series expansion. This is done as follows: The multiple polylogarithms satisfy the Hölder convolution [58]. For $z_1 \neq 1$ and $z_w \neq 0$ this identity reads

$$\begin{aligned} &G\left(z_1, ..., z_w; 1\right) \\ &= \sum_{j=0}^{w} (-1)^j \, G\left(1-z_j, 1-z_{j-1}, ..., 1-z_1; 1-\frac{1}{p}\right) G\left(z_{j+1}, ..., z_w; \frac{1}{p}\right). \end{aligned} \tag{9.17}$$

The Hölder convolution can be used to accelerate the convergence for the series representation of the multiple polylogarithms.

10 From Feynman integrals to multiple polylogarithms

In Section 3 we saw that the Feynman parameter integrals depend on two graph polynomials $\mathcal{U}$ and $\mathcal{F}$, which are homogeneous functions of the Feynman parameters. In this section we will discuss how multiple polylogarithms arise in the calculation of Feynman parameter integrals.

We will discuss two approaches. In the first approach one uses a Mellin-Barnes transformation and sums up residues. This leads to the sum representation of multiple polylogarithms. In the second approach one first derives a differential equation for the Feynman parameter integral, which is then solved by an ansatz in terms of the iterated integral representation of multiple polylogarithms.

Let us start with the first approach. Assume for the moment that the two graph polynomials $\mathcal{U}$ and $\mathcal{F}$ are absent from the Feynman parameter integral. In this case we have

$$\int\limits_0^1 \left(\prod\limits_{j=1}^{n} dx_j\, x_j^{\nu_j - 1} \right) \delta(1 - \sum\limits_{i=1}^{n} x_i) = \frac{\prod\limits_{j=1}^{n} \Gamma(\nu_j)}{\Gamma(\nu_1 + ... + \nu_n)}. \quad (10.1)$$

With the help of the Mellin-Barnes transformation we now reduce the general case to equation (10.1). The Mellin-Barnes transformation reads

$$(A_1 + A_2 + ... + A_n)^{-c} = \frac{1}{\Gamma(c)} \frac{1}{(2\pi i)^{n-1}} \int\limits_{-i\infty}^{i\infty} d\sigma_1 ... \int\limits_{-i\infty}^{i\infty} d\sigma_{n-1}$$
$$\times \Gamma(-\sigma_1) ... \Gamma(-\sigma_{n-1}) \Gamma(\sigma_1 + ... + \sigma_{n-1} + c)\; A_1^{\sigma_1} ... A_{n-1}^{\sigma_{n-1}} A_n^{-\sigma_1 - ... - \sigma_{n-1} - c}. \quad (10.2)$$

Each contour is such that the poles of $\Gamma(-\sigma)$ are to the right and the poles of $\Gamma(\sigma + c)$ are to the left. This transformation can be used to convert the sum of monomials of the polynomials $\mathcal{U}$ and $\mathcal{F}$ into a product, such that all Feynman parameter integrals are of the form of equation (10.1). As this transformation converts sums into products it is the "inverse" of Feynman parametrisation.

With the help of equation (10.1) and equation (10.2) we may exchange the Feynman parameter integrals against multiple contour integrals. A single contour integral is of the form

$$I = \frac{1}{2\pi i} \int\limits_{\gamma - i\infty}^{\gamma + i\infty} d\sigma \frac{\Gamma(\sigma + a_1) ... \Gamma(\sigma + a_m)}{\Gamma(\sigma + c_2) ... \Gamma(\sigma + c_p)} \frac{\Gamma(-\sigma + b_1) ... \Gamma(-\sigma + b_n)}{\Gamma(-\sigma + d_1) ... \Gamma(-\sigma + d_q)} x^{-\sigma}. \quad (10.3)$$

If $\max\left(\text{Re}(-a_1), ..., \text{Re}(-a_m)\right) < \min\left(\text{Re}(b_1), ..., \text{Re}(b_n)\right)$ the contour can be chosen as a straight line parallel to the imaginary axis with

$$\max\left(\text{Re}(-a_1), ..., \text{Re}(-a_m)\right) < \text{Re}\,\gamma < \min\left(\text{Re}(b_1), ..., \text{Re}(b_n)\right), \quad (10.4)$$

otherwise the contour is indented, such that the residues of $\Gamma(\sigma + a_1)$, ..., $\Gamma(\sigma + a_m)$ are to the right of the contour, whereas the residues of

$\Gamma(-\sigma + b_1)$, ..., $\Gamma(-\sigma + b_n)$ are to the left of the contour. The integral equation (10.3) is most conveniently evaluated with the help of the residuum theorem by closing the contour to the left or to the right. To sum up all residues which lie inside the contour it is useful to know the residues of the Gamma function:

$$\begin{aligned} \text{res}\ (\Gamma(\sigma + a), \sigma = -a - n) &= \frac{(-1)^n}{n!}, \\ \text{res}\ (\Gamma(-\sigma + a), \sigma = a + n) &= -\frac{(-1)^n}{n!}. \end{aligned} \tag{10.5}$$

In general there are multiple contour integrals, and as a consequence one obtains multiple sums. Having collected all residues, one then expands the Gamma-functions:

$$\begin{aligned} \Gamma(n + \varepsilon) = \Gamma(1+\varepsilon)\Gamma(n)\Big[1 + \varepsilon Z_1(n-1) + \varepsilon^2 Z_{11}(n-1) \\ + \varepsilon^3 Z_{111}(n-1) + ... + \varepsilon^{n-1} Z_{11...1}(n-1)\Big], \end{aligned} \tag{10.6}$$

where $Z_{m_1,...,m_k}(n)$ are Euler-Zagier sums defined by

$$Z_{m_1,...,m_k}(n) = \sum_{n \geq i_1 > i_2 > ... > i_k > 0} \frac{1}{{i_1}^{m_1}} \cdots \frac{1}{{i_k}^{m_k}}. \tag{10.7}$$

This motivates the following definition of a special form of nested sums, called Z-sums [68–71]:

$$Z(n; m_1, ..., m_k; x_1, ..., x_k) = \sum_{n \geq i_1 > i_2 > ... > i_k > 0} \frac{x_1^{i_1}}{{i_1}^{m_1}} \cdots \frac{x_k^{i_k}}{{i_k}^{m_k}}. \tag{10.8}$$

k is called the depth of the Z-sum and $w = m_1 + ... + m_k$ is called the weight. If the sums go to infinity ($n = \infty$) the Z-sums are multiple polylogarithms:

$$Z(\infty; m_1, ..., m_k; x_1, ..., x_k) = \text{Li}_{m_1,...,m_k}(x_1, ..., x_k). \tag{10.9}$$

For $x_1 = ... = x_k = 1$ the definition reduces to the Euler-Zagier sums [72–76]:

$$Z(n; m_1, ..., m_k; 1, ..., 1) = Z_{m_1,...,m_k}(n). \tag{10.10}$$

For $n = \infty$ and $x_1 = ... = x_k = 1$ the sum is a multiple ζ-value [58,77]:

$$Z(\infty; m_1, ..., m_k; 1, ..., 1) = \zeta_{m_1,...,m_k}. \tag{10.11}$$

The usefulness of the Z-sums lies in the fact, that they interpolate between multiple polylogarithms and Euler-Zagier sums. The Z-sums form a quasi-shuffle algebra. In this approach multiple polylogarithms appear through equation (10.9).

An alternative approach to the computation of Feynman parameter integrals is based on differential equations [65, 78–82]. To evaluate these integrals within this approach one first finds for each master integral a differential equation, which this master integral has to satisfy. The derivative is taken with respect to an external scale, or a ratio of two scales. An example for a one-loop four-point function is given by

$\frac{\partial}{\partial s_{123}}$ 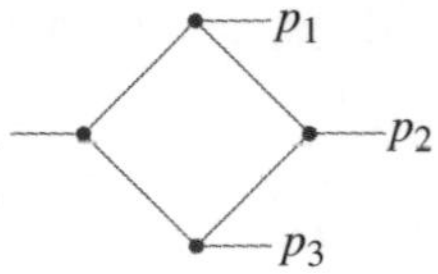 $= \frac{D-4}{2(s_{12}+s_{23}-s_{123})}$ 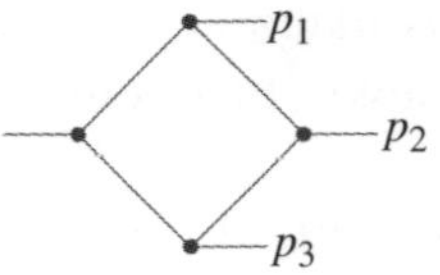

$+ \frac{2(D-3)}{(s_{123}-s_{12})(s_{123}-s_{12}-s_{23})} \left[\frac{1}{s_{123}} \right.$  $- \frac{1}{s_{12}}$ 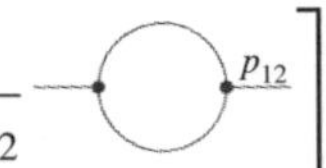 $\left. \right]$

$+ \frac{2(D-3)}{(s_{123}-s_{23})(s_{123}-s_{12}-s_{23})} \left[\frac{1}{s_{123}} \right.$ 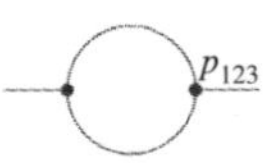 $- \frac{1}{s_{23}}$ 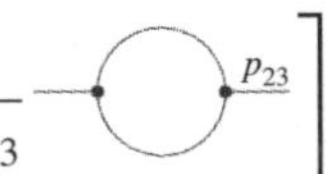 $\left. \right].$

The two-point functions on the right hand side are simpler and can be considered to be known. This equation is solved iteratively by an ansatz for the solution as a Laurent expression in ε. Each term in this Laurent series is a sum of terms, consisting of basis functions times some unknown (and to be determined) coefficients. This ansatz is inserted into the differential equation and the unknown coefficients are determined order by order from the differential equation. The basis functions are taken as a subset of multiple polylogarithms. In this approach the iterated integral representation of multiple polylogarithms is the most convenient form. This is immediately clear from the simple formula for the derivative as in equation (9.8).

11 Conclusions

In this talk we reported on mathematical properties of Feynman integrals. We first showed that under rather weak assumptions all the coefficients of the Laurent expansion of a multi-loop integral are periods. In the second part we focused on multiple polylogarithms and how they appear in the calculation of Feynman integrals.

References

[1] G. 'T HOOFT and M. J. G. VELTMAN, Nucl. Phys. **B44**, 189 (1972).

[2] C. G. BOLLINI and J. J. GIAMBIAGI, Nuovo Cim. **B12**, 20 (1972).

[3] G. M. CICUTA and E. MONTALDI, Nuovo Cim. Lett. **4**, 329 (1972).

[4] S. MOCH, J. A. M. VERMASEREN and A. VOGT, Nucl. Phys. **B688**, 101 (2004), hep-ph/0403192.

[5] A. VOGT, S. MOCH and J. A. M. VERMASEREN, Nucl. Phys. **B691**, 129 (2004), hep-ph/0404111.

[6] L. W. GARLAND, T. GEHRMANN, E. W. N. GLOVER, A. KOUKOUTSAKIS and E. REMIDDI, Nucl. Phys. **B627**, 107 (2002), hep-ph/0112081.

[7] L. W. GARLAND, T. GEHRMANN, E. W. N. GLOVER, A. KOUKOUTSAKIS and E. REMIDDI, Nucl. Phys. **B642**, 227 (2002), hep-ph/0206067.

[8] S. MOCH, P. UWER and S. WEINZIERL, Phys. Rev. **D66**, 114001 (2002), hep-ph/0207043.

[9] A. GEHRMANN-DE RIDDER, T. GEHRMANN, E. W. N. GLOVER and G. HEINRICH, Phys. Rev. Lett. **99**, 132002 (2007), 0707.1285.

[10] A. GEHRMANN-DE RIDDER, T. GEHRMANN, E. W. N. GLOVER and G. HEINRICH, JHEP **12**, 094 (2007), 0711.4711.

[11] A. GEHRMANN-DE RIDDER, T. GEHRMANN, E. W. N. GLOVER and G. HEINRICH, Phys. Rev. Lett. **100**, 172001 (2008), 0802.0813.

[12] A. GEHRMANN-DE RIDDER, T. GEHRMANN, E. W. N. GLOVER and G. HEINRICH, JHEP **05**, 106 (2009), 0903.4658.

[13] S. WEINZIERL, Phys. Rev. Lett. **101**, 162001 (2008), 0807.3241.

[14] S. WEINZIERL, JHEP **06**, 041 (2009), 0904.1077.

[15] S. WEINZIERL, Phys. Rev. **D80**, 094018 (2009), 0909.5056.

[16] S. BLOCH, H. ESNAULT and D. KREIMER, Comm. Math. Phys. **267**, 181 (2006), math.AG/0510011.

[17] S. BLOCH and D. KREIMER, Commun. Num. Theor. Phys. **2**, 637 (2008), 0804.4399.

[18] S. BLOCH, (2008), 0810.1313.

[19] F. BROWN, Commun. Math. Phys. **287**, 925 (2008), 0804.1660.

[20] F. BROWN, (2009), 0910.0114.

[21] F. BROWN and K. YEATS, (2009), 0910.5429.

[22] O. SCHNETZ, (2008), 0801.2856.

[23] O. SCHNETZ, (2009), 0909.0905.

[24] P. ALUFFI and M. MARCOLLI, Commun. Num. Theor. Phys. **3**, 1 (2009), 0807.1690.

[25] P. ALUFFI and M. MARCOLLI, (2008), 0811.2514.

[26] P. ALUFFI and M. MARCOLLI, (2009), 0901.2107.
[27] P. ALUFFI and M. MARCOLLI, (2009), 0907.3225.
[28] C. BERGBAUER, R. BRUNETTI and D. KREIMER, (2009), 0908.0633.
[29] S. LAPORTA, Phys. Lett. **B549**, 115 (2002), hep-ph/0210336.
[30] S. LAPORTA and E. REMIDDI, Nucl. Phys. **B704**, 349 (2005), hep-ph/0406160.
[31] S. LAPORTA, Int. J. Mod. Phys. **A23**, 5007 (2008), 0803.1007.
[32] D. H. BAILEY, J. M. BORWEIN, D. BROADHURST and M. L. GLASSER, (2008), 0801.0891.
[33] I. BIERENBAUM and S. WEINZIERL, Eur. Phys. J. **C32**, 67 (2003), hep-ph/0308311.
[34] O. V. TARASOV, Phys. Rev. **D54**, 6479 (1996), hep-th/9606018.
[35] O. V. TARASOV, Nucl. Phys. **B502**, 455 (1997), hep-ph/9703319.
[36] M. KONTSEVICH and D. ZAGIER, in: B. Engquis and W. Schmid, editors, Mathematics unlimited - 2001 and beyond , 771 (2001).
[37] M. YOSHINAGA, (2008), 0805.0349.
[38] B. FRIEDRICH, (2005), math.AG/0506113.
[39] P. BELKALE and P. BROSNAN, Int. Math. Res. Not. , 2655 (2003).
[40] C. BOGNER and S. WEINZIERL, J. Math. Phys. **50**, 042302 (2009), 0711.4863.
[41] K. HEPP, Commun. Math. Phys. **2**, 301 (1966).
[42] M. ROTH and A. DENNER, Nucl. Phys. **B479**, 495 (1996), hep-ph/9605420.
[43] T. BINOTH and G. HEINRICH, Nucl. Phys. **B585**, 741 (2000), hep-ph/0004013.
[44] T. BINOTH and G. HEINRICH, Nucl. Phys. **B680**, 375 (2004), hep-ph/0305234.
[45] C. BOGNER and S. WEINZIERL, Comput. Phys. Commun. **178**, 596 (2008), 0709.4092.
[46] A. V. SMIRNOV and M. N. TENTYUKOV, Comput. Phys. Commun. **180**, 735 (2009), 0807.4129.
[47] A. V. SMIRNOV and V. A. SMIRNOV, (2008), arXiv:0812.4700.
[48] H. HIRONAKA, Ann. Math. **79**, 109 (1964).
[49] M. SPIVAKOVSKY, Progr. Math. **36**, 419 (1983).
[50] S. ENCINAS and H. HAUSER, Comment. Math. Helv. **77**, 821 (2002).
[51] H. HAUSE, Bull. Amer. Math. Soc. **40**, 323 (2003).
[52] D. ZEILLINGER, Enseign. Math. **52**, 143 (2006).
[53] J. ECALLE, ARI/GARI. *La dimorphie et l'arithmétique des multizetas: un premier bilan*, Journal de Théorie des Nombres de Bordeaux **15** (2003), 411–478.

[54] C. REUTENAUER, *Free Lie Algebras*, Clarendon Press, Oxford, 1993.
[55] M. SWEEDLER, *Hopf Algebras*, Benjamin, New York, 1969.
[56] M. E. HOFFMAN, J. Algebraic Combin. **11**, 49 (2000), math. QA/9907173.
[57] L. GUO and W. KEIGHER, Adv. in Math. **150**, 117 (2000), math.RA/0407155.
[58] J. M. BORWEIN, D. M. BRADLEY, D. J. BROADHURST and P. LISONEK, Trans. Amer. Math. Soc. **353:3**, 907 (2001), math. CA/9910045.
[59] A. B. GONCHAROV, Math. Res. Lett. **5**, 497 (1998) (available at http://www.math.uiuc.edu/K-theory/0297).
[60] H. M. MINH, M. PETITOT and J. VAN DER HOEVEN, Discrete Math. **225:1-3**, 217 (2000).
[61] P. CARTIER, Séminaire Bourbaki, 885 (2001).
[62] G. RACINET, Publ. Math. Inst. Hautes Études Sci. **95**, 185 (2002), math.QA/0202142.
[63] N. NIELSEN, Nova Acta Leopoldina (Halle) **90**, 123 (1909).
[64] E. REMIDDI and J. A. M. VERMASEREN, Int. J. Mod. Phys. **A15**, 725 (2000), hep-ph/9905237.
[65] T. GEHRMANN and E. REMIDDI, Nucl. Phys. **B601**, 248 (2001), hep-ph/0008287.
[66] G. 'T HOOFT and M. J. G. VELTMAN, Nucl. Phys. **B153**, 365 (1979).
[67] J. VOLLINGA and S. WEINZIERL, Comput. Phys. Commun. **167**, 177 (2005), hep-ph/0410259.
[68] S. MOCH, P. UWER and S. WEINZIERL, J. Math. Phys. **43**, 3363 (2002), hep-ph/0110083.
[69] S. WEINZIERL, Comput. Phys. Commun. **145**, 357 (2002), math-ph/0201011.
[70] S. WEINZIERL, J. Math. Phys. **45**, 2656 (2004), hep-ph/0402131.
[71] S. MOCH and P. UWER, Comput. Phys. Commun. **174**, 759 (2006), math-ph/0508008.
[72] L. EULER, Novi Comm. Acad. Sci. Petropol. **20**, 140 (1775).
[73] D. ZAGIER, First European Congress of Mathematics, Vol. II, Birkhäuser, Boston, 497 (1994).
[74] J. A. M. VERMASEREN, Int. J. Mod. Phys. **A14**, 2037 (1999), hep-ph/9806280.
[75] J. BLÜMLEIN and S. KURTH, Phys. Rev. **D60**, 014018 (1999), hep-ph/9810241.
[76] J. BLÜMLEIN, Comput. Phys. Commun. **159**, 19 (2004), hep-ph/0311046.

[77] J. BLÜMLEIN, D. J. BROADHURST and J. A. M. VERMASEREN, Comput. Phys. Commun. **181**, 582 (2010), 0907.2557.
[78] A. V. KOTIKOV, Phys. Lett. **B254**, 158 (1991).
[79] A. V. KOTIKOV, Phys. Lett. **B267**, 123 (1991).
[80] E. REMIDDI, Nuovo Cim. **A110**, 1435 (1997), hep-th/9711188.
[81] T. GEHRMANN and E. REMIDDI, Nucl. Phys. **B580**, 485 (2000), hep-ph/9912329.
[82] T. GEHRMANN and E. REMIDDI, Nucl. Phys. **B601**, 287 (2001), hep-ph/0101124.

The flexion structure and dimorphy: flexion units, singulators, generators, and the enumeration of multizeta irreducibles

Jean Ecalle
with computational assistance from S. Carr

Abstract. We present a self-contained survey of the flexion structure and its core ARI//GARI. We explain why this pair algebra//group is uniquely suited to the generation, manipulation, description and illumination of double symmetries, and therefore conducive to an in-depth understanding of arithmetical dimorphy. Special emphasis is laid on the monogenous algebras generated by flexion units, their special bimoulds, and the corresponding singulators. We then attempt a broad-brush overview of the whole question of canonical irreducibles and introduce the promising subject of perinomal algebra. As a recreational aside, we also state, justify, and computationally check a refinement of the standard conjectures about the enumeration of multizeta irreducibles.

Contents **27**

1 Introduction and reminders . . . 30
1.1 Multizetas and dimorphy . . . 30
1.2 From scalars to generating series . . . 32
1.3 ARI//GARI and its dimorphic substructures . . . 35
1.4 Flexion units, singulators, double symmetries . . 36
1.5 Enumeration of multizeta irreducibles . . . 37
1.6 Canonical irreducibles and perinomal algebra . . 38
1.7 Purpose of the present survey . . . 38
2 Basic dimorphic algebras . . . 40
2.1 Basic operations . . . 40
2.2 The algebra *ARI* and its group *GARI* . . . 45
2.3 Action of the basic involution *swap* . . . 48
2.4 Straight symmetries and subsymmetries . . . 49
2.5 Main subalgebras . . . 52
2.6 Main subgroups . . . 53
2.7 The dimorphic algebra $ARI^{\underline{al}/\underline{al}} \subset ARI^{al/al}$ . . . 53
2.8 The dimorphic group $GARI^{\underline{as}/\underline{as}} \subset GARI^{as/as}$ . . 54
3 Flexion units and twisted symmetries . . . 54
3.1 The free monogenous flexion algebra $Flex(\mathfrak{E})$ . . 54

	3.2	Flexion units	57
	3.3	Unit-generated algebras $Flex(\mathfrak{E})$	61
	3.4	Twisted symmetries and subsymmetries in universal mode	63
	3.5	Twisted symmetries and subsymmetries in polar mode	67
4	Flexion units and dimorphic bimoulds		70
	4.1	Remarkable substructures of $Flex(\mathfrak{E})$	70
	4.2	The secondary bimoulds $\mathfrak{ess}^\bullet$ and $\mathfrak{esz}^\bullet$	79
	4.3	The related primary bimoulds $\mathfrak{es}^\bullet$ and $\mathfrak{ez}^\bullet$	87
	4.4	Some basic bimould identities	88
	4.5	Trigonometric and bitrigonometric bimoulds	89
	4.6	Dimorphic isomorphisms in universal mode	94
	4.7	Dimorphic isomorphisms in polar mode	95
5	Singulators, singulands, singulates		99
	5.1	Some heuristics. Double symmetries and imparity	99
	5.2	Universal singulators $senk(\mathfrak{ess}^\bullet)$ and $seng(\mathfrak{es}^\bullet)$	101
	5.3	Properties of the universal singulators	102
	5.4	Polar singulators: description and properties	104
	5.5	Simple polar singulators	105
	5.6	Composite polar singulators	105
	5.7	From $\underline{al}/\underline{al}$ to $\underline{al}/\underline{il}$. Nature of the singularities	106
6	A natural basis for $ALIL \subset ARI^{\underline{al}/\underline{il}}$		107
	6.1	Singulation-desingulation: the general scheme	107
	6.2	Singulation-desingulation up to length 2	111
	6.3	Singulation-desingulation up to length 4	112
	6.4	Singulation-desingulation up to length 6	112
	6.5	The basis $lama^\bullet/lami^\bullet$	116
	6.6	The basis $loma^\bullet/lomi^\bullet$	116
	6.7	The basis $luma^\bullet/lumi^\bullet$	117
	6.8	Arithmetical vs analytic smoothness	118
	6.9	Singulator kernels and "wandering" bialternals	119
7	A conjectural basis for $ALAL \subset ARI^{\underline{al}/\underline{al}}$. The three series of bialternals		120
	7.1	Basic bialternals: the enumeration problem	120
	7.2	The regular bialternals: *ekma, doma*	120
	7.3	The irregular bialternals: *carma*	121
	7.4	Main differences between regular and irregular bialternals	121
	7.5	The *pre-doma* potentials	123
	7.6	The *pre-carma* potentials	124
	7.7	Construction of the *carma* bialternals	126

7.8 Alternative approach 127
7.9 The global bialternal ideal and the universal 'restoration' mechanism 129
8 The enumeration of bialternals. Conjectures and computational evidence 130
8.1 Primary, sesquary, secondary algebras 130
8.2 The 'factor' algebra *EKMA* and its subalgebra *DOMA* 132
8.3 The 'factor' algebra *CARMA* 133
8.4 The total algebra of bialternals *ALAL* and the original BK-conjecture 133
8.5 The factor algebras and our sharper conjectures . 133
8.6 Cell dimensions for *ALAL* 135
8.7 Cell dimensions for *EKMA* 135
8.8 Cell dimensions for *DOMA*. 136
8.9 Cell dimensions for *CARMA* 136
8.10 Computational checks (Sarah Carr) 137
9 Canonical irreducibles and perinomal algebra 139
9.1 The general scheme 139
9.2 Arithmetical criteria 144
9.3 Functional criteria 144
9.4 Notions of perinomal algebra 147
9.5 The all-encoding perinomal mould $peri^{\bullet}$ 149
9.6 A glimpse of perinomal splendour 150
10 Provisional conclusion 152
10.1 Arithmetical and functional dimorphy 152
10.2 Moulds and bimoulds. The flexion structure . . . 154
10.3 *ARI*/*GARI* and the handling of double symmetries 158
10.4 What has already been achieved 160
10.5 Looking ahead: what is within reach and what beckons from afar 164
11 Complements 165
11.1 Origin of the flexion structure 165
11.2 From simple to double symmetries. The *scramble* transform 167
11.3 The bialternal tesselation bimould 168
11.4 Polar, trigonometric, bitrigonometric symmetries 171
11.5 The separative algebras $Inter(Qi_c)$ and $Exter(Qi_c)$ 175
11.6 Multizeta cleansing: elimination of unit weights 180
11.7 Multizeta cleansing: elimination of odd degrees . 189
11.8 $GARI_{\mathfrak{se}}$ and the two separation lemmas 192
11.9 Bisymmetrality of $\mathfrak{ess}^{\bullet}$: conceptual proof 193

11.10 Bisymmetrality of $\mathfrak{ess}^\bullet$: combinatorial proof . . 195
12 Tables, index, references 198
12.1 Table 1: basis for $Flex(\mathfrak{E})$ 198
12.2 Table 2: basis for $Flexin(\mathfrak{E})$ 202
12.3 Table 3: basis for $Flexinn(\mathfrak{E})$ 202
12.4 Table 4: the universal bimould $\mathfrak{ess}^\bullet$ 205
12.5 Table 5: the universal bimould $\mathfrak{esz}^\bullet_\sigma$ 206
12.6 Table 6: the bitrigonometric bimould $taal^\bullet/tiil^\bullet$. 207
12.7 Index of terms and notations 207
References 210

1 Introduction and reminders

1.1 Multizetas and dimorphy

Let us take as our starting point *arithmetical dimorphy*, which in its purest form manifests in the ring of *multizetas*. Some extremely important $\mathbb{Q}$-rings of transcendental numbers happen to be *dimorphic*, *i.e.* to possess two *natural* $\mathbb{Q}$-prebases[1] $\{\alpha_m\}$, $\{\beta_n\}$ with a simple *conversion rule* and two independent *multiplication tables*, all of which involve only rational coefficients and finite sums:

$$\alpha_m = \sum\nolimits^* H_m^n\, \beta_n \quad , \quad \beta_n = \sum\nolimits^* K_n^m\ \alpha_m \qquad (H_m^n\,,\ K_n^m \in \mathbb{Q})$$
$$\alpha_{m_1}\,\alpha_{m_2} = \sum\nolimits^* A^{m_3}_{m_1,m_2}\,\alpha_{m_3} \ ,\ \beta_{n_1}\,\beta_{n_2} = \sum\nolimits^* B^{n_3}_{n_1,n_2}\,\beta_{n_3} \ \ (A^{n_3}_{n_1,n_2},\ B^{n_3}_{n_1,n_2} \in \mathbb{Q}).$$

The simplest, most basic of all such rings is $\mathbb{Z}$eta, which is not only *multiplicatively* generated but also *linearly* spanned by the so-called *multizetas.*[2]

In the *first prebasis*, the multizetas are given by polylogarithmic integrals:

$$\mathrm{Wa}_*^{\alpha_1,\ldots,\alpha_l} := (-1)^{l_0} \int_0^1 \frac{dt_l}{(\alpha_l - t_l)} \cdots \int_0^{t_3} \frac{dt_2}{(\alpha_2 - t_2)} \int_0^{t_2} \frac{dt_1}{(\alpha_1 - t_1)} \tag{1.1}$$

with indices α_j that are either 0 or unit roots, and $l_0 := \sum_{\alpha_i=0} 1$.

---

[1] With some natural countable indexation $\{m\}$, $\{n\}$, not necessarily on $\mathbb{N}$ or $\mathbb{Z}$. We recall that a set $\{\alpha_m\}$ is a $\mathbb{Q}$-prebasis (or 'spanning subset') of a $\mathbb{Q}$-ring $\mathbb{D}$ if any $\alpha \in \mathbb{D}$ is expressible as a finite linear combination of the α_m's with rational coefficients. But the α_m's need not be $\mathbb{Q}$-independent. When they are, we say that $\{\alpha_m\}$ is a $\mathbb{Q}$-basis.

[2] Also known as MZV, short for *multiple zeta values.*

In the *second prebasis*, multizetas are expressed as "harmonic sums":

$$\mathrm{Ze}_*^{\binom{\epsilon_1 \,,\ldots,\, \epsilon_r}{s_1 \,,\ldots,\, s_r}} := \sum_{n_1 > \cdots > n_r > 0} n_1^{-s_1} \ldots n_r^{-s_r}\, e_1^{-n_1} \ldots e_r^{-n_r} \tag{1.2}$$

with $s_j \in \mathbb{N}^*$ and unit roots $e_j := \exp(2\pi i \epsilon_j)$ with 'logarithms' $\epsilon_j \in \mathbb{Q}/\mathbb{Z}$.

The stars $*$ means that the integrals or sums are provisionally assumed to be convergent or semi-convergent: for $Wa_*^{\boldsymbol{\alpha}}$ this means that $\alpha_1 \neq 0$ and $\alpha_l \neq 1$, and for $Ze_*^{\binom{\boldsymbol{\epsilon}}{\mathbf{s}}}$ this means that $\binom{\epsilon_1}{s_1} \neq \binom{0}{1}$ *i.e.* $\binom{e_1}{s_1} \neq \binom{1}{1}$.

The corresponding moulds $Wa_*^\bullet$ and $Ze_*^\bullet$ turn out to be respectively *symmetral* and *symmetrel*:[3]

$$\mathrm{Wa}_*^{\boldsymbol{\alpha}^1}\, \mathrm{Wa}_*^{\boldsymbol{\alpha}^2} = \sum_{\boldsymbol{\alpha} \in \mathrm{sha}(\boldsymbol{\alpha}^1, \boldsymbol{\alpha}^2)} \mathrm{Wa}_*^{\boldsymbol{\alpha}} \qquad \forall \boldsymbol{\alpha}^1, \forall \boldsymbol{\alpha}^2 \tag{1.3}$$

$$\mathrm{Ze}_*^{\binom{\boldsymbol{\epsilon}^1}{\mathbf{s}^1}}\, \mathrm{Ze}_*^{\binom{\boldsymbol{\epsilon}^2}{\mathbf{s}^2}} = \sum_{\binom{\boldsymbol{\epsilon}}{\mathbf{s}} \in \mathrm{she}\left(\binom{\boldsymbol{\epsilon}^1}{\mathbf{s}^1}, \binom{\boldsymbol{\epsilon}^2}{\mathbf{s}^2}\right)} \mathrm{Ze}_*^{\binom{\boldsymbol{\epsilon}}{\mathbf{s}}} \qquad \forall \binom{\boldsymbol{\epsilon}^1}{\mathbf{s}^1}, \forall \binom{\boldsymbol{\epsilon}^2}{\mathbf{s}^2}. \tag{1.4}$$

These are the so-called *quadratic relations*, which express multizeta dimorphy. As for the conversion rule, it reads:[4]

$$\mathrm{Wa}_*^{e_1, 0^{[s_1-1]}, \ldots, e_r, 0^{[s_r-1]}} := \mathrm{Ze}_*^{\binom{\epsilon_r \,,\, \epsilon_{r-1:r} \,,\ldots,\, \epsilon_{1:2}}{s_r \,,\, s_{r-1} \,,\ldots,\, s_1}} \tag{1.5}$$

$$\mathrm{Ze}_*^{\binom{\epsilon_1 \,,\, \epsilon_2 \,,\ldots,\, \epsilon_r}{s_1 \,,\, s_2 \,,\ldots,\, s_r}} =: \mathrm{Wa}_*^{e_1 \ldots e_r, 0^{[s_r-1]}, \ldots, e_1 e_2, 0^{[s_2-1]}, e_1, 0^{[s_1-1]}} \tag{1.6}$$

with $0^{[k]}$ denoting a subsequence of k zeros.

There happen to be unique extensions $Wa_*^\bullet \to Wa^\bullet$ and $Ze_*^\bullet \to Ze^\bullet$ that cover the divergent cases and keep our moulds symmetral or symmetrel while conforming to the 'initial conditions' $Wa^0 = Wa^1 = 0$ and $Ze^{\binom{0}{1}} = 0$. The only price to pay will be a slight modification of the conversion rule: see Section 1.2 *infra*.

Basic gradations/filtrations

Four parameters dominate the discussion:

– the *weight* $s := \sum s_i$ (in the $Ze^\bullet$-encoding) or $s := l$ (in the $Wa^\bullet$-encoding);

– the *length* or *"depth"* $r :=$ number of ϵ_i's or s_i's or non-zero α_i's;

[3] As usual, $sha(\boldsymbol{\omega}', \boldsymbol{\omega}'')$ denotes the set of all simple shufflings of the sequences $\boldsymbol{\omega}', \boldsymbol{\omega}''$, whereas in $she(\boldsymbol{\omega}', \boldsymbol{\omega}'')$ we allow (any number of) order-compatible contractions $\omega'_i + \omega''_j$.

[4] With the usual shorthand for differences: $\epsilon_{i:j} := \epsilon_i - \epsilon_j$.

– the *degree* $d := s - r$ = number of zero α_i's in the $Wa^{\bullet}$-encoding;[5]
– the *"coloration"* $p :=$ smallest p such that all root-related ϵ_i be in $\frac{1}{p}\mathbb{Z}/\mathbb{Z}$.
Only the weight s defines an (additive and multiplicative) gradation; the other parameters merely induce filtrations.

1.2 From scalars to generating series

The natural encodings $Wa^{\bullet}$ and $Ze^{\bullet}$ being unwieldy and too heterogeneous in their indexations, we must replace them by suitable *generating series*, so chosen as to preserve the simplicity of the two quadratic relations and of the conversion rule. This essentially *imposes* the following definitions:[6]

$$\mathrm{Zag}^{\left(\begin{smallmatrix} u_1 & ,\dots, & u_r \\ \epsilon_1 & ,\dots, & \epsilon_r \end{smallmatrix}\right)} := \sum_{1 \le s_j} \mathrm{Wa}^{e_1, 0^{[s_1-1]}, \dots, e_r, 0^{[s_r-1]}}\, u_1^{s_1-1} u_{1,2}^{s_2-1} \dots u_{1\dots r}^{s_r-1} \quad (1.7)$$

$$\mathrm{Zig}^{\left(\begin{smallmatrix} \epsilon_1 & ,\dots, & \epsilon_r \\ v_1 & ,\dots, & v_r \end{smallmatrix}\right)} := \sum_{1 \le s_j} \mathrm{Ze}^{\left(\begin{smallmatrix} \epsilon_1 & ,\dots, & \epsilon_r \\ s_1 & ,\dots, & s_r \end{smallmatrix}\right)}\, v_1^{s_1-1} \dots v_r^{s_r-1}. \quad (1.8)$$

The first series $Zag^{\bullet}$, via its Taylor coefficients, gives rise to yet another $\mathbb{Q}$-prebasis $\{Za^{\bullet}\}$ for the $\mathbb{Q}$-ring of multizetas. The mould $Za^{\bullet}$ is symmetral like $Wa^{\bullet}$ but quite distinct from it and much closer, in form and indexation, to the symmetrel mould $Ze^{\bullet}$:

$$\mathrm{Zag}^{\left(\begin{smallmatrix} u_1 & ,\dots, & u_r \\ \epsilon_1 & ,\dots, & \epsilon_r \end{smallmatrix}\right)} =: \sum_{1 \le s_j} \mathrm{Za}^{\left(\begin{smallmatrix} s_1 & ,\dots, & s_r \\ \epsilon_1 & ,\dots, & \epsilon_r \end{smallmatrix}\right)}\, u_1^{s_1-1} \dots u_r^{s_r-1}. \quad (1.9)$$

These power series are actually convergent: they define *generating functions*[7] that are meromorphic, with multiple poles at simple locations. These functions, in turn, verify simple difference equations, and admit an elementary mould factorisation (mark the exchange in the positions of *do* and *co*):

$$\mathrm{Zag}^{\bullet} := \lim_{k\to\infty} \mathrm{Zag}_k^{\bullet} = \lim_{k\to\infty} (\mathrm{doZag}_k^{\bullet} \times \mathrm{coZag}_k^{\bullet}) \quad (1.10)$$

$$\mathrm{Zig}^{\bullet} := \lim_{k\to\infty} \mathrm{Zig}_k^{\bullet} = \lim_{k\to\infty} (\mathrm{coZig}_k^{\bullet} \times \mathrm{doZig}_k^{\bullet}) \quad (1.11)$$

5 d is called *degree*, because under the correspondence *scalars* $\to$ *generating series*, the multizetas become coefficients of monomials of total degree d. See (2.19), (2.23).

6 With the usual abbreviations: $u_{i,j} = u_i + u_j$, $u_{i,j,k} = u_i + u_j + u_k$ etc.

7 Still denoted by the same symbols.

with dominant parts *doZag•/doZig•* that carry the $\boldsymbol{u}/\boldsymbol{v}$-dependence[8]:

$$\mathrm{doZag}_k^{\binom{u_1\ ,\dots,\ u_r}{\epsilon_1\ ,\dots,\ \epsilon_r}} := \sum_{1\le m_i\le k} e_1^{-m_1}\dots e_r^{-m_r} P(m_1-u_1)P(m_{1,2}-u_{1,2})\dots P(m_{1..r}-u_{1..r}) \tag{1.12}$$

$$\mathrm{doZig}_k^{\binom{\epsilon_1\ ,\dots,\ \epsilon_r}{v_1\ ,\dots,\ v_r}} := \sum_{k\ge n_1>n_2>\dots n_r\ge 1} e_1^{-n_1}\dots e_r^{-n_r} P(n_1-v_1)P(n_2-v_2)\dots P(n_r-v_r) \tag{1.13}$$

and corrective parts *coZag•/coZig•* that reduce to constants:

$$\mathrm{coZag}_k^{\binom{u_1\ ,\dots,\ u_r}{0\ ,\dots,\ 0}} := (-1)^r \sum_{1\le m_i\le k} P(m_1)P(m_{1,2})\dots P(m_{1\dots r}) \tag{1.14}$$

$$\mathrm{coZig}_k^{\binom{0\ ,\dots,\ 0}{v_1\ ,\dots,\ v_r}} := (-1)^r \sum_{k\ge n_1\ge n_2\ge\dots n_r\ge 1} \mu^{n_1,\dots,n_r} P(n_1)P(n_2)\dots P(n_r) \tag{1.15}$$

$$\mathrm{coZag}_k^{\binom{u_1\ ,\dots,\ u_r}{\epsilon_1\ ,\dots,\ \epsilon_r}} := 0 \quad \mathit{if} \quad (\epsilon_1,\dots,\epsilon_r)\neq(0,\dots,0) \tag{1.16}$$

$$\mathrm{coZig}_k^{\binom{\epsilon_1\ ,\dots,\ \epsilon_r}{v_1\ ,\dots,\ v_r}} := 0 \quad \mathit{if} \quad (\epsilon_1,\dots,\epsilon_r)\neq(0,\dots,0) \tag{1.17}$$

with $P(t) := 1/t$ (here and throughout) and with $\mu^{n_1,n_2,\dots,n_r} := \frac{1}{r_1!r_2!..r_l!}$ if the non-increasing sequence $(n_1,\dots,n_r)$ attains r_1 times its highest value, r_2 times its second highest value, etc.

Setting $\mathit{Mini}_k^\bullet := \mathit{Zig}_k^\bullet\|_{\mathbf{v}=0}$ we find:[9]

$$\mathrm{Mini}_k^{\binom{0\ ,\dots,\ 0}{v_1\ ,\dots,\ v_r}} := \sum_{\left[{2\le r_1\le r_2\dots\le r_l \atop r_1+r_2+\dots r_l=r}\right]}^{\left[{1\le l\le r/2 \atop 1\le n_i\le k}\right]} (-1)^{(r-l)} \mu^{r_1,\dots,r_l} \frac{(P(n_1))^{r_1}}{r_1}\dots\frac{(P(n_l))^{r_l}}{r_l} \tag{1.18}$$

$$\mathrm{Mini}_k^{\binom{\epsilon_1\ ,\dots,\ \epsilon_r}{v_1\ ,\dots,\ v_r}} := 0 \quad \mathit{if} \quad (\epsilon_1,\dots,\epsilon_r)\neq(0,\dots,0). \tag{1.19}$$

Let us now compare the bimoulds $\overline{C}_k^\bullet$ and $\underline{C}_k^\bullet$ thus defined:

$$\mathrm{swap.Zag}_k^\bullet = \overbrace{\mathrm{swap.coZag}_k^\bullet} \times \overbrace{\mathrm{swap.doZag}_k^\bullet} = \overline{A}_k^\bullet \times \overline{B}_k^\bullet =: \overline{C}_k^\bullet \tag{1.20}$$

[8] With the usual abbreviations $m_{i,j} := m_i+m_j$, $m_{i,j,k} := m_i+m_j+m_k$ etc.

[9] If we had no factor $\mu^{n_1,\dots,n_r}$ in (1.18), we would have $\mathit{Zig}_k^\bullet\|_{\mathbf{v}=0} = 0$ and therefore no $\mathit{Mini}_k^\bullet$ terms. But the mould $\mathit{Zig}_k^\bullet$ would fail to be *symmetril*, as required. Herein lies the origin of the corrective terms in the conversion rule.

$$(\mathrm{Mini}_k^\bullet)^{-1}\times\mathrm{Zig}_k^\bullet=\overbrace{(\mathrm{Mini}_k^\bullet)^{-1}\times\mathrm{coZig}_k^\bullet}\times\overbrace{\mathrm{doZig}_k^\bullet}=\underline{A}_k^\bullet\times\underline{B}_k^\bullet=:\underline{C}_k^\bullet \quad (1.21)$$

with $\times$ standing for ordinary mould or bimould multiplication[10]; with $(\mathrm{Mini}_k^\bullet)^{-1}$ denoting the multiplicative inverse of $(\mathrm{Mini}_k^\bullet)$; and with the involution *swap* defined as in (2.9) *infra*. Here, the $\boldsymbol{v}$-dependent factors $\overline{B}_k^{\binom{\epsilon}{v}}$ and $\underline{B}_k^{\binom{\epsilon}{v}}$ are both given by the finite sum

$$\sum e_1^{-n_1}\dots e_r^{-n_r}\,P(n_1-v_1)\dots P(n_r-v_r) \qquad (1.22)$$

with summation respectively over the domains $\overline{\mathcal{B}}_{r,k}$ and $\underline{\mathcal{B}}_{r,k}$

$$\overline{\mathcal{B}}_{r,k}:=\{k\geq n_r\geq 1,\,2k\geq n_{r-1}>n_r,\dots,\,(r-1)k\geq n_2>n_3,\,r\,k\geq n_1>n_2\}$$
$$\underline{\mathcal{B}}_{r,k}:=\{k\geq n_1>n_2>\dots n_{r-1}>n_r\geq 1\}.$$

Likewise, the $\boldsymbol{v}$-independent factors $\overline{A}_k^{\binom{\epsilon}{v}}$ and $\underline{A}_k^{\binom{\epsilon}{v}}$ vanish unless $\boldsymbol{\epsilon}=0$, in which case they are both given by the finite sum

$$\sum(-1)^r\,P(n_1)\dots P(n_r) \qquad (1.23)$$

with summation respectively over the domains $\overline{\mathcal{A}}_{r,k}$ and $\underline{\mathcal{A}}_{r,k}$

$$\overline{\mathcal{A}}_{r,k}:=\{k\geq n_1\geq 1,\,2k\geq n_2\geq n_1,\dots,(r-1)k\geq n_{r-1}\geq n_{r-2},\,rk\geq n_r\geq n_{r-1}\}$$
$$\underline{\mathcal{A}}_{r,k}:=\{k\geq n_r\geq n_{r-1}\geq\dots n_2\geq n_1\geq 1\}.$$

It easily follows from the above that for any compact $K\subset\mathbb{C}^r$ and k large enough, the difference $\overline{C}_k^{\binom{\epsilon}{v}}-\underline{C}_k^{\binom{\epsilon}{v}}$ is holomorphic on K, and that there exists a constant c_K such that:

$$\|\overline{C}_k^{\binom{\epsilon}{v}}-\underline{C}_k^{\binom{\epsilon}{v}}\|\leq(c_K)^r\,\frac{(\log k)^{r-1}}{k}\qquad(\boldsymbol{v}\in K\ ,\ k\ \text{large}). \qquad (1.24)$$

Summing up, we have an exact equivalence between old and new symmetries:[11]

$$\{\mathrm{Wa}^\bullet\ \text{symmetral}\}\Longleftrightarrow\{\mathrm{Zag}^\bullet\ \text{symmetral}\} \qquad (1.25)$$
$$\{\mathrm{Ze}^\bullet\ \text{symmetrel}\}\Longleftrightarrow\{\mathrm{Zig}^\bullet\ \text{symmetril}\} \qquad (1.26)$$

[10] In the case of bimoulds, $\times$ is often noted *mu* the better to distinguish it from the various other *flexion* products.

[11] *Symmetrility* is precisely defined in Section 3.5. Roughly, it mirrors *symmetrelity*, but with all contractions $M^{(\dots,\omega'_i+\omega''_j,\dots)}$ systematically replaced by $M^{\binom{\dots,\ u'_i+u''_j\ ,\dots}{\dots,\ \ v'_i\ \ ,\dots}}\,P(v'_i-v''_j)+M^{\binom{\dots,\ u'_i+u''_j\ ,\dots}{\dots,\ \ v''_j\ \ ,\dots}}\,P(v''_j-v'_i)$.

and the old conversion rule for scalar multizetas [12] becomes:

$$\mathrm{Zig}^\bullet = \mathrm{Mini}^\bullet \times \mathrm{swap(Zag)}^\bullet \tag{1.27}$$

$$(\iff \mathrm{swap}(\mathrm{Zig}^\bullet) = \mathrm{Zag}^\bullet \times \mathrm{Mana}^\bullet) \tag{1.28}$$

with elementary moulds $Mana^\bullet/Mini^\bullet := \lim_{k\to\infty} Mana_k^\bullet/Mini_k^\bullet$ whose only non-zero components:

$$\mathrm{Mana}^{\binom{u_1\ ,\dots,\ u_r}{0\ ,\dots,\ 0}} \equiv \mathrm{Mini}^{\binom{0\ ,\dots,\ 0}{v_1\ ,\dots,\ v_r}} \equiv \mathrm{Mono}_r \tag{1.29}$$

due to (1.18), may be expressed in terms of monozetas:

$$1 + \sum_{r\geq 2} \mathrm{Mono}_r\, t^r := \exp\left(\sum_{s\geq 2}(-1)^{s-1}\zeta(s)\,\frac{t^s}{s}\right). \tag{1.30}$$

To these relations one must add the so-called *self-consistency* relations:

$$\mathrm{Zag}^{\binom{u_1\ ,\dots,\ u_r}{q\epsilon_1\ ,\dots,\ q\epsilon_r}} \equiv \sum_{q\epsilon_i^*=q\epsilon_i} \mathrm{Zag}^{\binom{q u_1\ ,\dots,\ q u_r}{\epsilon_1^*\ ,\dots,\ \epsilon_r^*}}\ \forall q|p, \forall u_i\in\mathbb{C}, \forall\epsilon_i,\epsilon_i^*\in\frac{1}{p}\mathbb{Z}/\mathbb{Z} \tag{1.31}$$

which merely reflect trivial identities between unit roots of order p.

1.3 ARI//GARI and its dimorphic substructures

What is required at this point is an algebraic apparatus capable of accommodating Janus-like objects like $Zag^\bullet/Zig^\bullet$, *i.e.* an apparatus with operations that not only respect double symmetries and reproduce them under composition, but also construct them from scratch, *i.e.* from a few simple generators.

Such a machinery is at hand: it is the *flexion structure*, which arose in the early 90s in the context of *singularity analysis*, more precisely in the investigation of *parametric* or *"co-equational" resurgence*. Its objects are *bimoulds*, *i.e.* moulds $M^\bullet$ of the form

$$M^\bullet \in \mathrm{BIMU} \quad \iff \quad M^\bullet = \{M^{w_1,\dots,w_r} = M^{\binom{u_1\ \dots\ u_r}{v_1\ \dots\ v_r}}\} \tag{1.32}$$

with a double-layered indexation $w_i = \binom{u_i}{v_i}$. What makes these $M^\bullet$ into bimoulds, however, is not so much their double indexation as the very specific manner in which upper and lower indices transform and interact: all bimould operations can be expressed in terms of four elementary *flexions* that go by pairs, $\rceil$ with $\lfloor$ and $\rfloor$ with $\lceil$, and have the effect

[12] Namely, some modified form of the rules (2.16),(2.17), which apply in the *convergent* case.

of *adding together* several consecutive u_i and of *pairwise subtracting* several v_i, and that too in such a way as to conserve the scalar product $<\boldsymbol{u}, \boldsymbol{v}> := \sum u_i v_i$ and the symplectic form $d\boldsymbol{w} := \sum du_i \wedge dv_i$. Lastly, central to the flexion structure is a basic involution *swap* which acts on *BIMU* by turning the u_i's into differences of v_j's, and the v_i's into sums of u_j's (see Section 2.1 below).

The *flexion structure*, to put it loosely but tellingly, is the sum total of all interesting operations and structures that can be constructed on *BIMU* by deftly combining the four elementary flexions. It turns out that these *interesting structures* consist, up to isomorphism, of:
– seven + one Lie groups;
– seven + one Lie algebras (each with its pre-Lie structure);
– seven + one pre-Lie algebras.
In the three series, there exist exactly two triplets of type

group//algebra//superalgebra,

which "respect dimorphy", namely

GARI//ARI//SUARI and *GALI//ALI//SUALI*.

Moreover, when restricted to dimorphic bimoulds (*i.e.* bimoulds displaying a double symmetry), these two triplets actually coincide, thus sparing us the agony of choosing between them.

1.4 Flexion units, singulators, double symmetries

To understand dimorphy, and in particular to decompose the pair $Zag^\bullet/Zig^\bullet$ into the elementary building blocks capable of yielding the *multizeta irreducibles*, we require bimoulds $M^\bullet$ which combine three properties that do not sit well together:
— $M^\bullet$ must possess a given symmetry, say alternal or symmetral;
— swap.$M^\bullet$ must possess its own symmetry, which usually coincides with that of $M^\bullet$ or a variant thereof;
— $M^\bullet$ and swap.$M^\bullet$ must be *entire*, *i.e.* for a given length r their dependence on the complex indices (the u_i's in the case of $M^\bullet$ and the v_i's in the case of swap.$M^\bullet$) must be polynomial or holomorphic or a power series. That precludes, in particular, singularities at the origin.

The strange thing, however, is that in order to come to grips with *"entire dimorphy"* in the above sense, we cannot avoid making repeated use of bimoulds that are dimorphic alright, but with abundant *poles* at the origin. We must then get rid of these poles by subtracting suitable bimoulds, with exactly the same singular part, but without destroying the double symmetry. The only way to pull this off is by using very specific operators, the so-called *singulators*, whose basic ingredients are quite special dimorphic bimoulds, which:

— possess poles at the origin;
— lack the crucial parity property which most other dimorphic bimoulds possess and which ensures their stability under the *ARI* or *GARI* operations;
— are constructed from very elementary functions $\mathfrak{E}^{w_1} = \mathfrak{E}^{\binom{u_1}{v_1}}$, the so-called *flexion units*, of which there exist about a dozen. These *units* are odd in w_1 and verify an elementary functional equation, the *tripartite relation*, which is the most basic relation expressible in terms of flexions.

1.5 Enumeration of multizeta irreducibles

The $\mathbb{Q}$-ring $\mathbb{Z}eta$ of formal multizetas (*i.e.* of multizeta symbols subject only to the two *quadratic relations* (1.3), (1.4)) is known to be a polynomial ring, freely generated by a countable set of so-called *irreducibles.*[13] Hence the question: how many irreducibles (let us call that number $D_{d,r}$) must one pick in each cell of degree d and length r to get a complete and free system of irreducibles? The so-called BK-conjectures,[14] which were formulated in 1996 (they applied to the *genuine* rather than *formal* multizetas, and resulted from purely numerical tests) suggest a startlingly complicated formula for $D_{d,r}$ but no plausible rationale for its strange form. Soon after that, we published in [4] a convincing explanation for the formula, which however went largely unnoticed. We therefore return to the question in Section 5 and Section 7 in much greater detail. We actually enunciate four new conjectures which considerably improve on the original BK-formula, and in Section 8 we report on formal computations carried out by S. Carr to test these strengthened conjectures. But the key lies in the theoretical explanation: in our approach, the irreducibles correspond one-to-one to *polynomial bialternal* bimoulds, of which there exist two series: the regular and utterly simple $ekma^\bullet$ on the one hand, and the exceptional, highly intricate $carma^\bullet$ on the other. We explain in detail the mechanism responsible for the creation of these exceptional generators. That mechanism crucially involves the singulators mentioned in the preceding section.

[13] This fact is almost implicit in the (right) formalism. Indeed, with the notations of Section 9, the general bisymmetral, entire bimould $zag^\bullet$ factors as $zag^\bullet = gari\big(Zag_I, expari(ma^\bullet)\big)$ with $ma^\bullet = \sum_S \rho_S \, ma^\bullet_S$ denoting the general element of *ALIL*. Thus, to any linear basis $\{ma^\bullet_S\}$ of *ALIL*, there corresponds one-to-one a set $\{\rho_S\}$ of irreducibles, with the same countable indexation S, and a transparent formula for expressing the multizetas in terms of these irreducibles. A written exposition, resting on very similar ideas but couched in a quite different formalism, may be found in G. Racinet, *Doubles mélanges des polylogarithmes multiples aux racines de l'unité* , Publ. Math. IHES, 2002.

[14] See [1] and Section 8.4.

1.6 Canonical irreducibles and perinomal algebra

In Section 6 and Section 9 we move from the (d,r)-gradation to the more natural s-gradation, s being the weight. In that new setting, the irreducibles correspond to entire bimoulds which are no longer alternal/alternal (or *bialternal* for short) but alternal/alternil and which for that reason never reduce to a single component, as bialternals do. That may seem a complication, and it is, but it also brings a drastic simplification in its wake: instead of the dual system of generators $\{ekma^{\bullet}_{d}, carma^{\bullet}_{d,k}\}$ for the algebra $ALAL \subset ARI^{\mathrm{al/al}}_{ent}$ of entire bialternals, we now have a single system, either $\{lama^{\bullet}_{s}\}$ or $\{loma^{\bullet}_{s}\}$[15], of generators for the algebra $ALIL \subset ARI^{\mathrm{al/il}}_{ent}$ of all entire bimoulds of alternal/alternil type, with a transparent indexation by all odd weights $s = 3, 5, 7$ etc. Like $carma^{\bullet}$, but to an even greater extent, $lama^{\bullet}$ and $loma^{\bullet}$ depend for their construction on the repeated use of singulators, with parasitical poles being alternately *produced* and then *destroyed*. In Section 6.7 and Section 9 we also introduce a third system of generators for *ALIL*, namely $\{{}^{n}luma^{\bullet}\}$, with indices n now running through $\mathbb{N}^{*}$ and with *functional simplicity*[16] replacing *arithmetical simplicity*[17] as guiding principle. Just like with $lama^{\bullet}$ and $loma^{\bullet}$, the singulators are key to the construction of $luma^{\bullet}$, but under a quite different mechanism, which involves infinitely many (interrelated) linear representations of $Sl_r(\mathbb{Z})$. This is a whole new field unto itself, and a fascinating one at that, which we call *perinomal algebra*, and of which we try to give a foretaste.

1.7 Purpose of the present survey

A four-volume series (on the flexion structure and its applications) is 'in the works', but as often happens with fast-evolving subjects, centrifugal temptations are hard to resist, centripetal discipline difficult to maintain, and the whole bloated project shows more signs of expanding and mutating than of converging. To remedy this, we intend to post some of the accumulated material (including a library of Maple programmes for *ARI//GARI* calculations) online, on our Web-page, in the course of 2012.

[15] These are closely related variants.

[16] The components $luma^{w}$ are meromorphic functions with simple poles away from the origin.

[17] The components $lama^{w}$ and $loma^{w}$ have rational coefficients with "manageable" denominators.

But we feel that a compact Survey like the present one might also serve a purpose – not least that of fixing notations and nomenclature.[18]

Some of the subject-matter laid out here is fairly old – going back eight years in some cases – but unpublished for the most part.[19] There are novelties, too, the main one being perhaps the systematic use of *flexion units* as a means of introducing order into the theory's bewildering plethora of notions and objects: operations, symmetries, structures (algebras, groups) and substructures, bimoulds, bimould identities etc.

'The' flexion unit $\mathfrak{E}^\bullet$ is an unspecified function $\mathfrak{E}^{w_1}$ that is odd in $w_1 := \binom{u_1}{v_1}$ and verifies a bilinear, three-term relation[20] – the so-called *tripartite relation.* From $\mathfrak{E}^\bullet$ one then constructs a whole string of objects (bimoulds, symmetries, subalgebras of *ARI*, subgroups of *GARI*, etc.) which, despite their considerable complexity, owe all their properties to the *tripartite relation* verified by the seed-unit $\mathfrak{E}^\bullet$. As it happens, $\mathfrak{E}^\bullet$ is capable of a dozen or so *distinct realisations* as a concrete function of w_1, each of which automatically induces a realisation of the whole string of satellite objects (bimoulds, symmetries, etc.). The total effect is thus a drastic and welcome *'division by twelve'* of the flexion jungle.

Throughout, there is as much emphasis on the apparatus – the flexion structure and its special bimoulds – as on the applications to multizeta theory. We wind up with a sketch of perinomal algebra, in the hope of stimulating interest in this brand-new subject and of paving the way for a collective programme of exploration,[21] to start hopefully in the course of 2011.

One last word of caution: throughout this paper, the somewhat contentious word *canonical* is never used as a substitute for *unique* (when meaning *unique*, we say *unique*) but as a pointer to the existence, within a class of seemingly undistinguishable objects (like the many conceivable systems of *multizeta irreducibles*) of genuinely privileged representatives. To single out these representatives, esthetic considerations are *unavoidable*, with the residual (often minimal) fuzziness that this entails. But the subjectivity that attaches to the notion in no way detracts from its importance. Quite the opposite, in fact.

[18] Which up till now were still fluctuating from context to context in our various papers. Working out a coherent standardisation was, strangely, the hardest part in producing this survey.

[19] Although much of it was circulated as private notes and e-files, or taught at Orsay in two DEA courses.

[20] Involving the product $\mathfrak{E}^{w_1}\,\mathfrak{E}^{w_2}$ and two flexions thereof.

[21] Vast, multi-facetted, and very demanding in terms of computation, this field calls, or rather cries, for sustained teamwork.

2 Basic dimorphic algebras

2.1 Basic operations

Elementary flexions

In addition to ordinary, non-commutative mould multiplication *mu* (or $\times$):

$$A^\bullet = B^\bullet \times C^\bullet = \mathrm{mu}(B^\bullet, C^\bullet) \iff A^{\boldsymbol{w}} = \sum_{\boldsymbol{w}^1.\boldsymbol{w}^2=\boldsymbol{w}}^{r(\boldsymbol{w}^1),r(\boldsymbol{w}^2)\geq 0} B^{\boldsymbol{w}^1} C^{\boldsymbol{w}^2} \quad (2.1)$$

and its inverse *invmu*:

$$(\mathrm{invmu}.A)^{\boldsymbol{w}} = \sum_{1\leq s\leq r(\boldsymbol{w})} (-1)^s \sum_{\boldsymbol{w}^1\ldots\boldsymbol{w}^s=\boldsymbol{w}} A^{\boldsymbol{w}^1}\ldots A^{\boldsymbol{w}^s} \qquad (\boldsymbol{w}^i \neq \emptyset) \quad (2.2)$$

the bimoulds $A^\bullet$ in $BIMU = \oplus_{0\leq r} BIMU_r$ (see (1.32))[22] can be subjected to a host of specific operations, all constructed from four elementary *flexions* $\lfloor, \rceil, \lceil, \rfloor$ that are always defined relative to a given factorisation of the total sequence $\boldsymbol{w}$. The way the flexions act is apparent from the following examples:

$$\begin{array}{lll} \boldsymbol{w}=\boldsymbol{a}.\boldsymbol{b} & \boldsymbol{a} = \binom{u_1,\,u_2,\,u_3}{v_1,\,v_2,\,v_3} & \boldsymbol{b} = \binom{u_4,\,u_5,\,u_6}{v_4,\,v_5,\,v_6} \\ \Longrightarrow & \boldsymbol{a}\rfloor = \binom{u_1,\quad u_2,\quad u_3}{v_{1:4},\,v_{2:4},\,v_{3:4}} & \lceil\boldsymbol{b} = \binom{u_{1234},\,u_5,\,u_6}{v_4,\quad v_5,\quad v_6} \end{array}$$

$$\begin{array}{lll} \boldsymbol{w}=\boldsymbol{b}.\boldsymbol{c} & \boldsymbol{b} = \binom{u_1,\,u_2,\,u_3}{v_1,\,v_2,\,v_3} & \boldsymbol{c} = \binom{u_4,\,u_5,\,u_6}{v_4,\,v_5,\,v_6} \\ \Longrightarrow & \boldsymbol{b}\rceil = \binom{u_1,\,u_2,\,u_{3456}}{v_1,\,v_2,\quad v_3} & \lfloor\boldsymbol{c} = \binom{u_4,\quad u_5,\quad u_6}{v_{4:3},\,v_{5:3},\,v_{6:3}} \end{array}$$

$$\begin{array}{llll} \boldsymbol{w}=\boldsymbol{a}.\boldsymbol{b}.\boldsymbol{c} & \boldsymbol{a} = \binom{u_1,\,u_2,\,u_3}{v_1,\,v_2,\,v_3} & \boldsymbol{b} = \binom{u_4,\,u_5,\,u_6}{v_4,\,v_5,\,v_6} & \boldsymbol{c} = \binom{u_7,\,u_8,\,u_9}{v_7,\,v_8,\,v_9} \\ \Longrightarrow & \boldsymbol{a}\rfloor = \binom{u_1,\quad u_2,\quad u_3}{v_{1:4},\,v_{2:4},\,v_{3:4}} & \lceil\boldsymbol{b}\rceil = \binom{u_{1234},\,u_5,\,u_{6789}}{v_4,\quad v_5,\quad v_6} & \lfloor\boldsymbol{c} = \binom{u_7,\quad u_8,\quad u_9}{v_{7:6},\,v_{8:6},\,v_{9:6}} \end{array}$$

with the usual short-hand: $u_{i,\ldots,j} := u_i + \ldots + u_j$ and $v_{i:j} := v_i - v_j$. Here and throughout the sequel, we use boldface (with upper indexation) to denote sequences ($\boldsymbol{w}$, $\boldsymbol{w}^i$, $\boldsymbol{w}^j$ etc), and ordinary characters (with lower indexation) to denote single sequence elements (w_i, w_j etc), or sometimes sequences of length $r(\boldsymbol{w}) = 1$. Of course, the 'product' $\boldsymbol{w}^1.\boldsymbol{w}^2$ denotes the concatenation of the two factor sequences.

Short and long indexations on bimoulds

For bimoulds $M^\bullet \in BIMU_r$ it is sometimes convenient to switch from the usual *short indexation* (with r indices w_i's) to a more homogeneous

---

[22] $BIMU_r$ of course regroups all bimoulds whose components of length other than r vanish. These are often dubbed "length-r bimoulds" for short.

long indexation (with a redundant initial w_0 which gets bracketted for distinctiveness). The correspondence goes like this:

$$M^{\binom{u_1\ ,\dots,\ u_r}{v_1\ ,\dots,\ v_r}} \cong M^{\binom{[u_0^*],\ u_1^*\ ,\dots,\ u_r^*}{[v_0^*],\ v_1^*\ ,\dots,\ v_r^*}} \tag{2.3}$$

with the dual conditions on upper and lower indices:

$$\begin{aligned} &u_0^* = -u_{1\dots r} := -(u_1+\dots+u_r)\ ,\ u_i^* &&= u_i\ \forall i \geq 1 \\ &v_0^*\ \ \text{arbitrary} &&,\ v_i^* - v_0^* = v_i\ \forall i \geq 1 \end{aligned}$$

and of course $\sum_{1\leq i\leq r} u_i v_i \equiv \sum_{0\leq i\leq r} u_i^* v_i^*$.

Unary operations

The following linear transformations on *BIMU* are of constant use:[23]

$$B^\bullet = \text{minu}.A^\bullet \Rightarrow B^{w_1,\dots,w_r} = -\ A^{w_1,\dots,w_r} \tag{2.4}$$

$$B^\bullet = \text{pari}.A^\bullet \Rightarrow B^{w_1,\dots,w_r} = (-1)^r\ A^{w_1,\dots,w_r} \tag{2.5}$$

$$B^\bullet = \text{anti}.A^\bullet \Rightarrow B^{w_1,\dots,w_r} = A^{w_r,\dots,w_1} \tag{2.6}$$

$$B^\bullet = \text{mantar}.A^\bullet \Rightarrow B^{w_1,\dots,w_r} = (-1)^{r-1}\ A^{w_r,\dots,w_1} \tag{2.7}$$

$$B^\bullet = \text{neg}.A^\bullet \Rightarrow B^{w_1,\dots,w_r} = A^{-w_1,\dots,-w_r} \tag{2.8}$$

$$B^\bullet = \text{swap}.A^\bullet \Rightarrow B^{\binom{u_1\ ,\dots,\ u_r}{v_1\ ,\dots,\ v_r}} = A^{\binom{v_r\ \ ,\dots,\ v_{3:4}\ ,\ v_{2:3},\ v_{1:2}}{u_{1..r}\ ,\dots,\ u_{123},\ u_{12},\ \ u_1}} \tag{2.9}$$

$$B^\bullet = \text{pus}.A^\bullet \Rightarrow B^{\binom{u_1\ ,\dots,\ u_r}{v_1\ ,\dots,\ v_r}} = A^{\binom{u_r,\ u_1,\ u_2\ ,\dots,\ u_{r-1}}{v_r,\ v_1,\ v_2\ ,\dots,\ v_{r-1}}} \tag{2.10}$$

$$B^\bullet = \text{push}.A^\bullet \Rightarrow B^{\binom{u_1\ ,\dots,\ u_r}{v_1\ ,\dots,\ v_r}} = A^{\binom{-u_{1\dots r},\ u_1\ \ ,\ u_2\ \ ,\dots,\ u_{r-1}}{-v_r\ \ \ ,\ v_{1:r},\ v_{2:r}\ ,\dots,\ v_{r-1:r}}}. \tag{2.11}$$

All are involutions, save for *pus* and *push*, whose restrictions to each $BIMU_r$ reduce to circular permutations of order r respectively $r+1$:[24]

$$\text{push} = \text{neg}.\text{anti}.\text{swap}.\text{anti}.\text{swap} \tag{2.12}$$

$$\text{leng}_r = \text{push}^{r+1}.\text{leng}_r = \text{pus}^r.\text{leng}_r \tag{2.13}$$

with $leng_r$ standing for the natural projection of *BIMU* onto $BIMU_r$.

23 The reason for dignifying the humble sign change in (2.4) with the special name *minu* is that *minu* enters the definition of scores of operators acting on various algebras: the rule for forming the corresponding operators that act on the corresponding groups, is then simply to change the trivial, linear *minu*, which commutes with everybody, into the non-trivial, non-linear *invmu*, which commutes with practically nobody (see (2.2)). To keep the minus sign instead of *minu* (especially when it occurs twice and so cancels out) would be a sure recipe for getting the transposition wrong.

24 *Pus* respectively *push* is a circular permutation in the *short* respectively *long* indexation of bimoulds. Indeed: $(push.M)^{[w_0],w_1,\dots,w_r} = M^{[w_r],w_0,\dots,w_{r-1}}$.

Inflected derivations and automorphisms of *BIMU*

Let $BIMU_*$ respectively $BIMU^*$ denote the subset of all bimoulds $M^\bullet$ such that $M^\emptyset = 0$ respectively $M^\emptyset = 1$. To each pair $\mathcal{A}^\bullet = (\mathcal{A}_L^\bullet, \mathcal{A}_R^\bullet) \in BIMU_* \times BIMU_*$ respectively $BIMU^* \times BIMU^*$ we attach two remarkable operators:

$$\text{axit}(\mathcal{A}^\bullet) \in \text{Der}(BIMU) \qquad \text{respectively} \qquad \text{gaxit}(\mathcal{A}^\bullet) \in \text{Aut}(BIMU)$$

whose action on *BIMU* is given by:[25]

$$N^\bullet = \text{axit}(\mathcal{A}^\bullet).M^\bullet \Leftrightarrow N^w = \sum\nolimits^1 M^{a\lceil c}\mathcal{A}_L^{b\rfloor} + \sum\nolimits^2 M^{a\rfloor c}\mathcal{A}_R^{\lfloor b} \tag{2.14}$$

$$N^\bullet = \text{gaxit}(\mathcal{A}^\bullet).M^\bullet \Leftrightarrow N^w = \sum\nolimits^3 M^{\lceil b^1\rceil \dots \lceil b^s\rceil}\mathcal{A}_L^{a^1\rfloor}\dots\mathcal{A}_L^{a^s\rfloor}\mathcal{A}_R^{\lfloor c^1}\dots\mathcal{A}_R^{\lfloor c^s} \tag{2.15}$$

and verifies the identities:

$$\text{axit}(\mathcal{A}^\bullet).\text{mu}(M_1^\bullet, M_2^\bullet) \equiv \text{mu}(\text{axit}(\mathcal{A}^\bullet).M_1^\bullet, M_2^\bullet) + \text{mu}(M_1^\bullet, \text{axit}(\mathcal{A}^\bullet).M_2^\bullet) \tag{2.16}$$

$$\text{gaxit}(\mathcal{A}^\bullet).\text{mu}(M_1^\bullet, M_2^\bullet) \equiv \text{mu}(\text{gaxit}(\mathcal{A}^\bullet).M_1^\bullet, \text{gaxit}(\mathcal{A}^\bullet).M_2^\bullet). \tag{2.17}$$

The *BIMU*-derivations *axit* are stable under the Lie bracket for operators. More precisely, the identity holds:

$$[\text{axit}(\mathcal{B}^\bullet), \text{axit}(\mathcal{A}^\bullet)] = \text{axit}(\mathcal{C}^\bullet) \quad \text{with} \quad \mathcal{C}^\bullet = \text{axi}(\mathcal{A}^\bullet, \mathcal{B}^\bullet) \tag{2.18}$$

relative to a Lie law *axi* on $BIMU_* \times BIMU_*$ given by:

$$\mathcal{C}_L^\bullet := \text{axit}(\mathcal{B}^\bullet).\mathcal{A}_L^\bullet - \text{axit}(\mathcal{A}^\bullet).\mathcal{B}_L^\bullet + \text{lu}(\mathcal{A}_L^\bullet, \mathcal{B}_L^\bullet) \tag{2.19}$$

$$\mathcal{C}_R^\bullet := \text{axit}(\mathcal{B}^\bullet).\mathcal{A}_R^\bullet - \text{axit}(\mathcal{A}^\bullet).\mathcal{B}_R^\bullet - \text{lu}(\mathcal{A}_R^\bullet, \mathcal{B}_R^\bullet). \tag{2.20}$$

Here, *lu* denotes the standard (non-inflected) Lie law on *BIMU*:

$$\text{lu}(A^\bullet, B^\bullet) := \text{mu}(A^\bullet, B^\bullet) - \text{mu}(B^\bullet, A^\bullet). \tag{2.21}$$

Let *AXI* denote the Lie algebra consisting of all pairs $\mathcal{A}^\bullet \in BIMU_* \times BIMU_*$ under this law *axi*.

Likewise, the *BIMU*-automorphisms *gaxit* are stable under operator composition. More precisely:

$$\text{gaxit}(\mathcal{B}^\bullet).\text{gaxit}(\mathcal{A}^\bullet) = \text{gaxit}(\mathcal{C}^\bullet) \quad \text{with} \quad \mathcal{C}^\bullet = \text{gaxi}(\mathcal{A}^\bullet, \mathcal{B}^\bullet) \tag{2.22}$$

---

[25] The sum $\sum^1$ respectively $\sum^2$ extends to all sequence factorisations $\boldsymbol{w} = \boldsymbol{a}.\boldsymbol{b}.\boldsymbol{c}$ with $\boldsymbol{b} \neq \emptyset$, $\boldsymbol{c} \neq \emptyset$ respectively $\boldsymbol{a} \neq \emptyset$, $\boldsymbol{b} \neq \emptyset$. The sum $\sum^3$ extends to all factorisations $\boldsymbol{w} = \boldsymbol{a}^1.\boldsymbol{b}^1.\boldsymbol{c}^1.\boldsymbol{a}^2.\boldsymbol{b}^2.\boldsymbol{c}^2 \dots \boldsymbol{a}^s.\boldsymbol{b}^s.\boldsymbol{c}^s$ such that $s \geq 1$, $\boldsymbol{b}^i \neq \emptyset$, $\boldsymbol{c}^i.\boldsymbol{a}^{i+1} \neq \emptyset\ \forall i$. Note that the extreme factor sequences $\boldsymbol{a}^1$ and $\boldsymbol{c}^s$ may be $\emptyset$.

relative to a law *gaxi* on $BIMU^* \times BIMU^*$ given by:

$$\mathcal{C}^\bullet_L := \text{mu}(\text{gaxit}(\mathcal{B}^\bullet).\mathcal{A}^\bullet_L, \mathcal{B}^\bullet_L) \tag{2.23}$$

$$\mathcal{C}^\bullet_R := \text{mu}(\mathcal{B}^\bullet_R, \text{gaxit}(\mathcal{B}^\bullet).\mathcal{A}^\bullet_R). \tag{2.24}$$

Let *GAXI* denote the Lie group consisting of all pairs $\mathcal{A}^\bullet \in BIMU^* \times BIMU^*$ under this law *gaxi*. This group *GAXI* clearly admits *AXI* as its Lie algebra.

The mixed operations *amnit* = *anmit*

For $\mathcal{A}^\bullet := (A^\bullet, 0^\bullet)$ and $\mathcal{B}^\bullet := (0^\bullet, B^\bullet)$ the operators $axit(\mathcal{A}^\bullet)$ and $axit(\mathcal{B}^\bullet)$ reduce to $amit(A^\bullet)$ and $anit(B^\bullet)$ respectively (see (2.32) and (2.33) *infra*), and the, identity (2.18) becomes:

$$\text{amnit}(A^\bullet, B^\bullet) \equiv \text{anmit}(A^\bullet, B^\bullet) \qquad (\forall A^\bullet, B^\bullet \in \text{BIMU}_*) \tag{2.25}$$

with

$$\text{amnit}(A^\bullet, B^\bullet) := \text{amit}(A^\bullet).\text{anit}(B^\bullet) - \text{anit}(\text{amit}(A^\bullet).B^\bullet) \tag{2.26}$$

$$\text{anmit}(A^\bullet, B^\bullet) := \text{anit}(B^\bullet).\text{amit}(A^\bullet) - \text{amit}(\text{anit}(B^\bullet).A^\bullet). \tag{2.27}$$

When one of the two arguments $(A^\bullet, B^\bullet)$ vanishes, the definitions reduce to:

$$\text{amnit}(A^\bullet, 0^\bullet) = \text{anmit}(A^\bullet, 0^\bullet) := \text{amit}(A^\bullet) \tag{2.28}$$

$$\text{amnit}(0^\bullet, B^\bullet) = \text{anmit}(0^\bullet, B^\bullet) = \text{anit}(B^\bullet). \tag{2.29}$$

Moreover, when *amnit* operates on a one-component bimould $M^\bullet \in BIMU_1$ (such as the *flexion units* $\mathfrak{E}^\bullet$, see Section 3.1 and Section 3.3 *infra*), its action drastically simplifies:

$$\begin{aligned} N^\bullet &:= \text{amnit}(A^\bullet, B^\bullet).M^\bullet \equiv \text{anmit}(A^\bullet, B^\bullet).M^\bullet \Leftrightarrow N^{\boldsymbol{w}} \\ &:= \sum_{\boldsymbol{a}\, w_i\, \boldsymbol{b}=\boldsymbol{w}} A^{\boldsymbol{a}\rfloor} M^{\lceil w_i \rceil} B^{\lfloor \boldsymbol{b}}. \end{aligned} \tag{2.30}$$

Unary substructures

We have two obvious subalgebras//subgroups of *AXI*//*GAXI*, answering to the conditions:

$$\text{AMI} \subset \text{AXI} : \mathcal{A}^\bullet_R = 0^\bullet \ , \quad \text{GAMI} \subset \text{GAXI} : \mathcal{A}^\bullet_R = 1^\bullet$$

$$\text{ANI} \subset \text{AXI} : \mathcal{A}^\bullet_L = 0^\bullet \ , \quad \text{GANI} \subset \text{GAXI} : \mathcal{A}^\bullet_L = 1^\bullet$$

but we are more interested in the *mixed* unary substructures, consisting of elements of the form:

$$\mathcal{A}^\bullet = (\mathcal{A}^\bullet_L, \mathcal{A}^\bullet_R) \ \text{ with } \ \mathcal{A}^\bullet_R \equiv \text{h}(\mathcal{A}^\bullet_L) \ \text{ and h a fixed involution} \tag{2.31}$$

with everything expressible in terms of the left element $\mathcal{A}_L^\bullet$ of the pair $\mathcal{A}^\bullet$. There exist, up to isomorphism, exactly seven such mixed unary substructures:

algebra	h	*swap*	algebra	h
......	....	....	......	....
ARI	*minu*	$\leftrightarrow$	IRA	*minu.push*
ALI	*anti.pari*	$\leftrightarrow$	ILA	*anti.pari.neg*
ALA	$anti.pari.neg_u$	$\leftrightarrow$	ALA	$anti.pari.neg_u$
ILI	$anti.pari.neg_v$	$\leftrightarrow$	ILI	$anti.pari.neg_v$
AWI	*anti.neg*	$\leftrightarrow$	IWA	*anti*
AWA	$anti.neg_u$	$\leftrightarrow$	AWA	$anti.neg_u$
IWI	$anti.neg_v$	$\leftrightarrow$	IWI	$anti.neg_v$

group	h	*swap*	group	h
......	....	....	......	....
GARI	*invmu*	$\leftrightarrow$	GIRA	*push.swap.invmu.swap*
GALI	*anti.pari*	$\leftrightarrow$	GILA	*anti.pari.neg*
GALA	$anti.pari.neg_u$	$\leftrightarrow$	GALA	$anti.pari.neg_u$
GILI	$anti.pari.neg_v$	$\leftrightarrow$	GILI	$anti.pari.neg_v$
GAWI	*anti.neg*	$\leftrightarrow$	GIWA	*anti*
GAWA	$anti.neg_u$	$\leftrightarrow$	GAWA	$anti.neg_u$
GIWI	$anti.neg_v$	$\leftrightarrow$	GIWI	$anti.neg_v$.

Each algebra in the first table (*e.g. ARI*) is of course *the* Lie algebra of the like-named group (*e.g. GARI*). Conversely, each Lie group in the second table is essentially determined by its eponymous Lie algebra *and* the condition of left-linearity.[26]

Dimorphic substructures
Among all seven pairs of substructures, only two respect dimorphy, namely *ARI//GARI* and *ALI//GALI*. Moreover, when restricted to dimorphic objects, they actually coincide:

$$\mathrm{ARI}^{\underline{\mathrm{al}}/\underline{\mathrm{al}}} = \mathrm{ALI}^{\underline{\mathrm{al}}/\underline{\mathrm{al}}} \text{ with } \{\underline{\mathrm{al}}/\underline{\mathrm{al}}\} = \{\text{alternal/alternal and even}\}$$
$$\mathrm{GARI}^{\underline{\mathrm{as}}/\underline{\mathrm{as}}} = \mathrm{GALI}^{\underline{\mathrm{as}}/\underline{\mathrm{as}}} \text{ with} \{\underline{\mathrm{as}}/\underline{\mathrm{as}}\} = \{\text{symmetral/symmetral and even}\}.$$

We shall henceforth work with the pair *ARI//GARI*, whose definition involves a simpler involution h (it dispenses with the sequence inversion *anti*: see above table).

26 Meaning that the group operation (like $A^\bullet, B^\bullet \mapsto gari(A^\bullet, B^\bullet)$ in our example) is linear in $A^\bullet$ but highly non-linear in $B^\bullet$.

2.2 The algebra *ARI* and its group *GARI*

Basic anti-actions

The proper way to proceed is to define the anti-actions (on *BIMU*, with its uninflected product *mu* and bracket *lu*) first of the lateral pairs *AMI//GAMI*, *ANI//GANI* and then of the mixed pair *ARI//GARI*:

$$N^\bullet = \mathrm{amit}(A^\bullet).M^\bullet \Leftrightarrow N^w = \sum\nolimits^1 M^{a\lceil c} A^{b\rfloor} \tag{2.32}$$

$$N^\bullet = \mathrm{anit}(A^\bullet).M^\bullet \Leftrightarrow N^w = \sum\nolimits^2 M^{a\rfloor c} A^{\lfloor b} \tag{2.33}$$

$$N^\bullet = \mathrm{arit}(A^\bullet).M^\bullet \Leftrightarrow N^w = \sum\nolimits^1 M^{a\lceil c} A^{b\rfloor} - \sum\nolimits^2 M^{a\rfloor c} A^{\lfloor b} \tag{2.34}$$

with sums $\sum^1$ (respectively $\sum^2$) ranging over all sequence factorisations $\boldsymbol{w} = \boldsymbol{abc}$ such that $\boldsymbol{b} \neq \emptyset$, $\boldsymbol{c} \neq \emptyset$ (respectively $\boldsymbol{a} \neq \emptyset$, $\boldsymbol{b} \neq \emptyset$).

$$N^\bullet = \mathrm{gamit}(A^\bullet).M^\bullet \Leftrightarrow N^w = \sum\nolimits^1 M^{\lceil b^1 \dots \lceil b^s} A^{a^1\rfloor} \dots A^{a^s\rfloor} \tag{2.35}$$

$$N^\bullet = \mathrm{ganit}(A^\bullet).M^\bullet \Leftrightarrow N^w = \sum\nolimits^2 M^{b^1\rceil \dots b^s\rceil} A^{\lfloor c^1} \dots A^{\lfloor c^s} \tag{2.36}$$

$$N^\bullet = \mathrm{garit}(A^\bullet).M^\bullet \Leftrightarrow N^w = \sum\nolimits^3 M^{\lceil b^1\rceil \dots \lceil b^s\rceil} A^{a^1\rfloor} \dots A^{a^s\rfloor} A_*^{\lfloor c^1} \dots A_*^{\lfloor c^s} \tag{2.37}$$

with $A_*^\bullet := \mathrm{invmu}(A^\bullet)$ and with sums $\sum^1, \sum^2, \sum^3$ ranging respectively over all sequence factorisations of the form:

$$\begin{array}{lll} \boldsymbol{w} = \boldsymbol{a}^1\boldsymbol{b}^1 \dots \boldsymbol{a}^s\boldsymbol{b}^s & (s \geq 1\ ,\ \text{only } \boldsymbol{a}^1 \text{ may be } \emptyset) \\ \boldsymbol{w} = \boldsymbol{b}^1\boldsymbol{c}^1 \dots \boldsymbol{b}^s\boldsymbol{c}^s & (s \geq 1\ ,\ \text{only } \boldsymbol{c}^s \text{ may be } \emptyset) \\ \boldsymbol{w} = \boldsymbol{a}^1\boldsymbol{b}^1\boldsymbol{c}^1 \dots \boldsymbol{a}^s\boldsymbol{b}^s\boldsymbol{c}^s & (s \geq 1\ ,\ \text{with } \boldsymbol{b}^i \neq \emptyset \text{ and } \boldsymbol{c}^i\boldsymbol{a}^{i+1} \neq \emptyset). \end{array}$$

More precisely, in $\sum^3$ two *inner* neighbour factors $\boldsymbol{c}^i$ and $\boldsymbol{a}^{i+1}$ may vanish separately but not simultaneously, whereas the *outer* factors $\boldsymbol{a}^1$ and $\boldsymbol{c}^s$ may of course vanish separately or even simultaneously.

Lie brackets and group laws

We can now concisely express the Lie algebra brackets *ami, ani, ari* and the group products *gami, gani, gari*:

$$\mathrm{ami}(A^\bullet, B^\bullet) := \mathrm{amit}(B^\bullet).A^\bullet - \mathrm{amit}(A^\bullet).B^\bullet + \mathrm{lu}(A^\bullet, B^\bullet) \tag{2.38}$$

$$\mathrm{ani}(A^\bullet, B^\bullet) := \mathrm{anit}(B^\bullet).A^\bullet - \mathrm{anit}(A^\bullet).B^\bullet - \mathrm{lu}(A^\bullet, B^\bullet) \tag{2.39}$$

$$\mathrm{ari}(A^\bullet, B^\bullet) := \mathrm{arit}(B^\bullet).A^\bullet - \mathrm{arit}(A^\bullet).B^\bullet + \mathrm{lu}(A^\bullet, B^\bullet) \tag{2.40}$$

$$\mathrm{gami}(A^\bullet, B^\bullet) := \mathrm{mu}(\mathrm{gamit}(B^\bullet).A^\bullet), B^\bullet) \tag{2.41}$$

$$\mathrm{gani}(A^\bullet, B^\bullet) := \mathrm{mu}(B^\bullet, \mathrm{ganit}(B^\bullet).A^\bullet)) \tag{2.42}$$

$$\mathrm{gari}(A^\bullet, B^\bullet) := \mathrm{mu}(\mathrm{garit}(B^\bullet).A^\bullet), B^\bullet). \tag{2.43}$$

Pre-Lie products ('pre-brackets')

Parallel with the three Lie brackets, we have three pre-Lie brackets:

$$\mathrm{preami}(A^\bullet, B^\bullet) := \mathrm{amit}(B^\bullet).A^\bullet + \mathrm{mu}(A^\bullet, B^\bullet) \tag{2.44}$$
$$\mathrm{preani}(A^\bullet, B^\bullet) := \mathrm{anit}(B^\bullet).A^\bullet - \mathrm{mu}(A^\bullet, B^\bullet) \quad \text{(sign!)} \tag{2.45}$$
$$\mathrm{preari}(A^\bullet, B^\bullet) := \mathrm{arit}(B^\bullet).A^\bullet + \mathrm{mu}(A^\bullet, B^\bullet) \tag{2.46}$$

with the usual relations:

$$\mathrm{ari}(A^\bullet, B^\bullet) \equiv \mathrm{preari}(A^\bullet, B^\bullet) - \mathrm{preari}(B^\bullet, A^\bullet) \tag{2.47}$$
$$\mathrm{assopreari}(A^\bullet, B^\bullet, C^\bullet) \equiv \mathrm{assopreari}(A^\bullet, C^\bullet, B^\bullet) \tag{2.48}$$

with *assopreari* denoting the *associator*[27] of the pre-bracket *preari*. The same holds of course for *ami* and *ani*.

Exponentiation from *ARI* to *GARI*

Provided we properly define the multiple pre-Lie brackets, *i.e.* from left to right:

$$\mathrm{preari}(A_1^\bullet, \ldots, A_s^\bullet) = \mathrm{preari}(\mathrm{preari}(A_1^\bullet, \ldots, A_{s-1}^\bullet), A_s^\bullet) \tag{2.49}$$

we have a simple expression for the exponential mapping from a Lie algebra to its group. Thus, the exponential *expari* : *ARI* → *GARI* can be expressed as a series of pre-brackets:

$$\mathrm{expari}(A^\bullet) = \sum_{0\leq n} \frac{1}{n!}\,\mathrm{preari}(\overbrace{A^\bullet, \ldots, A^\bullet}^{n\ times}) = 1^\bullet + \sum_{0<n} \frac{1}{n!}\,\mathrm{preari}(\ldots) \tag{2.50}$$

or, what amounts to the same, as a mixed *mu+arit*-expansion:

$$\mathrm{expari}(A^\bullet) = 1^\bullet + \sum_{1\leq r\,,\,1\leq n_i} \mathrm{Ex}^{n_1,\ldots,n_r}\,\mathrm{mu}(A_{n_1}^\bullet, \ldots, A_{n_r}^\bullet) \tag{2.51}$$

with $A_n^\bullet := \big(arit(A^\bullet)\big)^{n-1}.A^\bullet$ and with the symmetral mould $Ex^\bullet$:

$$\mathrm{Ex}^{n_1,\ldots,n_r} := \frac{1}{(n_1-1)!}\,\frac{1}{(n_2-1)!}\cdots\frac{1}{(n_r-1)!}\,\frac{1}{n_{1\ldots r}\,n_{2\ldots r}\ldots n_r} \tag{2.52}$$

The operation from *GARI* to *ARI* that inverses *expari* shall be denoted as *logari*. It, too, can be expressed as a series of multiple *pre-ari* products, but in a much less straightforward manner than (2.50).

[27] Here, the associator *assobin* of a binary operation *bin* is straightforwardly defined as $assobin(a, b, c) := bin(bin(a, b), c) - bin(a, bin(b, c))$. Nothing to do with the Drinfeld associators of the sequel!

For any *alternal* mould $L^\bullet$ we also have the identities:

$$\sum_{\sigma\subset\mathfrak{S}(r)} L^{\omega_{\sigma(1)},\dots,\omega_{\sigma(r)}}\text{preari}(A^\bullet_{\sigma(1)},\dots,A^\bullet_{\sigma(r)}) \equiv \frac{1}{r}\sum_{\sigma\subset\mathfrak{S}(r)} L^{\omega_{\sigma(1)},\dots,\omega_{\sigma(r)}}\text{ari}(A^\bullet_{\sigma(1)},\dots,A^\bullet_{\sigma(r)}) \qquad (\forall A^\bullet_1,\dots,A^\bullet_r) \tag{2.53}$$

which actually characterise *preari*.

Adjoint actions

We shall require the adjoint actions, *adgari* and *adari*, of *GARI* on *GARI* and *ARI* respectively. The definitions are straightforward:

$$\text{adgari}(A^\bullet).B^\bullet := \text{gari}(A^\bullet, B^\bullet, \text{invgari}.A^\bullet) \quad (A^\bullet, B^\bullet \in \text{GARI}) \tag{2.54}$$

$$\text{adari}(A^\bullet).B^\bullet := \text{logari}(\text{adgari}(A^\bullet).\text{expari}(B^\bullet)) \tag{2.55}$$

$$:= \text{fragari}(\text{preari}(A^\bullet, B^\bullet), A^\bullet) \; (A^\bullet \in \text{GARI},\ B^\bullet \in \text{ARI}) \tag{2.56}$$

except for definition (2.56), which results from (2.55) and (2.43) and uses the *pre-ari* product[28] defined as in (2.46) *supra* and the *gari*-quotient[29] defined as in (2.60) *infra*.

Definition (2.56) has over the equivalent definition (2.55) the advantage of bringing out the $B^\bullet$-linearity of $adari(A^\bullet).B^\bullet$ and of leading to much simpler calculations.[30]

The centers of *ARI* and *GARI*

The sets *Center(ARI)* respectively *Center(GARI)* consist of all bimoulds $M^\bullet$ that verify

(i) $M^\emptyset = 0$ respectively $M^\emptyset = 1$;

(ii) $M^{\binom{u_1,\dots,u_r}{0,\dots,0}} = m_r \in \mathbb{C} \qquad \forall u_i$;

(iii) $M^{\binom{u_1,\dots,u_r}{v_1,\dots,v_r}} = 0$ unless $0 = v_1 = \dots = v_r$.

---

28 Properly speaking, *preari* applies only to elements $M^\bullet$ of *ARI*, *i.e.* such that $M^\emptyset = 0$. Here, however, only $B^\bullet$ is in *ARI*, whilst $A^\bullet$ is in *GARI* and therefore $A^\emptyset = 1$. But this is no obstacle to applying the rule (2.46).

29 Properly speaking, *fragari* applies only to arguments $S^\bullet_1$, $S^\bullet_2$ in *GARI*, *i.e.* such that $S^\emptyset_i = 1$. Here, however, only $S^\bullet_2 := A^\bullet$ is in *GARI*, whilst $S^\bullet_1 := preari(A^\bullet, B^\bullet)$ is in *ARI* and therefore $S^\emptyset_1 = 0$. But this is no obstacle to applying the rule:

$$\text{fragari}(S^\bullet_1, S^\bullet_2) := \text{mu}(\text{garit}(S^\bullet_2)^{-1}.S^\bullet_1, \text{invgari}.S^\bullet_2) = \text{mu}(\text{garit}(\text{invgari}.S^\bullet_2).S^\bullet_1, \text{invgari}.S^\bullet_2).$$

30 Despite the spontaneous occurence of the *pre-ari* product in (2.56), it should be noted that $adari(A^\bullet)$ is an automorphisms of *ARI* but *not* of *PREARI*.

Moreover, in view of (2.43), *gari*-multiplication by a central element $C^\bullet$ amounts to ordinary post-multiplication by that same $C^\bullet$:

$$\text{gari}(C^\bullet, A^\bullet) \equiv \text{gari}(A^\bullet, C^\bullet) \equiv \text{mu}(A^\bullet, C^\bullet) \quad (C^\bullet \in \text{Center(GARI)}) \quad (2.57)$$

Relatedness of the four main group inversions
Lastly, we may note that the inversions relative to the four group laws *mu, gari, gami, gani* are not totally unrelated, but verify the rather unexpected identity:

$$\text{invmu} = \text{invgari.invgami.invgani} = \text{invgani.invgami.invgari}. \qquad (2.58)$$

In fact, the group generated by these four involutions is isomorphic to the group with presentation $< a, b, c, d > /\{a^2, b^2, c^2, d^2, abcd\}$.

Complexity of the flexion operations
Compared with the uninflected mould operations, the flexion operations on bimoulds tend to be staggeringly complex. Here is the natural complexity ranking for some of the main *unary* operations:

$$\textit{invgami} \sim \textit{invgani} \ll \textit{invgari} \ll \textit{logari} \ll \textit{expari}$$

and here is the number of summands involved[31] in $\textit{invgari}(A^\bullet)$ or $\textit{expari}(A^\bullet)$ as the length r increases:

length r	1	2	3	4	5	6	7	8	...
#(*invgari*)	1	4	20	112	672	4224	27459	183040	...
#(*expari*)	1	4	21	126	818	5594	39693	289510	...

Fortunately, the whole field is so strongly and harmoniously structured, and offers so many props to intuition, that this underlying complexity remains manageable. While formal computation is often indispensable at the exploratory stage, the patterns and properties that it brings to light tend to yield rather readily to rigorous proof.

2.3 Action of the basic involution *swap*

Dimorphy is a property that bears on a bimould and its *swappee*. However, even the group product most respectful of dimorphy, *i.e. gari*, does not commute with the involution *swap*. But if we set

$$\text{gira}(A^\bullet, B^\bullet) := \text{swap.gari}(\text{swap}.A^\bullet, \text{swap}.B^\bullet) \qquad (2.59)$$

$$\text{fragari}(A^\bullet, B^\bullet) := \text{gari}(A^\bullet, \text{invgari}.B^\bullet) \qquad (2.60)$$

$$\text{fragira}(A^\bullet, B^\bullet) := \text{gira}(A^\bullet, \text{invgira}.B^\bullet) \qquad (2.61)$$

[31] Each of these *inflected* summands, taken in isolation, is fairly complex!

the operation *gari//gira* and *fragari//fragira*, though distinct, can be expressed in terms of each other

$$\mathrm{gira}(A^\bullet, B^\bullet) \equiv \mathrm{ganit}(\mathrm{rash}.B^\bullet).\mathrm{gari}(A^\bullet, \mathrm{ras}.B^\bullet) \tag{2.62}$$
$$\mathrm{gari}(A^\bullet, B^\bullet) \equiv \mathrm{ganit}(\mathrm{rish}.B^\bullet).\mathrm{gira}(A^\bullet, \mathrm{ris}.B^\bullet) \tag{2.63}$$
$$\mathrm{fragira}(A^\bullet, B^\bullet) \equiv \mathrm{ganit}(\mathrm{crash}.B^\bullet).\mathrm{fragari}(A^\bullet, B^\bullet) \tag{2.64}$$
$$\mathrm{fragari}(A^\bullet, B^\bullet) \equiv \mathrm{ganit}(\mathrm{crish}.B^\bullet).\mathrm{fragira}(A^\bullet, B^\bullet) \tag{2.65}$$

via the anti-action $\mathrm{ganit}(B_*^\bullet)$ and with inputs $B_*^\bullet$ related to $B^\bullet$ through one of the following, highly non-linear operations

$$\mathrm{ras}.B^\bullet := \mathrm{invgari.swap.invgari.swap}.B^\bullet \tag{2.66}$$
$$\mathrm{rash}.B^\bullet := \mathrm{mu}(\mathrm{push.swap.invmu.swap}.B^\bullet, B^\bullet) \tag{2.67}$$
$$\mathrm{crash}.B^\bullet := \mathrm{rash.swap.invgari.swap}.B^\bullet \tag{2.68}$$

$$\mathrm{ris} := \mathrm{ras}^{-1} = \mathrm{swap.invgari.swap.invgari} \tag{2.69}$$
$$\mathrm{rish} := \mathrm{invgani.rash.ris} \tag{2.70}$$
$$\mathrm{crish} := \mathrm{invgani.crash} = \mathrm{rish.invgari} \tag{2.71}$$

2.4 Straight symmetries and subsymmetries

• **alternality and symmetrality**.
Like a mould, a bimould $A^\bullet$ is said to be *alternal* (respectively *symmetral*) if it verifies

$$\sum_{\boldsymbol{w}\in \mathrm{sha}(\boldsymbol{w}', \boldsymbol{w}'')} A^{\boldsymbol{w}} \equiv 0 \;\left(\text{respectively } \equiv A^{\boldsymbol{w}'} A^{\boldsymbol{w}''}\right) \quad \forall \boldsymbol{w}' \neq \emptyset, \forall \boldsymbol{w}'' \neq \emptyset \tag{2.72}$$

with $\boldsymbol{w}$ running through the set $sha(\boldsymbol{w}', \boldsymbol{w}'')$ of all shufflings of $\boldsymbol{w}'$ and $\boldsymbol{w}''$.

• **{alternal} $\Longrightarrow$ {mantar-invariant, pus-neutral}.**
Alternality implies *mantar*-invariance, with *mantar* = *minu.pari.anti* defined as in (2.7).

It also implies *pus*-neutrality, which means this:

$$\left(\sum_{1\le l\le r(\bullet)} \mathrm{pus}^l\right).A^\bullet \equiv 0 \quad i.e. \quad \sum_{\boldsymbol{w}' \overset{\mathrm{circ}}{\sim} \boldsymbol{w}} A^{\boldsymbol{w}'} \equiv 0 \qquad (\text{if } r(\boldsymbol{w}) \ge 2) \tag{2.73}$$

• **{symmetral} $\Longrightarrow$ {gantar-invariant, gus-neutral}.**
Symmetrality implies likewise *gantar*-invariance, with

$$\mathrm{gantar} := \mathrm{invmu.anti.pari} \tag{2.74}$$

as well as *gus*-neutrality, which means $\big(\sum_{1\leq l\leq r(\bullet)} \mathrm{pus}^l\big).\mathrm{logmu}.A^\bullet \equiv 0$ *i.e.*

$$\sum_{1\leq k\leq r(\boldsymbol{w})} (-1)^{k-1} \sum_{\boldsymbol{w}^1\ldots\boldsymbol{w}^k \overset{\mathrm{circ}}{\sim} \boldsymbol{w}} A^{\boldsymbol{w}^1}\ldots A^{\boldsymbol{w}^k} \equiv 0 \qquad (\text{if } r(\boldsymbol{w})\geq 2) \quad (2.75)$$

• **{bialternal}** $\overset{\mathrm{ess^{ly}}}{\Longrightarrow}$ **{neg-invariant, push-invariant}.**
Bialternality implies not only invariance under *neg.push* but also separate *neg*-invariance and *push*-invariance for any $A^\bullet \in BIMU_r$ but the implication holds only if $r > 1$, since on $BIMU_1$ we have *neg=push*. So *neg.push=id*, meaning that there is no constraint at all on elements of $BIMU_1$. But we must nonetheless impose *neg*-invariance on $BIMU_1$ (or what amounts to the same, *push*-invariance) to ensure the stability of bialternals under the *ari*-bracket: see Section 2.7.

• **{bisymmetral}** $\overset{\mathrm{ess^{ly}}}{\Longrightarrow}$ **{neg-invariant, gush-invariant}.**
Bisymmetrality implies not only invariance under *neg.gush*, with

$$\mathrm{gush} := \mathrm{neg.gantar.swap.gantar.swap} \qquad (2.76)$$

but also separate *neg*-invariance and *gush*-invariance, but only if we assume *neg*-invariance for the component of length 1. If we do not make that assumption, every bisymmetral bimould in *GARI* splits into two bisymmetral factors: a regular right factor (invariant under *neg*) and an irregular left factor (invariant under *pari.neg*).

Let us now examine the *stable* combinations of alternality or 'subalternality' (respectively symmetrality or 'subsymmetrality'), *i.e.* the combinations that are preserved under at least *some* flexion operations and give rise to interesting algebras or groups.

Primary and secondary subalgebras and subgroups
Broadly speaking, simple symmetries or subsymmetries (*i.e.* those that bear only on bimoulds or their swappees but not both) tend to be stable under a vast range of binary operations, both uninflected (like the *lu*-bracket or the *mu*-product) or inflected (like *ari/gari* or *ali/gali*). The corresponding algebras or groups are called *primary*. On the other hand, double symmetries or subsymmetries (*i.e.* those that bear simultaneously on bimoulds *and* their swappees) are only stable – when at all – under (suitable) inflected operations. We speak in this case of *secondary* algebras or groups.

"Finitary" and "infinitary" constraints

Another important distinction lies in the character – "finitary" or otherwise – of the contraints corresponding to each set of symmetries of subsymmetries. These constraints always assume the form

$$0=\sum_{\tau}\epsilon(\tau)\,M^{\tau(\boldsymbol{w})}+\sum_{\sigma}\epsilon(\sigma,\boldsymbol{w})\,M^{\sigma(\boldsymbol{w})} \tag{2.77}$$

$$\text{with } \boldsymbol{w}=(w_1,\ldots,w_r);\ \epsilon(\tau)\in\mathbb{Z},\ \epsilon(\sigma,\boldsymbol{w})\in\mathbb{C},\ \tau\in\mathrm{Gl}_r(\mathbb{Z}) \tag{2.78}$$

with a first sum involving a finite number of sequences $\tau(\boldsymbol{w})$ (respectively $\sigma(\boldsymbol{w})$) that are linearly dependent on $\boldsymbol{w}$ and of equal (respectively lesser) length. What really matters is the subgroup $<\tau>_r$ of $\mathrm{Gl}_r(\mathbb{Z})$ generated by the τ in the first sum and unambiguously determined (up to isomorphism) by the constraints. When $<\tau>_r$ is finite[32] we speak of *finitary* constraints. The corresponding algebras or groups are always easy to investigate; the algebras in particular split into 'cells', or component subspaces in $BIMU_r$, whose dimensions are readily calculated by using standard invariant theory. When $<\tau>_r$ is infinite [33] things can of course get much trickier, but the important point to note is this: whereas simple symmetries (like alternality) are always *finitary*, and full double symmetries (like bialternality) always infinitary, there exists a very useful intermediary class – that namely of *finitary double symmetries.* The prototypal case is the (*ari*-stable) combination of alternality and *push*-invariance.[34]

We can now proceed to catalogue all the *basic* symmetry-induced algebras and groups – *basic* in the sense that all others can be derived from them by intersection.

Throughout, we adopt the following convenient notations. For any set $E\subset BIMU$:

(i) E^h or $E^{h/*}$ denotes the subset of all bimoulds $M^\bullet$ with the property h;
(ii) $E^{h/k}$ denotes the subset of all bimoulds such that $M^\bullet$ has the property h and $swap.M^\bullet$ has the property k;
(iii) if h or k is a unary operation, the property in question should be taken to mean h- or k-invariance;
(iv) $\overline{pusnu}$ or $\overline{gusnu}$ denote *pus* - or *gus* -neutrality (see Section 2.4);

[32] Like with the alternality constraints, in which case $<\tau>_r\sim\mathfrak{S}_r$.

[33] Like with the bialternality constraints, in which case $<\tau>_r$ is generated by two distinct finite subgroups of $\mathrm{Gl}_r(\mathbb{Z})$, which we may denote as $\mathfrak{S}_r$ and $swap.\mathfrak{S}_r.swap$.

[34] That combination is indeed a double symmetry, since a bimould's *push*-invariance is a consequence of its *and* its swappee's alternality or at least *mantar*-invariance.

(v) the underlining (as in $\underline{al}/\underline{al}$ or $\underline{as}/\underline{as}$) always signals the *parity condition* for the length-1 component;
(vi) boldface **ARI** or **GARI** is used to distinguish the few *infinitary* subalgebras or subgroups of *ARI* or *GARI*.

The only *infinitary* algebras are:

$$\mathbf{ARI}^{\underline{\mathrm{al}}/\underline{\mathrm{al}}}\,,\ \mathbf{ARI}^{\overline{\mathrm{pusnu}}/\overline{\mathrm{pusnu}}}\,,\ \mathbf{ARI}^{\overline{\mathrm{pusnu}}/\overline{\mathrm{pusnu}}}_{\mathrm{mantar}/.} := \mathbf{ARI}^{\overline{\mathrm{pusnu}}/\overline{\mathrm{pusnu}}} \cap ARI^{\mathrm{mantar}/.}$$

As for the intersection $\mathbf{ARI}^{\overline{\mathrm{pusnu}}/\overline{\mathrm{pusnu}}} \cap ARI^{\mathrm{push}}$, it can be shown to coincide with $\mathbf{ARI}^{\underline{\mathrm{al}}/\underline{\mathrm{al}}}$ deprived of its length-one component. The same pattern holds the groups.

2.5 Main subalgebras

$la^{\bullet}$		$li^{\bullet} := \mathrm{swap}(la^{\bullet})$		subalgebra
....	..	....	...	....
push-invariant	$\Leftrightarrow$	push-invariant	...	ARI^{push}
pus-neutral		....	...	$ARI^{\mathrm{pusnu}/*}$
....		pus-neutral (strictly)	...	$ARI^{*/\overline{\mathrm{pusnu}}}$
pus-neutral (strictly)		pus-neutral (strictly)	...	$\mathbf{ARI}^{\overline{\mathrm{pusnu}}/\overline{\mathrm{pusnu}}}$
push-neutral	$\Leftrightarrow$	push-neutral	...	unstable
pus-invariant		....	...	unstable
....		pus-invariant	...	unstable
mantar-invariant		....	...	$ARI^{\mathrm{mantar}/*}$
....		mantar-invariant	...	unstable
mantar-invariant		mantar-invariant	...	*unstable*
mantar-invariant		mantar-invariant	neg	$ARI^{\underline{\mathrm{mantar}}/\underline{\mathrm{mantar}}}$
push-invariant		mantar-invariant	...	$ARI^{\mathrm{push}/\mathrm{mantar}}$
mantar-invariant		push-invariant	...	$ARI^{\mathrm{mantar}/\mathrm{push}}$
alternal		....	...	$ARI^{\mathrm{al}/*}$
....		alternal	...	unstable
alternal		alternal	...	unstable
alternal		alternal	neg	$\mathbf{ARI}^{\underline{\mathrm{al}}/\underline{\mathrm{al}}}$
alternal		mantar-invariant	...	unstable
alternal		mantar-invariant	neg	$ARI^{\underline{\mathrm{al}}/\underline{\mathrm{mantar}}}$
alternal		push-invariant	...	$ARI^{\mathrm{al}/\mathrm{push}}$
mantar-invariant		alternal	...	unstable
mantar-invariant		alternal	neg	$ARI^{\underline{\mathrm{mantar}}/\underline{\mathrm{al}}}$
push-invariant		alternal	...	$ARI^{\mathrm{push}/\mathrm{al}}$.

2.6 Main subgroups

$\mathrm{ga}^\bullet$		$\mathrm{gi}^\bullet := \mathrm{swap}(\mathrm{ga}^\bullet)$		subgroup
………………	..	………………	…	………………
gush-invariant	$\Leftrightarrow$	gush-invariant	…	$\mathrm{GARI}^{\mathrm{gush}}$
gus-neutral		………………	…	$\mathrm{GARI}^{\mathrm{gusnu}/*}$
………………		gus-neutral (strictly)	…	$\mathrm{GARI}^{*/\overline{\mathrm{gusnu}}}$
gus-neutral (strictly)		gus-neutral (strictly)	…	$\mathbf{GARI}^{\overline{\mathrm{gusnu}}/\overline{\mathrm{gusnu}}}$
gush-neutral	$\Leftrightarrow$	gush-neutral	…	unstable
gus-invariant		………………	…	unstable
………………		gus-invariant	…	unstable
gantar-invariant		………………	…	$\mathrm{GARI}^{\mathrm{gantar}/*}$
………………		gantar-invariant	…	unstable
gantar-invariant		gantar-invariant	…	unstable
gantar-invariant		gantar-invariant	neg	$\mathrm{GARI}^{\underline{\mathrm{gantar}}/\underline{\mathrm{gantar}}}$
gush-invariant		gantar-invariant	…	$\mathrm{GARI}^{\mathrm{gush}/\mathrm{gantar}}$
gantar-invariant		gush-invariant	…	$\mathrm{GARI}^{\mathrm{gantar}/\mathrm{gush}}$
alternal		………………	…	$\mathrm{GARI}^{\mathrm{as}/*}$
………………		symmetral	…	unstable
symmetral		symmetral	…	unstable
symmetral		symmetral	neg	$\mathbf{GARI}^{\underline{\mathrm{as}}/\underline{\mathrm{as}}}$
symmetral		gantar-invariant	…	*unstable*
symmetral		gantar-invariant	neg	$\mathrm{GARI}^{\underline{\mathrm{as}}/\underline{\mathrm{gantar}}}$
symmetral		gush-invariant	…	$\mathrm{GARI}^{\mathrm{as}/\mathrm{gush}}$
gantar-invariant		symmetral	…	unstable
gantar-invariant		symmetral	neg	$\mathrm{GARI}^{\underline{\mathrm{gantar}}/\underline{\mathrm{as}}}$
gush-invariant		symmetral	…	$\mathrm{GARI}^{\mathrm{gush}/\mathrm{as}}$.

2.7 The dimorphic algebra $ARI^{\underline{al}/\underline{al}} \subset ARI^{al/al}$.

The space $ARI^{\underline{al}/\underline{al}}$ of *bialternal* and *even* bimoulds is a subalgebra of *ARI*. The total space $ARI^{al/al}$ of *all* bialternals is only marginally larger, since

$$\mathrm{ARI}^{\mathrm{al/al}} = \mathrm{ARI}^{\dot{\mathrm{al}}/\dot{\mathrm{al}}} \oplus \mathrm{ARI}^{\underline{\mathrm{al}}/\underline{\mathrm{al}}} \tag{2.79}$$

with a complement space $ARI^{\dot{a}l/\dot{a}l} := BIMU_{1,odd}$ that simply consists of all *odd* bimoulds with a single non-zero component of length 1. The total space $ARI^{al/al}$ is not an algebra, but there is some additional structure on it, in the form of a bilinear mapping *oddari* of $ARI^{\dot{a}l/\dot{a}l}$ into $ARI^{\underline{al}/\underline{al}}$:

$$\text{oddari}: \quad (\text{ARI}^{\dot{a}l/\dot{a}l}, \text{ARI}^{\dot{a}l/\dot{a}l}) \longrightarrow \text{ARI}^{\underline{al}/\underline{al}} \qquad (\text{oddari} \neq \text{ari}) \tag{2.80}$$

with

$$C^\bullet = \text{oddari}(A^\bullet.B^\bullet) \implies \tag{2.81}$$

$$\begin{aligned} C^{\binom{u_1\,,\,u_2}{v_1\,,\,v_2}} := &+A^{\binom{u_1}{v_1}} B^{\binom{u_2}{v_2}} + A^{\binom{-u_1-u_2}{-v_2}} B^{\binom{u_1}{v_1-v_2}} + A^{\binom{u_2}{v_2-v_1}} B^{\binom{-u_1-u_2}{-v_1}} \\ &-B^{\binom{u_1}{v_1}} A^{\binom{u_2}{v_2}} - B^{\binom{-u_1-u_2}{-v_2}} A^{\binom{u_1}{v_1-v_2}} - B^{\binom{u_2}{v_2-v_1}} A^{\binom{-u_1-u_2}{-v_1}} \end{aligned}$$

Remark. Although *swap* doesn't act as an automorphism on *ARI*, it does on $ARI^{\underline{al}/\underline{al}}$, essentially because all elements of $ARI^{\underline{al}/\underline{al}}$ are *push* invariant.

2.8 The dimorphic group $GARI^{\underline{as}/\underline{as}} \subset GARI^{as/as}$

The set $GARI^{\underline{al}/\underline{al}}$ of *bisymmetral* and *even* bimoulds is a subgroup of *GARI*. The total set $GARI^{as/as}$ of *all* bisymmetrals is only marginally larger, since we have the factorisation

$$\text{GARI}^{as/as} = \text{gari}(\text{GARI}^{\dot{a}s/\dot{a}s}, \text{GARI}^{\underline{as}/\underline{as}}) \tag{2.82}$$

$$\text{GARI}^{\dot{a}s/\dot{a}s} = \bigcup_{\mathfrak{E}} \mathfrak{ess}^\bullet_{\mathfrak{E}} \quad (\mathfrak{E} = \text{flexion unit},\ \mathfrak{ess}^\bullet_{\mathfrak{E}} \text{ bisymmetral}) \tag{2.83}$$

with a left factor $\text{GARI}^{\dot{a}s/\dot{a}s}$ consisting of bisymmetral bimoulds that are invariant under *pari.neg* (rather than *neg*) and correspond one-to-one to very special bimoulds of $BIMU_1$, the so-called *flexion units* (see Section 3.2 and Section 3.5). Of course, the union $\bigcup_{\mathfrak{E}^\bullet}$ extends to the *vanishing* unit $\mathfrak{E}^\bullet = 0^\bullet$, to which there corresponds $\mathfrak{ess}_{\mathfrak{E}} = id_{GARI}$. The total set $GARI^{as/as}$ is not a group, but the above decomposition makes it clear that it is stable under postcomposition by $GARI^{\underline{as}/\underline{as}}$:

$$\text{gari}(\text{GARI}^{as/as}, \text{GARI}^{\underline{as}/\underline{as}}) = \text{GARI}^{as/as} \tag{2.84}$$

Remark. Although *swap* doesn't act as an automorphism on *GARI*, it does on $GARI^{\underline{as}/\underline{as}}$, essentially because all elements of $GARI^{\underline{as}/\underline{as}}$ are *gush* invariant. In fact, for $B^\bullet$ in $GARI^{\underline{as}/\underline{as}}$, formula (2.62) reads $gira(A^\bullet, B^\bullet) = gari(A^\bullet, B^\bullet)$ since in that case $rash(B^\bullet) = 1^\bullet$ and $ras(B^\bullet) = B^\bullet$.

3 Flexion units and twisted symmetries

3.1 The free monogenous flexion algebra $Flex(\mathfrak{E})$

To any $\mathfrak{E}^\bullet \in BIMU_1$ of a given parity type $\binom{s_1}{s_2}$, *i.e.* such that

$$\mathfrak{E}^{\binom{\epsilon u_1}{\eta v_1}} \equiv \epsilon^{s_1}\, \eta^{s_2}\, \mathfrak{E}^{\binom{u_1}{v_1}} \quad \text{with } s_1, s_2 \in \{0,1\};\ \epsilon, \eta \in \{+,-\};\ \forall u_1, v_1 \tag{3.1}$$

let us attach the space *Flex*($\mathfrak{E}$) of all bimoulds generated by $\mathfrak{E}^\bullet$ under *all* flexion operations, unary or binary[35]. *Flex*($\mathfrak{E}$) thus contains subalgebras not just of *ARI* but of all 7+1 distinct flexion algebras, and subgroups not just of *GARI* but of all 7+1 distinct flexion groups. Moreover, for truly random generators $\mathfrak{E}^\bullet$, all realisations *Flex*($\mathfrak{E}$) are clearly isomorphic: they depend only on the parity type $\binom{s_1}{s_2}$. Lastly, for all four parity types, we have the same universal decomposition of *Flex*($\mathfrak{E}$) into cells $Flex_r(\mathfrak{E}) \subset BIMU_r$ whose dimensions are as follows:

$$\mathrm{Flex}(\mathfrak{E}) = \bigoplus_{r\geq 0} \mathrm{Flex}_r(\mathfrak{E}) \quad \text{with} \quad \dim(\mathrm{Flex}_r(\mathfrak{E})) = \frac{(3r)!}{r!\,(2r+1)!}. \tag{3.2}$$

The reason is that $Flex_r(\mathfrak{E})$ can be *freely* generated by just two operations, namely *mu* and *amnit*:

$$A_i^\bullet \in Flex_{r_i}(\mathfrak{E}) \Longrightarrow \mathrm{mu}(A_1,\ldots,A_s) \in Flex_{r_1+..r_s}(\mathfrak{E}) \tag{3.3}$$
$$A_i^\bullet \in Flex_{r_i}(\mathfrak{E}) \Longrightarrow \mathrm{amnit}(A_1,A_2).\mathfrak{E}^\bullet \in Flex_{1+r_1+r_2}(\mathfrak{E}). \tag{3.4}$$

As a consequence, each cell $Flex_r(\mathfrak{E})$ can be shown to possess four natural bases of exactly the required cardinality, namely $\{\mathfrak{e}_{\boldsymbol{t}}^\bullet\} \sim \{\mathfrak{e}_{\boldsymbol{p}}^\bullet\} \sim \{\mathfrak{e}_{\boldsymbol{o}}^\bullet\} \sim \{\mathfrak{e}_{\boldsymbol{g}}^\bullet\}$. Theses bases are actually *one*, and merely differ by the indexation:
1) $\boldsymbol{t}$ runs through all r-node ternary trees;
2) $\boldsymbol{p}$ runs through all r-fold arborescent parenthesisings;
3) $\boldsymbol{o}$ runs through all arborescent, coherent orders on $\{1,\ldots,r\}$;
4) $\boldsymbol{g}$ runs through all pairs $\boldsymbol{g} = (\boldsymbol{ga}, \boldsymbol{gi})$ of r-edged, non-overlapping graphs.

The basis $\{\mathfrak{e}_{\boldsymbol{t}}^\bullet\}$

The free generation of $Flex_r(\mathfrak{E})$ under the operations (3.3) and (3.4) produces an indexation by trees $\boldsymbol{\theta}$ of a definite sort which, though not ternary, stand in one-to-one correspondence with ternary trees $\boldsymbol{t}$. We need not bother with that here.

The basis $\{\mathfrak{e}_{\boldsymbol{g}}^\bullet\}$

We fix r and puncture the unit circle at all points Si_k and Sa_k of the form

$$Si_k := \exp\left(2\pi i \frac{k}{r+1}\right), \quad Sa_k := \exp\left(2\pi i \frac{k+\frac{1}{2}}{r+1}\right) \qquad (k \in \mathbb{Z}/(r+1)\mathbb{Z})$$

[35] Other than *swap*, which exchanges the u_i's and v_i's, and *pus* (see (2.10)) which, we recall, doesn't qualify as a proper flexion operation. But *push* is allowed, as well as all algebra and group operations.

Let G_r be the set of all $\frac{(3r)!}{r!\,(2r+1)!}$ pairs $g = (ga, gi)$ such that:
(i) ga is a connected graph with vertices at each Sa_j and with exactly r straight, non-intersecting edges;
(ii) gi is a connected graph with vertices at each Si_j and with exactly r straight, non-intersecting edges;
(iii) ga and gi are 'orthogonal' in the sense that each edge of one intersects exactly one edge of the other.[36]

To each such $g = (ga, gi)$ we attach the bimould $\mathfrak{e}_g^\bullet \in Flex_r(\mathfrak{E})$ defined by

$$\mathfrak{e}_g^{\binom{u_1,\dots,u_r}{v_1,\dots,v_r}} := \prod_{x \in ga \cap gi} \mathfrak{E}^{\binom{u(x)}{v(x)}} \qquad \text{(exactly } r \text{ factors)} \tag{3.5}$$

with

$$u(x) := \sum_{[\mathrm{Si}_0 < \mathrm{Sa}_{m_1} < \mathrm{Sa}_n < \mathrm{Sa}_{m_2}]^{\mathrm{circ}}} u_n \qquad (\text{with } 1 \le n \le r)$$

$$v(x) := v_{n_2} - v_{n_1} \qquad (n_2 \neq 0;\; v_{n_1} = 0 \text{ if } n_1 = 0)$$

with Sa_{m_1}, Sa_{m_2} (respectively Si_{n_1}, Si_{n_2}) denoting the end-points of the edge of ga (respectively gi) going through x and with the indexation order so chosen as to ensure

$$[\mathrm{Si}_0 < \mathrm{Sa}_{m_1} < \mathrm{Sa}_{m_2}]^{\mathrm{circ}} \quad \text{and} \quad [\mathrm{Si}_{n_1} < \mathrm{Sa}_{m_1} < \mathrm{Si}_{n_2} < \mathrm{Sa}_{m_2}]^{\mathrm{circ}}.$$

The basis $\{\mathfrak{e}_o^\bullet\}$

A partial order o on $\{1, \dots, r\}$ is *arborescent* if each i in $\{1, \dots, r\}$ has at most one direct o-antecedent i_-, and it is *coherent* if the following implication (which involves both the natural order $\le$ and the o-order $\preceq$) holds:

$$\{i_1 \le i_2 \le i_3 \;,\; i \preceq i_1 \;,\; i \preceq i_3\} \Longrightarrow \{i \preceq i_2\} \tag{3.6}$$

This amounts to saying that the set of all j such that $i \preceq j$ has to be an *interval* $i^- \le j \le i^+$ for the natural order. The basis elements are then defined as follows

$$\mathfrak{e}_o^{\binom{u_1,\dots,u_r}{v_1,\dots,v_r}} := \prod_{1 \le i \le r} \mathfrak{E}^{\binom{u(i)}{v(i)}} \quad \text{with } u(i) := \sum_{i \preceq j} u_j = \sum_{j=i^-}^{j=i^+} u_j, \quad v(i) := v_i - v_{i_-}$$

If i has no o-antecedent i_- we must of course set $v(i) := v_i$.

---

36 Each ga verifying (i) has *one* orthogonal gi verifying (ii) and *vice versa*. We are told that these objects are known as *non-crossing trees* in combinatorics.

The basis $\{\mathfrak{e}^\bullet_p\}$

The set $\mathcal{P}_r$ of all r-fold arborescent parenthesisings may be visualised as consisting of non-commutative words $\boldsymbol{p}$ made up of r letters a ("opening parentheses"), r letters b ("inter-parenthesis content") and r letters c ("closing parentheses"). These words, in turn, are defined by a simple induction: each non-prime $\boldsymbol{p}$ admits a unique factorisation into prime factors $\boldsymbol{p}_i$, and each prime $\boldsymbol{p}$ admits a unique expression of the form

$$\boldsymbol{p} = a.\boldsymbol{p}_1.b.\boldsymbol{p}_2.c \qquad\qquad (\boldsymbol{p}_1, \boldsymbol{p}_2 \in \mathcal{P}) \tag{3.7}$$

with factors $\boldsymbol{p}_1$, $\boldsymbol{p}_2$ that need not be prime, and one of which may be empty.[37] Thus $\mathcal{P}_1 = \{abc\}$, $\mathcal{P}_2 = \{aabcbc, ababcc, abcabc\}$, etc.

To define the correspondance between the $\boldsymbol{p}$- and $\boldsymbol{o}$-indexations, we assimilate each i in $\{1, \dots, r\}$ to the i-th letter b in the words $\boldsymbol{p} \in \mathcal{P}_r$ and set

$$h(i) := \alpha - \gamma = \gamma' - \alpha' \tag{3.8}$$

if that i-th letter b is preceded in $\boldsymbol{p}$ by α letters a and γ letters c or, what amounts to the same, followed by α' letters a and γ' letters c. We then define the order $\boldsymbol{o}$ on $\{1, \dots, r\}$ by decreeing that $i \prec j$ iff $h(i) < h(j)$ and $h(i) < h(k)$ for all k between i and j.[38]

3.2 Flexion units

As it happens, the most useful monogenous algebras $Flex(\mathfrak{E})$ are not those spawned by 'random' generators $\mathfrak{E}$ but on the contrary by very special ones – the so-called *flexion units*.

Exact flexion units. The tripartite relation

A *flexion unit* is a bimould $\mathfrak{E}^\bullet \in BIMU_1$ that is *odd* in w_1 and verifies the *tripartite relation* below. More precisely:

$$\begin{aligned} \mathfrak{E}^{-w_1} &\equiv -\mathfrak{E}^{w_1}, & \mathfrak{E}^{w_1}\,\mathfrak{E}^{w_2} &\equiv \mathfrak{E}^{w_1\rfloor}\,\mathfrak{E}^{\lceil w_2} + \mathfrak{E}^{w_1\rceil}\,\mathfrak{E}^{\lfloor w_2} \quad i.e \\ \mathfrak{E}^{\left(\begin{smallmatrix} -u_1 \\ -v_1 \end{smallmatrix}\right)} &\equiv -\mathfrak{E}^{\left(\begin{smallmatrix} u_1 \\ v_1 \end{smallmatrix}\right)}, & \mathfrak{E}^{\left(\begin{smallmatrix} u_1 \\ v_1 \end{smallmatrix}\right)}\mathfrak{E}^{\left(\begin{smallmatrix} u_2 \\ v_2 \end{smallmatrix}\right)} &\equiv \mathfrak{E}^{\left(\begin{smallmatrix} u_1 \\ v_{1:2} \end{smallmatrix}\right)}\mathfrak{E}^{\left(\begin{smallmatrix} u_{12} \\ v_2 \end{smallmatrix}\right)} + \mathfrak{E}^{\left(\begin{smallmatrix} u_{12} \\ v_1 \end{smallmatrix}\right)}\mathfrak{E}^{\left(\begin{smallmatrix} u_2 \\ v_{2:1} \end{smallmatrix}\right)}. \end{aligned} \tag{3.9}$$

In view of the imparity of $\mathfrak{E}^\bullet$ the tripartite identity may also be written in more symmetric form:

$$\mathfrak{E}^{\left(\begin{smallmatrix} u_1 \\ v_{1:0} \end{smallmatrix}\right)}\mathfrak{E}^{\left(\begin{smallmatrix} u_2 \\ v_{2:0} \end{smallmatrix}\right)} + \mathfrak{E}^{\left(\begin{smallmatrix} u_2 \\ v_{2:1} \end{smallmatrix}\right)}\mathfrak{E}^{\left(\begin{smallmatrix} u_0 \\ v_{0:1} \end{smallmatrix}\right)} + \mathfrak{E}^{\left(\begin{smallmatrix} u_0 \\ v_{0:2} \end{smallmatrix}\right)}\mathfrak{E}^{\left(\begin{smallmatrix} u_1 \\ v_{1:2} \end{smallmatrix}\right)} \equiv 0 \;\; \forall u_i, \forall v_i \text{ with } u_0+u_1+u_2=0$$

[37] Or even both, if $\boldsymbol{p} \in \mathcal{P}_1$.

[38] As a consequence, if the i-th and j-th letters b fall into distinct prime factors of $\boldsymbol{p}$, then i and j are non-comparable.

Another way of characterising flexion units is via the *push*-neutrality of their powers $mu^n(\mathfrak{E}^\bullet)$. Indeed, if we set:

$$\mathrm{mu}^n(\mathfrak{E}^\bullet) = \mathrm{mu}(\overbrace{\mathfrak{E}^\bullet, \ldots, \mathfrak{E}^\bullet}^{n\ times}) \tag{3.10}$$

then $\mathfrak{E}$ is a flexion unit *iff* $mu^1(\mathfrak{E}^\bullet)$ and $mu^2(\mathfrak{E}^\bullet)$ are *push*-neutral, in which case it can be shown that *all* powers $mu^n(\mathfrak{E}^\bullet)$ are automatically *push*-neutral:

$$\{\mathfrak{E} \text{ is a flexion unit}\} \Leftrightarrow \left\{\left(\sum_{0\le k\le n} \mathrm{push}^k\right).\mathrm{mu}^n(\mathfrak{E}^\bullet) = 0,\ \ \forall n \in \mathbb{N}^*\right\}. \tag{3.11}$$

If two units $\mathfrak{E}^\bullet$ and $\mathfrak{O}^\bullet$ are *constant* respectively in v_1 and u_1, then the sum $\mathfrak{E}^\bullet + \mathfrak{O}^\bullet$ is also a unit.

Lastly, if $\mathfrak{E}^\bullet$ is a unit, then for each $\alpha, \beta, \gamma, \delta \in \mathbb{C}$ the relation

$$\mathfrak{E}_{[\alpha,\beta,\gamma,\delta]}^{\binom{u_1}{v_1}} := \delta\, e^{\gamma\, u_1 v_1}\, \mathfrak{E}^{\binom{u_1/\alpha}{v_1/\beta}} \tag{3.12}$$

defines a new unit $\mathfrak{E}^\bullet_{[\alpha,\beta,\gamma,\delta]}$.

Conjugate units

If $\mathfrak{E}^\bullet$ is a unit, then the relation $\mathfrak{O}^{\binom{u_1}{v_1}} := \mathfrak{E}^{\binom{v_1}{u_1}}$ define another unit $\mathfrak{O}^\bullet$ – the so-called *conjugate* of $\mathfrak{E}^\bullet$. Indeed, setting $(u_1, u_2) := (v_1', v_2' - v_1')$, $(v_1, v_2) := (u_1' + u_2', u_2')$, then using the *imparity* of $\mathfrak{E}^\bullet$ and re-ordering the terms, we find that (3.9) becomes:

$$\mathfrak{O}^{\binom{u_1'}{v_1'}}\mathfrak{O}^{\binom{u_2'}{v_2'}} \equiv \mathfrak{O}^{\binom{u_1'}{v_{1:2}'}}\mathfrak{O}^{\binom{u_{12}'}{v_2'}} + \mathfrak{O}^{\binom{u_{12}'}{v_1'}}\mathfrak{O}^{\binom{u_2'}{v_{2:1}'}} \qquad \text{with} \quad \mathfrak{O}^{\binom{u_1}{v_1}} := \mathfrak{E}^{\binom{v_1}{u_1}}$$

i.e. conserves its form.

Let us now mention the most useful flexion units, some *exact* and others only *approximate*. Throughout the sequel, we shall set:

$$P(t) := \frac{1}{t}, \qquad Q(t) := \frac{1}{\tan(t)}, \qquad Q_c(t) := \frac{c}{\tan(c\,t)} \tag{3.13}$$

Polar units

They consist purely of poles at the origin:

$$\mathrm{Pa}^{w_1} = P(u_1) \tag{3.14}$$

$$\mathrm{Pi}^{w_1} = P(v_1) \tag{3.15}$$

$$\mathrm{Pai}^{w_1}_{\alpha,\beta} = P\left(\frac{u_1}{\alpha}\right) + P\left(\frac{v_1}{\beta}\right) = \frac{\alpha}{u_1} + \frac{\beta}{v_1} \tag{3.16}$$

$Pa^\bullet$, $Pi^\bullet$, $Pai^\bullet_{\alpha,\beta}$ are *exact* units.

Trigonometric units
They are 'periodised' variants of the polar units:

$$\mathrm{Qa}_c^{w_1} = \mathcal{Q}_c(u_1) = \frac{c}{\tan(c\,u_1)} \tag{3.17}$$

$$\mathrm{Qi}_c^{w_1} = \mathcal{Q}_c(v_1) = \frac{c}{\tan(c\,v_1)} \tag{3.18}$$

$$\mathrm{Qai}_{c,\alpha,\beta}^{w_1} = \mathcal{Q}_c\left(\frac{u_1}{\alpha}\right) + \mathcal{Q}_c\left(\frac{v_1}{\beta}\right) = \frac{c}{\tan\left(\frac{c\,u_1}{\alpha}\right)} + \frac{c}{\tan\left(\frac{c\,v_1}{\beta}\right)} \tag{3.19}$$

$$\mathrm{Qaih}_{c,\alpha,\beta}^{w_1} = \mathcal{Q}_c\left(\frac{u_1}{\alpha}\right) - \mathcal{Q}_{c\,i}\left(\frac{v_1}{\beta}\right) = \frac{c}{\tan\left(\frac{c\,u_1}{\alpha}\right)} - \frac{c}{\tanh\left(\frac{c\,v_1}{\beta}\right)} \tag{3.20}$$

$Qa_c^{\bullet}$, $Qi_c^{\bullet}$ are *approximate* units but $Qai_{c,\alpha,\beta}^{\bullet}$, $Qaih_{c,\alpha,\beta}^{\bullet}$ are *exact.*

Elliptic units (after C. Brembilla)
Let $\sigma(z\,;\ g_2, g_3)$ be the classical Weierstrass sigma function:

$$\sigma(z\,;\ g_2, g_3) = z - \frac{g_2}{2^4.3.5}\,z^5 - \frac{g_3}{2^3.3.5.7}\,z^7 + \mathcal{O}(z^9) \qquad \text{with}$$

$$\sigma(z\,;\ g_2, g_3) \equiv -\sigma(-z\,;\ g_2, g_3) \equiv t\,\sigma(z\,t^{-1}\,;\ g_2\,t^4, g_3\,t^6) \qquad (\forall t).$$

Then for all $g_2, g_3, \alpha, \beta, \gamma, \delta \in \mathbb{C}$ $(\alpha\beta \neq 0)$, the relation

$$\mathfrak{E}_{g_2,g_3}^{\binom{u_1}{v_1}} := \frac{\sigma(u_1 + v_1\,;\ g_2,\ g_3)}{\sigma(u_1\,;\ g_2,\ g_3)\,\sigma(v_1\,;\ g_2,\ g_3)} \tag{3.21}$$

defines a two-parameter family of exact flexion units, which in turn, under the standard parameter saturation of (3.12), give rise to:

$$\mathfrak{E}_{g_2,g_3,\alpha,\beta,\gamma,\delta}^{\binom{u_1}{v_1}} := \delta\, e^{\gamma\,u_1\,v_1}\ \mathfrak{E}_{g_2,g_3}^{\binom{u_1/\alpha}{v_1/\beta}} \tag{3.22}$$

$$\mathfrak{E}_{g_2,g_3,\alpha,\beta,\gamma,\delta}^{\bullet} \equiv \mathfrak{E}_{g_2\,t^4,\,g_3\,t^6,\,\alpha\,t,\,\beta\,t,\,\gamma,\,\delta\,t^{-1}}^{\bullet} \qquad (\forall t). \tag{3.23}$$

This six-parameter, five-dimensional complex variety of flexion units contains all previously listed *exact units* (polar or trigonometric) as limit cases. In fact, it would seem (the matter is still under investigation) that it exhausts *all* flexion units meromorphic in both u_1 and v_1.

We must now examine further units, exact or approximate, that fail to be meromorphic in one of these variables, or both.

Bitrigonometric units
$Qaa_c^{w_1}$ (respectively $Qii_c^{w_1}$) is defined for $u_1 \in \mathbb{C}$ and $v_1 \in \mathbb{Q}/\mathbb{Z}$ (respectively vice versa):

$$\begin{aligned}
\mathrm{Qaa}_c^{\binom{u_1}{v_1}} &:= \sum_{n_1 \in \mathbb{Z}} \frac{c\, e^{-2\pi i n_1 v_1}}{\pi n_1 + c u_1} \\
&= \sum_{1 \le n_1 \le \mathrm{den}(v_1)} \frac{c\, e^{-2\pi i n_1 v_1}}{\mathrm{den}(v_1)}\, \mathcal{Q}_c\left(\frac{\pi\, n_1 + c\, u_1}{\mathrm{den}(v_1)}\right) \\
\mathrm{Qii}_c^{\binom{u_1}{v_1}} &:= \sum_{n_1 \in \mathbb{Z}} \frac{c\, e^{-2\pi i n_1 u_1}}{\pi n_1 + c v_1} \\
&= \sum_{1 \le n_1 \le \mathrm{den}(u_1)} \frac{c\, e^{-2\pi i n_1 u_1}}{\mathrm{den}(u_1)}\, \mathcal{Q}_c\left(\frac{\pi\, n_1 + c\, v_1}{\mathrm{den}(u_1)}\right) = \mathrm{Qaa}_c^{\binom{v_1}{u_1}}
\end{aligned} \tag{3.24}$$

with *den* denoting the denominator (of a rational number). $Qaa_c^\bullet$ and $Qii_c^\bullet$ are both *approximate* units (see (3.30), (3.31) below).

Flat units
Let σ be the sign function on $\mathbb{R}$, *i.e.* $\sigma(\mathbb{R}^\pm) = \pm 1$ and $\sigma(0) = 0$. Then set:

$$\mathrm{Sa}^{w_1} = \sigma(u_1), \quad \mathrm{Si}^{w_1} = \sigma(v_1), \quad \mathrm{Sai}^{w_1} = \sigma(u_1) + \sigma(v_1) \tag{3.25}$$

$Sa^\bullet$, $Si^\bullet$ are *approximate* units but $Sai^\bullet$ is *exact*.[39]

Mixed units

$$\mathrm{Qas}_{c,\pm}^{w_1} = \mathrm{Q}_c(u_1) \pm c\, i\, \sigma(v_1), \quad \mathrm{Qis}_{c,\pm}^{w_1} = \mathrm{Q}_c(v_1) \pm c\, i\, \sigma(u_1) \tag{3.26}$$

$Qas_{c,\pm}^\bullet$, $Qis_{c,\pm}^\bullet$ are *exact* units.

“False” units

$$\mathrm{Qi}_{c,\pm}^{w_1} = Qi_c^{w_1} \pm c\, i = c\, Q(c\, v_1) \pm c\, i = \pm 2\, c\, i \frac{e^{\pm 2 c i v_1}}{e^{\pm 2 c i v_1} - 1} \tag{3.27}$$

$Qi_{c,+}^\bullet$ and $Qi_{c,-}^\bullet$ verify the exact *tripartite relation* but not the *imparity* condition.[40]

[39] When viewed as a distribution or as an almost-everywhere defined function on $\mathbb{R}$. But when viewed as a function on $\mathbb{Z}$, it becomes an approximate unit.

[40] In terms of applications, the failure of imparity has more disruptive consequences than the failure to verify the exact *tripartite equation*, because it means that $\mathfrak{E}$ has no proper conjugate $\mathfrak{O}$, which in turn prevents it from serving as building block for dimorphic bimoulds such as $\mathfrak{ess}^\bullet$ etc.

Approximate flexion units. Tweaking the tripartite relation
The approximate flexion units listed above verify *tweaked* variants of the tripartite relation:

$$\mathrm{Qa}_c^{w_1}\,\mathrm{Qa}_c^{w_2} \equiv \mathrm{Qa}_c^{w_1\rfloor}\,\mathrm{Qa}_c^{\lceil w_2} + \mathrm{Qa}_c^{w_1\rceil}\,\mathrm{Qa}_c^{\lfloor w_2} + c^2 \tag{3.28}$$

$$\mathrm{Qi}_c^{w_1}\,\mathrm{Qi}_c^{w_2} \equiv \mathrm{Qi}_c^{w_1\rfloor}\,\mathrm{Qi}_c^{\lceil w_2} + \mathrm{Qi}_c^{w_1\rceil}\,\mathrm{Qi}_c^{\lfloor w_2} - c^2 \tag{3.29}$$

$$\mathrm{Qaa}_c^{w_1}\,\mathrm{Qaa}_c^{w_2} \equiv \mathrm{Qaa}_c^{w_1\rfloor}\,\mathrm{Qaa}_c^{\lceil w_2} + \mathrm{Qaa}_c^{w_1\rceil}\,\mathrm{Qaa}_c^{\lfloor w_2} + c^2\,\delta(v_1)\,\delta(v_2) \tag{3.30}$$

$$\mathrm{Qii}_c^{w_1}\,\mathrm{Qii}_c^{w_2} \equiv \mathrm{Qii}_c^{w_1\rfloor}\,\mathrm{Qii}_c^{\lceil w_2} + \mathrm{Qii}_c^{w_1\rceil}\,\mathrm{Qii}_c^{\lfloor w_2} - c^2\,\delta(u_1)\,\delta(u_2) \tag{3.31}$$

$$\mathrm{Sa}^{w_1}\,\mathrm{Sa}^{w_2} \equiv \mathrm{Sa}^{w_1\rfloor}\,\mathrm{Sa}^{\lceil w_2} + \mathrm{Sa}^{w_1\rceil}\,\mathrm{Sa}^{\lfloor w_2} - 1 + \delta(u_1)\,\delta(u_2) \tag{3.32}$$

$$\mathrm{Si}^{w_1}\,\mathrm{Si}^{w_2} \equiv \mathrm{Si}^{w_1\rfloor}\,\mathrm{Si}^{\lceil w_2} + \mathrm{Si}^{w_1\rceil}\,\mathrm{Si}^{\lfloor w_2} + 1 - \delta(v_1)\,\delta(v_2). \tag{3.33}$$

In the last four relations, $\delta(t) := 1$ if $t = 0$ and $\delta(t) := 0$ otherwise.

3.3 Unit-generated algebras $Flex(\mathfrak{E})$

For an *exact* flexion unit $\mathfrak{E}^\bullet$ the monogenous flexion algebra $Flex(\mathfrak{E})$, also known as *eumonogeneous*[41] algebra, is richer in interesting bimoulds, though much smaller in size than in the case of a random generator $\mathfrak{E}^\bullet$. The total algebra $Flex(\mathfrak{E})$ can still, as in Section 3.1, be freely-canonically generated, but under the sole operation *amnit* and *without* mould multiplication *mu*. In other words, we retain only the steps (3.4) and forego the steps (3.3). As a consequence, $Flex(\mathfrak{E})$ decomposes into cells $Flex_r(\mathfrak{E}) \subset BIMU_r$ whose dimensions are given by the Catalan numbers and whose inductive construction goes like this:

$$\mathrm{Flex}(\mathfrak{E}) = \bigoplus_{r\geq 0} \mathrm{Flex}_r(\mathfrak{E}) \quad \text{with} \quad \dim(\mathrm{Flex}_r(\mathfrak{E})) = \frac{(2r)!}{r!\,(r+1)!} \tag{3.34}$$

$$\mathrm{Flex}_r(\mathfrak{E}) = \bigoplus_{\substack{r_1+r_2=r-1\\ r_1,r_2\geq 0}} \mathrm{amnit}(\mathrm{Flex}_{r_1}(\mathfrak{E}), \mathrm{Flex}_{r_2}(\mathfrak{E}))\,.\,\mathfrak{E}^\bullet. \tag{3.35}$$

The new basis $\{\mathfrak{e}_t^\bullet\}$
It follows from (3.35) that $Flex_r(\mathfrak{E})$ has a natural basis $\{\mathfrak{e}_t^\bullet\}$ indexed by all r-node binary trees $\boldsymbol{t}$. The construction is by induction on r:

$$\mathfrak{e}_t^\bullet = \mathrm{amnit}(\mathfrak{e}_{t_1}^\bullet, \mathfrak{e}_{t_2}^\bullet)\,.\,\mathfrak{E}^\bullet = \mathrm{anmit}(\mathfrak{e}_{t_1}^\bullet, \mathfrak{e}_{t_2}^\bullet)\,.\,\mathfrak{E}^\bullet \tag{3.36}$$

[41] With *eu* standing for *good*. For the polar respectively trigonometric specialisations of the unit, $Flex(\mathfrak{E})$ is known as the *eupolar* respectively *eutrigonometric* algebra. In the eutrigonometric case, though, the basis elements are more numerous than in the eupolar case, and *amnit* is no longer sufficient to generate everything. See the last table in Section 12.1.

where $\boldsymbol{t_1}, \boldsymbol{t_2}$ denote the left and right subtrees (one of them possibly empty) attached to the root of the binary tree $\boldsymbol{t}$.

This new basis $\{\mathfrak{e}^\bullet_t\}$ is a natural subset of the analogous basis of Section 3.1, which was indexed by *ternary* trees.

The new basis $\{\mathfrak{e}^\bullet_g\}$

It coincides with the analogous system in Section 3.1, but restricted to the pairs $\boldsymbol{g} = (\boldsymbol{ga}, \boldsymbol{gi})$ meeting either of these two equivalent conditions:
(i) the graph $\boldsymbol{ga}$ has no pair of edges issuing from the same vertex and containing Si_0 in the angle so defined.
(ii) the graph $\boldsymbol{gi}$ has no pair of edges with end-points (Si_p, Si_k), (Si_{k+1}, Si_q) disposed in the circular order $0 \leq p < k < k+1 < q \leq r+1$.

The new basis $\{\mathfrak{e}^\bullet_o\}$

It coincides with its prototype in of Section 3.1, but under restriction to the *separative* orders $\boldsymbol{o}$, *i.e.* to orders such that:

$$\{i - j = 1\} \Longrightarrow \{i \preceq j\} \text{ } or \text{ } \{j \preceq i\} \tag{3.37}$$

In other words, elements that are *consecutive* in the natural order must be *comparable* in the $\boldsymbol{o}$-order. This implies that $\boldsymbol{o}$ has a *smallest* element. It also implies that if i, j are not $\boldsymbol{o}$-comparable, then the intervals $[i^-, i^+]$ and $[j^-, j^+]$ cannot be contiguous (which justifies calling the order $\boldsymbol{o}$ *"separative"*).

The new basis $\{\mathfrak{e}^\bullet_p\}$

It coincides with the analogous system in Section 3.1, but restricted to the words $\boldsymbol{p}$ constructed from the sole induction rule (3.7), without recourse to word concatenation. These less numerous $\boldsymbol{p}$ are necessarily prime, and can be compactly represented by sequences $\boldsymbol{h} = [h(1), \ldots, h(r)]$, with $h(i)$ denoting the height of the i-th letter b in $\boldsymbol{p}$, as defined in (3.8). For the lengths $r \leq 3$ we have thus:

$$\begin{array}{ll}
\mathcal{H}_1 = \{[1]\} & \longleftrightarrow \mathcal{P}_1 = \{abc\} \\
\mathcal{H}_2 = \{[1,2],[2,1]\} & \longleftrightarrow \mathcal{P}_2 = \{ababcc,\, aabcbc\} \\
\mathcal{H}_3 = \{[1,2,3],[1,3,2],[2,1,2],[2,3,1],[3,2,1]\} & \longleftrightarrow \mathcal{P}_3 = \{abababccc, \ldots\}
\end{array}$$

The involution *syap* between conjugate flexion structures

All monogenous structures $Flex(\mathfrak{E})$ generated by the exact flexion units listed in Section 3.2 are actually isomorphic. In the case of two conjugate units, the isomorphism becomes an involution, denoted *syap*:

$$\text{syap}: \quad \text{Flex}_r(\mathfrak{E}) \leftrightarrow \text{Flex}_r(\mathfrak{O}), \quad \mathfrak{e}^\bullet_t \leftrightarrow \mathfrak{o}^\bullet_t \quad (\mathfrak{E}, \mathfrak{O} \text{ conjugate}). \tag{3.38}$$

The involution *syap*, being defined only on monogenous structures, is quite distinct from the universal involution *swap*, which applies to the whole of *BIMU*. On the other hand, *syap* is more regular: it commutes with *all* flexion operations, whether unary or binary, whereas *swap* commutes only with a few, such as *ami*//*gami*.

The involution *sap* on each flexion structure

Both mappings *swap* and *syap* exchange $Flex(\mathfrak{E})$ and $Flex(\mathfrak{O})$. Since these two involutions actually commute, their product *sap* is also a linear involution, with eigenspaces $\{\pm 1\}$ of approximately equal size:

$$\text{syap}: \quad \text{Flex}_r(\mathfrak{E}) \leftrightarrow \text{Flex}_r(\mathfrak{O}) \tag{3.39}$$

$$\text{swap}: \quad \text{Flex}_r(\mathfrak{E}) \leftrightarrow \text{Flex}_r(\mathfrak{O}) \tag{3.40}$$

$$\text{sap}: \quad \text{Flex}_r(\mathfrak{E}) \leftrightarrow \text{Flex}_r(\mathfrak{E}),\ \ \text{Flex}_r(\mathfrak{O}) \leftrightarrow \text{Flex}_r(\mathfrak{O}) \tag{3.41}$$

$$\text{with} \quad \text{sap} := \quad \text{syap.swap} = \text{swap.syap}. \tag{3.42}$$

For r even, the dimensions $d_r^{\pm}$ of *sap*'s eigenspaces of eigenvalues ± 1 are equal, but for r odd d_r^+ is slightly larger than d_r^-. In fact, computational evidence supports the following conjectures[42]:

$$d_{2r}^+ - d_{2r}^- = 0 \qquad (\forall r) \tag{3.43}$$

$$d_{2r+1}^+ - d_{2r+1}^- = \frac{(2r)!}{r!(r+1)!} = d_r^+ + d_r^- \qquad (\forall r) \tag{3.44}$$

Polar specialisation and graphic interpretation

In the specal case $(\mathfrak{E}^\bullet, \mathfrak{O}^\bullet) = (Pa^\bullet, Pi^\bullet)$, both the canonical basis and the involution *syap* have a simple interpretation, as shown on the polygonal diagrams in Section 12.1, with the *dotted* respectively *full* lines representing the variables $\boldsymbol{u}$ respectively $\boldsymbol{v}$.

3.4 Twisted symmetries and subsymmetries in universal mode

To every exact flexion unit $\mathfrak{E}$ there correspond *twisted* variants of all *straight* symmetries and subsymmetries listed in Section 2.4. But before defining these, we must introduce two elementary bimoulds $\mathfrak{ez}^\bullet$ and $\underline{\mathfrak{ez}}^\bullet = \text{pari}.\mathfrak{ez}^\bullet$:

$$\mathfrak{ez}^{w_1,\dots,w_r} := \mathfrak{E}^{w_1}\dots\mathfrak{E}^{w_r}, \qquad \underline{\mathfrak{ez}}^{w_1,\dots,w_r} := (-1)^r\ \mathfrak{E}^{w_1}\dots\mathfrak{E}^{w_r} \tag{3.45}$$

as well as the *symmetral* bimould $\mathfrak{es}^\bullet := sap.\mathfrak{ez}^\bullet$. (See also (4.70)).

[42] They have been verified up to $r = 8$.

- **$\mathfrak{E}$-alternality and $\mathfrak{E}$-symmetrality**

The simplest characterisation of the $\mathfrak{E}$-twisted symmetries is by means of the equivalence:

$$\{B^\bullet \;\; \mathfrak{E}\text{-alternal resp. } \mathfrak{E}\text{-symmetral}\} \Longleftrightarrow \{A^\bullet \text{ alternal resp. symmetral}\}$$

with $B^\bullet = \text{ganit}(\mathfrak{ez}^\bullet).A^\bullet$ or $B^\bullet = \text{gamit}(\mathfrak{ez}^\bullet).A^\bullet$, on choice.[43]

As for the analytic expression of the twisted symmetries, it reproduces that of the *straight* symmetries on which they are patterned, except for the systematic occurence of inflected pairs (w_i, w_j), with w_i, w_j not in the same factor sequence. Let us illustrate the $\mathfrak{E}$-alternality (respectively $\mathfrak{E}$-symmetrality) relations for two sequences $\boldsymbol{w}'$, $\boldsymbol{w}''$ first of length 1:

$$B^{w_1,w_2} + B^{w_2,w_1} + B^{w_1\rceil}\underline{\mathfrak{ez}}^{\lfloor w_2} + B^{\lceil w_2}\underline{\mathfrak{ez}}^{w_1\rfloor} = 0 \text{ (resp. } B^{w_1}B^{w_2}) \quad i.e$$

$$B^{\binom{u_1\,,\,u_2}{v_1\,,\,v_2}} + B^{\binom{u_2\,,\,u_1}{v_2\,,\,v_1}} - B^{\binom{u_{12}}{v_1}}\mathfrak{E}^{\binom{u_2}{v_{2:1}}} - B^{\binom{u_{12}}{v_2}}\mathfrak{E}^{\binom{u_1}{v_{1:2}}} = 0 \text{ (resp. } B^{\binom{u_1}{v_1}}B^{\binom{u_2}{v_2}})$$

and then of length 2:

$$\begin{aligned}
&B^{w_1,w_2,w_3,w_4} + B^{w_1,w_3,w_2,w_4} + B^{w_3,w_1,w_2,w_4} + B^{w_1,w_3,w_4,w_2} + B^{w_3,w_1,w_4,w_2} + B^{w_3,w_4,w_1,w_2}\\
&+B^{w_1\rceil,w_2,w_4}\,\underline{\mathfrak{ez}}^{\lfloor w_3} + B^{\lceil w_3,w_2,w_4}\,\underline{\mathfrak{ez}}^{w_1\rfloor} + B^{w_1\rceil,w_4,w_2}\,\underline{\mathfrak{ez}}^{\lfloor w_3} + B^{\lceil w_3,w_4,w_2}\,\underline{\mathfrak{ez}}^{w_1\rfloor}\\
&+B^{w_3,w_1\rceil,w_2}\,\underline{\mathfrak{ez}}^{\lfloor w_4} + B^{w_3,\lceil w_4,w_2}\,\underline{\mathfrak{ez}}^{w_1\rfloor} + B^{w_1,w_2\rceil,w_4}\,\underline{\mathfrak{ez}}^{\lfloor w_3} + B^{w_1,\lceil w_3,w_4}\,\underline{\mathfrak{ez}}^{w_2\rfloor}\\
&+B^{w_1,w_3,w_2\rceil}\,\underline{\mathfrak{ez}}^{\lfloor w_4} + B^{w_1,w_3,\lceil w_4}\,\underline{\mathfrak{ez}}^{w_2\rfloor} + B^{w_3,w_1,w_2\rceil}\,\underline{\mathfrak{ez}}^{\lfloor w_4} + B^{w_3,w_1,\lceil w_4}\,\underline{\mathfrak{ez}}^{w_2\rfloor}\\
&+B^{w_1\rceil,w_2\rceil}\,\underline{\mathfrak{ez}}^{\lfloor w_3,\lfloor w_4} + B^{\lceil w_3,w_2\rceil}\,\underline{\mathfrak{ez}}^{w_1\rfloor,\lfloor w_4} + B^{w_1\rceil,\lceil w_4}\,\underline{\mathfrak{ez}}^{\lfloor w_3,w_2\rfloor} + B^{\lceil w_3,\lceil w_4}\,\underline{\mathfrak{ez}}^{w_1\rfloor,w_2\rfloor}\\
&= 0 \quad \text{(respectively } B^{w_1,w_2}B^{w_3,w_4}).
\end{aligned}$$

These two examples should suffice to make the pattern clear. Remarkably, when $\mathfrak{E}$ runs through the set of all flexion units, the corresponding $\mathfrak{E}$-symmetralities essentially exhaust all commutative *flexion products*[44] that may be defined on *BIMU*.

Like their *straight* models, the *twisted* symmetries induce important subsymmetries, which we must now sort out.

- **$\{\mathfrak{E}$-alternal$\} \Longrightarrow \{\mathfrak{E}$-mantar-invariant, $\mathfrak{E}$-pus-neutral$\}$**

$\mathfrak{E}$-*mantar* is a linear operator conjugate to *mantar*:

$$\mathfrak{E}\text{-mantar} := \text{ganit}(\mathfrak{ez}^\bullet).\text{mantar}.\text{ganit}(\mathfrak{ez}^\bullet)^{-1} \tag{3.46}$$

---

[43] $\text{ganit}(\mathfrak{ez}^\bullet)$ and $\text{gamit}(\mathfrak{ez}^\bullet)$ define two distinct mappings $A^\bullet \mapsto B^\bullet$, but both result in the same transformation of symmetries.

[44] Provided we include the approximate flexion units, for which the twisted symmetries become more intricate. For the trigonometric case, see Section 11.4.

and with explicit action:

$$((\mathfrak{E}\text{-mantar}).B)^{\boldsymbol{w}} = (-1)^{r-1} \sum_{\prod_i \boldsymbol{a}^i b_i \boldsymbol{c}^i = \tilde{\boldsymbol{w}}} B^{\lceil b_1 \rceil \ldots \lceil b_s \rceil} \prod_i \underline{\mathfrak{e}_3}^{\boldsymbol{a}^i \rfloor} \prod_i \underline{\mathfrak{e}_3}^{\lfloor \boldsymbol{c}^i} \quad (3.47)$$

(Note that $\tilde{\boldsymbol{w}}$ always denotes the sequence $\boldsymbol{w}$ in reverse order).

$\mathfrak{E}$-*pus*-neutrality also is derived from straight *pus*-neutrality:

$$\left(\sum_{1 \le l \le r(\bullet)} \text{pus}^l \right) .\text{ganit}(\mathfrak{e}_3)^{-1}.B^\bullet \equiv 0$$

and admits a simpler direct expression:

$$\sum_{\boldsymbol{w}' \overset{\text{circ}}{\sim} \boldsymbol{w}} B^{\boldsymbol{w}'} + (-1)^{r(\boldsymbol{w})} \sum_{\boldsymbol{a}^i w_i \boldsymbol{b}^i = \boldsymbol{w}} B^{\lceil w_i \rceil} \; \underline{\mathfrak{e}_3}^{\boldsymbol{a}^i \rfloor} \; \underline{\mathfrak{e}_3}^{\lfloor \boldsymbol{b}^i} \equiv 0. \quad (3.48)$$

- **{$\mathfrak{E}$-symmetral} $\Longrightarrow$ {$\mathfrak{E}$-gantar-invariant, $\mathfrak{E}$-gus-neutral}**

$\mathfrak{E}$-*gantar* is a non-linear operator conjugate to *gantar*:

$$\begin{aligned}
\mathfrak{E}\text{-gantar} &:= \text{ganit}(\mathfrak{e}_3{}^\bullet).\text{gantar}.\text{ganit}(\mathfrak{e}_3{}^\bullet)^{-1} \\
&= \text{ganit}(\mathfrak{e}_3{}^\bullet).\text{invmu}.\text{anti}.\text{pari}.\text{ganit}(\mathfrak{e}_3{}^\bullet)^{-1} \\
&= \text{ganit}(\mathfrak{e}_3{}^\bullet).\text{invmu}.\text{anti}.\text{pari}.\text{minu}.\text{ganit}(\mathfrak{e}_3{}^\bullet)^{-1}\text{minu} \\
&= \text{invmu}.\text{ganit}(\mathfrak{e}_3{}^\bullet).\text{anti}.\text{pari}.\text{minu}.\text{ganit}(\mathfrak{e}_3{}^\bullet)^{-1}\text{minu} \\
&= \text{invmu}.(\mathfrak{E}\text{-mantar}).\text{minu}.
\end{aligned}$$

To establish the above sequence, we used the commutation of *ganit*($M^\bullet$) with both *minu* and *invmu*, and the mutual commutation of *minu, anti, pari*.

Using the last identity, we see that the action of $\mathfrak{E}$-*gantar* is given by:

$$\begin{aligned}
&((\mathfrak{E}\text{-gantar}).B)^{\boldsymbol{w}} \\
&= \sum_{\prod_i \boldsymbol{a}^i b_i \boldsymbol{c}^i = \tilde{\boldsymbol{w}}} \; \sum_{\prod_j \boldsymbol{b}^j = \prod_i \lceil b_i \rceil} (-1)^{r-s} \prod_{1 \le j \le s} B^{\boldsymbol{b}^j} \prod_i \underline{\mathfrak{e}_3}^{\boldsymbol{a}^i \rfloor} \prod_i \underline{\mathfrak{e}_3}^{\lfloor \boldsymbol{c}^i} \qquad (3.49)
\end{aligned}$$

$\mathfrak{E}$-*gus*-neutrality also is derived from straight *gus*-neutrality:

$$\left(\sum_{1 \le l \le r(\bullet)} \text{gus}^l \right) .\text{ganit}(\mathfrak{e}_3)^{-1}.B^\bullet \equiv 0$$

and admits a simpler direct expression:

$$\sum_{1 \le s} (-1)^s \sum_{\boldsymbol{w}^1 \ldots \boldsymbol{w}^s \overset{\text{circ}}{\sim} \boldsymbol{w}} B^{\boldsymbol{w}^1} \ldots B^{\boldsymbol{w}^s} \equiv (-1)^{r(\boldsymbol{w})} \sum_{\boldsymbol{a}^i w_i \boldsymbol{b}^i = \boldsymbol{w}} B^{\lceil w_i \rceil} \; \underline{\mathfrak{e}_3}^{\boldsymbol{a}^i \rfloor} \; \underline{\mathfrak{e}_3}^{\lfloor \boldsymbol{b}^i}. \quad (3.50)$$

One should take care to interpret the circular sums correctly, *i.e.* without repetitions. Thus, if $\boldsymbol{w}$ has length 4, on the left-hand side of (3.50) the terms $B^{w_1,w_2}B^{w_3,w_4}$ and $B^{w_2,w_3}B^{w_4,w_1}$ occur *once* rather than *twice*, and the term $B^{w_1}B^{w_2}B^{w_3}B^{w_4}$ also occurs *once*, not *four times*.

• **{alternal//$\mathfrak{O}$-alternal}** $\overset{\text{ess}^{\text{ly}}}{\Longrightarrow}$ **{$\mathfrak{E}$-neg-invariant, $\mathfrak{E}$-push-invariant}**
As mentioned in Section 2.4, bialternality implies invariance not just under *negpush= mantar.swap.mantar.swap* but also[45] separate invariance under *neg* and *push*. Likewise, given any pair of conjugate flexion units $(\mathfrak{E}, \mathfrak{O})$, a bimould $B^\bullet$ of type $\underline{\text{al}}/\underline{\text{ol}}$ (*i.e.* alternal and with a $\mathfrak{O}$-alternal swappee) is ipso facto invariant not just under $\mathfrak{E}$-*negpush* but also[46] separately so under $\mathfrak{E}$-*neg* and $\mathfrak{E}$-*push*. The definitions of these operators run parallel to those of the straight case[47]:

$$\mathfrak{E}\text{-negpush} := \text{mantar.swap.}(\mathfrak{E}\text{-mantar}).\text{swap} \tag{3.51}$$
$$\mathfrak{E}\text{-neg} := \text{neg.adari}(\mathfrak{es}^\bullet) = \text{adari}(\text{pari.}\mathfrak{es}^\bullet).\text{neg} \tag{3.52}$$
$$\mathfrak{E}\text{-push} := (\mathfrak{E}\text{-neg}).\text{mantar.swap.}(\mathfrak{E}\text{-mantar}).\text{swap}. \tag{3.53}$$

In fact, invariance under $\mathfrak{E}$-*push* is equivalent to invariance under a distinct and simpler operator $\mathfrak{E}$-$push_*$, which is defined as follows:

$$\mathfrak{E}\text{-push}_* := (\mathfrak{E}\text{-ter})^{-1}.\text{push.mantar.}(\mathfrak{E}\text{-ter}).\text{mantar} \tag{3.54}$$

with

$$((\mathfrak{E}\text{-ter}).B^\bullet)^{w_1,\ldots,w_r} := B^{w_1,\ldots,w_r} - B^{w_1,\ldots,w_{r-1}}\mathfrak{E}^{w_r} + B^{w_1,\ldots,w_{r-1}\rceil}\mathfrak{E}^{\lfloor w_r} \tag{3.55}$$
$$((\mathfrak{E}\text{-ter})^{-1}.B^\bullet)^{w_1,\ldots,w_r} := \sum_{\boldsymbol{a}.\boldsymbol{b}.\boldsymbol{c}\,=\,\boldsymbol{w}=(w_1,\ldots,w_r)} B^{\boldsymbol{a}\rceil}\,\mathfrak{mues}^{\lfloor \boldsymbol{b}}\,\mathfrak{es}^{\boldsymbol{c}} \tag{3.56}$$

and with $\mathfrak{mues}^\bullet := \text{invmu.}\mathfrak{es}^\bullet = \text{pari.anti.}\mathfrak{es}^\bullet$ and $\mathfrak{es}^\bullet$ as in (4.70).

The reason for this equivalence is the identity:

$$(\text{id} - \mathfrak{E}\text{-push}_*).B^\bullet \equiv \text{swamu}(\mathfrak{es}^\bullet, (\text{id} - \mathfrak{E}\text{-push}).B^\bullet) \qquad \forall B^\bullet \tag{3.57}$$

with *swamu* defined as the *swap*-conjugate of *mu*.[48]

---

[45] Provided we assume (as assume we must, to ensure *ari*-stability) the component of length 1 to be *even*.

[46] Again, assuming parity for the length-1 component.

[47] See (2.12) for *push* and also (4.70) for $\mathfrak{es}^\bullet$.

[48] *I.e.* $\text{swamu}(M_1^\bullet, M_2^\bullet) := \text{swap.mu}(\text{swap.}M_1^\bullet, \text{swap.}M_2^\bullet)$

The notable advantage of $\mathfrak{E}$-push$_*$-invariance over $\mathfrak{E}$-push-invariance is that it leads straightaway to the so-called *senary* relation:[49]

$$(\mathfrak{E}\text{-ter}).B^\bullet = \text{push.mantar.}(\mathfrak{E}\text{-ter}).\text{mantar.}B^\bullet \qquad (3.58)$$

which is the simplest way of expressing the $\mathfrak{E}$-*push*-invariance of $B^\bullet$.

• **{symmetral//$\mathfrak{O}$-symmetral} $\overset{\text{ess}^{\text{ly}}}{\Longrightarrow}$ {$\mathfrak{E}$-geg-invariant, $\mathfrak{E}$-gush-invariant}**
Here, the first induced subsymmetry is the same as above, namely invariance under the linear operator $\mathfrak{E}$-*geg*, defined as $\mathfrak{E}$-*neg* in (3.52) but with *adari* replaced by *adgari*:

$$\mathfrak{E}\text{-geg} := \text{neg.adgari}(\mathfrak{es}^\bullet) = \text{adgari}(\text{pari.}\mathfrak{es}^\bullet)\text{.neg.} \qquad (3.59)$$

The second induced subsymmetry is $\mathfrak{E}$-*gush*-invariance, with:

$$\mathfrak{E}\text{-gush} := (\mathfrak{E}\text{-geg}).\text{gantar.swap.}(\mathfrak{E}\text{-gantar}).\text{swap.} \qquad (3.60)$$

The only moot point is whether $\mathfrak{E}$-*gush*-invariance is equivalent to invariance under some simpler operator $\mathfrak{E}$-*gush*$_*$ defined along the same lines as (3.54). Even though the existence of a senary relation, or for that matter of a relation of *finite* arity is unlikely, it ought to be possible to improve considerably on $\mathfrak{E}$-*gush*.

3.5 Twisted symmetries and subsymmetries in polar mode

Let us now restate the above results for the most important unit specialisation, which is the polar specialisation $(\mathfrak{E}^\bullet, \mathfrak{O}^\bullet) = (Pa^\bullet, Pi^\bullet)$. The transposition goes like this:

$\mathfrak{E}$-alternal	→ alternul	(∗)	;	$\mathfrak{O}$-alternal	→	alternil	
$\mathfrak{E}$-symmetral	→ symmetrul	(∗)	;	$\mathfrak{O}$-symmetral	→	symmetril	
$\mathfrak{E}$-mantar	→ mantur	(∗)	;	$\mathfrak{O}$-mantar	→	mantir	
$\mathfrak{E}$-gantar	→ gantur	(∗)	;	$\mathfrak{O}$-gantar	→	gantir	
$\mathfrak{E}$-pus	→ pusu	(∗)	;	$\mathfrak{O}$-mantar	→	pusi	
$\mathfrak{E}$-gus	→ gusu	(∗)	;	$\mathfrak{O}$-gus	→	gusi	
$\mathfrak{E}$-push	→ pushu		;	$\mathfrak{O}$-push	→	pushi	(∗)
$\mathfrak{E}$-gush	→ gushu		;	$\mathfrak{O}$-gush	→	gushi	(∗)
$\mathfrak{E}$-neg	→ negu		;	$\mathfrak{O}$-neg	→	negi	(∗)
$\mathfrak{E}$-geg	→ gegu		;	$\mathfrak{O}$-geg	→	gegi	(∗)
$\mathfrak{E}$-ter	→ teru		;	$\mathfrak{O}$-ter	→	teri	(∗)

[49] So-called because it involves only six terms – three on the left-hand side and three on the right.

And of course:

$$\begin{aligned} &\text{alternal}/\mathfrak{O}\text{-}\mathit{alternal} \;\rightarrow\; \text{alternal/alternil}\\ &\text{alternal}/\mathfrak{E}\text{-}\mathit{alternal} \;\rightarrow\; \text{alternal/alternul} \;\; (*). \end{aligned}$$

In the above tables, the stars $(*)$ accompany all symmetry types that are *incompatible* with *entireness*. For further details, see Section 4.7.

- **Alternility and symmetrility**

Let us write down the alternility (respectively symmetrility) relations for two sequences $\boldsymbol{w}'$, $\boldsymbol{w}''$ first of length (1,1):

$$B^{\binom{u_1,\,u_2}{v_1,\,v_2}} + B^{\binom{u_1,\,u_2}{v_1,\,v_2}} - B^{\binom{u_{12}}{v_1}} P^{v_{2:1}} - B^{\binom{u_{12}}{v_2}} P^{v_{1:2}} = 0 \;\; \Big(\text{resp. } B^{\binom{u_1}{v_1}} B^{\binom{u_2}{v_2}}\Big)$$

then of length (1,2):

$$\begin{aligned} &B^{\binom{u_1,\,u_2,\,u_3}{v_1,\,v_2,\,v_3}} + B^{\binom{u_2,\,u_1,\,u_3}{v_2,\,v_1,\,v_3}} + B^{\binom{u_2,\,u_3,\,u_1}{v_2,\,v_3,\,v_1}} - B^{\binom{u_{12},\,u_3}{v_1,\,v_3}} P^{v_{2:1}} - B^{\binom{u_{12},\,u_3}{v_2,\,v_3}} P^{v_{1:2}}\\ &- B^{\binom{u_2,\,u_{13}}{v_2,\,v_1}} P^{v_{3:1}} - B^{\binom{u_2,\,u_{13}}{v_2,\,v_3}} P^{v_{1:3}} = 0 \;\; \Big(\text{respectively } B^{\binom{u_1}{v_1}} B^{\binom{u_2,\,u_3}{v_2,\,v_3}}\Big) \end{aligned}$$

and then of length (2,2):

$$\begin{aligned} &B^{\binom{u_1,\,u_2,\,u_3,\,u_4}{v_1,\,v_2,\,v_3,\,v_4}} + B^{\binom{u_1,\,u_3,\,u_2,\,u_4}{v_1,\,v_3,\,v_2,\,v_4}} + B^{\binom{u_3,\,u_1,\,u_2,\,u_4}{v_3,\,v_1,\,v_2,\,v_4}}\\ &+ B^{\binom{u_1,\,u_3,\,u_4,\,u_2}{v_1,\,v_3,\,v_4,\,v_2}} + B^{\binom{u_3,\,u_1,\,u_4,\,u_2}{v_3,\,v_1,\,v_4,\,v_2}} + B^{\binom{u_3,\,u_4,\,u_1,\,u_2}{v_3,\,v_4,\,v_1,\,v_2}}\\ &- B^{\binom{u_{13},\,u_2,\,u_4}{v_1,\,v_2,\,v_4}} P^{v_{3:1}} - B^{\binom{u_{13},\,u_2,\,u_4}{v_3,\,v_2,\,v_4}} P^{v_{1:3}} - B^{\binom{u_{13},\,u_4,\,u_2}{v_1,\,v_4,\,v_2}} P^{v_{3:1}} - B^{\binom{u_{13},\,u_4,\,u_2}{v_3,\,v_4,\,v_2}} P^{v_{1:3}}\\ &- B^{\binom{u_{13},\,u_1,\,u_2}{v_3,\,v_1,\,v_2}} P^{v_{4:1}} - B^{\binom{u_{13},\,u_{14},\,u_2}{v_3,\,v_4,\,v_2}} P^{v_{1:4}} - B^{\binom{u_1,\,u_{23},\,u_4}{v_1,\,v_2,\,v_4}} P^{v_{3:2}} - B^{\binom{u_1,\,u_{23},\,u_4}{v_1,\,v_3,\,v_4}} P^{v_{2:3}}\\ &- B^{\binom{u_1,\,u_3,\,u_{24}}{v_1,\,v_3,\,v_2}} P^{v_{4:2}} - B^{\binom{u_1,\,u_3,\,u_{24}}{v_1,\,v_3,\,v_4}} P^{v_{2:4}} - B^{\binom{u_3,\,u_1,\,u_{24}}{v_3,\,v_1,\,v_2}} P^{v_{4:2}} - B^{\binom{u_3,\,u_1,\,u_{24}}{v_3,\,v_1,\,v_4}} P^{v_{2:4}}\\ &+ B^{\binom{u_{13},\,u_{24}}{v_1,\,v_2}} P^{v_{3:1}} P^{v_{4:2}} + B^{\binom{u_{13},\,u_{24}}{v_3,\,v_2}} P^{v_{1:3}} P^{v_{4:2}}\\ &+ B^{\binom{u_{13},\,u_{24}}{v_1,\,v_4}} P^{v_{3:1}} P^{v_{2:4}} + B^{\binom{u_{13},\,u_{24}}{v_3,\,v_4}} P^{v_{1:3}} P^{v_{2:4}}\\ &= 0 \;\Big(\text{respectively } \;\; B^{\binom{u_1,\,u_2}{v_1,\,v_2}} B^{\binom{u_3,\,u_4}{v_3,\,v_4}}\Big). \end{aligned}$$

Here and in all such formulas, we set $P^{v_i} := P(v_i) := 1/v_i$, purely for typographical coherence.

• **{alternil} $\Longrightarrow$ {mantir-invariant, pusi-neutral}**
For length $r = 1, 2, 3$ the *mantir* operator acts thus:[50]

$$
\begin{aligned}
(\mathrm{mantir}.B)^{\binom{u_1}{v_1}} &= +B^{\binom{u_1}{v_1}} \\
(\mathrm{mantir}.B)^{\binom{u_1,\, u_2}{v_1,\, v_2}} &= -B^{\binom{u_2,\, u_1}{v_2,\, v_1}} + B^{\binom{u_{12}}{v_1}} P^{v_{2:1}} + B^{\binom{u_{12}}{v_2}} P^{v_{1:2}} \\
(\mathrm{mantir}.B)^{\binom{u_1,\, u_2,\, u_3}{v_1,\, v_2,\, v_3}} &= +B^{\binom{u_3,\, u_2,\, u_1}{v_3,\, v_2,\, v_1}} \\
&\quad -B^{\binom{u_{23},\, u_1}{v_3,\, v_1}} P^{v_{2:3}} - B^{\binom{u_{23},\, u_1}{v_2,\, v_1}} P^{v_{3:2}} - B^{\binom{u_3,\, u_{12}}{v_3,\, v_2}} P^{v_{1:2}} - B^{\binom{u_3,\, u_{12}}{v_3,\, v_1}} P^{v_{2:1}} \\
&\quad +B^{\binom{u_{123}}{v_1}} P^{v_{2:1}} P^{v_{3:1}} + B^{\binom{u_{123}}{v_2}} P^{v_{1:2}} P^{v_{3:2}} + B^{\binom{u_{123}}{v_3}} P^{v_{1:3}} P^{v_{2:3}}
\end{aligned}
$$

and *pusi*-neutrality means this:

$$
\sum_{\mathrm{circ}} B^{\binom{u_1,\, u_2}{v_1,\, v_2}} = +B^{\binom{u_{12}}{v_1}} P^{v_{2:1}} + B^{\binom{u_{12}}{v_2}} P^{v_{1:2}}
$$

$$
\sum_{\mathrm{circ}} B^{\binom{u_1,\, u_2,\, u_3}{v_1,\, v_2,\, v_3}} = +B^{\binom{u_{123}}{v_1}} P^{v_{2:1}} P^{v_{3:1}} + B^{\binom{u_{123}}{v_2}} P^{v_{1:2}} P^{v_{3:2}} + B^{\binom{u_{123}}{v_3}} P^{v_{1:3}} P^{v_{2:3}}
$$

• **{symmetril} $\Longrightarrow$ {gantir-invariant, gusi-neutral}**
For length $r = 1, 2, 3$ the *gantir* operator acts thus:

$$
\begin{aligned}
(\mathrm{gantir}.B)^{\binom{u_1}{v_1}} &= +B^{\binom{u_1}{v_1}} \\
(\mathrm{gantir}.B)^{\binom{u_1,\, u_2}{v_1,\, v_2}} &= -B^{\binom{u_2,\, u_1}{v_2,\, v_1}} + B^{\binom{u_2}{v_2}} B^{\binom{u_1}{v_1}} + B^{\binom{u_{12}}{v_1}} P^{v_{2:1}} + B^{\binom{u_{12}}{v_2}} P^{v_{1:2}} \\
(\mathrm{gantir}.B)^{\binom{u_1,\, u_2,\, u_3}{v_1,\, v_2,\, v_3}} &= +B^{\binom{u_3,\, u_2,\, u_1}{v_3,\, v_2,\, v_1}} + B^{\binom{u_3}{v_3}} B^{\binom{u_2}{v_2}} B^{\binom{u_1}{v_1}} - B^{\binom{u_3\ u_2}{v_3\ v_2}} B^{\binom{u_1}{v_1}} - B^{\binom{u_3}{v_3}} B^{\binom{u_2\ u_1}{v_2\ v_1}} \\
&\quad - B^{\binom{u_{23},\, u_1}{v_3,\, v_1}} P^{v_{2:3}} - B^{\binom{u_{23},\, u_1}{v_2,\, v_1}} P^{v_{3:2}} - B^{\binom{u_3,\, u_{12}}{v_3,\, v_2}} P^{v_{1:2}} - B^{\binom{u_3,\, u_{12}}{v_3,\, v_1}} P^{v_{2:1}} \\
&\quad +B^{\binom{u_{23}}{v_3}} B^{\binom{u_1}{v_1}} P^{v_{2:3}} + B^{\binom{u_{23}}{v_2}} B^{\binom{u_1}{v_1}} P^{v_{3:2}} + B^{\binom{u_3}{v_3}} B^{\binom{u_{12}}{v_2}} P^{v_{1:2}} + B^{\binom{u_3}{v_3}} B^{\binom{u_{12}}{v_1}} P^{v_{2:1}} \\
&\quad + B^{\binom{u_{123}}{v_1}} P^{v_{2:1}} P^{v_{3:1}} + B^{\binom{u_{123}}{v_2}} P^{v_{1:2}} P^{v_{3:2}} + B^{\binom{u_{123}}{v_3}} P^{v_{1:3}} P^{v_{2:3}}.
\end{aligned}
$$

As for *gusi*-neutrality, it has the same expression as *pusi*-neutrality, but with left-hand side replaced for $r = 2, 3$, etc., respectively by:

$$
B^{\binom{u_1,\, u_2}{v_1,\, v_2}} + B^{\binom{u_2,\, u_1}{v_2,\, v_1}} - B^{\binom{u_1}{v_1}} B^{\binom{u_3}{v_3}}
$$

$$
\begin{gathered}
B^{\binom{u_1, u_2, u_3}{v_1, v_2, v_3}} + B^{\binom{u_2, u_3, u_1}{v_2, v_3, v_1}} + B^{\binom{u_3, u_1, u_2}{v_3, v_1, v_2}} \\
-B^{\binom{u_1, u_2}{v_1, v_2}} B^{\binom{u_3}{v_3}} - B^{\binom{u_2, u_3}{v_2, v_3}} B^{\binom{u_1}{v_1}} - B^{\binom{u_3, u_1}{v_3, v_1}} B^{\binom{u_2}{v_2}}
\end{gathered}
$$

etc.

• **{alternal//alternil} $\Longrightarrow$ {negu-invariant, pushu-invariant}**
The first induced subsymmetry here is invariance under *negu*, with

$$
\mathrm{negu} := \mathrm{neg.adari}(\mathrm{paj}^{\bullet}) = \mathrm{adari}(\mathrm{pari.paj}^{\bullet}).\mathrm{neg} \tag{3.61}
$$

[50] To get the general formula, one simply transposes (3.47).

and with $paj^{\bullet}$ defined as in (4.72). The second induced subsymmetry is invariance under *pushu*,with

$$\text{pushu} := \text{negu.mantar.swap.mantir.swap} \tag{3.62}$$

with *mantar* as in (2.7) and *mantir* as above; and it is in fact equivalent to invariance under the simpler operator $pushu_*$:

$$\text{pushu}_* := \text{teru}^{-1}\text{.push.mantar.teru.mantar} \tag{3.63}$$

whose main ingredient is the arity-3 operator *teru* and its inverse:[51]

$$\begin{aligned} (\text{teru}.B^{\bullet})^{w_1,\dots,w_r} &:= B^{w_1,\dots,w_r} - B^{w_1,\dots,w_{r-1}}\, \mathfrak{E}^{w_r} + B^{w_1,\dots,w_{r-1}\rceil}\, Pa^{\lfloor w_r} \\ (\text{teru}^{-1}.B^{\bullet})^{w_1,\dots,w_r} &:= \sum_{\boldsymbol{a}.\boldsymbol{b}.\boldsymbol{c} = \boldsymbol{w} = (w_1,\dots,w_r)} B^{\boldsymbol{a}\rceil}\, \text{mupaj}^{\lfloor \boldsymbol{b}}\, \text{paj}^{\boldsymbol{c}} \end{aligned}$$

leading to the linear *senary relation*:

$$\text{teru}.B^{\bullet} = \text{push.mantar.teru.mantar}.B^{\bullet} \tag{3.64}$$

• **{symmetral//symmetril} $\Longrightarrow$ {negu-invariant, gushu-invariant}** Here, the first induced subsymmetry is *gegu*-invariance, with *gegu* defined as *negu* in(3.61), but with *adari* replaced by *adgari*:

$$\text{gegu} := \text{neg.adgari}(\text{paj}^{\bullet}) = \text{adgari}(\text{pari.paj}^{\bullet})\text{.neg} \tag{3.65}$$

and the second is *gushu*-invariance, with

$$\text{gushu} := \text{gegu.gantar.swap.gantir.swap} \tag{3.66}$$

with *gantar* as in (2.74) and *gantir* as above.

4 Flexion units and dimorphic bimoulds

4.1 Remarkable substructures of $Flex(\mathfrak{E})$

We shall now use the flexion units to construct two objects of pivotal importance: two very special *secondary* or *dimorphic* bimoulds (*i.e.* bimoulds with a double symmetry) which are, uncharacteristically, inva-

[51] The inverse $teru^{-1}$ is not of finite arity, of course, but its main ingredient is the mould $mupaj^{\bullet} :=$ $invmu.paj^{\bullet}$ which, due to symmetrality, has the simple form $pari.anti.paj^{\bullet}$.

riant under *pari.neg* rather than *neg*, and which, owing to that rare property, will prove helpful:
– in bridging the gap between *straight* and *twisted* double symmetries;
– in connecting $GARI^{as/as}$ with $GARI^{\underline{as}/\underline{as}}$;
– in constructing the *singulators* on which all the deeper results rest.
To do this, however, we must proceed step by step, and begin by constructing some important subspaces of $Flex(\mathfrak{E})$ and some remarkable bimould families like the $\mathfrak{re}^\bullet_r$ which, though not exactly dimorphic, come very close.

The subspaces $Flexinn(\mathfrak{E}) \subset Flexin(\mathfrak{E}) \subset Flex(\mathfrak{E})$
For each integer sequence $\boldsymbol{r} := (r_1, \ldots, r_s)$ let us define inductively the three bimoulds $\mathfrak{me}^\bullet_{\boldsymbol{r}}$, $\mathfrak{ne}^\bullet_{\boldsymbol{r}}$, $\mathfrak{re}^\bullet_{\boldsymbol{r}}$:[52]

$$\begin{aligned}
\mathfrak{me}^\bullet_1 &:= \mathfrak{E}^\bullet\ ;\ \mathfrak{me}^\bullet_r := \mathrm{amit}(\mathfrak{me}^\bullet_{r-1}).\mathfrak{E}^\bullet\ ;\ \mathfrak{me}^\bullet_{r_1,\ldots,r_s} := \mathrm{mu}(\mathfrak{me}^\bullet_{r_1},\ldots,\mathfrak{me}^\bullet_{r_s})\\
\mathfrak{ne}^\bullet_1 &:= \mathfrak{E}^\bullet\ ;\ \mathfrak{ne}^\bullet_r := \mathrm{anit}(\mathfrak{ne}^\bullet_{r-1}).\mathfrak{E}^\bullet\ ;\ \mathfrak{ne}^\bullet_{r_1,\ldots,r_s} := \mathrm{mu}(\mathfrak{ne}^\bullet_{r_1},\ldots,\mathfrak{ne}^\bullet_{r_s})\\
\mathfrak{re}^\bullet_1 &:= \mathfrak{E}^\bullet\ ;\ \mathfrak{re}^\bullet_r := \mathrm{arit}(\mathfrak{re}^\bullet_{r-1}).\mathfrak{E}^\bullet\ ;\ \mathfrak{re}^\bullet_{r_1,\ldots,r_s} := \mathrm{mu}(\mathfrak{re}^\bullet_{r_1},\ldots,\mathfrak{re}^\bullet_{r_s}).
\end{aligned}$$

Clearly, $\mathfrak{me}^\bullet_{\boldsymbol{r}}, \mathfrak{ne}^\bullet_{\boldsymbol{r}}, \mathfrak{re}^\bullet_{\boldsymbol{r}}$ are in $Flex_r(\mathfrak{E})$ with $r := \|\boldsymbol{r}\| = \sum r_i$. In fact, one can show that all three sets: $\{\mathfrak{me}^\bullet_{\boldsymbol{r}}, \|\boldsymbol{r}\| = r\}$, $\{\mathfrak{ne}^\bullet_{\boldsymbol{r}}, \|\boldsymbol{r}\| = r\}$, $\{\mathfrak{re}^\bullet_{\boldsymbol{r}}, \|\boldsymbol{r}\| = r\}$ span one and the same[53] subspace $Flexin_r(\mathfrak{E})$ of $Flex_r(\mathfrak{E})$, with dimension 2^{r-1}.

These three bases of $Flexin_r(\mathfrak{E})$ are connected by six simple matrices (two of them rational-valued, the other four entire-valued). Indeed:

$$\begin{aligned}
\mathfrak{me}^\bullet_{r_0} &= \sum_{1\le s}\ \sum_{\sum r_i = r_0} (-1)^{s+r}\ \mathfrak{ne}^\bullet_{r_1,\ldots,r_s}\\
\mathfrak{ne}^\bullet_{r_0} &= \sum_{1\le s}\ \sum_{\sum r_i = r_0} (-1)^{s+r}\ \mathfrak{me}^\bullet_{r_1,\ldots,r_s}\\
\mathfrak{re}^\bullet_{r_0} &= \sum_{1\le s}\ \sum_{\sum r_i = r_0} (-1)^{s+1}\, r_s\ \mathfrak{me}^\bullet_{r_1,\ldots,r_s}\\
\mathfrak{re}^\bullet_{r_0} &= \sum_{1\le s}\ \sum_{\sum r_i = r_0} (-1)^{s+r}\, r_1\ \mathfrak{ne}^\bullet_{r_1,\ldots,r_s}\\
\mathfrak{me}^\bullet_{r_0} &= \sum_{1\le s}\ \sum_{\sum r_i = r_0} \frac{1}{r_1 r_{12} \ldots r_{12\ldots s}}\ \mathfrak{re}^\bullet_{r_1,\ldots,r_s}\\
\mathfrak{ne}^\bullet_{r_0} &= \sum_{1\le s}\ \sum_{\sum r_i = r_0} \frac{(-1)^{s+r}}{r_{12\ldots s} \ldots r_{s-1,s} r_s}\ \mathfrak{re}^\bullet_{r_1,\ldots,r_s}
\end{aligned}$$

with $r_{i,j,\ldots}$ or even $r_{ij\ldots}$ standing as usual for $r_i + r_j + \ldots$

[52] For their analytical expressions, see Section 12.2.

[53] This would no longer be the case if $\mathfrak{E}^\bullet$ were not a flexion unit.

If we now denote by $\{r_1, \ldots, r_s\}$ any non-ordered integer set with repetitions allowed (or 'partition', if you prefer) and if we set:[54]

$$\mathfrak{re}^\bullet_{\{r_1,\ldots,r_s\}} = \sum_{\{r'_1,\ldots,r'_s\}=\{r_1,\ldots,r_s\}} \frac{1}{s!}\ \mathrm{preari}(\mathfrak{re}^\bullet_{r'_1}, \ldots, \mathfrak{re}^\bullet_{r'_s})$$

$$\mathfrak{se}^\bullet_{\{r_1,\ldots,r_s\}} = \sum_{\{r'_1,\ldots,r'_s\}=\{r_1,\ldots,r_s\}} \frac{1}{r'_1 r'_{12} \ldots r'_{12\ldots s}}\ \mathrm{preari}(\mathfrak{re}^\bullet_{r'_1}, \ldots, \mathfrak{re}^\bullet_{r'_s})$$

then it can be shown that, despite the very different summation weights, the two sets $\{\mathfrak{re}^\bullet_{\{\boldsymbol{r}\}}, \|\boldsymbol{r}\| = r\}$, $\{\mathfrak{se}^\bullet_{\{\boldsymbol{r}\}}, \|\boldsymbol{r}\| = r\}$ span one and the same[55] subspace $Flexinn_r(\mathfrak{E})$ of $Flexin_r(\mathfrak{E})$, with dimension $p(r)$ equal to the number of partitions of r. Summing up, we have:

$$Flexinn(\mathfrak{E}) = \oplus Flexinn_r(\mathfrak{E}) \subset Flexin(\mathfrak{E}) = \oplus Flexin_r(\mathfrak{E}) \subset Flex(\mathfrak{E}) = \oplus Flex_r(\mathfrak{E})$$

$$\dim(\mathrm{Flexinn}_r(\mathfrak{E})) = p(r);\ \dim(\mathrm{Flexin}_r(\mathfrak{E})) = 2^{r-1};\ \dim(\mathrm{Flex}_r(\mathfrak{E})) = \frac{(2r)!}{r!\,(r+1)!}$$

(i) $Flex(\mathfrak{E})$ is stable under all flexion operations.
(ii) $Flexin(\mathfrak{E})$ is stable under *mu*, *lu*, and $arit(\mathfrak{re}^\bullet_{r_0})$ $(\forall r_0)$.
(iii) $Flexinn(\mathfrak{E})$ is stable under nothing much, but crucial nonetheless.

Action of $arit(\mathfrak{re}^\bullet_r)$ on $Flexin(\mathfrak{E})$

It is neatly encapsulated in the formulas:

$$\mathrm{arit}(\mathfrak{re}^\bullet_q).\mathfrak{me}^\bullet_p = \sum_{s\geq 1} \sum_{\sum r_i = p+q\,,\, r_1 \geq p} (-1)^{1+s}\, r_s\ \mathfrak{me}^\bullet_{r_1,\ldots,r_s} \tag{4.1}$$

$$\mathrm{arit}(\mathfrak{re}^\bullet_q).\mathfrak{ne}^\bullet_p = \sum_{s\geq 1} \sum_{\sum r_i = p+q\,,\, r_s \geq p} (-1)^{1+s+q}\, r_1\ \mathfrak{ne}^\bullet_{r_1,\ldots,r_s} \tag{4.2}$$

$$\mathrm{arit}(\mathfrak{re}^\bullet_q).\mathfrak{re}^\bullet_p = p\ \mathfrak{re}^\bullet_{p+q} + \sum_{i\leq q} \mathrm{lu}(\mathfrak{re}^\bullet_i, \mathfrak{re}^\bullet_{p+q-i}) \tag{4.3}$$

$$= p\ \mathfrak{re}^\bullet_{p+q} + \sum_{i< p} \mathrm{lu}(\mathfrak{re}^\bullet_i, \mathfrak{re}^\bullet_{p+q-i}) \tag{4.4}$$

The algebra $ARI_{<\mathfrak{re}>}$ and its group $GARI_{<\mathfrak{se}>}$

Of the three bases of $Flex(\mathfrak{E})$, the first two are simplest, in the sense that

---

[54] With the multiple pre-brackets *preari* taken, as usual, *from left to right*.

[55] The simplest way to show that $\{\mathfrak{re}^\bullet_{\{\boldsymbol{r}\}}\}$ and $\{\mathfrak{se}^\bullet_{\{\boldsymbol{r}\}}\}$ span the same space and to find the conversion rule between the two bases, is to equate the expansions (4.11) and (4.12) for $\mathfrak{Se}^\bullet_f$ while expressing the coefficients α_n of the infinitesimal generator and the coefficients γ_n of the infinitesimal dilator in terms of each other.

here we have *atomic* basis elements $\mathfrak{me}^\bullet_{r_1,\dots,r_s}$ or $\mathfrak{ne}^\bullet_{r_1,\dots,r_s}$, *i.e.* elements that reduce to single products of the form $\mathfrak{E}^{w'_1}\dots\mathfrak{E}^{w'_r}$ for suitably inflected w'_i. With the third basis, on the other hand, we have *molecular* basis elements that can only be expressed as superpositions of at least $\prod r_i$ atoms. But the individual $\mathfrak{re}^\bullet_r$ ($r \in \mathbb{N}^*$), whose definition we recall:

$$\begin{aligned} \mathfrak{re}_r^{w_1,\dots,w_s} &:= 0 && \textit{if } r \neq s \\ \mathfrak{re}_1^{w_1} := \mathfrak{E}^{w_1}, \quad \mathfrak{re}^\bullet_r &:= \mathrm{arit}(\mathfrak{re}^\bullet_{r-1}).\mathfrak{re}^\bullet_1 && \textit{if } r \geq 2 \end{aligned} \tag{4.5}$$

more than make up for their 'molecularity' by possessing three essential properties:
(i) the bimoulds $\mathfrak{re}^\bullet_r$ thus defined are alternal;
(ii) when suitably combined, they exhibit traces of dimorphy, since the bimould $\mathfrak{sre}^\bullet$:

$$\mathfrak{sre}^\bullet := \frac{1}{2}\mathfrak{re}^\bullet_1 + \frac{1}{6}\mathfrak{re}^\bullet_2 + \frac{1}{12}\mathfrak{re}^\bullet_3 + \cdots = \sum_{r\geq 1} \frac{1}{r(r+1)}\,\mathfrak{re}^\bullet_r \quad \in \mathrm{ARI}^{\mathrm{al}/\mathfrak{ol}} \tag{4.6}$$

is not only alternal, but has a $\mathfrak{D}$-alternal swappee $\mathfrak{sr\ddot{o}}^\bullet$;
(iii) but the real importance of the $\mathfrak{re}^\bullet_r$ derives from the remarkable identities:

$$\mathrm{ari}(\mathfrak{re}^\bullet_{r_1}, \mathfrak{re}^\bullet_{r_2}) = (r_1 - r_2)\,\mathfrak{re}^\bullet_{r_1+r_2} \qquad \forall r_1, r_2 \geq 1 \tag{4.7}$$

which lead straightaway to the following commutative diagram:

$$\begin{array}{llll} \mathrm{GIFF}_{<x>} \overset{\text{isom.}}{\longrightarrow} \mathrm{GARI}_{<\mathfrak{se}>} & \subset \mathrm{GARI}^{\mathrm{as}} & \| \ \mathrm{se}_r(x) = \frac{x}{(1-x^r)^{\frac{1}{r}}} & \longrightarrow \mathfrak{se}^\bullet_r \\ \uparrow \exp \qquad\qquad \uparrow \mathrm{expari} & & \| \uparrow \exp & \qquad \uparrow \mathrm{expari} \\ \mathrm{DIFF}_{<x>} \overset{\text{isom.}}{\longrightarrow} \mathrm{ARI}_{<\mathfrak{re}>} & \subset \mathrm{ARI}^{\mathrm{al}} & \| \ \mathrm{re}_r(x) = x^{r+1}\partial_x & \longrightarrow \mathfrak{re}^\bullet_r \end{array}$$

Here, $GIFF_{<x>}$ denotes the group of (formal, one-dimensional) identity-tangent mappings of the form:

$$f := x \mapsto x.\left(1 + \sum_{1\leq r} a_r\, x^r\right) \tag{4.8}$$

and $DIFF_{<x>}$ denotes its infinitesimal algebra, whose elements may be represented as sums $\sum_{1\leq r} a_r\, x^{r+1}\,\partial_x$, provided we *change the sign* before their natural bracket.

Of course, since the Lie algebra $ARI_{<\mathfrak{re}>}$ contains only *alternal* bimoulds, its exponential, the group $GARI_{<\mathfrak{se}>}$, contains only *symmetral* bimoulds. Moreover, since elements of $ARI_{<\mathfrak{re}>}$ also possess traces of dimorphy, so too will their images in $GARI_{<\mathfrak{se}>}$. In the case of two remarkable bimoulds, $\mathfrak{ess}^\bullet$ and $\mathfrak{esz}^\bullet$ of $GARI_{<\mathfrak{se}>}$, we shall even get exact dimorphy rather than 'traces'.

But rather than jumping ahead, let us first explicate the isomorphisms $f \leftrightarrow \mathfrak{Se}^\bullet_f$ between the classical group $GIFF_{<x>}$ and its counterpart $GARI_{<\mathfrak{se}>}$ in the flexion structure. We begin with the easier direction, *i.e.* from *flexion* to *classical.*

The isomorphism $GARI_{<\mathfrak{se}>} \to GIFF_{<x>}$ made explicit
Let $\mathfrak{Se}^\bullet_f$ in $GARI_{<\mathfrak{se}>}$ be the image of some $f(x) := x\,(1 + \sum a_r\, x^r)$ in $GIFF_{<x>}$. How do we read the coefficients a_r directly off the bimould $\mathfrak{Se}^\bullet_f$ itself, without going through the costly operation *logari*? The answer is given by the bilinear operator *gepar*:

$$\text{gepar}.H^\bullet := \text{mu}(\text{anti.swap}.H^\bullet, \text{swap}.H^\bullet) \tag{4.9}$$

and by the formula:

$$(\text{gepar}.\mathfrak{Se}_f)^{w_1,\dots,w_r} \equiv (r+1)\, a_r \mathfrak{O}^{w_1} \dots \mathfrak{O}^{w_r} \quad \text{with } \mathfrak{O} \text{ conjugate to } \mathfrak{E} \tag{4.10}$$

The isomorphism $GIFF_{<x>} \to GARI_{<\mathfrak{se}>}$ made explicit
The isomorphism from *classical* to *flexion* is more difficult but also more interesting to unravel. We may of course transit through $DIFF_{<x>}$ and $ARI_{<\mathfrak{te}>}$ in the above diagram, but that involves performing the 'costly' operation *expari* and leads, in the course of the calculations, to rational coefficients with large denominators, which vanish in the end result. Concretely, that means forming the infinitesimal generator f_* of f (see (4.13), (4.15)) and inserting its coefficients ϵ_n into (4.11). Fortunately, there exists a much more direct scheme, which involves only integer coefficients: this time, we form the infinitesimal dilator $f_\#$ of f, which is a far more accessible object than f_* (see (4.14), (4.16)) and inject its coefficients γ_n into (4.12).

$$\mathfrak{Se}^\bullet_f = \sum_{\{r\}} \mathfrak{te}^\bullet_{\{r\}}\, \epsilon_{\{r\}} \qquad \text{with} \qquad \epsilon_{\{r_1,\dots,r_s\}} := \epsilon_{r_1} \dots \epsilon_{r_s} \tag{4.11}$$

$$\mathfrak{Se}^\bullet_f = \sum_{\{r\}} \mathfrak{se}^\bullet_{\{r\}}\, \gamma_{\{r\}} \qquad \text{with} \qquad \gamma_{\{r_1,\dots,r_s\}} := \gamma_{r_1} \dots \gamma_{r_s} \tag{4.12}$$

$$f_*(x) = x \sum_{1 \le k} \epsilon_k\, x^k = \text{infinitesimal generator of } f \tag{4.13}$$

$$f_\#(x) = x \sum_{1 \le k} \gamma_k\, x^k = x - \frac{f(x)}{f'(x)} = \text{infinitesimal dilator of } f \tag{4.14}$$

$$\big(\exp(f_*(x)\,\partial_x)\big).x = f(x) \tag{4.15}$$

$$\big(f \circ (id + \epsilon\, f_\#)\big)(x) = x + \sum_{1\le n}(1+\epsilon\, n)\, a_n\, x^{n+1} + \mathcal{O}(\epsilon^2). \tag{4.16}$$

Ultimately, of course, the coefficients γ_n of $f_\#$ have to be expressed in terms of those of f itself. Here, however, we have the choice between the three main representations of $GIFF_{<x>}$:

$$\begin{array}{llll} x \mapsto & f(x) = x + \sum_{1\le n} a_n\, x^{1+n} & & (x \sim 0) \\ y \mapsto & \underline{f}(y) = y + \sum_{1\le n} b_n\, y^{1-n} & = 1/f(y^{-1}) & (y \sim 0) \\ z \mapsto & \underline{\underline{f}}(z) = z + \sum_{1\le n} c_n\, e^{n\,z} & = \log f(e^z) & (z \sim 0) \end{array}$$

leading for γ_n to three rather similar expressions:

$$\begin{aligned} \sum \gamma_n\, x^n &\equiv \frac{\sum n\, a_n\, x^n}{1+\sum(n+1)\, a_n\, x^n} \\ &\equiv \frac{-\sum n\, b_n\, x^n}{1-\sum(n-1)\, b_n\, x^n} \equiv \frac{\sum n\, c_n\, x^n}{1+\sum n\, c_n\, x^n}. \end{aligned} \tag{4.17}$$

Under closer examination, it turns out that the coefficients $\{a_n\}$, $\{b_n\}$, $\{c_n\}$ of $f, \underline{f}, \underline{\underline{f}}$ are well-suited for expressing $\mathfrak{Se}^\bullet_f$ in the bases $\{\mathfrak{ne}^\bullet_r\}$,$\{\mathfrak{me}^\bullet_r\}$,$\{\mathfrak{re}^\bullet_r\}$ respectively (mark the order!), leading to three expansions:

$$\mathfrak{Se}^\bullet_f = \sum_r \mathbf{A}^r\, \mathfrak{ne}^\bullet_r = \sum_r \mathbf{B}^r\, \mathfrak{me}^\bullet_r = \sum_r \mathbf{C}^r\, \mathfrak{re}^\bullet_r. \tag{4.18}$$

To get a complete grip on the situation, we must calculate the moulds $\mathbf{A}^\bullet$, $\mathbf{B}^\bullet$, $\mathbf{C}^\bullet$ in terms of the a_n, b_n, c_n. To this end, we lift the infinitesimal dilation identity (4.16) from $GIFF_{<x>}$ to $GARI_{<\mathfrak{se}>}$. We find:

$$r(\bullet).\mathfrak{Se}^\bullet_f = \mathrm{arit}(\mathfrak{Te}^\bullet_f).\mathfrak{Se}^\bullet_f + \mathrm{mu}(\mathfrak{Se}^\bullet_f, \mathfrak{Te}^\bullet_f) \text{ with } \mathfrak{Te}^\bullet_f := \sum_{1\le r}\gamma_r\, \mathfrak{te}^\bullet_r \tag{4.19}$$

or more compactly:

$$r(\bullet).\, \mathfrak{Se}^\bullet_f = \mathrm{preari}(\mathfrak{Se}^\bullet_f, \mathfrak{Te}^\bullet_f) \quad \text{with} \quad \mathfrak{Te}^\bullet_f := \sum_{1\le r}\gamma_r\, \mathfrak{te}^\bullet_r. \tag{4.20}$$

In view of the formulas (4.1), (4.2), (4.3) for the action of $arit(\mathfrak{te}^\bullet_r)$ on $Flexin(\mathfrak{E})$, the identity (4.19) immediately translates into these three simple induction rules for the calculation of $\mathbf{A}^\bullet, \mathbf{B}^\bullet, \mathbf{C}^\bullet$:

$$\|\boldsymbol{r}\|\mathbf{A}^{\boldsymbol{r}} = \sum_{\boldsymbol{r}^1\boldsymbol{r}^2=\boldsymbol{r}} \mathbf{A}^{\boldsymbol{r}^1}\mathcal{N}_0^{\boldsymbol{r}^2} + \sum_{\boldsymbol{r}^1\boldsymbol{r}^2\boldsymbol{r}^3=\boldsymbol{r}} \sum_{1\le r_0<\|\boldsymbol{r}^2\|} \mathbf{A}^{\boldsymbol{r}^1 r_0 \boldsymbol{r}^3}\mathcal{N}_{r_0}^{\boldsymbol{r}^2} \tag{4.21}$$

$$\|\boldsymbol{r}\|\mathbf{B}^{\boldsymbol{r}} = \sum_{\boldsymbol{r}^1\boldsymbol{r}^2=\boldsymbol{r}} \mathbf{B}^{\boldsymbol{r}^1}\mathcal{M}_0^{\boldsymbol{r}^2} + \sum_{\boldsymbol{r}^1\boldsymbol{r}^2\boldsymbol{r}^3=\boldsymbol{r}} \sum_{1\le r_0<\|\boldsymbol{r}^2\|} \mathbf{B}^{\boldsymbol{r}^1 r_0 \boldsymbol{r}^3}\mathcal{M}_{r_0}^{\boldsymbol{r}^2} \tag{4.22}$$

$$\|\boldsymbol{r}\|\mathbf{C}^{\boldsymbol{r}} = \sum_{\boldsymbol{r}^1\boldsymbol{r}^2=\boldsymbol{r}} \mathbf{C}^{\boldsymbol{r}^1}\mathcal{R}_0^{\boldsymbol{r}^2} + \sum_{\boldsymbol{r}^1\boldsymbol{r}^2\boldsymbol{r}^3=\boldsymbol{r}} \sum_{1\le r_0<\|\boldsymbol{r}^2\|} \mathbf{C}^{\boldsymbol{r}^1 r_0 \boldsymbol{r}^3}\mathcal{R}_{r_0}^{\boldsymbol{r}^2}. \tag{4.23}$$

The auxiliary moulds $\mathcal{N}^\bullet, \mathcal{M}^\bullet, \mathcal{R}^\bullet$ are defined as follows:

$$\mathcal{N}_{r_0}^{\boldsymbol{r}} := \gamma_{\|\boldsymbol{r}\|-r_0}(\boldsymbol{a})\ (-1)^{1+s+\|\boldsymbol{r}\|-r_0}\, r_1\, (\Xi_{0<r_0\le r_s} - \Xi_{0=r_0}) \tag{4.24}$$

$$\mathcal{M}_{r_0}^{\boldsymbol{r}} := \gamma_{\|\boldsymbol{r}\|-r_0}(\boldsymbol{b})\ (-1)^{1+s}\, r_s\, (\Xi_{0\le r_0\le r_1}) \tag{4.25}$$

$$\mathcal{R}_{r_0}^{r_1} := \gamma_{\|\boldsymbol{r}\|-r_0}(\boldsymbol{c})\ (\Xi_{0=r_0=r_1} + r_0\, \Xi_{0<r_0<r_1}) \tag{4.26}$$

$$\mathcal{R}_{r_0}^{r_1,r_2} := \gamma_{\|\boldsymbol{r}\|-r_0}(\boldsymbol{c})\ (\Xi_{r_1<r_0\le r_2} - \Xi_{r_2<r_0\le r_1}) \tag{4.27}$$

$$\mathcal{R}_{r_0}^{r_1,\ldots,r_s} := 0 \qquad \textit{if}\ \ s \ge 3 \tag{4.28}$$

(i) with $\boldsymbol{r} := (r_1, \ldots, r_s)$ for any $s, r_i \in \mathbb{N}^*$;
(ii) with $\Xi_{\mathcal{S}}$ denoting the characteristic function of any given set $\mathcal{S}$;
(iii) with $\gamma_n(\boldsymbol{a}), \gamma_n(\boldsymbol{b}), \gamma_n(\boldsymbol{c})$ denoting the coefficients of the infinitesimal dilator $f_\#$ expressed (via the formulas (4.17)) in terms of the coefficients a_i, b_i, c_i respectively.

The main facts here are these:
(i) The moulds $\mathbf{B}^\bullet$ and $\mathbf{A}^\bullet$ are *symmetrel* whereas $\mathbf{C}^\bullet$ is *symmetral*;
(ii) $\mathbf{A}^{\boldsymbol{r}}, \mathbf{B}^{\boldsymbol{r}}, \mathbf{C}^{\boldsymbol{r}}$ are homogeneous polynomials of total degree $\|\boldsymbol{r}\|$ in the variables a_i, b_i, c_i respectively, but whereas $\mathbf{C}^{\boldsymbol{r}}$ has (predictably) rational coefficients, $\mathbf{A}^{\boldsymbol{r}}$ and $\mathbf{B}^{\boldsymbol{r}}$ have (unexpectedly) entire coefficients;
(iii) These rational (respectively entire) coefficients display remarkable symmetry properties: see (4.31), (4.32) below.

Here are the first structure polynomials $\mathbf{A}^{\boldsymbol{r}}, \mathbf{B}^{\boldsymbol{r}}, \mathbf{C}^{\boldsymbol{r}}$ up to $\|\boldsymbol{r}\| = 4$:

$$\mathbf{B}^1 = -b_1 \quad \mathbf{A}^1 = a_1 \quad \mathbf{C}^1 = c_1$$

$$\begin{array}{llll}
\mathbf{B}^2 = -2b_2 + b_1^2 & \mathbf{A}^2 = -2a_2 + a_1^2 & \mathbf{C}^2 = +c_2 \\
\mathbf{B}^{1,1} = +b_2 & \mathbf{A}^{1,1} = +a_2 & \mathbf{C}^{1,1} = +\frac{1}{2}c_1^2
\end{array}$$

$$\begin{array}{lll}
\mathbf{B}^3 = -3b_3 + 3b_1b_2 - b_1^3 & \mathbf{A}^3 = +3a_3 - 3a_1a_2 + a_1^3 & \mathbf{C}^3 = +c_3 \\
\mathbf{B}^{1,2} = +2b_3 & \mathbf{A}^{1,2} = -a_3 - a_1a_2 + a_1^3 & \mathbf{C}^{1,2} = +c_1c_2 - \frac{1}{6}c_1^3 \\
\mathbf{B}^{2,1} = +b_3 - b_1b_2 & \mathbf{A}^{2,1} = -2a_3 + 2a_1a_2 - a_1^3 & \mathbf{C}^{2,1} = +\frac{1}{6}c_1^3 \\
\mathbf{B}^{1,1,1} = -b_3 & \mathbf{A}^{1,1,1} = +a_3 & \mathbf{C}^{1,1,1} = +\frac{1}{6}c_1^3
\end{array}$$

$$
\begin{array}{lllllll}
\mathbf{B}^{4} & = -4\,b_4 & +4\,b_1\,b_3 & +2\,b_2^2 & -4\,b_1^2\,b_2 & +b_1^4 \\
\mathbf{B}^{1,3} & = +3\,b_4 & & & & \\
\mathbf{B}^{3,1} & = \ \ +b_4 & -b_1\,b_3 & -2\,b_2^2 & +b_1^2\,b_2 & \\
\mathbf{B}^{2,2} & = +2\,b_4 & -2\,b_1\,b_3 & +b_2^2 & & \\
\mathbf{B}^{1,1,2} & = -2\,b_4 & & -b_2^2 & & \\
\mathbf{B}^{1,2,1} & = \ \ -b_4 & & +b_2^2 & & \\
\mathbf{B}^{2,1,1} & = \ \ -b_4 & +b_1\,b_3 & & & \\
\mathbf{B}^{1,1,1,1} & = \ \ +b_4 & & & &
\end{array}
$$

$$
\begin{array}{lllllll}
\mathbf{A}^{4} & = -4\,a_4 & +4\,a_1\,a_3 & +2\,a_2^2 & -4\,a_1^2\,a_2 & +a_1^4 \\
\mathbf{A}^{1,3} & = \ \ +a_4 & +2\,a_1\,a_3 & +a_2^2 & -5\,a_1^2\,a_2 & +2\,a_1^4 \\
\mathbf{A}^{3,1} & = +3\,a_4 & -3\,a_1\,a_3 & -3\,a_2^2 & +6\,a_1^2\,a_2 & -2\,a_1^4 \\
\mathbf{A}^{2,2} & = +2\,a_4 & -2\,a_1\,a_3 & +a_2^2 & & \\
\mathbf{A}^{1,1,2} & = \ \ -a_4 & & -a_2^2 & +a_1^2\,a_2 & \\
\mathbf{A}^{1,2,1} & = \ \ -a_4 & -a_1\,a_3 & & +2\,a_1^2\,a_2 & -a_1^4 \\
\mathbf{A}^{2,1,1} & = -2\,a_4 & +2\,a_1\,a_3 & +a_2^2 & -3\,a_1^2\,a_2 & +a_1^4 \\
\mathbf{A}^{1,1,1,1} & = \ \ +a_4 & & & &
\end{array}
$$

$$
\begin{array}{lllllll}
\mathbf{C}^{4} & = +c_4 & & & & \\
\mathbf{C}^{1,3} & = & +c_1\,c_3 & +\frac{1}{2}\,c_2^2 & -\frac{1}{2}\,c_1^2\,c_2 & +\frac{1}{24}\,c_1^4 \\
\mathbf{C}^{3,1} & = & & -\frac{1}{2}\,c_2^2 & +\frac{1}{2}\,c_1^2\,c_2 & -\frac{1}{24}\,c_1^4 \\
\mathbf{C}^{2,2} & = & & +\frac{1}{2}\,c_2^2 & & \\
\mathbf{C}^{1,1,2} & = & & & +\frac{1}{2}\,c_1^2\,c_2 & -\frac{1}{8}\,c_1^4 \\
\mathbf{C}^{1,2,1} & = & & & & +\frac{1}{12}\,c_1^4 \\
\mathbf{C}^{2,1,1} & = & & & & +\frac{1}{24}\,c_1^4 \\
\mathbf{C}^{1,1,1,1} & = & & & & +\frac{1}{24}\,c_1^4.
\end{array}
$$

For any *unordered* integer sequence $\{\boldsymbol{r}\} := \{r_1, \ldots, r_s\}$, with *repetitions* allowed, we set:

$$
\boldsymbol{a}_{\{r\}} := \prod_i a_{r_i}; \qquad \boldsymbol{b}_{\{r\}} := \prod_i b_{r_i}; \qquad \boldsymbol{c}_{\{r\}} := \prod_i c_{r_i}. \tag{4.29}
$$

There exist efficient algorithms for calculating the three series of structure coefficients $A^{\bullet,\{\bullet\}}, B^{\bullet,\{\bullet\}}, C^{\bullet,\{\bullet\}}$ which occur in the above tables:

$$
\mathbf{A}^{r} = \sum_{\{r''\}} A^{r,\{r''\}}\,\boldsymbol{a}_{\{r''\}} \quad \mathbf{B}^{r} = \sum_{\{r''\}} B^{r,\{r''\}}\,\boldsymbol{b}_{\{r''\}} \quad \mathbf{C}^{r} = \sum_{\{r''\}} C^{r,\{r''\}}\,\boldsymbol{c}_{\{r''\}}
$$

and which encode, each in their way, all the information about the mapping from $GIFF_{<x>}$ to $GARI_{<\mathfrak{se}>}$. These structure coefficients have many properties, some of which are still imperfectly understood. We mention here but two of them. Consider the regularised coefficients $B^{\{\bullet\},\{\bullet\}}, A^{\{\bullet\},\{\bullet\}}$ defined by:[56]

$$A^{\{r'\},\{r''\}} = \sum_{r \in \{r'\}} A^{r,\{r''\}}; \qquad B^{\{r'\},\{r''\}} = \sum_{r \in \{r'\}} B^{r,\{r''\}}. \tag{4.30}$$

We then have the remarkable symmetry properties:

$$A^{\{r'\},\{r''\}} = A^{\{r''\},\{r'\}}; \qquad B^{\{r'\},\{r''\}} = B^{\{r''\},\{r'\}} \tag{4.31}$$

together with the identity:

$$B^{\{r'\},\{r''\}} = (-1)^r \, A^{\{r'\},\{r''\}} \qquad \text{with} \qquad r := \sum r'_i = \sum r''_i. \tag{4.32}$$

The following tables give $A^{\{r'\},\{r''\}}$ up to $r = 6$. The entries left vacant correspond to zeros.

	2	$\mathbf{1^2}$
2	−2	+1
$\mathbf{1^2}$	+1	

	3	**1.2**	$\mathbf{1^3}$
3	+3	−3	+1
1.2	−3	+1	
$\mathbf{1^3}$	+1		

	4	**1.3**	$\mathbf{2^2}$	$\mathbf{1^2.2}$	$\mathbf{1^4}$
4	−4	+4	+2	−4	+1
1.3	+4	−1	−2	+1	
$\mathbf{2^2}$	+2	−2	+1		
$\mathbf{1^2.2}$	−4	+1			
$\mathbf{1^4}$	+1				

	5	**1.4**	**2.3**	$\mathbf{1^2.3}$	$\mathbf{1.2^2}$	$\mathbf{1^3.2}$	$\mathbf{1^5}$
5	+5	−5	−5	+5	+5	−5	+1
1.4	−5	+1	+5	−1	−3	+1	
2.3	−5	+5	−1	−2	+1		
$\mathbf{1^2.3}$	+5	−1	−2	+1			
$\mathbf{1.2^2}$	+5	−3	+1				
$\mathbf{1^3.2}$	−5	+1					
$\mathbf{1^5}$	+1						

[56] The two sums in (4.30) range over all ordered sequences $\boldsymbol{r}$ that coincide, up to order, with the unordered sets $\{\boldsymbol{r}'\}$.

	6	**1.5**	**2.4**	$\mathbf{3^2}$	$\mathbf{1^2.4}$	**1.2.3**	$\mathbf{1^3.3}$	$\mathbf{2^3}$	$\mathbf{1^2.2^2}$	$\mathbf{1^4.2}$	$\mathbf{1^6}$
1.5	+6	−1	−6	−3	+1	+7	−1	+2	−4	+1	
2.4	+6	−6	+2	−3	+2	+4	−2	−2	+1		
3.3	+3	−3	−3	+3	+3	−3	0	+1			
$\mathbf{1^2.4}$	−6	+1	+2	+3	−1	−3	+1				
1.2.3	−12	+7	+4	−3	−3	+1					
$\mathbf{1^3.3}$	+6	−1	−2	0	+1						
$\mathbf{2^3}$	−2	+2	−2	+1							
$\mathbf{1^2.2^2}$	+9	−4	+1								
$\mathbf{1^4.2}$	−6	+1									
$\mathbf{1^6}$	+1										

If now, following (4.30), we set $C^{\{r'\},\{r''\}} = \sum_{r\in\{r'\}} C^{r,\{r''\}}$, we are saddled with rational numbers, but the symmetry relation becomes even more striking than with $A^{\{\bullet\},\{\bullet\}}$ and $B^{\{\bullet\},\{\bullet\}}$. Indeed:

$$C^{\{r'\},\{r''\}} = C^{\{r''\},\{r'\}} = 0 \quad \text{if} \quad \{r'\} \neq \{r''\} \tag{4.33}$$

$$C^{\{r\},\{r\}} = \frac{C_{r_1}^{s_1}}{s_1!}\,\frac{C_{r_2}^{s_2}}{s_2!}\ldots \quad \text{if } \{r\} = \{\overbrace{r_1,\ldots,r_1}^{s_1\ times},\ \overbrace{r_2,\ldots,r_2}^{s_2\ times},\ \ldots\}. \tag{4.34}$$

4.2 The secondary bimoulds $\mathfrak{ess}^\bullet$ and $\mathfrak{esz}^\bullet$

Dimorphic elements of $GARI_{<\mathfrak{se}>}$

We are now, at long last, in a position to construct the two main dimorphic bimoulds $\mathfrak{ess}_\sigma^\bullet$ and $\mathfrak{esz}_\sigma^\bullet$ of $GARI_{<\mathfrak{se}>}$, simply by taking the images of two well-chosen elements f_σ and g_σ of $GIFF_{<x>}$. In the last section, we mentioned the *economical* way of taking such images, without transiting through the algebras. Here, for the sake of expediency, we plump for the *theoretical* way, via the infinitesimal generators:

$$\begin{array}{lll|lll}
\mathfrak{f}_\sigma(x) & \longrightarrow & \mathfrak{ess}_\sigma^\bullet & \mathfrak{g}_\sigma(x) & \longrightarrow & \mathfrak{esz}_\sigma^\bullet \\
\uparrow \exp & & \uparrow \text{expari} & \uparrow \exp & & \uparrow \text{expari} \\
\mathfrak{f}_{*\sigma}(x) & \longrightarrow & \mathfrak{less}_\sigma^\bullet & \mathfrak{g}_{*\sigma}(x) & \longrightarrow & \mathfrak{lesz}_\sigma^\bullet .
\end{array}$$

The above diagram immediately translates into the formulas:

$$\mathfrak{ess}_\sigma^\bullet := \text{expari}\left(\sum_{r\geq 1} \sigma^r\, \epsilon_r\, \mathfrak{te}_r^\bullet\right) \longleftrightarrow f_\sigma(x) := \frac{1-e^{-\sigma x}}{\sigma} \tag{4.35}$$

$$\mathfrak{esz}_\sigma^\bullet := \text{expari}\left(\sum_{r\geq 1} \eta_{\sigma,r}\, \mathfrak{te}_r^\bullet\right) \longleftrightarrow g_\sigma(x) := \frac{1-(1-x)^{1-2\sigma}}{1-2\sigma} \tag{4.36}$$

with rational coefficients ϵ_r and $\eta_{\sigma,r}$ determined by:

$$\left(\exp\left(\left(\sum_{r\geq 1}\sigma^r \epsilon_r\, x^r\right).x.\partial_x\right)\right).x = f_\sigma(x) = x\left(1+\sum_{r\geq 1}\sigma^r c_r\, x^r\right) = x - \frac{\sigma}{2}x^2 + \ldots \tag{4.37}$$

$$\left(\exp\left(\left(\sum_{r\geq 1}\eta_{\sigma,r}\, x^r\right).x.\partial_x\right)\right).x = g_\sigma(x) = x\left(1+\sum_{r\geq 1} d_{\sigma,r}\; x^r\right) = x + \sigma\, x^2 + \ldots \tag{4.38}$$

Thus:

$$\begin{aligned}
&\epsilon_1 = -\frac{1}{2},\ \epsilon_2 = -\frac{1}{12},\ \epsilon_3 = -\frac{1}{48},\ \epsilon_4 = -\frac{1}{180},\\
&\epsilon_5 = -\frac{11}{8640},\ \epsilon_6 = -\frac{1}{6720}\ldots\\
&\eta_{\sigma,1} = s,\ \eta_{\sigma,2} = \frac{1}{3}\sigma\,(1-\sigma),\ \eta_{\sigma,3} = \frac{1}{6}\sigma\,(1-\sigma)^2,\\
&\eta_{\sigma,4} = \frac{1}{90}\sigma\,(1-\sigma)(3-4\,\sigma)(3-2\,\sigma)\ldots
\end{aligned}$$

Main property: The bimoulds $\mathfrak{ess}^\bullet_\sigma$ are bisymmetral (*i.e.* of type as/as) whilst the bimoulds $\mathfrak{esz}^\bullet_\sigma$ are symmetral/$\mathfrak{O}$-symmetral (*i.e.* of type $as/\mathfrak{os}$). Here, $\mathfrak{O}$ denotes as usual the flexion unit conjugate to $\mathfrak{E}$.

Remark 1: This is a survey, dedicated to *stating* rather than *proving*. However, the double symmetries of $\mathfrak{ess}^\bullet_\sigma$ and $\mathfrak{esz}^\bullet_\sigma$ are so essential that we must pause to justify them. The symmetrality of these two bimoulds is easy enough: it simply results from their being, by construction, elements of $GARI_{<\mathfrak{se}>}$. But what about their *swappees*? The way the operator *gepar* is defined (see (4.1)), it is clear that if $\mathfrak{ess}^\bullet_\sigma$ is to be symmetral, then $gepar.\mathfrak{ess}^\bullet_\sigma$ too has to be symmetral. Similarly, if $\mathfrak{esz}^\bullet_\sigma$ is to be $\mathfrak{O}$-symmetral, then $gepar.\mathfrak{esz}^\bullet_\sigma$ too has to be $\mathfrak{O}$-symmetral. Now, in view of (4.10) and (4.37), (4.38), we can see that

$$(\text{gepar}.\mathfrak{ess}_\sigma)^{w_1,\ldots,w_r} = \mathcal{S}_{\sigma,r}\, \mathfrak{O}^{w_1}\ldots\mathfrak{O}^{w_r} \tag{4.39}$$

$$(\text{gepar}.\mathfrak{esz}_\sigma)^{w_1,\ldots,w_r} = \mathcal{Z}_{\sigma,r}\, \mathfrak{O}^{w_1}\ldots\mathfrak{O}^{w_r} \tag{4.40}$$

with

$$\mathcal{S}_{\sigma,r} = (r+1)\,\sigma^r\, c_r = \frac{(-\sigma)^r}{r!} \tag{4.41}$$

$$\mathcal{Z}_{\sigma,r} = (r+1)\, d_{\sigma,r} = \frac{1}{r!}\prod_{0\leq j\leq r-1}(2\,\sigma+j). \tag{4.42}$$

Now, it is an easy matter to check that the above coefficients $\mathcal{S}_{\sigma,r}$ respectively $\mathcal{Z}_{\sigma,r}$ are *the only ones* that can make the bimoulds defined by the right-hand sides of (4.39) respectively (4.40) *symmetral* respectively $\mathfrak{O}$-*symmetral*. Thus, $gepar.\mathfrak{ess}^\bullet_\sigma$ and $gepar.\mathfrak{esz}^\bullet_\sigma$ do possess the right symmetries, and from there it is but a short step to check that their constituent factors, namely $swap.\mathfrak{ess}^\bullet_\sigma$ and $anti.swap.\mathfrak{ess}^\bullet_\sigma$ respectively $swap.\mathfrak{esz}^\bullet_\sigma$ and $anti.swap.\mathfrak{esz}^\bullet_\sigma$, also possess the right symmetries: see Section 11.9 and Section 11.10.

Remark 2: Whereas the bimoulds $\mathfrak{esz}^\bullet_\sigma$ really differ when σ varies, the bimoulds $\mathfrak{ess}^\bullet_\sigma$ merely undergo dilatation – an elementary transform that commutes with all flexion operations. So all these $\mathfrak{ess}^\bullet_\sigma$ essentially reduce to their prototype $\mathfrak{ess}^\bullet := \mathfrak{ess}^\bullet_1$, which we shall henceforth call *the* bisymmetral element of $Flex(\mathfrak{E})$.

Remark 3: By continuity in σ, we see that $g_{1/2}(x) = -\log(1-x)$. Thus $f_1 \circ g_{1/2} = id$ and therefore $gari(\mathfrak{ess}^\bullet_1, \mathfrak{esz}^\bullet_{1/2}) = id_{GARI} = 1^\bullet$, which shows that invgari.$\mathfrak{ess}^\bullet_1$ and by implication all invgari.$\mathfrak{ess}^\bullet_\sigma$ are *not* bisymmetral.

Remark 4: odd-even factorisations of the bisymmetrals.
The pre-image $f(x) := 1 - e^{-x}$ of $\mathfrak{ess}^\bullet$ in $GIFF_{<x>}$ factors as $f = f_\diamond \circ f_{\diamond\diamond}$, with an elementary first factor and a second factor that carries only even-indexed coefficients:

$$f_\diamond(x) := \frac{x}{1+\frac{1}{2}x} = \left(\exp\left(-\frac{1}{2}x^2\,\partial_x\right)\right).x \tag{4.43}$$

$$f_{\diamond\diamond}(x) := x\left(1 + \sum_{1\leq n} a^{\diamond\diamond}_{2n}\,x^{2n}\right). \tag{4.44}$$

For $\mathfrak{ess}^\bullet$ this immediately translates into the factorisation (4.45) in $GARI_{<\mathfrak{se}>}$. For $swap.\mathfrak{ess}^\bullet =: \ddot{\mathfrak{o}}\mathfrak{ss}^\bullet$, it translates, though less immediately, into the factorisation (4.46) in *BIMU*. Mark the order inversion, though, and note that $swap.\mathfrak{ess}^\bullet_* \neq \ddot{\mathfrak{o}}\mathfrak{ss}^\bullet_\star$, $swap.\mathfrak{ess}^\bullet_{**} \neq \ddot{\mathfrak{o}}\mathfrak{ss}^\bullet_{\star\star}$.

$$\mathfrak{ess}^\bullet = \mathrm{gari}(\mathfrak{ess}^\bullet_*, \mathfrak{ess}^\bullet_{**}) \quad \text{with } \mathfrak{ess}^\bullet_*, \mathfrak{ess}^\bullet_{**} \text{ symmetral} \tag{4.45}$$

$$\mathrm{swap}.\mathfrak{ess}^\bullet = \ddot{\mathfrak{o}}\mathfrak{ss}^\bullet = \mathrm{mu}(\ddot{\mathfrak{o}}\mathfrak{ss}^\bullet_{\star\star}, \ddot{\mathfrak{o}}\mathfrak{ss}^\bullet_\star) \quad \text{with } \ddot{\mathfrak{o}}\mathfrak{ss}^\bullet_\star, \ddot{\mathfrak{o}}\mathfrak{ss}^\bullet_{\star\star} \text{ symmetral.} \tag{4.46}$$

All four factor bimoulds are symmetral. The single-starred ones are elementary:

$$\mathfrak{ess}^{\bullet}_{*} := \mathrm{expari}\left(-\frac{1}{2}\mathfrak{E}^{\bullet}\right) \Rightarrow \mathfrak{ess}_{*}^{\binom{u_1\ ,\dots,\ u_r}{v_1\ ,\dots,\ v_r}} = \frac{(-1)^r}{2^r}\,\mathfrak{E}^{\binom{u_1}{v_{1:2}}}\mathfrak{E}^{\binom{u_{12}}{v_{2:3}}}\dots\mathfrak{E}^{\binom{u_{1\dots r}}{v_r}} \tag{4.47}$$

$$\ddot{\mathfrak{o}}\mathfrak{ss}^{\bullet}_{\star} := \mathrm{expmu}\left(-\frac{1}{2}\mathfrak{O}^{\bullet}\right) \Rightarrow \ddot{\mathfrak{o}}\mathfrak{ss}_{\star}^{\binom{u_1\ ,\dots,\ u_r}{v_1\ ,\dots,\ v_r}} = \frac{(-1)^r}{2^r\,r!}\,\mathfrak{O}^{\binom{u_1}{v_1}}\mathfrak{O}^{\binom{u_2}{v_2}}\dots\mathfrak{O}^{\binom{u_r}{v_r}}. \tag{4.48}$$

The double-starred factors, though non-elementary, carry only (non-zero) components of *even* length:

$$\mathfrak{ess}^{\bullet}_{*}\,,\ \ddot{\mathfrak{o}}\mathfrak{ss}^{\bullet}_{\star}\ \in\ \mathrm{BIMU}^{\mathrm{as}}_{\mathrm{neg.pari}} \tag{4.49}$$

$$\mathfrak{ess}^{\bullet}_{**}\,,\ \ddot{\mathfrak{o}}\mathfrak{ss}^{\bullet}_{\star\star}\ \in\ \mathrm{BIMU}^{\mathrm{as}}_{\mathrm{neg}} \cap \mathrm{BIMU}^{\mathrm{as}}_{\mathrm{pari}}. \tag{4.50}$$

As a consequence:

$$\mathrm{anti}.\ddot{\mathfrak{o}}\mathfrak{ss}^{\bullet}_{\star} = \ddot{\mathfrak{o}}\mathfrak{ss}^{\bullet}_{\star}; \qquad \mathrm{anti}.\mathfrak{oss}^{\bullet}_{\star\star} = \mathrm{invmu}.\mathfrak{oss}^{\bullet}_{\star} \tag{4.51}$$

and therefore:

$$\begin{aligned}\mathrm{gepar}.\mathfrak{ess}^{\bullet} &= \mathrm{mu}(\mathrm{anti}.\ddot{\mathfrak{o}}\mathfrak{ss}^{\bullet}_{\star}, \mathrm{anti}.\ddot{\mathfrak{o}}\mathfrak{ss}^{\bullet}_{\star\star}, \ddot{\mathfrak{o}}\mathfrak{ss}^{\bullet}_{\star\star}, \ddot{\mathfrak{o}}\mathfrak{ss}^{\bullet}_{\star})\\ &= \mathrm{mu}(\mathfrak{oss}^{\bullet}_{\star}, \mathfrak{oss}^{\bullet}_{\star}) = \mathrm{expmu}(-\mathfrak{O}^{\bullet}).\end{aligned} \tag{4.52}$$

Remark 5: induction for the calculation of $\mathfrak{ess}$, $\mathfrak{ess}_{**}$ and $\mathfrak{oss}$, $\ddot{\mathfrak{o}}\mathfrak{ss}_{\star\star}$. The source diffeos for $\mathfrak{ess}^{\bullet}$ and $\mathfrak{ess}^{\bullet}_{**}$ are f and f, with infinitesimal dilators:

$$f_{\#}(x) = 1 + x - e^x; \qquad f_{\diamond\diamond\#}(x) = x - \cosh(x) \tag{4.53}$$

to which there answer the following elements of $ARI_{<\mathfrak{re}>}$ and $IRA_{<\mathfrak{r}\ddot{\mathfrak{o}}>}$:

$$\mathfrak{ett}^{\bullet} := -\sum_{1\le n}\frac{1}{(n+1)!}\,\mathfrak{re}^{\bullet}_{n}\ ;\quad \mathfrak{ett}^{\bullet}_{**} := -\sum_{1\le n}\frac{1}{(2n+1)!}\,\mathfrak{re}^{\bullet}_{2n}$$

$$\ddot{\mathfrak{o}}\mathfrak{tt}^{\bullet} := -\sum_{1\le n}\frac{1}{(n+1)!}\,\mathfrak{r}\ddot{\mathfrak{o}}^{\bullet}_{n}\ ;\quad \ddot{\mathfrak{o}}\mathfrak{tt}^{\bullet}_{\star\star} := -\sum_{1\le n}\frac{1}{(2n+1)!}\,\mathfrak{r}\ddot{\mathfrak{o}}^{\bullet}_{2n}$$

which in turn lead to these linear and highly effective inductive formulas[57] for the calculation of our four bimoulds:

$$r(\bullet)\,\mathfrak{ess}^\bullet = \mathrm{preari}(\mathfrak{ess}^\bullet, \mathfrak{ett}^\bullet) \tag{4.54}$$

$$r(\bullet)\,\mathfrak{ess}^\bullet_{**} = \mathrm{preari}(\mathfrak{ess}^\bullet_{**}, \mathfrak{ett}^\bullet_{**}) \tag{4.55}$$

$$r(\bullet)\,\ddot{\mathfrak{o}}\mathfrak{ss}^\bullet = \mathrm{preira}(\ddot{\mathfrak{o}}\mathfrak{ss}^\bullet, \ddot{\mathfrak{o}}\mathfrak{tt}^\bullet) \tag{4.56}$$

$$r(\bullet)\,\ddot{\mathfrak{o}}\mathfrak{ss}^\bullet_{\star\star} = \mathrm{preira}(\ddot{\mathfrak{o}}\mathfrak{ss}^\bullet_{\star\star}, \ddot{\mathfrak{o}}\mathfrak{tt}^\bullet_{\star\star}) + \frac{1}{2}\,\mathrm{mu}(\ddot{\mathfrak{o}}\mathfrak{ss}^\bullet_{\star\star}, \ddot{\mathfrak{o}}\mathfrak{tt}^\bullet_{\star}) \tag{4.57}$$

with

$$\ddot{\mathfrak{o}}\mathfrak{tt}^\bullet_\star := \mathrm{coshmu}(\mathfrak{O}^\bullet) := \frac{1}{2}\big(\mathrm{expmu}(\mathfrak{O}^\bullet) + \mathrm{expmu}(-\mathfrak{O}^\bullet)\big).$$

In (4.54) and (4.55), *preari* may be replaced by *preali* or *preawi*; and in (4.56) and (4.57), *preira* may be replaced by *preila* or *preiwa*, since the involutions h_1, h_2, h_3 that define the algebras *ARI*, *ALI*, *AWI* (see Section 2.1 towards the end) have the same effects on the basic alternals $\mathfrak{te}^\bullet_n$:

$$h_1\,\mathfrak{te}^\bullet_n \equiv h_2\,\mathfrak{te}^\bullet_n \equiv h_3\,\mathfrak{te}^\bullet_n \quad ; \quad h_1^*\,\mathfrak{t}\ddot{\mathfrak{o}}^\bullet_n \equiv h_2^*\,\mathfrak{t}\ddot{\mathfrak{o}}^\bullet_n \equiv h_3^*\,\mathfrak{t}\ddot{\mathfrak{o}}^\bullet_n. \tag{4.58}$$

Whatever the pre-bracket chosen, the induction algorithm yields the same result, but expressed in very different bases. For the direct bimoulds, the best choice is *preari* or *preali*[58]; and for the swappees, it is *preiwa*[59] along with the following expression of $\mathfrak{t}\ddot{\mathfrak{o}}^\bullet$:

$$\mathfrak{t}\ddot{\mathfrak{o}}^{\binom{u_1\ ,\ldots,\ u_r}{v_1\ ,\ldots,\ v_r}} = \sum_i (r+1-i)\ \mathfrak{O}^{\binom{u_{1\ldots r}}{v_i}} \prod_{j\neq i} \mathfrak{O}^{\binom{u_j}{v_{j:i}}} \tag{4.59}$$

Comparing $\mathfrak{ess}^\bullet$ and $\ddot{\mathfrak{e}}\mathfrak{ss}^\bullet := sap.\mathfrak{ess}^\bullet$

The bimould $\mathfrak{ess}^\bullet$ belongs to the group $GARI_{<\mathfrak{se}>}$ whereas its image $\ddot{\mathfrak{e}}\mathfrak{ss}^\bullet$ under the involution *sap=swap.syap* belongs to $swap.GARI_{<syap.\mathfrak{se}>}$ *i.e.* to $swap.GARI_{<\mathfrak{so}>}$, which is not a group – only the *swappee* of one. Nevertheless, $\mathfrak{ess}^\bullet$ and $\ddot{\mathfrak{e}}\mathfrak{ss}^\bullet$ have much in common, since they:
– belong both to $Flex(\mathfrak{E})$ and are both bisymmetral, *i.e.* in $GARI^{as/as}$;

[57] These are true induction, since the sought-after bimoulds occur only once, with length r, on the left-hand side; and several times on the right-hand side, but with lengths at most $r-1$ (respectively $r-2$) in (4.54), (4.56) (respectively (4.55), (4.57)).

[58] Since h_1 and h_2, unlike h_3, involve no sign changes.

[59] Since h_3^*, unlike h_1^* and h_2^*,, involves no sign changes.

– are both invariant under *pari.neg*;
– have both the same length-one component: $\mathfrak{ess}^{w_1} = \ddot{\mathfrak{e}}\mathfrak{ss}^{w_1}$.
This is enough for them to be exchanged under *gari*-postcomposition by a bimould $\mathfrak{s}\ddot{\mathfrak{e}}\mathfrak{es}^\bullet$ that is not only bisymmetral, but also *even*[60], *i.e.* in $GARI^{\underline{as}/\underline{as}}$. It is therefore the exponential of an element $\mathfrak{l}\ddot{\mathfrak{e}}\mathfrak{el}^\bullet$ of $ARI^{\underline{al}/\underline{al}}$. In other words:

$$\mathfrak{ess}^\bullet = \mathrm{gari}(\ddot{\mathfrak{e}}\mathfrak{ss}^\bullet, \mathfrak{s}\ddot{\mathfrak{e}}\mathfrak{es}^\bullet) = \mathrm{gari}(\ddot{\mathfrak{e}}\mathfrak{ss}^\bullet, \mathrm{expari}(\mathfrak{l}\ddot{\mathfrak{e}}\mathfrak{el}^\bullet)) \tag{4.60}$$

But since both $\mathfrak{s}\ddot{\mathfrak{e}}\mathfrak{es}^\bullet$ and $\mathfrak{l}\ddot{\mathfrak{e}}\mathfrak{el}^\bullet$ are invariant under *neg* and *pari.neg*, they are invariant under *pari*. All their non-vanishing components are therefore of *even* length; or more precisely of even length $r \geq 4$, since an initial, length-2 component of $\mathfrak{s}\ddot{\mathfrak{e}}\mathfrak{es}^\bullet$ would have to be a bialternal element of $Flex_2(\mathfrak{E})$, and no such element exists.

Up to length $r = 14$, the bialternal subalgebra $Flex^{\underline{al}/\underline{al}}(\mathfrak{E})$ of $ARI^{\underline{al}/\underline{al}}$ is *freely* generated by the non-vanishing components of $\mathfrak{l}\ddot{\mathfrak{e}}\mathfrak{el}^\bullet$, *i.e.*

$$\mathfrak{l}\ddot{\mathfrak{e}}\mathfrak{el}^\bullet_4\ ,\ \mathfrak{l}\ddot{\mathfrak{e}}\mathfrak{el}^\bullet_6\ ,\ \mathfrak{l}\ddot{\mathfrak{e}}\mathfrak{el}^\bullet_8\ ,\ \mathfrak{l}\ddot{\mathfrak{e}}\mathfrak{el}^\bullet_{10}\ ,\ \mathfrak{l}\ddot{\mathfrak{e}}\mathfrak{el}^\bullet_{12}\ ,\ \mathfrak{l}\ddot{\mathfrak{e}}\mathfrak{el}^\bullet_{14}\ \ldots \tag{4.61}$$

or, alternatively, by the series of *singulates* $\mathfrak{lel}^\bullet_{2r}$ (see Section 4.2 below):

$$\mathfrak{lel}^\bullet_{2r} := \mathrm{senk}_{2r}(\mathfrak{ess}^\bullet).\mathfrak{E}^\bullet \qquad (r \geq 2) \tag{4.62}$$

but after 14 this no longer holds. As of now, for large values of r, the exact dimension of $Flex^{\underline{al}/\underline{al}}_{2r}(\mathfrak{E})$ is not known.

If we now repeat the above construction but with $\mathfrak{E}$ replaced by the conjugate unit $\mathfrak{O}$, identity (4.60) becomes, with self-explanatory notations:

$$\mathfrak{oss}^\bullet = \mathrm{gari}(\ddot{\mathfrak{o}}\mathfrak{ss}^\bullet, \mathfrak{s}\ddot{\mathfrak{o}}\mathfrak{os}^\bullet) = \mathrm{gari}(\ddot{\mathfrak{o}}\mathfrak{ss}^\bullet, \mathrm{expari}(\mathfrak{l}\ddot{\mathfrak{o}}\mathfrak{ol}^\bullet)). \tag{4.63}$$

So far, so predictable. The remarkable thing, however, is that the components $\mathfrak{l}\ddot{\mathfrak{e}}\mathfrak{el}^\bullet_{2r}$ and $\mathfrak{l}\ddot{\mathfrak{o}}\mathfrak{ol}^\bullet_{2r}$ of the rightmost bimoulds in (4.60) and (4.63) get exchanged, up to sign, under the involutions *swap* and *syap* (see (Section 3.3)). As a consequence, each one of them is, again up to sign, invariant under the involution *sap*.

Polar and trigonometric specialisations

Let us now consider the three polar and the three trigonometric specialisations of $\mathfrak{E}^\bullet$, along with the corresponding bisymmetrals and their

[60] *I.e.* invariant under *neg* rather than *pari.neg*.

swappees:

Flexion units	$\mathfrak{E}^\bullet$	:	$\mathrm{Pa}^\bullet$	$\mathrm{Pi}^\bullet$	$\mathrm{Pai}^\bullet_{\alpha,\beta}$	$\mathrm{Qa}^\bullet_c$	$\mathrm{Qi}^\bullet_c$	$\mathrm{Qai}^\bullet_{c,\alpha,\beta}$	
bisymmetrals	$\mathfrak{ess}^\bullet$	:	$\mathrm{par}^\bullet$	$\mathrm{pil}^\bullet$	$\mathrm{pail}^\bullet_{\alpha,\beta}$	$\ldots$	$\mathrm{til}^\bullet_c$	$\mathrm{tail}^\bullet_{c,\alpha,\beta}$	
swappees	$\ddot{\mathfrak{o}}\mathfrak{ss}^\bullet$	:	$\mathrm{pir}^\bullet$	$\mathrm{pal}^\bullet$	$\mathrm{pial}^\bullet_{\alpha,\beta}$	$\ldots$	$\mathrm{tal}^\bullet_c$	$\mathrm{tial}^\bullet_{c,\alpha,\beta}$	
type as/$\mathfrak{os}$	$\mathfrak{esz}^\bullet_\sigma$	:	$\mathrm{bar}^\bullet_\sigma$	$\mathrm{bil}^\bullet_\sigma$	$\mathrm{bail}^\bullet_{c,\alpha,\beta}$	$\ldots$	$\ldots$	$\mathrm{dail}^\bullet_{\sigma,c,\alpha,\beta}$	
swappees	$\ddot{\mathfrak{o}}\mathfrak{sz}^\bullet_\sigma$	:	$\mathrm{bir}^\bullet_\sigma$	$\mathrm{bal}^\bullet_\sigma$	$\mathrm{bial}^\bullet_{\sigma,\alpha,\beta}$	$\ldots$	$\ldots$	$\mathrm{dial}^\bullet_{\sigma,c,\alpha,\beta}$.	

All these unit specialisations are *exact*, except $Qa^\bullet_c$, which generates no bisymmetral, and $Qi^\bullet_c$, which does.[61]

Let D^t be the dilation operator:

$$(D^t.M)^{\binom{u_1\ ,\dots,\ u_r}{v_1\ ,\dots,\ v_r}} := M^{\binom{u_1/t\ ,\dots,\ u_r/t}{v_1/t\ ,\dots,\ v_r/t}}.$$

It clearly respects bialternality and bisymmetrality. Due to the general identities:

$$\begin{aligned} A^\bullet\ \boldsymbol{v}\textit{-constant}\text{ and }B^\bullet\ \boldsymbol{u}\textit{-constant} &\quad \Longrightarrow \\ \mathrm{swap.mu}(A^\bullet, B^\bullet) \quad &\equiv \quad \mathrm{mu}(\mathrm{swap}.B^\bullet, \mathrm{swap}.A^\bullet) \end{aligned}$$

we can form new bisymmetrals:

$$\begin{aligned} \mathrm{vipail}^\bullet_{\alpha,\beta} &:= \mathrm{mu}(D^\alpha.\mathrm{pal}^\bullet, D^\beta.\mathrm{pil}^\bullet) \subset \mathrm{GARI}^{\mathrm{as/as}} \\ \mathrm{vipair}^\bullet_{\alpha,\beta} &:= \mathrm{mu}(D^\alpha.\mathrm{par}^\bullet, D^\beta.\mathrm{pir}^\bullet) \subset \mathrm{GARI}^{\mathrm{as/as}} \\ \mathrm{vitail}^\bullet_{c,\alpha,\beta} &:= \mathrm{mu}(D^\alpha.\mathrm{tal}^\bullet_c, D^\beta.\mathrm{til}^\bullet_c) \quad \subset \mathrm{GARI}^{\mathrm{as/as}}. \end{aligned}$$

The next four identities are special cases of (4.60) when $\mathfrak{E}$ specialises respectively to Pa, Pi, $Pai_{\alpha,\beta}$, $Qai_{c,\alpha,\beta}$:

$$\begin{array}{llll} \mathrm{par}^\bullet & \equiv \mathrm{gari}(\mathrm{pal}^\bullet, \mathrm{lar}^\bullet) & \text{with } \mathrm{lar}^\bullet & \subset \mathrm{GARI}^{\underline{\mathrm{as}}/\underline{\mathrm{as}}} \\ \mathrm{pil}^\bullet & \equiv \mathrm{gari}(\mathrm{pir}^\bullet, \mathrm{ril}^\bullet) & \text{with } \mathrm{ril}^\bullet & \subset \mathrm{GARI}^{\underline{\mathrm{as}}/\underline{\mathrm{as}}} \\ \mathrm{pail}^\bullet_{\alpha,\beta} & \equiv \mathrm{gari}(\mathrm{pial}^\bullet_{\beta,\alpha}, \mathrm{lappil}^\bullet_{\alpha,\beta}) & \text{with } \mathrm{lappil}^\bullet_{\alpha,\beta} & \subset \mathrm{GARI}^{\underline{\mathrm{as}}/\underline{\mathrm{as}}} \\ \mathrm{tail}^\bullet_{c,\alpha,\beta} & \equiv \mathrm{gari}(\mathrm{tial}^\bullet_{c,\beta,\alpha}, \mathrm{lattil}^\bullet_{c,\alpha,\beta}) & \text{with } \mathrm{lattil}^\bullet_{c,\alpha,\beta} & \subset \mathrm{GARI}^{\underline{\mathrm{as}}/\underline{\mathrm{as}}} \\ \mathrm{vipail}^\bullet_{\alpha,\beta} & \equiv \mathrm{gari}(\mathrm{pail}^\bullet_{\alpha,\beta}, \mathrm{paiv}^\bullet_{\alpha,\beta}) & \text{with } \mathrm{paiv}^\bullet_{\alpha,\beta} & \subset \mathrm{GARI}^{\underline{\mathrm{as}}/\underline{\mathrm{as}}} \\ \mathrm{vitail}^\bullet_{c,\alpha,\beta} & \equiv \mathrm{gari}(\mathrm{tail}^\bullet_{c,\alpha,\beta}, \mathrm{taiv}^\bullet_{c,\alpha,\beta}) & \text{with } \mathrm{taiv}^\bullet_{c,\alpha,\beta} & \subset \mathrm{GARI}^{\underline{\mathrm{as}}/\underline{\mathrm{as}}} \end{array}$$

while the last two identities provide yet other examples of elements of $GARI^{as/as}$ sharing the same first component and related under postcomposition by an element of $GARI^{\underline{as}/\underline{as}}$.

[61] But of course with an elementary corrective factor $mini^\bullet_c \in center(GARI)$ in the connection formula: $swap.til^\bullet_c = gari(mana^\bullet_c, tal_c{}^\bullet) = gari(tal_c{}^\bullet, mana^\bullet_c)$.

Difference between even and non-even bissymmetrals

To bring out the sharp difference between *even* and *non-even* bisymmetrals, we introduce two distinct copies $\mathfrak{E}_1, \mathfrak{E}_2$ of the universal unit $\mathfrak{E}$, and define their *blend* as follows:

$$\mathfrak{sse}^{\bullet}_{1,2} = \mathrm{blend}(\mathfrak{E}^{\bullet}_1, \mathfrak{E}^{\bullet}_2) \qquad \Longleftrightarrow$$
$$\mathfrak{sse}_{1,2}^{\binom{u_1,\ldots,u_r}{v_1,\ldots,v_r}} = \mathfrak{E}_1^{\binom{u_1}{v_{1:2}}} \mathfrak{E}_1^{\binom{u_{12}}{v_{2:3}}} \mathfrak{E}_1^{\binom{u_{123}}{v_{3:4}}} \ldots \mathfrak{E}_1^{\binom{u_{1..r}}{v_r}} \mathfrak{E}_2^{\binom{u_1}{v_1}} \mathfrak{E}_2^{\binom{u_2}{v_2}} \mathfrak{E}_2^{\binom{u_3}{v_3}} \ldots \mathfrak{E}_2^{\binom{u_r}{v_r}}. \tag{4.64}$$

The blend $\mathfrak{sse}^{\bullet}_{1,2}$ is obviously *even*. It is also easily seen to be *symmetral*. In fact, since, up to order, *blend* commutes with *swap*:

$$\mathrm{blend}(\mathfrak{E}^{\bullet}_1, \mathfrak{E}^{\bullet}_2) \overset{\mathrm{swap}}{\longleftrightarrow} \mathrm{blend}(\mathrm{swap}.\mathfrak{E}^{\bullet}_2, \mathrm{swap}.\mathfrak{E}^{\bullet}_1) \tag{4.65}$$

and since the swappee of an exact flexion unit $\mathfrak{E}$ coincides with the conjugate unit $\mathfrak{O}$, the *blend* is actually *bisymmetral*.

Moreover, we have a remarkable (non-elementary) identity for expressing the *gari*-inverse of the *blend* of two flexion units: it is itself a *blend*, but preceded by *pari* and with the two arguments arguments exchanged. Therefore, under *invgari*, the two entries of (4.65) become:

$$\mathrm{pari.blend}(\mathfrak{E}^{\bullet}_2, \mathfrak{E}^{\bullet}_1) \overset{\mathrm{swap}}{\longleftrightarrow} \mathrm{pari.blend}(\mathrm{swap}.\mathfrak{E}^{\bullet}_1, \mathrm{swap}.\mathfrak{E}^{\bullet}_2) \tag{4.66}$$

and are still connected by *swap*.

As a consequence, for the *even*[62] bisymmetral $\mathfrak{ess}^{\bullet}_{1,2}$ we have this commutative diagram,[63] with self-explanatory notations:

$$\begin{array}{lccl}
\text{(symmetral)} & \mathfrak{sse}^{\bullet}_{1,2} & \overset{\mathrm{swap}}{\longleftrightarrow} & \mathfrak{sso}^{\bullet}_{2,1} \quad \text{(symmetral)} \\
& \mathrm{invgari} \updownarrow & & \updownarrow \mathrm{invgari} \\
\text{(symmetral)} & \mathrm{pari}.\mathfrak{sse}^{\bullet}_{2,1} & \overset{\mathrm{swap}}{\longleftrightarrow} & \mathrm{pari}.\mathfrak{sso}^{\bullet}_{1,2} \quad \text{(symmetral!)}.
\end{array}$$

In sharp contrast, with the *non-even*[64] bisymmetral $\mathfrak{ess}^{\bullet}$ constructed in (4.35), the diagram's commutativity breaks down:

$$\begin{array}{lccccl}
\text{(symmetral)} & \mathfrak{ess}^{\bullet} & \overset{\mathrm{swap}}{\longleftrightarrow} & \ddot{\mathfrak{o}}\mathfrak{ss}^{\bullet} & & \text{(symmetral)} \\
& \mathrm{invgari} \updownarrow & & & \searrow \mathrm{invgari} & \\
\text{(symmetral)} & \mathfrak{ess}^{\bullet}_{*} & \overset{\mathrm{swap}}{\longleftrightarrow} & \ddot{\mathfrak{o}}\mathfrak{ss}^{\bullet}_{*} & \neq & \ddot{\mathfrak{o}}\mathfrak{ss}^{\bullet}_{**} \quad \text{(non symmetral!)}.
\end{array}$$

[62] *I.e.* *neg*-invariant.

[63] We would of course have similarly commutative diagrams (only with less explicit *gari*-inverses) if we replaced $\mathfrak{sse}^{\bullet}$ by any element of $GARI^{\underline{\mathrm{as}}/\underline{\mathrm{as}}}$, since on that subgroup *swap* acts as an automorphism, just as it does on $ARI^{\underline{\mathrm{al}}/\underline{\mathrm{al}}}$.

[64] More precisely: $\mathfrak{ess}^{\bullet}$ is *pari.neg*-invariant instead of *neg*-invariant.

4.3 The related primary bimoulds $\mathfrak{es}^\bullet$ and $\mathfrak{ez}^\bullet$

After constructing the *secondary* bimoulds $\mathfrak{ess}^\bullet$, $\mathfrak{esz}^\bullet_\sigma$ (non-elementary, with a double symmetry), we must now define the much simpler, yet closely related *primary* bimoulds $\mathfrak{es}^\bullet$, $\mathfrak{ez}^\bullet$ (elementary, with a single symmetry):

$$\mathfrak{es}^\bullet := \mathrm{expari}(\mathfrak{E}^\bullet) \tag{4.67}$$

$$\mathfrak{ez}^\bullet := \mathrm{invmu}(1^\bullet - \mathfrak{E}^\bullet) \tag{4.68}$$

$$\mathfrak{es}^\bullet \overset{\mathrm{sap}}{\leftrightarrow} \mathfrak{ez}^\bullet \tag{4.69}$$

This leads to the more explicit formulas:[65]

$$\mathfrak{es}^{\binom{u_1\ ,\dots,\ u_r}{v_1\ ,\dots,\ v_r}} := \mathfrak{E}^{\binom{u_1}{v_{1:2}}}\mathfrak{E}^{\binom{u_{12}}{v_{2:3}}}\mathfrak{E}^{\binom{u_{123}}{v_{3:4}}}\dots\mathfrak{E}^{\binom{u_{1\dots r}}{v_r}} \quad \text{(symmetral)} \tag{4.70}$$

$$\mathfrak{ez}^{\binom{u_1\ ,\dots,\ u_r}{v_1\ ,\dots,\ v_r}} := \mathfrak{E}^{\binom{u_1}{v_1}}\mathfrak{E}^{\binom{u_2}{v_2}}\mathfrak{E}^{\binom{u_3}{v_3}}\dots\mathfrak{E}^{\binom{u_r}{v_r}} \quad (\mathfrak{E}\text{-symmetral}). \tag{4.71}$$

The symmetrality of $\mathfrak{es}^\bullet$ respectively $\mathfrak{E}$-symmetrality of $\mathfrak{ez}^\bullet$ relies entirely on $\mathfrak{E}$ being an exact flexion unit, but the definitions also extend, albeit at the cost of significant complications, to approximate units.

Let us now consider the three polar and the three trigonmetric specialisations of $\mathfrak{E}^\bullet$ and the corresponding incarnations of $\mathfrak{es}^\bullet$ and $\mathfrak{ez}^\bullet$:

$$\begin{array}{rl} \text{Flexion units} \ \ \mathfrak{E}^\bullet & : \mathrm{Pa}^\bullet \ \ \mathrm{Pi}^\bullet \ \ \mathrm{Pai}^\bullet_{\alpha,\beta} \ \ \mathrm{Qa}^\bullet_c \ \ \mathrm{Qi}^\bullet_c \ \ \mathrm{Qai}^\bullet_{c,\alpha,\beta} \\ \text{symmetrals} \ \ \mathfrak{es}^\bullet & : \mathrm{paj}^\bullet \ \ \mathrm{pij}^\bullet \ \ \mathrm{paij}^\bullet_{\alpha,\beta} \ \ \mathrm{taj}_c \ \ \mathrm{tij}^\bullet_c \ \ \mathrm{taij}^\bullet_{c,\alpha,\beta} \\ \mathfrak{E}\text{-symmetrals} \ \ \mathfrak{ez}^\bullet & : \mathrm{pac}^\bullet \ \ \mathrm{pic}^\bullet \ \ \mathrm{paic}^\bullet_{\alpha,\beta} \ \ \mathrm{tac}^\bullet_c \ \ \mathrm{tic}^\bullet_c \ \ \mathrm{taic}^\bullet_{c,\alpha,\beta}. \end{array}$$

The definitions of the new bimoulds are straightforward for the exact units, but less so for the approximate units Qa_c and Qi_c. In those two cases, we mention only the elementary part (mod. c^2), which conforms entirely to the general formulas (4.70) and (4.71), and refer to Section 3.9 for the corrective terms.

$$\mathrm{paj}^{w_1,\dots,w_r} := \prod_{1\le j\le r} P(u_1+\dots+u_j) \tag{4.72}$$

$$\mathrm{pij}^{w_1,\dots,w_r} := \prod_{1\le j\le r}^{\mathrm{circ}} P(v_j - v_{j+1}) \tag{4.73}$$

$$\mathrm{paij}^{w_1,\dots,w_r}_{\alpha,\beta} := \prod_{1\le j\le r}^{\mathrm{circ}} \left(P\left(\frac{u_1+\dots+u_j}{\alpha}\right) + P\left(\frac{v_j - v_{j+1}}{\beta}\right)\right) \tag{4.74}$$

[65] To derive (4.70) from (4.67), one must use the fact that $\mathfrak{E}$ is a flexion unit.

$$\mathrm{taj}_c^{w_1,\dots,w_r} := \prod_{1\le j\le r} Q_c(u_1+\dots+u_j) \qquad \text{(modulo } c^2\text{)} \tag{4.75}$$

$$\mathrm{tij}_c^{w_1,\dots,w_r} := \prod_{1\le j\le r}^{\mathrm{circ}} Q_c(v_j - v_{j+1}) \qquad \text{(modulo } c^2\text{)} \tag{4.76}$$

$$\mathrm{taij}_{c,\alpha,\beta}^{w_1,\dots,w_r} := \prod_{1\le j\le r}^{\mathrm{circ}} \left(Q_c\left(\frac{u_1+\dots+u_j}{\alpha}\right) + Q_c\left(\frac{v_j - v_{j+1}}{\beta}\right)\right) \text{ (exactly)}. \tag{4.77}$$

In the above products, *circ* means that the (non-existing) variable v_{r+1} should be construed as $v_0 = 0$ whenever it occurs. No such precaution is required for the following specialisations of $\mathfrak{ez}^\bullet$.

$$\mathrm{pac}^{w_1,\dots,w_r} := \prod_{1\le j\le r} P(u_j) \tag{4.78}$$

$$\mathrm{pic}^{w_1,\dots,w_r} := \prod_{1\le j\le r} P(v_j) \tag{4.79}$$

$$\mathrm{paic}_{\alpha,\beta}^{w_1,\dots,w_r} := \prod_{1\le j\le r} \left(P\left(\frac{u_j}{\alpha}\right) + P\left(\frac{v_j}{\beta}\right)\right) \tag{4.80}$$

$$\mathrm{tac}_c^{w_1,\dots,w_r} := \prod_{1\le j\le r} Q_c(u_j) \qquad \text{(modulo } c^2\text{)} \tag{4.81}$$

$$\mathrm{tic}_c^{w_1,\dots,w_r} := \prod_{1\le j\le r} Q_c(v_j) \qquad \text{(modulo } c^2\text{)} \tag{4.82}$$

$$\mathrm{taic}_{c,\alpha,\beta}^{w_1,\dots,w_r} := \prod_{1\le j\le r} \left(Q_c\left(\frac{u_j}{\alpha}\right) + Q_c\left(\frac{v_j}{\beta}\right)\right) \qquad \text{(exactly)} \tag{4.83}$$

4.4 Some basic bimould identities

Let us list, first in universal mode, the main relations between the primary bimoulds:

$\mathrm{invmu}.\mathfrak{es}^\bullet$	$=$	$\mathrm{pari.anti}.\mathfrak{es}^\bullet$	‖	$\mathrm{invmu}.\mathfrak{ez}^\bullet$	$=$	$1^\bullet - \mathfrak{E}^\bullet$
$\mathrm{invgami}.\mathfrak{es}^\bullet$	$=$	$\mathrm{pari}.\mathfrak{ez}^\bullet$	‖	$\mathrm{invgami}.\mathfrak{ez}^\bullet$	$=$	$\mathrm{pari}.\mathfrak{es}^\bullet$
$\mathrm{invgani}.\mathfrak{es}^\bullet$	$=$	*unremarkable*	‖	$\mathrm{invgani}.\mathfrak{ez}^\bullet$	$=$	$\mathrm{pari.anti}.\mathfrak{es}^\bullet$
$\mathrm{invgari}.\mathfrak{es}^\bullet$	$=$	$\mathrm{pari}.\mathfrak{es}^\bullet$	‖	$\mathrm{invgari}.\mathfrak{ez}^\bullet$	$=$	*unremarkable*.

The relations that really matter, however, are the ones linking primary and secondary bimoulds. To state them, we require a highly non-linear operator *slash* which measures, in terms of *GARI*, the *un-evennness* of a bimould:

$$\mathrm{slash}.B^\bullet := \mathrm{fragari}(\mathrm{neg}.B^\bullet, B^\bullet) = \mathrm{gari}(\mathrm{neg}.B^\bullet, \mathrm{invgari}.B^\bullet). \tag{4.84}$$

We can now write down the two secondary-to-primary identities:

$$\text{slash}.\mathfrak{ess}^\bullet = \mathfrak{es}^\bullet \quad \text{with} \quad \mathfrak{ess}^\bullet := \mathfrak{ess}_1^\bullet \tag{4.85}$$

$$\text{sap}.\mathfrak{esz}_0^\bullet = \mathfrak{ez}^\bullet \quad \text{with} \quad \text{sap} := \text{syap.swap} = \text{swap.syap}. \tag{4.86}$$

To conclude this section, let us reproduce some of the above identities in the polar and trigonometric specialisations – for definiteness, and also to show which relations survive and which don't when $\mathfrak{E}$ specialises to the *approximate* flexion units like $Qa_c^\bullet$ and $Qi_c^\bullet$.

$$\begin{aligned} &\text{slash.pal}^\bullet &&= \text{paj}^\bullet &&, \quad \text{slash.tal}_c^\bullet &&= \text{taj}_c^\bullet \\ &\text{slash.pil}^\bullet &&= \text{pij}^\bullet &&, \quad \text{slash.til}_c^\bullet &&= \text{tij}_c^\bullet \\ &\text{slash.pail}_{\alpha,\beta}^\bullet &&= \text{paij}_{\alpha,\beta}^\bullet &&, \quad \text{slash.tail}_{c,\alpha,\beta}^\bullet &&= \text{taij}_{c,\alpha,\beta}^\bullet \end{aligned}$$

$$\begin{aligned} &\text{paj}^\bullet &&= \text{expari.Pa}^\bullet &&, \quad \text{taj}_c^\bullet &&= \text{expari.Qa}_c^\bullet \\ &\text{pij}^\bullet &&= \text{expari.Pi}^\bullet &&, \quad \text{tij}_c^\bullet &&\neq \text{expari.Qi}_c^\bullet \\ &\text{paij}_{\alpha,\beta}^\bullet &&= \text{expari.Pai}_{\alpha,\beta}^\bullet &&, \quad \text{taij}_{c,\alpha,\beta}^\bullet &&= \text{expari.Qai}_{c,\alpha,\beta}^\bullet \end{aligned}$$

$$\begin{aligned} &\text{invgami.paj}^\bullet &&\overset{\text{trivially}}{=} \text{invgani.anti.paj}^\bullet &&\equiv \text{pari.pac}^\bullet \\ &\text{invgami.pij}^\bullet &&\overset{\text{trivially}}{=} \text{invgani.anti.pij}^\bullet &&\equiv \text{pari.pic}^\bullet \\ &\text{invgami.paij}_{\alpha,\beta}^\bullet &&\overset{\text{trivially}}{=} \text{invgani.anti.paij}_{\alpha,\beta}^\bullet &&\equiv \text{pari.paic}_{\alpha,\beta}^\bullet \\ &\text{invgami.taj}_c^\bullet &&\overset{\text{trivially}}{=} \text{invgani.anti.taj}_c^\bullet &&\not\equiv \text{pari.tac}_c^\bullet \\ &\text{invgami.tij}_c^\bullet &&\overset{\text{trivially}}{=} \text{invgani.anti.tij}_c^\bullet &&\not\equiv \text{pari.tic}_c^\bullet \\ &\text{invgami.taij}_{c,\alpha,\beta}^\bullet &&\overset{\text{trivially}}{=} \text{invgani.anti.taij}_{c,\alpha,\beta}^\bullet &&\equiv \text{pari.taic}_{c,\alpha,\beta}^\bullet . \end{aligned}$$

4.5 Trigonometric and bitrigonometric bimoulds

Correspondence between polar and trigonometric

Polar bimoulds of a given type may have one trigonometric equivalent, or several, or none. The reverse correspondence, however, is always straightforward: when c goes to 0, (Qa_c, Qi_c) goes to (Pa, Pi) and the various trigonometric bimoulds, whenever they exist, go to their polar namesakes.

Correspondence between trigonometric and bitrigonometric

The correspondence, here, is always one-to-one. This may come as a surprise, since the bitrigonometric units Qaa_c, Qii_c are far more complex than their trigonometric counterparts Qa_c, Qi_c. To turn a trigonomeric

bimould of a given type into a bitrigonometric one of the same type, the recipe is:
– to change Qa_c respectively Qi_c into Qaa_c respectively Qii_c;
– to change c^{2s} into $c^{2s}\delta(lin_1^{\boldsymbol{w}})\ldots\delta(lin_{2s}^{\boldsymbol{w}})$ with discrete diracs δ defined as in Section 3.2 (see after (3.33)) and with their arguments $lin_j^{\boldsymbol{w}}$ denoting suitable differences of v_i's or sums of u_i's, as the case may be. There are simple rules for picking, in each instance, the right inputs $lin_j^{\boldsymbol{w}}$, which alone preserve the symmetries. We shall see examples in the last para of the present section, when explicating the passage from *trigo* to *bitrigo* for the primary bimoulds.

The secondary bimoulds $tal_c^\bullet/til_c^\bullet$ and $taal_c^\bullet/tiil_c^\bullet$

Of all the bimoulds constructed so far, these are the most important,[66] but also the most difficult to construct and describe. We can do no more here than state the main facts:
– the secondary bimoulds $\mathfrak{es}\mathfrak{z}_\sigma^\bullet$ have no trigonometric specialisation, whether under $\mathfrak{E} = Qa_c$ or $\mathfrak{E} = Qi_c$;
– the secondary bimould $\mathfrak{ess}^\bullet$ has no trigonometric specialisation under $\mathfrak{E} = Qa_c$, but it has one under $\mathfrak{E} = Qi_c$, namely $til_c^\bullet$, with $tal_c^\bullet$ as *swappee*.

In other words, while the polar pair $par^\bullet/pir^\bullet$ has no trigonometric, and therefore no bitrigonometric, counterpart, the polar pair $pal^\bullet/pil^\bullet$ does possess exact, though far more complex analogues, namely $tal_c^\bullet/til_c^\bullet$ and $taal_c^\bullet/tiil_c^\bullet$.

For illustration, the pair $taal_c^\bullet/tiil_c^\bullet$ has been tabulated in Section 12.6 up to length $r = 4$. The simpler pair $tal_c^\bullet/til_c^\bullet$ can be deduced from it, simply by recalibrating the flexion units and by changing all δ's into 1's.

Like $pil^\bullet$ in the polar case, the bisymmetral $til_c^\bullet$ and its *gari*-inverse $ritil_c^\bullet$ possess the important property of *separativity*: under the *gepar* transform[67] they turn into polynomials of c and the $Q_c(u_i)$ (all strict u_i-sums vanish!), with a particularly simple expression in the case of $ritil_c^\bullet$:

$$(\mathrm{gepar.til_c})^{w_1,\ldots,w_r} = \textit{homog. polynomial in}\,(c, Q_c(u_1),\ldots,Q_c(u_r)) \quad (4.87)$$

$$(\mathrm{gepar.ritil_c})^{w_1,\ldots,w_r} = \sum_{0\le s\le \frac{r}{2}} \frac{(-1)^s c^{2s}}{2s+1}\mathrm{sym}_{r-2s}(Q_c(u_1),\ldots,Q_c(u_r)) \quad (4.88)$$

with $sym_k(x_1,\ldots,x_r)$ denoting the k-th symmetric function of the x_i.[68]

[66] Because it is the main part of the first factor $Zag_I^\bullet$ in the trifactorisation of $Zag^\bullet$ and also the main ingredient of the canonical-rational associator.

[67] We recall that $gepar.S^\bullet := mu(anti.swap.S^\bullet, swap.S^\bullet)$.

[68] sym_0 is $\equiv 1$; sym_1 is the sum; sym_r is the product.

The primary bimoulds: trigonometric specialisation
To explicate the primary bimoulds, we require six series of coefficients that are best defined by their generating series:

$$\alpha(t) = \arctan(t) = \sum_{s\geq 0} \alpha_n\, t^{n+1} = t - \frac{1}{3}t^3 + \frac{1}{5}t^5 - \frac{1}{7}t^7 \ldots$$

$$\beta(t) = \tan(t) = \sum_{s\geq 0} \beta_n\, t^{n+1} = t + \frac{1}{3}t^3 + \frac{2}{15}t^5 + \frac{17}{315}t^7 \ldots$$

$$\hat{\alpha}(t) = \frac{t}{(1+t^2)^{1/2}} = t\,(\alpha'(t))^{\frac{1}{2}} = \sum_{s\geq 0} \hat{\alpha}_n\, t^{n+1} = t - \frac{1}{2}t^3 + \frac{3}{8}t^5 - \frac{5}{16}t^7 \ldots$$

$$\hat{\beta}(t) = \frac{t}{\cos(t)} = t\,(\beta'(t))^{\frac{1}{2}} = \sum_{s\geq 0} \hat{\beta}_n\, t^{n+1} = t + \frac{1}{2}t^3 + \frac{5}{24}t^5 + \frac{61}{720}t^7 \ldots$$

$$\check{\alpha}(t) = \frac{\arctan(t)}{(1+t^2)^{-1/2}} = \alpha(t)(\alpha'(t))^{-\frac{1}{2}} = \sum_{s\geq 0} \check{\alpha}_n\, t^{n+1} = t + \frac{1}{6}t^3 - \frac{11}{120}t^5 + \frac{103}{1680}t^7 \ldots$$

$$\check{\beta}(t) = \sin(t) = \beta(t)(\beta'(t))^{-\frac{1}{2}} = \sum_{s\geq 0} \check{\beta}_n\, t^{n+1} = t - \frac{1}{6}t^3 + \frac{1}{120}t^5 - \frac{1}{5040}t^7 \ldots$$

As in the polar case, the basic primary bimoulds $taj_c^\bullet$, $tij_c^\bullet$ (symmetral) derive from the secondary bimoulds $tal_c^\bullet$, $til_c^\bullet$ (bisymmetral) under the *slash*-tranform[69] and are best expressed via their *swappees*. To the polar pair $pac^\bullet/pic^\bullet$, however, there now correspond two trigonometric pairs, namely $tac_c^\bullet$, $tic_c^\bullet$ and the "correction" $tak_c^\bullet$, $tik_c^\bullet$ which will be needed to reproduce all the exact relations between primary bimoulds that obtained in the polar case. Let us begin with the definitions. We have:

$$\mathrm{swap.taj}_c^{\boldsymbol{w}} = \sum_{s\geq 0} \quad \sum_{\boldsymbol{w}^0 w_{n_1} \boldsymbol{w}^1 \ldots w_{n_s} \boldsymbol{w}^s = \boldsymbol{w}} \mathrm{Ji}_*^{\boldsymbol{w}^0}\, \mathrm{Qi}_c^{w_{n_1}}\, \mathrm{Ji}^{\boldsymbol{w}^1} \ldots \mathrm{Qi}_c^{w_{n_s}}\, \mathrm{Ji}^{\boldsymbol{w}^s}$$

$$\mathrm{swap.tij}_c^{\boldsymbol{w}} = \sum_{s\geq 0} \hat{\alpha}_{r-s} \sum_{\boldsymbol{w}^0 w_{n_1} \boldsymbol{w}^1 \ldots w_{n_s} \boldsymbol{w}^s = \boldsymbol{w}} \mathrm{Ja}^{\boldsymbol{w}^0}\mathrm{Qa}_c^{w_{n_1}}\mathrm{Ja}^{\boldsymbol{w}^1} \ldots \mathrm{Qa}_c^{w_{n_s}}\mathrm{Ja}^{\boldsymbol{w}^s}$$

$$\mathrm{tac}_c^{\boldsymbol{w}} = \sum_{s\geq 0} \alpha_{r-s} \sum_{\boldsymbol{w}^0 w_{n_1} \boldsymbol{w}^1 \ldots w_{n_s} \boldsymbol{w}^s = \boldsymbol{w}} \mathrm{Ca}^{\boldsymbol{w}^0}\mathrm{Qa}_c^{w_{n_1}}\mathrm{Ca}^{\boldsymbol{w}^1} \ldots \mathrm{Qa}_c^{w_{n_s}}\mathrm{Ca}^{\boldsymbol{w}^s}$$

$$\mathrm{tic}_c^{\boldsymbol{w}} = \sum_{s\geq 0} \quad \sum_{\boldsymbol{w}^0 w_{n_1} \boldsymbol{w}^1 \ldots w_{n_s} \boldsymbol{w}^s = \boldsymbol{w}} \mathrm{Ci}^{\boldsymbol{w}^0}\, \mathrm{Qi}_c^{w_{n_1}}\, \mathrm{Ci}^{\boldsymbol{w}^1} \ldots \mathrm{Qi}_c^{w_{n_s}}\, \mathrm{Ci}^{\boldsymbol{w}^s}$$

$$\mathrm{tak}_c^{\boldsymbol{w}} = \sum_{s\geq 0} \check{\alpha}_{r-s} \sum_{\boldsymbol{w}^0 w_{n_1} \boldsymbol{w}^1 \ldots w_{n_s} \boldsymbol{w}^s = \boldsymbol{w}} \mathrm{Ka}^{\boldsymbol{w}^0}\mathrm{Qa}_c^{w_{n_1}}\mathrm{Ka}^{\boldsymbol{w}^1} \ldots \mathrm{Qa}_c^{w_{n_s}}\mathrm{Ka}_*^{\boldsymbol{w}^s}$$

$$\mathrm{tik}_c^{\boldsymbol{w}} = \check{\beta}_r\, c^r \qquad \text{if } \boldsymbol{w} = (w_1, \ldots, w_r)$$

[69] We recall that $slash.S^\bullet := gari(neg.S^\bullet, invgari.S^\bullet)$.

with auxiliary building blocks themselves defined by:

$$\begin{array}{llllll}
\mathrm{Ja}^{w_1,\dots,w_r} & = & \mathrm{Ca}^{w_1,\dots,w_r} & := c^r & & (\forall r \geq 0)\\
\mathrm{Ji}^{w_1,\dots,w_r} & = & \mathrm{Ci}^{w_1,\dots,w_r} & := c^r\,\beta_r & & (\forall r \geq 0)\\
\mathrm{Ji}_*^{w_1,\dots,w_r} & := & c^r\,\hat{\beta}_r & & & (\forall r \geq 0)\\
\mathrm{Ka}^{w_1,\dots,w_r} & := & c^r & & & (\forall r \geq 0)\\
\mathrm{Ka}_*^{w_1,\dots,w_r} & := & c^r & (\forall r \geq 1) & \text{but} & \mathrm{Ka}_*^{\emptyset} := 0.
\end{array}$$

Here are some of the main trigonometric identities that are exact transpositions of their polar prototypes:

$$\mathrm{slash.tal}_c^\bullet = \mathrm{taj_c}^\bullet \tag{4.89}$$
$$\mathrm{slash.til}_c^\bullet = \mathrm{tij_c}^\bullet \tag{4.90}$$
$$\mathrm{invgani.tac}_c^\bullet = \mathrm{anti.swap.anti.pari.tic}_c^\bullet \tag{4.91}$$
$$\mathrm{invgani.tic}_c^\bullet = \mathrm{anti.swap.anti.pari.tac}_c^\bullet \tag{4.92}$$
$$\mathrm{invgami.taj}_c^\bullet = \mathrm{invgani.anti.taj}_c^\bullet \tag{4.93}$$
$$\mathrm{invgami.tij}_c^\bullet = \mathrm{invgani.anti.tij}_c^\bullet. \tag{4.94}$$

And here is an example when polar identities:

$$\mathrm{invmu.paj}^\bullet \overset{\text{trivially}}{=} \mathrm{pari.anti.paj}^\bullet = \mathrm{invgani.pac}^\bullet \tag{4.95}$$
$$\mathrm{invmu.pij}^\bullet \overset{\text{trivially}}{=} \mathrm{pari.anti.pij}^\bullet = \mathrm{invgani.pic}^\bullet \tag{4.96}$$

require a corrective term in the trigonometric transposition:

$$\mathrm{invmu.taj}_c^\bullet \overset{\text{trivially}}{=} \mathrm{pari.anti.taj}_c^\bullet = \mathrm{fragani}(\mathrm{tak}_c^\bullet, \mathrm{tac}_c^\bullet) \tag{4.97}$$
$$\mathrm{invmu.tij}_c^\bullet \overset{\text{trivially}}{=} \mathrm{pari.anti.tij}_c^\bullet = \mathrm{fragani}(\mathrm{tik}_c^\bullet, \mathrm{tic}_c^\bullet). \tag{4.98}$$

The abbreviation *fragani* denotes of course the *gani*-fraction:

$$\mathrm{fragani}(A^\bullet, B^\bullet) := \mathrm{gani}(A^\bullet, \mathrm{invgani}.B^\bullet)$$

and the relations (4.97), (4.98) basically reflect the functional identities:

$$\hat{\beta} = \check{\alpha} \circ \alpha; \qquad \hat{\alpha} = \check{\beta} \circ \beta.$$

Here is another example. The important polar identity:

$$\mathrm{pij}^\bullet = \mathrm{expari.Pi}^\bullet$$

doesn't transpose to $\mathrm{tij}_c^\bullet = \mathrm{expari.Qi}_c^\bullet$ but to the variant:

$$\mathrm{tijj}_c^\bullet = \mathrm{expari.Qi}_c^\bullet \qquad (\mathrm{anti.swap.tijj}^\bullet =: \mathrm{astajj})$$

with a bimould $tijj_c^\bullet$ best defined via its *anti.swap*-transform $astajj_c^\bullet$, for which the following remarkable expansion holds:

$$\mathrm{astajj}_c^{w_1,\ldots,w_r} = \sum_{0\le t\le \frac{r}{2}} (-1)^t\, c^{2t} \sum_{\boldsymbol{w}^0 w_{n_1} \boldsymbol{w}^1 \ldots w_{n_s} \boldsymbol{w}^s = \boldsymbol{w}}^{s=r-2t} \mathrm{Ta}^{m_1,m_2\ldots,m_{2t}}\, Qa^{w_{n_1}} \ldots Qa^{w_{n_s}}$$

with

$$[m_1, m_2, \ldots, m_{2t}] := [1, 2, \ldots, r] \dot{-} [n_1, n_2, \ldots, n_s]$$

and

$$\mathrm{Ta}^{m_1,m_2\ldots,m_{2t}} := \frac{m_1}{m_2}\frac{m_3}{m_4} \ldots \frac{m_{2t-1}}{m_{2t}}.$$

Primary bimoulds: bitrigonometric specialisation
The bimoulds of the preceding para become:

$$\mathrm{swap.taaj}_c^{\boldsymbol{w}} = \sum_{s\ge 0} \sum_{\boldsymbol{w}^0 w_{n_1} \boldsymbol{w}^1 \ldots w_{n_s} \boldsymbol{w}^s = \boldsymbol{w}} \mathrm{Jii}_*^{\boldsymbol{w}^0}\, \mathrm{Qii}_c^{w_{n_1}}\, \mathrm{Jii}^{\boldsymbol{w}^1} \ldots \mathrm{Qii}_c^{w_{n_s}}\, \mathrm{Jii}^{\boldsymbol{w}^s}$$

$$\mathrm{swap.tiij}_c^{\boldsymbol{w}} = \sum_{s\ge 0} \hat{\alpha}_{r-s} \sum_{\boldsymbol{w}^0 w_{n_1} \boldsymbol{w}^1 \ldots w_{n_s} \boldsymbol{w}^s = \boldsymbol{w}} \mathrm{Jaa}^{\boldsymbol{w}^0}\, \mathrm{Qaa}_c^{w_{n_1}}\, \mathrm{Jaa}^{\boldsymbol{w}^1} \ldots \mathrm{Qaa}_c^{w_{n_s}}\, \mathrm{Jaa}^{\boldsymbol{w}^s}$$

$$\mathrm{taac}_c^{\boldsymbol{w}} = \sum_{s\ge 0} \alpha_{r-s} \sum_{\boldsymbol{w}^0 w_{n_1} \boldsymbol{w}^1 \ldots w_{n_s} \boldsymbol{w}^s = \boldsymbol{w}} \mathrm{Caa}^{\boldsymbol{w}^0}\, \mathrm{Qaa}_c^{w_{n_1}}\, \mathrm{Caa}^{\boldsymbol{w}^1} \ldots \mathrm{Qaa}_c^{w_{n_s}}\, \mathrm{Caa}^{\boldsymbol{w}^s}$$

$$\mathrm{tiic}_c^{\boldsymbol{w}} = \sum_{s\ge 0} \sum_{\boldsymbol{w}^0 w_{n_1} \boldsymbol{w}^1 \ldots w_{n_s} \boldsymbol{w}^s = \boldsymbol{w}} \mathrm{Cii}^{\boldsymbol{w}^0}\, \mathrm{Qii}_c^{w_{n_1}}\, \mathrm{Cii}^{\boldsymbol{w}^1} \ldots \mathrm{Qii}_c^{w_{n_s}}\, \mathrm{Cii}^{\boldsymbol{w}^s}$$

$$\mathrm{taak}_c^{\boldsymbol{w}} = \sum_{s\ge 0} \check{\alpha}_{r-s} \sum_{\boldsymbol{w}^0 w_{n_1} \boldsymbol{w}^1 \ldots w_{n_s} \boldsymbol{w}^s = \boldsymbol{w}} \mathrm{Kaa}^{\boldsymbol{w}^0}\, \mathrm{Qaa}_c^{w_{n_1}}\, \mathrm{Kaa}^{\boldsymbol{w}^1} \ldots \mathrm{Qaa}_c^{w_{n_s}}\, \mathrm{Kaa}_*^{\boldsymbol{w}^s}$$

$$\mathrm{tiik}_c^{\boldsymbol{w}} = \check{\beta}_r\, c^r\, \delta(u_1) \ldots \delta(u_r) \qquad \text{if } \boldsymbol{w} = (w_1, \ldots, w_r)$$

with elementary building blocks defined by:

$$\begin{array}{llll}
\mathrm{Caa}^{w_1,\ldots,w_r} = \mathrm{Jaa}^{w_1,\ldots,w_r} & := & c^r\, \delta(v_1) \ldots \delta(v_r) & (\forall r \ge 0) \\
\mathrm{Cii}^{w_1,\ldots,w_r} = \mathrm{Jii}^{w_1,\ldots,w_r} & := & \beta_r\, c^r\, \delta(u_1) \ldots \delta(u_r) & (\forall r \ge 0) \\
\mathrm{Jii}^{\emptyset} := 0 \;,\; \mathrm{Jii}_*^{w_1,\ldots,w_r} & := & \hat{\beta}_r\, c^r\, \delta(u_1) \ldots \delta(u_r) & (\forall r \ge 1) \\
\mathrm{Kaa}^{\emptyset} := 1 \;,\; \mathrm{Kaa}^{w_1,\ldots,w_r} & := & c^r\, \delta(v_1) \ldots \delta(v_r) & (\forall r \ge 1) \\
\mathrm{Kaa}_*^{\emptyset} := 0 \;,\; \mathrm{Kaa}_*^{w_1,\ldots,w_r} & := & c^r\, \delta(v_1) \ldots \delta(v_r) & (\forall r \ge 1)
\end{array}$$

Remark 1: though there is one and only one 'proper' way of 'filling in' the trigonometric formulas with δ's to get the bitrigonometric equivalents, the procedure is non-trivial. Indeed, the arguments inside the δ's are not always single u_i's or v_i's but often non-trivial sums or differences.[70]

[70] As with $taaj_c^\bullet$ and $tiij_c^\bullet$, once we carry out the *swap* transform in the above definitions.

Remark 2: The even-odd factorisations (4.45), (4.46) have their exact counterpart here. Thus, in trigonometric mode:

$$til^\bullet = \mathrm{gari}(til^\bullet_*, til^\bullet_{**}) \qquad \text{with } til^\bullet_*, til^\bullet_{**} \text{ symmetral} \quad (4.99)$$

$$tal^\bullet = \mathrm{mu}(tal^\bullet_{\star\star}, tal^\bullet_\star) \qquad \text{with } tal^\bullet_\star, tal^\bullet_{\star\star} \text{ symmetral} \quad (4.100)$$

with elementary factors $tal^\bullet_\star$, $til^\bullet_*$ alongside non-elementary factors $tal^\bullet_{\star\star}$, $til^\bullet_{**}$ that carry only even-lengthed components.

4.6 Dimorphic isomorphisms in universal mode

We can now enunciate the main statement of the whole section, namely that there exists a canonical isomorphism between *straight* dimorphic structures (algebras or groups) and their *twisted* counterparts.[71] But before that, we must begin with the less remarkable isomorphisms which connect *straight* or *twisted* monomorphic structures[72] and exchange only *one* symmetry with another.

All these results are summarised in the following diagrams:
- with various groups in the upper lines;
- with various Lie algebras in the lower lines;
- with horizontal arrows that stand for (algebra or group) isomorphisms;
- with vertical arrows representing the natural exponential mapping of each Lie algebra into its group.

Basic diagrams of monomorphic transport

$$\begin{array}{ccc|ccc}
\mathrm{MU}^{\mathrm{as}} & \stackrel{\mathrm{ganit}(\mathfrak{ez}^\bullet)}{\longrightarrow} & \mathrm{MU}^{\mathfrak{es}} & \mathrm{MU}^{\mathrm{as}} & \stackrel{\mathrm{ganit}(\mathrm{pari.anti}.\mathfrak{es}^\bullet)}{\longleftarrow} & \mathrm{MU}^{\mathfrak{es}} \\
\uparrow \mathrm{expmu} & & \uparrow \mathrm{expmu} & \uparrow \mathrm{expmu} & & \uparrow \mathrm{expmu} \\
\mathrm{LU}^{\mathrm{al}} & \stackrel{\mathrm{ganit}(\mathfrak{ez}^\bullet)}{\longrightarrow} & \mathrm{LU}^{\mathfrak{el}} & \mathrm{LU}^{\mathrm{al}} & \stackrel{\mathrm{ganit}(\mathrm{pari.anti}.\mathfrak{es}^\bullet)}{\longleftarrow} & \mathrm{LU}^{\mathfrak{el}}
\end{array}$$

$$\begin{array}{ccc|ccc}
\mathrm{MU}^{\mathrm{as}} & \stackrel{\mathrm{gamit}(\mathfrak{ez}^\bullet)}{\longrightarrow} & \mathrm{MU}^{\mathfrak{es}} & \mathrm{MU}^{\mathrm{as}} & \stackrel{\mathrm{gamit}(\mathrm{pari}.\mathfrak{es}^\bullet)}{\longleftarrow} & \mathrm{MU}^{\mathfrak{es}} \\
\uparrow \mathrm{expmu} & & \uparrow \mathrm{expmu} & \uparrow \mathrm{expmu} & & \uparrow \mathrm{expmu} \\
\mathrm{LU}^{\mathrm{al}} & \stackrel{\mathrm{gamit}(\mathfrak{ez}^\bullet)}{\longrightarrow} & \mathrm{LU}^{\mathfrak{el}} & \mathrm{LU}^{\mathrm{al}} & \stackrel{\mathrm{gamit}(\mathrm{pari}.\mathfrak{es}^\bullet)}{\longleftarrow} & \mathrm{LU}^{\mathfrak{el}}.
\end{array}$$

[71] Or, more properly, "half-twisted", since the first symmetry remains straight, and only the second gets twisted.

[72] *I.e.* subgroups of $MU := \{BIMU^*, mu\}$ or subalgebras of $LU := \{BIMU_*, lu\}$.

Basic diagram of dimorphic transport

$$\begin{array}{ccc}
\mathrm{GARI}^{\underline{as}/\underline{as}} & \xrightarrow{\mathrm{adgari}(\mathfrak{ess}^\bullet)} & \mathrm{GARI}^{\underline{as}/\underline{\mathfrak{os}}} \\
\mathrm{logari} \downarrow\uparrow \mathrm{expari} & & \mathrm{logari} \downarrow\uparrow \mathrm{expari} \\
\mathrm{ARI}^{\underline{al}/\underline{al}} & \xrightarrow{\mathrm{adari}(\mathfrak{ess}^\bullet)} & \mathrm{ARI}^{\underline{al}/\underline{\mathfrak{ol}}}.
\end{array}$$

Dimorphic subsymmetries

The subsymmetries listed below are by no means the only ones[73] but they are the ones that matter most and also (whether coincidentally or not) the only ones that are properly dimorphic.[74]

$$\begin{array}{llcrcr}
A^\bullet \in \mathrm{ARI}^{\underline{al}/\underline{al}} & \Longrightarrow A^\bullet & = & \mathrm{neg}.A^\bullet & = & \mathrm{push}.A^\bullet \\
A^\bullet \in \mathrm{GARI}^{\underline{as}/\underline{as}} & \Longrightarrow A^\bullet & = & \mathrm{neg}.A^\bullet & = & \mathrm{gush}.A^\bullet \\
A^\bullet \in \mathrm{ARI}^{\underline{al}/\underline{\mathfrak{ol}}} & \Longrightarrow A^\bullet & = & \mathfrak{D}\text{-}\mathrm{neg}.A^\bullet & = & \mathfrak{D}\text{-}\mathrm{push}.A^\bullet \\
A^\bullet \in \mathrm{GARI}^{\underline{as}/\underline{\mathfrak{os}}} & \Longrightarrow A^\bullet & = & \mathfrak{D}\text{-}\mathrm{geg}.A^\bullet & = & \mathfrak{D}\text{-}\mathrm{gush}.A^\bullet.
\end{array}$$

As noted earlier, $\mathfrak{D}$-*neg*-invariance is expressible in terms of an elementary *primary* bimould $\mathfrak{es}^\bullet := slash.\mathfrak{ess}^\bullet$, and $\mathfrak{D}$-*push*-invariance also is equivalent to the much simpler *senary relation*.

4.7 Dimorphic isomorphisms in polar mode

Diagrams of monomorphic transport

For the specialisation $\mathfrak{E} = \mathrm{Pa}$, the first universal diagrams of monomorphic transport become:

$$\begin{array}{ccc|ccc}
\mathrm{MU}^{as} & \xrightarrow{\mathrm{ganit}(\mathrm{pac}^\bullet)} & \mathrm{MU}^{us} & \mathrm{MU}^{as} & \xleftarrow{\mathrm{ganit}(\mathrm{pari.anti.paj}^\bullet)} & \mathrm{MU}^{us} \\
\uparrow \mathrm{expmu} & & \uparrow \mathrm{expmu} & \uparrow \mathrm{expmu} & & \uparrow \mathrm{expmu} \\
\mathrm{LU}^{al} & \xrightarrow{\mathrm{ganit}(\mathrm{pac}^\bullet)} & \mathrm{LU}^{ul} & \mathrm{LU}^{al} & \xleftarrow{\mathrm{ganit}(\mathrm{pari.anti.paj}^\bullet)} & \mathrm{LU}^{ul}.
\end{array}$$

For the specialisation $\mathfrak{E} = \mathrm{Pi}$, they become:

$$\begin{array}{ccc|ccc}
\mathrm{MU}^{as} & \xrightarrow{\mathrm{ganit}(\mathrm{pic}^\bullet)} & \mathrm{MU}^{is} & \mathrm{MU}^{as} & \xleftarrow{\mathrm{ganit}(\mathrm{pari.anti.pij}^\bullet)} & \mathrm{MU}^{is} \\
\uparrow \mathrm{expmu} & & \uparrow \mathrm{expmu} & \uparrow \mathrm{expmu} & & \uparrow \mathrm{expmu} \\
\mathrm{LU}^{al} & \xrightarrow{\mathrm{ganit}(\mathrm{pic}^\bullet)} & \mathrm{LU}^{il} & \mathrm{LU}^{al} & \xleftarrow{\mathrm{ganit}(\mathrm{pari.anti.pij}^\bullet)} & \mathrm{LU}^{il}.
\end{array}$$

[73] See Section 3.4.

[74] In the sense that it takes *two* symmetries, not *one*, to induce them.

Diagrams of dimorphic transport
For the specialisation $(\mathfrak{E}, \mathfrak{O}) = (\text{Pa}, \text{Pi})$, the diagram of dimorphic transport becomes:

$$\begin{array}{ccc} \text{GARI}^{\underline{\text{as}}/\underline{\text{as}}} & \stackrel{\text{adgari(pal}^\bullet)}{\longrightarrow} & \text{GARI}^{\underline{\text{as}}/\underline{\text{is}}} \\ \text{logari} \downarrow\uparrow \text{expari} & & \text{logari} \downarrow\uparrow \text{expari} \\ \text{ARI}^{\underline{\text{al}}/\underline{\text{al}}} & \stackrel{\text{adari(pal}^\bullet)}{\longrightarrow} & \text{ARI}^{\underline{\text{al}}/\underline{\text{il}}} \end{array}$$

and the dimorphic subsymmetries become:

$$\begin{array}{lclcl} A^\bullet \in \text{ARI}^{\underline{\text{al}}/\underline{\text{il}}} \Longrightarrow A^\bullet & = & \text{negu}.A^\bullet & = & \text{pushu}.A^\bullet \\ A^\bullet \in \text{GARI}^{\underline{\text{as}}/\underline{\text{is}}} \Longrightarrow A^\bullet & = & \text{gegu}.A^\bullet & = & \text{gushu}.A^\bullet. \end{array}$$

For the 'conjugate' specialisation $(\mathfrak{E}, \mathfrak{O}) = (\text{Pi}, \text{Pa})$, the diagram becomes:

$$\begin{array}{ccc} \text{GARI}^{\underline{\text{as}}/\underline{\text{as}}} & \stackrel{\text{adgari(pil}^\bullet)}{\longrightarrow} & \text{GARI}^{\underline{\text{as}}/\underline{\text{us}}} \\ \text{logari} \downarrow\uparrow \text{expari} & & \text{logari} \downarrow\uparrow \text{expari} \\ \text{ARI}^{\underline{\text{al}}/\underline{\text{al}}} & \stackrel{\text{adari(pil}^\bullet)}{\longrightarrow} & \text{ARI}^{\underline{\text{al}}/\underline{\text{ul}}} \end{array}$$

and the dimorphic subsymmetries become:

$$\begin{array}{lclcl} A^\bullet \in \text{ARI}^{\underline{\text{al}}/\underline{\text{ul}}} \Longrightarrow A^\bullet & = & \text{negi}.A^\bullet & = & \text{pushi}.A^\bullet \\ A^\bullet \in \text{GARI}^{\underline{\text{as}}/\underline{\text{us}}} \Longrightarrow A^\bullet & = & \text{gegi}.A^\bullet & = & \text{gushi}.A^\bullet. \end{array}$$

The matter of 'entireness'
A few comments are in order here, regarding the preservation, or otherwise, of the *entire* character of bimoulds.[75]
(i) The simple symmetries *al* and *as* are compatible with entireness, and so are the double symmetries $\underline{al}/\underline{al}$ and $\underline{as}/\underline{as}$.
(ii) The twisted symmetries *il* and *is* are compatible with entireness, but *ul* and *us* are not.
(iii) However, even in second monomorphic diagram, when all four structures contain *entire* bimoulds and the isomorphism *ganit*($pic^\bullet$) might conceivably preserve *entireness*, it *does not.* The same holds when *ganit*($pic^\bullet$) is replaced by *gamit*($pic^\bullet$).
(iv) The (important) twisted double symmetries $\underline{al}/\underline{il}$ and $\underline{as}/\underline{is}$ are compatible with *entireness*, but the (less important) double symmetries $\underline{al}/\underline{ul}$ and $\underline{as}/\underline{us}$ are not.

[75] *I.e.* their being *polynomials* or *entire functions* or *formal power series* of their $\boldsymbol{u}$-variables.

(v) However, even in the first dimorphic diagram, where all four structures do contain *entire* bimoulds and when the isomorphism $adari(pal^\bullet)$ might conceivably preserve *entireness*, it *does not.*
(vi) The dimorphic subsymmetries induced by $\underline{al}/\underline{il}$ and $\underline{as}/\underline{is}$ (*i.e. negu-* and *pushu-* or *gushu*-invariance), despite the massive involvement of 'poles', are compatible with *entireness*, whereas the dimorphic subsymmetries induced by $\underline{al}/\underline{ul}$ and $\underline{as}/\underline{us}$ (*i.e. negi-* and *pushi-* or *gushi*-invariance), are not. For the first dimorphic subsymmetries (of the *'neg'* sort), both the compatibility and incompatibility may be checked on the formulas:

$$\mathrm{negu}.B^\bullet = \mathrm{neg.adari(paj^\bullet)}.B^\bullet = \mathrm{adari(pari.paj^\bullet).neg}.B^\bullet \qquad (4.101)$$

$$\mathrm{negi}.B^\bullet = \mathrm{neg.adari(pij^\bullet)}.B^\bullet = \mathrm{adari(pari.pij^\bullet).neg}.B^\bullet. \qquad (4.102)$$

For the first dimorphic subsymmetries (of the *'push'* sort), the compatibility respectively incompatibility may be checked on the *senary* relations:

$$\mathrm{teru}.B^\bullet = \mathrm{push.mantar.teru.mantar}.B^\bullet \qquad (4.103)$$

$$\mathrm{teri}.B^\bullet = \mathrm{push.mantar.teri.mantar}.B^\bullet \qquad (4.104)$$

which express *pushu*-invariance, respectively *pushi*-invariance, in much simpler form, and involve the elementary, linear operators:

$$C^\bullet = \mathrm{teru}.B^\bullet \Longleftrightarrow C^{w_1,\dots,w_r} = B^{w_1,\dots,w_r} - B^{w_1,\dots,w_{r-1}}\,\mathrm{Pa}^{w_r} + B^{w_1,\dots,w_{r-1}\rceil}\,\mathrm{Pa}^{\lfloor w_r}$$

$$C^\bullet = \mathrm{teri}.B^\bullet \Longleftrightarrow C^{w_1,\dots,w_r} = B^{w_1,\dots,w_r} - B^{w_1,\dots,w_{r-1}}\,\mathrm{Pi}^{w_r} + B^{w_1,\dots,w_{r-1}\rceil}\,\mathrm{Pi}^{\lfloor w_r}.$$

The six entire structures

All the above remarks still hold, mutatis mutandis, when we replace the polar symmetries by their trigonometric counterparts (to be precisely defined in Section 11.4). Thus, whereas for the six fundamental structures we have the following commutative diagram, with all horizontal arrows denoting either group or algebra isomorphisms:

$$\begin{array}{ccccccc}
\mathrm{GARI}^{\underline{as}/\underline{as}} & \stackrel{\mathrm{adgari(pal^\bullet)}}{\longrightarrow} & \mathrm{GARI}^{\underline{as}/\underline{is}} & \stackrel{\mathrm{adgari(Zag_I{}^\bullet)}}{\longrightarrow} & \mathrm{GARI}^{\underline{as}/\underline{iis}} & \stackrel{\mathrm{adgari(tal^\bullet)}}{\longleftarrow} & \mathrm{GARI}^{\underline{as}/\underline{as}} \\
\uparrow \mathrm{expari} & & \uparrow \mathrm{expari} & & \uparrow \mathrm{expari} & & \uparrow \mathrm{expari} \\
\mathrm{ARI}^{\underline{al}/\underline{al}} & \stackrel{\mathrm{adari(pal^\bullet)}}{\longrightarrow} & \mathrm{ARI}^{\underline{al}/\underline{il}} & \stackrel{\mathrm{adari(Zag_I{}^\bullet)}}{\longrightarrow} & \mathrm{ARI}^{\underline{al}/\underline{iil}} & \stackrel{\mathrm{adari(tal^\bullet)}}{\longleftarrow} & \mathrm{ARI}^{\underline{al}/\underline{al}}
\end{array}$$

the picture changes when we add the requirement of entireness: the straight and twisted structures are no longer isomorphic[76] and only the middling isomorphism $adari(Zag_I^\bullet)$ between the twisted structures (polar and

[76] Neither under $adari(pal^\bullet)$, $adari(tal^\bullet)$, nor any conceivable replacement.

trigonometric) survives, as pictured in the following diagram:

$$\begin{array}{ccccccc}
\mathrm{GARI}_{\mathrm{ent}}^{\underline{\mathrm{as}}/\underline{\mathrm{as}}} & \overset{\mathrm{adgari(pal^\bullet)}}{\not\rightarrow} & \mathrm{GARI}_{\mathrm{ent}}^{\underline{\mathrm{as}}/\underline{\mathrm{is}}} & \overset{\mathrm{adgari(Zag_I{}^\bullet)}}{\longrightarrow} & \mathrm{GARI}_{\mathrm{ent}}^{\underline{\mathrm{as}}/\underline{\mathrm{iis}}} & \overset{\mathrm{adgari(tal^\bullet)}}{\not\leftarrow} & \mathrm{GARI}_{\mathrm{ent}}^{\underline{\mathrm{as}}/\underline{\mathrm{as}}} \\
\uparrow \mathrm{expari} & & \uparrow \mathrm{expari} & & \uparrow \mathrm{expari} & & \uparrow \mathrm{expari} \\
\mathrm{ARI}_{\mathrm{ent}}^{\underline{\mathrm{al}}/\underline{\mathrm{al}}} & \overset{\mathrm{adari(pal^\bullet)}}{\not\rightarrow} & \mathrm{ARI}_{\mathrm{ent}}^{\underline{\mathrm{al}}/\underline{\mathrm{il}}} & \overset{\mathrm{adari(Zag_I{}^\bullet)}}{\longrightarrow} & \mathrm{ARI}_{\mathrm{ent}}^{\underline{\mathrm{al}}/\underline{\mathrm{iil}}} & \overset{\mathrm{adari(tal^\bullet)}}{\not\leftarrow} & \mathrm{ARI}_{\mathrm{ent}}^{\underline{\mathrm{al}}/\underline{\mathrm{al}}}.
\end{array}$$

The six entire and $\boldsymbol{v}$-constant structures

This applies in particular to the six important substructures below, whose bimoulds:

– are power series of the upper indices u_i;

– are constant in the lower indices v_i.

Here is the diagram, with self-explanatory notations:

$$\begin{array}{ccccccc}
\mathrm{ASAS} & \overset{\mathrm{adgari(pal^\bullet)}}{\not\rightarrow} & \mathrm{ASIS} & \overset{\mathrm{adgari(Zag_I{}^\bullet)}}{\longrightarrow} & \mathrm{ASIIS} & \overset{\mathrm{adgari(tal^\bullet)}}{\not\leftarrow} & \mathrm{ASAS} \\
\uparrow \mathrm{expari} & & \uparrow \mathrm{expari} & & \uparrow \mathrm{expari} & & \uparrow \mathrm{expari} \\
\mathrm{ALAL} & \overset{\mathrm{adari(pal^\bullet)}}{\not\rightarrow} & \mathrm{ALIL} & \overset{\mathrm{adari(Zag_I{}^\bullet)}}{\longrightarrow} & \mathrm{ALIIL} & \overset{\mathrm{adari(tal^\bullet)}}{\not\leftarrow} & \mathrm{ALAL}.
\end{array}$$

The projector *cut*:

$$(\mathrm{cut}.M)^{\left(\begin{smallmatrix} u_1 & ,\dots, & u_r \\ v_1 & ,\dots, & v_r \end{smallmatrix}\right)} := M^{\left(\begin{smallmatrix} u_1 & ,\dots, & u_r \\ 0 & ,\dots, & 0 \end{smallmatrix}\right)} \tag{4.105}$$

clearly defines epimorphisms of

$$ARI^{\underline{\mathrm{al}}/\underline{\mathrm{al}}},\ \ GARI^{\underline{\mathrm{as}}/\underline{\mathrm{as}}},\ \ ARI^{\underline{\mathrm{al}}/\underline{\mathrm{il}}},\ \ GARI^{\underline{\mathrm{as}}/\underline{\mathrm{is}}},\ \ ARI^{\underline{\mathrm{al}}/\underline{\mathrm{iil}}},\ \ GARI^{\underline{\mathrm{as}}/\underline{\mathrm{iis}}}$$

respectively onto

$$ALAL,\ \ ASAS,\ \ ALIL,\ \ ASIS,\ \ ALIIL,\ \ ASIIS.$$

Now, all the bimoulds associated with *colourless multizetas*, happen to have lower indices v_i that are all $= 0$ as elements of $\mathbb{Q}/\mathbb{Z}$. We shall take advantage of the above property of *cut* to identify these bimoulds with their *cuttees*, *i.e.* to view them as $\boldsymbol{v}$-constant.

Central corrections

For structures with a *twisted* double symmetry, instead of demanding that the exact *swappee* should display the second symmetry, we often relax the condition and simply demand that the *swappee* corrected[77] by a suitable *central element* should display that symmetry. Thus, under these

[77] *Additively* in the case of algebras; *multiplicatively* in the case of groups.

relaxed conditions:

$$A^\bullet \in \mathrm{ARI}^{\underline{\mathrm{al}}/\underline{\mathrm{il}}} \Leftrightarrow \{A^\bullet \in \mathit{alternal};\quad \mathrm{swap}(A^\bullet + C_A^\bullet) \in \mathit{alternil}\}$$
$$S^\bullet \in \mathrm{GARI}^{\underline{\mathrm{al}}/\underline{\mathrm{il}}} \Leftrightarrow \{S^\bullet \in \mathit{symmetral};\quad \mathrm{swap}(\mathrm{gari}(A^\bullet, C_S^\bullet)) \in \mathit{symmetril}\}$$

with $C_A^\bullet \in \mathrm{Center(ARI)}$ and $C_S^\bullet \in \mathrm{Center(GARI)}$.

The sets thus defined are still algebras or groups, albeit larger ones. In the case of the $\boldsymbol{v}$-constant family *ALIL*, *ASIS*, *ALIIL*, *ASIIS*, we shall *always* assume this relaxed definition for without the central corrections these sets would be *empty*.[78] Besides, the bimoulds $Zag^\bullet$ associated with the (coloured or uncoloured) multizetas also require a *central correction* to display their double symmetry.

5 Singulators, singulands, singulates

At this point, we already have a valuable tool at our disposal, namely the operator $adari(pal^\bullet)$, which acts as an algebra isomorphism and respects double symmetries. What it doesn't do, though, is respect entireness: when applied to entire bimoulds of type, say, $\underline{al}/\underline{al}$, it produces bimoulds that have the right type, in this case $\underline{al}/\underline{il}$, but with singularities at the origin. To remove these without destroying the double symmetry $\underline{al}/\underline{il}$, we require a universal machinery capable, roughly speaking, of producing all possible singularities of type $\underline{al}/\underline{il}$. Such a machinery is at hand. It consists of *singulators, singulands*, and *singulates*. The *singulators* are quite complex linear operators. The *singulands* are arbitrary entire bimoulds subject only to simple parity constraints. Lastly, when acting on singulands, the singulators turn them into *singulates*, which are bimoulds of type $\underline{al}/\underline{il}$ and with singularities at the origin that are, so to speak, 'made to order', and capable of neutralising, by subtraction, any given, unwanted singularity of type $\underline{al}/\underline{il}$.

After some heuristics (destined to divest our construction of its 'contrived' character), we shall examine the singulators, first in universal mode, then in the relevant polar specialisation.

5.1 Some heuristics. Double symmetries and imparity

Analytical definition of ***sen***

Let us first introduce a mapping $sen : (A^\bullet, S^\bullet) \mapsto B^\bullet$ that is:
– linear in $S^\bullet \in BIMU_1$;
– quadrilinear in $A^\bullet \in BIMU^*$;

[78] For the structures *ALAL* and *ASAS*, on the other hand, central corrections are not required. In fact, allowing such corrections makes no difference at all, which again shows that the pairs *ALAL*//*ASAS* and *ALIL*//*ASIS* cannot be isomorphic.

– which turns *group-like* properties of $A^\bullet$ into *algebra-like* properties of $B^\bullet$;
– whose action strongly depends on the *parity* properties of $A^\bullet$, $S^\bullet$.
Here goes the definition:

$$B^\bullet = \mathrm{sen}(A^\bullet).S^\bullet \Leftrightarrow 2\,B^{\boldsymbol{w}} = \sum_{w_i \boldsymbol{w}^1 \boldsymbol{w}^2 w_j \boldsymbol{w}^3 \boldsymbol{w}^4 \overset{\mathrm{circ}}{=} \boldsymbol{w}^*} A_1^{\boldsymbol{w}^1} A_2^{\boldsymbol{w}^2 \rfloor} S^{\lceil w_j \rceil} A_3^{\lfloor \boldsymbol{w}^3} A_4^{\boldsymbol{w}^4}$$

$$\text{with} \qquad \boldsymbol{w}^* = \mathrm{augment}(\boldsymbol{w}) \qquad \text{and} \tag{5.1}$$

$$A_1^\bullet = \mathrm{anti}.A^\bullet,\ A_2^\bullet = A^\bullet,\ A_3^\bullet = \mathrm{pari.anti}.A^\bullet,\ A_4^\bullet = \mathrm{pari}.A^\bullet$$

with the *augment* $\boldsymbol{w}^*$ defined in the usual way:

$$\boldsymbol{w} = \begin{pmatrix} u_1\,, \ldots, \,u_r \\ v_1\,, \ldots, \,v_r \end{pmatrix} \quad \Rightarrow \quad \boldsymbol{w}^* = \begin{pmatrix} [u_0],\, u_1\,, \ldots, \,u_r \\ [v_0]\,,\, v_1\,, \ldots, \,v_r \end{pmatrix}$$

with the redundant additional component w_0:

$$u_0 := -u_1 - u_2 \cdots - u_r, \qquad v_0 = 0$$

and with the circular summation rule amounting to the double summation

$$\sum_{0 \le i \le r} \quad \sum_{w_i \boldsymbol{w}^1 \boldsymbol{w}^2 w_j \boldsymbol{w}^3 \boldsymbol{w}^4 = w_i w_{i+1} \ldots w_r w_0 w_1 \ldots w_{i-1}} \quad . \tag{5.2}$$

Main properties of ***sen***

Let $B^\bullet := sen(A^\bullet).S^\bullet$.

$\boldsymbol{P_1}$: If r is even and $S^{-w_1} = S^{w_1}$ then $B^{w_1,\ldots,w_r} = 0$.

$\boldsymbol{P_2}$: If r is odd and $S^{-w_1} = -S^{w_1}$ then $B^{w_1,\ldots,w_r} = 0$.

$\boldsymbol{P_3}$: If $neg.A^\bullet = pari.A^\bullet$, then *sen* essentially commutes with *swap*:

$$\mathrm{swap.sen}(A^\bullet).S^\bullet = -\mathrm{pari.sen}(\mathrm{swap}.A^\bullet).\mathrm{swap}.S^\bullet \tag{5.3}$$
$$= +\mathrm{sen}(\mathrm{pari.swap}.A^\bullet).\mathrm{swap}.S^\bullet \tag{5.4}$$

$\boldsymbol{P_4}$: If $A^\bullet$ is *gantar*-invariant, then $B^\bullet$ is *mantar*-invariant.[79]

$\boldsymbol{P_5}$: If $A^\bullet$ is symmetral, then $B^\bullet$ is alternal.

$\boldsymbol{P_6}$: If $neg.A^\bullet = pari.A^\bullet$ and $A^\bullet$ is bisymmetral, then $B^\bullet$ is bialternal.

[79] We recall that *mantar* := −*pari.anti* and *gantar* := *invmu.pari.anti* with *invmu* denoting *inversion* with respect to the mould product *mu*.

Compact definition of *sen*

Lastly, we may note that, for $A^\bullet$ symmetral, the analytical definition (5.1) of $sen(A^\bullet).S^\bullet$ can be rewritten in compact form as:

$$2\,\mathrm{sen}(A^\bullet).S^\bullet = \mathrm{pushinvar.mut}(\mathrm{pari}.A^\bullet).\mathrm{garit}(A^\bullet).S^\bullet \quad (\forall A^\bullet \in as) \quad (5.5)$$

– with the linear mapping *pushinvar* of $\oplus_r BIMU_r$ onto $\oplus_r BIMU_r^{push}$:

$$\mathrm{pushinvar}.M^\bullet := \sum_{0\le k\le r} \mathrm{push}^k.M^\bullet \quad \text{if} \;\; M^\bullet \in \mathrm{BIMU_r}$$

– with the anti-action $mut(A^\bullet)$ of *MU* on *BIMU*:

$$\mathrm{mut}(A^\bullet).M^\bullet := \mathrm{mu}(A_*^\bullet, M^\bullet, A^\bullet) \quad \text{with} \quad A_*^\bullet = \mathrm{invmu}(A^\bullet) \qquad (5.6)$$

– with the anti-action $garit(A^\bullet)$ of *GARI* on *BIMU*, which is given by 2.37 but simplifies when $M^\bullet$ is of length 1:

$$(\mathrm{garit}(A^\bullet).M)^{\boldsymbol{w}} = \sum_{\boldsymbol{w}^1 w_2 \boldsymbol{w}^3 = \boldsymbol{w}} A^{\boldsymbol{w}^1\rfloor} M^{\lceil w_2 \rceil} A_*^{\lfloor \boldsymbol{w}^3} \quad \text{if } M^\bullet \in \mathrm{BIMU_1}. \quad (5.7)$$

(Pay attention to the position of $A_*^\bullet$ on the *left* in the definition of $mut(A^\bullet)$ and on the *right* in that of $garit(A^\bullet)$. Nonetheless, we have *anti-actions* in both cases.)

5.2 Universal singulators *senk*($\mathfrak{ess}^\bullet$) and *seng*($\mathfrak{es}^\bullet$)

Let $\mathfrak{E}$ be the universal (exact) flexion unit, and let $\mathfrak{es}^\bullet$ (respectively $\mathfrak{ess}^\bullet$) be the primary (respectively secondary) bimould attached to $\mathfrak{E}$. Further, let us set:

$$\mathrm{neginvar} := \mathrm{id} + \mathrm{neg} \qquad (5.8)$$

$$\mathrm{pushinvar} := \sum_{0\le r} (\mathrm{id} + \mathrm{push} + \mathrm{push}^2 + \ldots \mathrm{push}^r).\mathrm{leng}_r \qquad (5.9)$$

(with $leng_r$ denoting the projector from *BIMU* onto $BIMU_r$) and let us define *mut* as in (5.6) above, and *ganit*, *garit*, *adari* [80] as in Section 2.2.

One can then prove that the following two identities define one and the same operator $senk(\mathfrak{ess}^\bullet)$:

$$2\,\mathrm{senk}(\mathfrak{ess}^\bullet).S^\bullet := \mathrm{neginvar}.(\mathrm{adari}(\mathfrak{ess}^\bullet))^{-1}.\mathrm{mut}(\mathfrak{es}^\bullet).S^\bullet \qquad (5.10)$$

$$2\,\mathrm{senk}(\mathfrak{ess}^\bullet).S^\bullet := \mathrm{pushinvar.mut}(\mathrm{neg}.\mathfrak{ess}^\bullet).\mathrm{garit}(\mathfrak{ess}^\bullet).S^\bullet \qquad (5.11)$$

[80] *Adari* alone is an action; all the others are anti actions.

and, likewise, that the following two identities define one and the same operator $seng(\mathfrak{es}^\bullet)$:

$$2\,\mathrm{seng}(\mathfrak{es}^\bullet).S^\bullet := \big(\mathrm{id} + \mathrm{neg.adari}(\mathfrak{es}^\bullet)\big).\mathrm{mut}(\mathfrak{es}^\bullet).S^\bullet \tag{5.12}$$

$$\begin{aligned} 2\,\mathrm{seng}(\mathfrak{es}^\bullet).S^\bullet := &\ \mathrm{mut}(\mathfrak{es}^\bullet).S^\bullet + \mathrm{garit}(\mathfrak{es}^\bullet).\mathrm{neg}.S^\bullet \\ &-\mathrm{arit}\big(\mathrm{garit}(\mathfrak{es}^\bullet).\mathrm{neg}.S^\bullet\big).\mathrm{logari}(\mathfrak{es}^\bullet). \end{aligned} \tag{5.13}$$

The next identity shows how the two basic singulators $senk(\mathfrak{ess}^\bullet)$ and $seng(\mathfrak{es}^\bullet)$ are related; and the other two describe their near-commutation with the basic involution *swap*.

$$\mathrm{seng}(\mathfrak{es}^\bullet) \equiv \mathrm{adari}(\mathfrak{ess}^\bullet).\mathrm{senk}(\mathfrak{ess}^\bullet) \tag{5.14}$$

$$\mathrm{swap.senk}(\mathfrak{ess}^\bullet) \equiv \mathrm{senk}(\mathrm{neg.swap}.\mathfrak{ess}^\bullet).\mathrm{swap} \tag{5.15}$$

$$\mathrm{swap.seng}(\mathfrak{es}^\bullet) \equiv \mathrm{ganit}(\mathrm{syap}.\mathfrak{e}\mathfrak{z}^\bullet).\mathrm{seng}(\mathrm{syap}.\mathfrak{es}^\bullet).\mathrm{neg.swap}. \tag{5.16}$$

Thus, basically, under the impact of the involution *swap*, the inner argument of the singulators also undergoes an involution, namely *neg.swap* in the case of *senk*, and *syap* in the case of *seng*.

Without going into tedious details, let us point out that most of the properties listed above follow:
(i) from the the properties of *sen* (see Section 4.1);
(ii) from the fact that $senk(\mathfrak{ess}^\bullet).S^\bullet$, as defined by (5.11), is none other than $sen(\mathfrak{ess}^\bullet).S^\bullet$, as defined by (5.1) or (5.5);[81]
(iii) from the following identity, valid for any *push*-invariant bimould $M^\bullet$:

$$\mathrm{swap.adari}(\mathfrak{ess}^\bullet).M^\bullet \equiv \mathrm{ganit}(\mathrm{syap}.\mathfrak{e}\mathfrak{z}^\bullet).\mathrm{adari}(\mathrm{swap}.\mathfrak{ess}^\bullet).\mathrm{swap}.M^\bullet. \tag{5.17}$$

5.3 Properties of the universal singulators

The singulators $senk(\mathfrak{ess}^\bullet)$ and $seng(\mathfrak{es}^\bullet)$ do not yield remarkable results when acting on general bimoulds of *BIMU*, but they turn bimoulds of $BIMU_1$ into dimorphic bimoulds of type $\underline{al}/\underline{al}$ and $\underline{al}/\underline{\mathfrak{ol}}$ respectively. Thus:

$$\mathrm{senk}(\mathfrak{ess}^\bullet).S^\bullet \in \mathrm{ARI}^{\underline{al}/\underline{al}} \qquad \forall S^\bullet \in \mathrm{BIMU}_1 \tag{5.18}$$

$$\mathrm{seng}(\mathfrak{es}^\bullet).S^\bullet \in \mathrm{ARI}^{\underline{al}/\underline{\mathfrak{ol}}} \qquad \forall S^\bullet \in \mathrm{BIMU}_1. \tag{5.19}$$

For $senk(\mathfrak{ess}^\bullet)$, this follows from $senk(\mathfrak{ess}^\bullet) = sen(\mathfrak{ess}^\bullet)$ (because $\mathfrak{ess}^\bullet$ is symmetral, indeed bisymmetral) and then from (5.15). For $seng(\mathfrak{es}^\bullet)$, this follows from (5.14) or (5.16), on choice.

[81] Hint: use the fact that $\mathfrak{ess}^\bullet$ is on the one hand invariant under *pari.neg* and on the other of alternal (even bialternal) type, so that $invmu.\mathfrak{ess}^\bullet = pari.anti.\mathfrak{ess}^\bullet$.

These two operators, however, are in a sense too 'global'. To really generate all possible *'dimorphic derivatives'* of bimoulds $S^\bullet$ in $BIMU_1$, we need to split $senk(\mathfrak{ess}^\bullet)$ and $seng(\mathfrak{es}^\bullet)$ into separate components with the help of the projectors $leng_r$ of *BIMU* onto $BIMU_r$.

$$\mathrm{senk}(\mathfrak{ess}^\bullet)=\sum_{1\le r}\mathrm{senk}_r(\mathfrak{ess}^\bullet) \tag{5.20}$$

$$\mathrm{seng}(\mathfrak{es}^\bullet)=\sum_{1\le r}\mathrm{seng}_r(\mathfrak{ess}^\bullet)\quad (\text{mark : first } \mathfrak{es}^\bullet,\ \text{then } \mathfrak{ess}^\bullet\ !) \tag{5.21}$$

with

$$\mathrm{senk}_r(\mathfrak{ess}^\bullet):=\mathrm{leng}_r.\mathrm{senk}(\mathfrak{ess}^\bullet) \tag{5.22}$$

$$\mathrm{seng}_r(\mathfrak{ess}^\bullet):=\mathrm{adari}(\mathfrak{ess}^\bullet).\mathrm{senk}_r(\mathfrak{ess}^\bullet) \tag{5.23}$$

$$=\mathrm{adari}(\mathfrak{ess}^\bullet).\mathrm{leng}_r.\mathrm{adari}(\mathfrak{ess}^\bullet)^{-1}.\mathrm{seng}(\mathfrak{es}^\bullet). \tag{5.24}$$

Although the decomposition runs on different lines[82] in both cases, the resulting components share the same dimorphy-inducing properties:

$$\mathrm{senk}_r(\mathfrak{ess}^\bullet).S^\bullet \in \mathrm{ARI}^{\underline{\mathrm{al}}/\underline{\mathrm{al}}}\cap \mathrm{BIMU}_r \qquad \forall S^\bullet \in \mathrm{BIMU}_1 \tag{5.25}$$

$$\mathrm{seng}_r(\mathfrak{ess}^\bullet).S^\bullet \in \mathrm{ARI}^{\underline{\mathrm{al}}/\underline{\mathrm{ol}}}\cap \mathrm{BIMU}_{r\le} \qquad \forall S^\bullet \in \mathrm{BIMU}_1 \tag{5.26}$$

with

$$\mathrm{BIMU}_{r\le} := \oplus_{r\le r'}\ \mathrm{BIMU}_{r'}. \tag{5.27}$$

But beware: the r-indexation is slightly confusing since, as an operator acting on $BIMU_1$, $\mathrm{senk}_r(\mathfrak{ess}^\bullet)$ is $(r-1)$-linear in $\mathfrak{E}$. Moreover, $\mathfrak{E}^{w_1}$ is odd in w_1. As a consequence, $\mathrm{senk}_r(\mathfrak{ess}^\bullet).S^\bullet$ and therefore $\mathrm{seng}_r(\mathfrak{ess}^\bullet).S^\bullet$ automatically vanish in exactly two cases: when S^{w_1} and r are both *even* or both *odd*.[83]

Dimorphic elements in the monogenous algebra $Flex(\mathfrak{E})$

The above results also apply, of course, within $Flex(\mathfrak{E})$, but since the only singuland in $Flex_1(\mathfrak{E})$ is, up to scalar multiplication, the unit $\mathfrak{E}^\bullet$, which is *odd*, we only get bialternal singulates in $Flex_{2r}(\mathfrak{E})$. Moreover, the singulate in $Flex_2(\mathfrak{E})$ vanishes, because it essentially reduces to $oddari(\mathfrak{E}^\bullet, \mathfrak{E}^\bullet)$ (see 2.80). To sum up:

$$\mathrm{senk}_{2r-1}(\mathfrak{ess}^\bullet).\mathfrak{E}^\bullet=0\quad \forall r;\qquad \mathrm{senk}_2(\mathfrak{ess}^\bullet).\mathfrak{E}^\bullet=0 \tag{5.28}$$

$$\mathrm{senk}_{2r}(\mathfrak{ess}^\bullet).\mathfrak{E}^\bullet \in \mathrm{ARI}^{\underline{\mathrm{al}}/\underline{\mathrm{al}}}\cap \mathrm{Flex}_{2r}(\mathfrak{E})\ \text{ and }\ \neq 0\ \ \textit{if}\ \ r\ge 2. \tag{5.29}$$

[82] The components $seng_r(\mathfrak{ess}^\bullet)$ fully depend on $\mathfrak{ess}^\bullet$ whereas the global operator $seng(\mathfrak{es}^\bullet)$ only depends on $\mathfrak{es}^\bullet = slash.\mathfrak{ess}^\bullet$.

[83] In the obvious sense: *i.e.* H^{w_1} as a function of w_1, and r as an integer.

5.4 Polar singulators: description and properties

There is little point in considering the unit specialisation $\mathfrak{E} \mapsto Pi$, since it leads to the symmetry types $\underline{al}/\underline{ul}$ and $\underline{as}/\underline{us}$ which, as already pointed out, are not compatible with entireness. That leaves the specialisation $\mathfrak{E} \mapsto Pa$ and the symmetry types $\underline{al}/\underline{il}$ and $\underline{as}/\underline{is}$ that go with it. For the bisymmetral bimould, it induces the straightforward specialisation $\mathfrak{ess}^\bullet \mapsto par^\bullet$, but instead of $par^\bullet$ we may also consider $pal^\bullet$, which in fact turns out to be more convenient. This, however, has no impact on the specialisation *sang* of $seng(\mathfrak{es}^\bullet) = seng(slash.\mathfrak{ess}^\bullet)$ since $slash.par^\bullet = slash.pal^\bullet = paj^\bullet$. The definitions of Section 4.2 become:

$$2\,\mathrm{sang}.S^\bullet := \big(\mathrm{id} + \mathrm{neg.adari}(\mathrm{paj}^\bullet)\big).\mathrm{mut}(\mathrm{paj}^\bullet).S^\bullet \tag{5.30}$$

$$\begin{aligned} &= \mathrm{mut}(\mathrm{paj}^\bullet).S^\bullet + \mathrm{garit}(\mathrm{paj}^\bullet).\mathrm{neg}.S^\bullet \\ &\quad -\mathrm{arit}\big(\mathrm{garit}(\mathrm{paj}^\bullet).\mathrm{neg}.S^\bullet\big).\mathrm{logari}(\mathrm{paj}^\bullet) \end{aligned} \tag{5.31}$$

and the equivalence between these two definitions is relatively easy to check, based on the fact that the bimoulds $vipaj^\bullet$ and $vimupaj^\bullet$ thus defined:

$$\mathrm{vipaj}^\bullet := \mathrm{adari}(\mathrm{paj}^\bullet).\mathrm{paj}^\bullet \;,\quad \mathrm{vimupaj}^\bullet := \mathrm{adari}(\mathrm{paj}^\bullet).\mathrm{mupaj}^\bullet$$

admit the following expressions:

$$\begin{aligned} \mathrm{vipaj}^{w_1,\dots,w_r} &= (-1)^{r-1}\mathrm{mupaj}^{w_1,\dots,w_{r-1}}\, P(u_1 + \ldots + u_r) \\ \mathrm{vimupaj}^{w_1,\dots,w_r} &= (-1)^{r} \quad \mathrm{paj}^{w_2,\dots,w_r}\, P(u_1 + \ldots + u_r). \end{aligned}$$

This in turn enables us to recast definition (5.31) in more direct form:

$$\begin{aligned} 2\,(\mathrm{sang}.S)^{\boldsymbol{w}} &= + \sum_{\boldsymbol{a}\, w_i\, \boldsymbol{b} = \boldsymbol{w}} \mathrm{mupaj}^{\boldsymbol{a}}\, S^{w_i}\, \mathrm{paj}^{\boldsymbol{b}} \\ &\quad + \sum_{\boldsymbol{a}\, w_i\, \boldsymbol{b} = \boldsymbol{w}} \mathrm{paj}^{\boldsymbol{a}\rfloor}\, (\mathrm{neg}.S)^{\lceil w_i \rceil}\, \mathrm{mupaj}^{\lfloor \boldsymbol{b}} \\ &\quad + \sum_{\boldsymbol{a}\, w_i\, \boldsymbol{b}\, w_r = \boldsymbol{w}} \mathrm{paj}^{\boldsymbol{a}\rfloor}\, (\mathrm{neg}.S)^{\lceil w_i \rceil}\, \mathrm{mupaj}^{\lfloor \boldsymbol{b}}\, P(|\boldsymbol{u}|) \\ &\quad - \sum_{w_1\, \boldsymbol{a}\, w_i\, \boldsymbol{b} = \boldsymbol{w}} \mathrm{paj}^{\boldsymbol{a}\rfloor}\, (\mathrm{neg}.S)^{\lceil w_i \rceil}\, \mathrm{mupaj}^{\lfloor \boldsymbol{b}}\, P(|\boldsymbol{u}|). \end{aligned}$$

For the singulator $senk(\mathfrak{ess}^\bullet)$, however, we get two distinct specialisations *slank* and *srank*, based respectively on $pal^\bullet$ and $par^\bullet$:

$$2\,\mathrm{slank}.S^\bullet := \mathrm{neginvar}.(\mathrm{adari}(\mathrm{pal}^\bullet))^{-1}.\mathrm{mut}(\mathrm{pal}^\bullet).S^\bullet \tag{5.32}$$

$$= \mathrm{pushinvar.mut}(\mathrm{neg.pal}^\bullet).\mathrm{garit}(\mathrm{pal}^\bullet).S^\bullet \tag{5.33}$$

$$2\,\mathrm{srank}.S^\bullet := \mathrm{neginvar}.(\mathrm{adari}(\mathrm{par}^\bullet))^{-1}.\mathrm{mut}(\mathrm{par}^\bullet).S^\bullet \tag{5.34}$$

$$= \mathrm{pushinvar.mut}(\mathrm{neg.par}^\bullet).\mathrm{garit}(\mathrm{par}^\bullet).S^\bullet. \tag{5.35}$$

Both *slank* and *srank* relate to *sang* under the predictable formulas:

$$\text{sang} = \text{adari}(\text{pal}^\bullet).\text{slank} = \text{adari}(\text{par}^\bullet).\text{srank} \tag{5.36}$$

and both *slank* and *srank* (respectively *sang*) turn arbitrary singulands $S^\bullet \in BIMU_1$ into dimorphic singulates of type $\underline{al}/\underline{al}$ (respectively $\underline{al}/\underline{il}$).

5.5 Simple polar singulators

The polar singulators, like their universal models, have to be broken down into their constituent parts. For *slank* and *srank*, the formulas are straightforward:

$$\text{slank}_r := \text{leng}_r.\text{slank} \tag{5.37}$$

$$\text{srank}_r := \text{leng}_r.\text{srank}. \tag{5.38}$$

For *sang*, the decomposition is more roundabout, and depends on the choice of either $pal^\bullet$ or $par^\bullet$:

$$\text{slang}_r := \text{adari}(\text{pal}^\bullet).\text{leng}_r.\big(\text{adari}(\text{pal}^\bullet)\big)^{-1}.\text{sang} \tag{5.39}$$

$$= \text{adari}(\text{pal}^\bullet).\text{slank}_r \;\neq\; \text{leng}_r.\text{sang} \tag{5.40}$$

$$\text{srang}_r := \text{adari}(\text{par}^\bullet).\text{leng}_r.\big(\text{adari}(\text{par}^\bullet)\big)^{-1}.\text{sang} \tag{5.41}$$

$$= \text{adari}(\text{par}^\bullet).\text{srank}_r \;\neq\; \text{leng}_r.\text{sang}. \tag{5.42}$$

Thus, despite the similar-looking identities

$$\text{slank} = \sum_{r\geq 1}\text{slank}_r, \quad \text{srank} = \sum_{r\geq 1}\text{srank}_r, \quad \text{sang} = \sum_{r\geq 1}\text{slang}_r = \sum_{r\geq 1}\text{srang}_r$$

there is no way we can avoid *secondary* bimoulds (in this case, the bisymmetral $pal^\bullet$ or $par^\bullet$) even in the decomposition of the 'primary-looking' singulator *sang*.

5.6 Composite polar singulators

To produce all possible dimorphic singularities, we require not just the singulator components, but also their Lie brackets. For reasons that shall be spelt out in Section 4.7, we settle for the choice $pal^\bullet$ and the corresponding singulators, and we set, for any arguments $S_1^\bullet, \ldots, S_l^\bullet$ in $BIMU_1$:

$$\text{slank}_{[r_1,\ldots,r_l]} \cdot \text{mu}(S_1^\bullet, \ldots, S_l^\bullet) := \text{ari}(\text{slank}_{r_1}.S_1^\bullet, \ldots, \text{slank}_{r_l}.S_l^\bullet) \in \text{ARI}_r^{\underline{\text{al}}/\underline{\text{al}}}$$

$$\text{slang}_{[r_1,\ldots,r_l]} \cdot \text{mu}(S_1^\bullet, \ldots, S_l^\bullet) := \text{ari}(\text{slang}_{r_1}.S_1^\bullet, \ldots, \text{slang}_{r_l}.S_l^\bullet) \in \text{ARI}_{r\leq}^{\underline{\text{al}}/\underline{\text{il}}}$$

with $r := r_1 + \ldots + r_l$ and of course:

$$\mathrm{ARI}_r^{\underline{\mathrm{al}}/\underline{\mathrm{al}}} := \mathrm{ARI}^{\underline{\mathrm{al}}/\underline{\mathrm{al}}} \cap \mathrm{BIMU}_r; \quad \mathrm{ARI}_{r\leq}^{\underline{\mathrm{al}}/\underline{\mathrm{il}}} := \mathrm{ARI}^{\underline{\mathrm{al}}/\underline{\mathrm{il}}} \cap \left(\oplus_{r\leq r'} \mathrm{BIMU}_{r'} \right)$$

and with the multiple *ari*-braket defined from left to right. By multilinearity, the above actions extend to mappings:

$$\mathrm{slank}_{[r_1,\ldots,r_l]} : S^\bullet \mapsto \Sigma^\bullet; \ \mathrm{BIMU}_l \to \mathrm{ARI}_r^{\underline{\mathrm{al}}/\underline{\mathrm{al}}} \tag{5.43}$$

$$\mathrm{slang}_{[r_1,\ldots,r_l]} : S^\bullet \mapsto \Sigma^\bullet; \ \mathrm{BIMU}_l \to \mathrm{ARI}_{r\leq}^{\underline{\mathrm{al}}/\underline{\mathrm{il}}}. \tag{5.44}$$

It is sometimes convenient, nay indispensable,[84] to consider also the pre-Lie brackets of the singulator components. The formulas read:

$$\mathrm{slank}_{r_1,\ldots,r_l} . \mathrm{mu}(S_1^\bullet, \ldots, S_l^\bullet) := \mathrm{preari}(\mathrm{slank}_{r_1}.S_1^\bullet, \ldots, \mathrm{slank}_{r_l}.S_l^\bullet) \tag{5.45}$$

$$\mathrm{slang}_{r_1,\ldots,r_l} . \mathrm{mu}(S_1^\bullet, \ldots, S_l^\bullet) := \mathrm{preari}(\mathrm{slang}_{r_1}.S_1^\bullet, \ldots, \mathrm{slang}_{r_l}.S_l^\bullet) \tag{5.46}$$

with the multiple *pre-ari*-braket defined again from left to right, as in (2.49). By multilinearity, the above actions extend to mappings:

$$\mathrm{slank}_{r_1,\ldots,r_l} : S^\bullet \mapsto \Sigma^\bullet; \ \mathrm{BIMU}_l \to \mathrm{ARI}_r^{\mathrm{al}} \tag{5.47}$$

$$\mathrm{slang}_{r_1,\ldots,r_l} : S^\bullet \mapsto \Sigma^\bullet; \ \mathrm{BIMU}_l \to \mathrm{ARI}_{r\leq}^{\mathrm{al}}. \tag{5.48}$$

Here, the resulting singulates $\Sigma^\bullet$ are of course alternal, but their *swappees* exhibit no distinctive symmetry. In practical applications, however, these multiple singulators based on *preari* always occur in sums $\sum Q^\bullet slank_\bullet$ or $\sum Q^\bullet slang_\bullet$, with scalar moulds $Q^\bullet$ that are alternal (respectively symmetral), and these new composite operators *do* produce dimorphy: they turn arbitrary singulands $S^\bullet$ into singulates $\Sigma^\bullet$ of type $\underline{al}/\underline{al}$ or $\underline{al}/\underline{il}$ (respectively $\underline{as}/\underline{as}$ or $\underline{as}/\underline{is}$).

5.7 From $\underline{al}/\underline{al}$ to $\underline{al}/\underline{il}$. Nature of the singularities

The reason for preferring the singulator *slank* (built from $pal^\bullet$) to the singulator *srank* (built from $par^\bullet$) is that it leads to simpler denominators. Indeed, for a singuland S^{w_1} regular at the origin and 'random', although the bialternal singulates $slank_r.S^{\boldsymbol{w}}$ and $srank_r.S^{\boldsymbol{w}}$, as functions of $\boldsymbol{w} = (w_1, \ldots, w_r)$, have both multipoles of order $r-1$ at the origin, the total number of factors differs sharply. After common denominator reduction,

[84] For example in *perinomal algebra*: see Section 6 and Section 8.

$slank_r.S^{\boldsymbol{w}}$ has only $r+1$ factors on its denominator, whereas $srank_r.S^{\boldsymbol{w}}$ has $r\,(r+1)/2$. More precisely:

$$\text{denom}(\text{slank}_r.S^{\boldsymbol{w}}) = u_0\,u_1\ldots u_{r-1}\,u_r \quad \text{with } u_0 := -(u_1+\cdots+u_r)$$
$$\text{denom}(\text{srank}_r.S^{\boldsymbol{w}}) = \prod_{1\le i\le j\le r}\;\sum_{i\le k\le j} u_k.$$

The results are slightly more complex for the singulates of type $\underline{al}/\underline{il}$, namely $slang_r.S^\bullet$ and $srang_r.S^\bullet$, since these, as a rule, possess non-vanishing components of any length $r'\ge r$, but here again the first choice leads to simpler denominators.

Another reason for preferring the $pal^\bullet$-based choice to the $par^\bullet$-based one is that $pal^\bullet$ possesses a trigonometric counterpart $tal_c^\bullet$ whereas $par^\bullet$ doesn't.

6 A natural basis for $ALIL \subset ARI^{\underline{al}/\underline{il}}$

6.1 Singulation-desingulation: the general scheme

This section is devoted to the construction of bimoulds $l\o ma^\bullet$ in $ALIL$. In other words:
– $l\o ma^{\boldsymbol{w}}$ should be $\boldsymbol{u}$-entire, *i.e.* in $\mathbb{C}[[u_1,\ldots,u_r]]$;
– $l\o ma^{\boldsymbol{w}}$ should be $\boldsymbol{v}$-constant;
– $l\o ma^\bullet$ should be *alternal*;
– $l\o mi^\bullet := swap.l\o ma^\bullet$ should be *alternil* modulo *Center(ALIL)*.
But we also add two key conditions:
(i) $l\o ma^{\boldsymbol{w}}$ should be in $\mathbb{Q}[[u_1,\ldots,u_r]]$, *i.e.* carry rational Taylor coefficients;
(ii) the first component should be of the form:

$$l\o ma^{w_1} = u_1^2\,(1-u_1^2)^{-1} = u_1^2+u_1^4+u_1^6+u_1^8+\ldots \tag{6.1}$$

Condition (ii) is there to ensure that in the iso-weight decomposition:

$$\text{l\o ma}^\bullet = \text{l\o ma}_3^\bullet + \text{l\o ma}_5^\bullet + \text{l\o ma}_7^\bullet + \text{l\o ma}_9^\bullet + \ldots \tag{6.2}$$

the part $l\o ma_s^\bullet$ of weight s be non-zero[85] and start with $l\o ma_s^{w_1} = u_1^{s-1}$, with the ultimate objective of getting a basis $\{l\o ma_s^\bullet\,;\ s \text{ odd} \ge 3\}$ of $ALIL$.

[85] s is odd ≥ 3. $l\o ma_s^\bullet$ (respectively $l\o mi_s^\bullet$) carries exactly $s-1$ (respectively s) nonzero components of length $r\in[1,s-1]$ (respectively $r\in[1,s]$) and degree $d=s-r$. Indeed, the last components are $l\o ma_s^{w_1,\ldots,w_s}=0$ and $l\o mi_s^{w_1,\ldots,w_s}=1/s$.

The 'central correction' formula reads:

$$\mathrm{lomi}_s^\bullet = \mathrm{swap}(\mathrm{loma}_s^\bullet + \mathrm{Ca}_s^\bullet); \quad \mathrm{Ca}_s^\bullet \in \mathrm{Center(ALIL)} \qquad (6.3)$$

with a central bimould $Ca_s^\bullet$ which, due to condition (6.1), can be shown to be of the form:

$$\mathrm{Ca}_s^{w_1,\dots,w_s} = \frac{1}{s} \ (\forall w_i); \quad \mathrm{Ca}_s^{w_1,\dots,w_r} = 0 \ \text{ if } \ r \neq s \ \ (\forall w_i) \qquad (6.4)$$

Expanding $l\o ma^\bullet$ into series of singulates

Before decomposing $l\o ma_s^\bullet$ weight-by-weight, we must *construct* it as a series of singulates. There are actually two variants:

$$\mathrm{l\o ma}^\bullet = \overbrace{\Sigma^\bullet_{[1]}}^{r\le 2} + \overbrace{\Sigma^\bullet_{[1,2]}}^{r\le 4} + \overbrace{\Sigma^\bullet_{[1,4]} + \Sigma^\bullet_{[2,3]} + \Sigma^\bullet_{[1,1,3]} + \Sigma^\bullet_{[2,1,2]} + \Sigma^\bullet_{[1,1,1,2]}}^{r\le 6} + \dots \qquad (6.5)$$

$$\begin{aligned}\mathrm{l\o ma}^\bullet = &\overbrace{\Sigma^\bullet_{1}}^{r\le 2} + \overbrace{\Sigma^\bullet_{1,2} + \Sigma^\bullet_{2,1}}^{r\le 4} + \overbrace{\Sigma^\bullet_{1,4} + \Sigma^\bullet_{4,1} + \Sigma^\bullet_{2,3} + \Sigma^\bullet_{3,2}}^{r\le 6}\\ &+ \overbrace{\Sigma^\bullet_{1,1,3} + \Sigma^\bullet_{1,3,1} + \Sigma^\bullet_{3,1,1} + \Sigma^\bullet_{2,2,1} + \Sigma^\bullet_{2,1,2} + \Sigma^\bullet_{1,2,2}}^{r\le 6} \qquad (6.6)\\ &+ \overbrace{\Sigma^\bullet_{1,1,1,2} + \Sigma^\bullet_{1,1,2,1} + \Sigma^\bullet_{1,2,1,1} + \Sigma^\bullet_{2,1,1,1}}^{r\le 6} + \dots\end{aligned}$$

with

$$\Sigma^\bullet_{[r_1,\dots,r_l]} := \mathrm{slang}_{[r_1,\dots,r_l]}.S^\bullet_{[r_1,\dots,r_l]} \qquad (6.7)$$

$$\Sigma^\bullet_{r_1,\dots,r_l} := \mathrm{slang}_{r_1,\dots,r_l}.S^\bullet_{r_1,\dots,r_l}. \qquad (6.8)$$

The *singulates* $\Sigma^\bullet_{[r_1,\dots,r_l]}$ are going to be in $\mathrm{ARI}^{\mathrm{al/il}}_{r\le}$ but the *singulates* $\Sigma^\bullet_{r_1,\dots,r_l}$ only in $\mathrm{ARI}^{\mathrm{al}}_{r\le}$. As for the *singulands* $S_{[r_1,\dots,r_l]}$ and $S_{r_1,\dots,r_l}$, they are merely in BIMU_l, but with a definite parity in each x_i, which is exactly opposite to the parity of r_i. Moreover, we can without loss of generality assume that they vanish as soon as one of the x_i's vanishes. Then again, they may be sought either in the form of power series or of meromorphic functions of a quite specific type:

$$S^{x_1,\dots,x_l}_{[r_1,\dots,r_l]} \in x_1^{\nu_1}\dots x_l^{\nu_l}\,\mathbb{C}[[x_1^2,\dots,x_l^2]] \qquad \text{(power series)} \qquad (6.9)$$

$$S^{x_1,\dots,x_l}_{[r_1,\dots,r_l]} = \sum_{n_i\in\mathbb{Z}^*} R^{n_1,\dots,n_l}_{[r_1,\dots,r_l]} P(n_1+x_1)\dots P(n_l+x_l) \qquad \text{(merom. funct.)} \qquad (6.10)$$

with $\nu_i = 1$ (respectively 2) if r_i is even (respectively odd).

Both expansions (6.5) and (6.6) lead to the same results. The first expansion (6.5) relies on *ari*-brackets and has the advantage of involving fewer summands. The downside is that it forces us to choose a basis in the Lie algebra generated by the *simple* singulates $\Sigma^{\bullet}_{r_i}$ and that there exist no clear canonical choices for such bases. This arbitrariness, though, manifests only during the construction and doesn't show in the final result.

The second expansion (6.6) relies on *pre-ari*-brackets, and here the position is exactly the reverse: we have uniqueness and canonicity at every construction step, but more numerous summands.

Altogether, the *ari*-expansion is to be preferred in calculations, whereas the *pre-ari*-expansion is theoretically more appealing. In perinomal algebra, its use will even become mandatory (see Section 9). In any case, the conversion rules for changing from the one to the other are simple enough. Thus, up to length $r = 5$, we find:

$$S_{1,2}^{x_1,x_2} = +S_{[1,2]}^{x_1,x_2}; \quad S_{2,1}^{x_1,x_2} = -S_{[1,2]}^{x_2,x_1}$$

$$S_{1,4}^{x_1,x_2} = +S_{[1,4]}^{x_1,x_2}; \quad S_{4,1}^{x_1,x_2} = -S_{[1,4]}^{x_2,x_1}$$

$$S_{2,3}^{x_1,x_2} = +S_{[2,3]}^{x_1,x_2}; \quad S_{3,2}^{x_1,x_2} = -S_{[3,2]}^{x_2,x_1}$$

$$S_{1,1,3}^{x_1,x_2,x_3} = +S_{[1,1,3]}^{x_1,x_2,x_3}; \quad S_{1,3,1}^{x_1,x_2,x_3} = -S_{[1,1,3]}^{x_1,x_3,x_2} - S_{[1,1,3]}^{x_3,x_1,x_2}; \quad S_{3,1,1}^{x_1,x_2,x_3} = +S_{[1,1,3]}^{x_3,x_2,x_1}$$

$$S_{2,2,1}^{x_1,x_2,x_3} = -S_{[2,1,2]}^{x_1,x_3,x_2}; \quad S_{2,1,2}^{x_1,x_2,x_3} = +S_{[2,1,2]}^{x_1,x_2,x_3} + S_{[2,1,2]}^{x_3,x_2,x_1}; \quad S_{1,2,2}^{x_1,x_2,x_3} = -S_{[2,1,2]}^{x_3,x_1,x_2}$$

$$S_{1,1,1,2}^{x_1,x_2,x_3,x_4} = +S_{[1,1,1,2]}^{x_1,x_2,x_3,x_4}$$

$$S_{1,1,2,1}^{x_1,x_2,x_3,x_4} = -S_{[1,1,1,2]}^{x_1,x_2,x_4,x_3} - S_{[1,1,1,2]}^{x_1,x_4,x_2,x_3} - S_{[1,1,1,2]}^{x_4,x_1,x_2,x_3}$$

$$S_{1,2,1,1}^{x_1,x_2,x_3,x_4} = +S_{[1,1,1,2]}^{x_1,x_4,x_3,x_2} + S_{[1,1,1,2]}^{x_4,x_1,x_3,x_2} + S_{[1,1,1,2]}^{x_4,x_3,x_1,x_2}$$

$$S_{2,1,1,1}^{x_1,x_2,x_3,x_4} = -S_{[1,1,1,2]}^{x_4,x_3,x_2,x_1}.$$

In the above table, as indeed throughout the sequel, we write down only the upper indices of the singulands (since, in the *colourless case* with which we are concerned here, the lower indices don't matter). Moreover, we write these upper indices of the singulands as "x_i" rather than "u_i", the better to bring out their independence from the u_i's that serve as upper indices for the singulates. Indeed, when expressing the *entireness condition* for the sums of singulates (see Section 6.3, Section 6.4 below), we may work either with $\Theta^{\bullet}_{r_*}$ itself or *swap*.$\Theta^{\bullet}_{r_*}$, and the distinct but equivalent constraints on the singulands which both approaches yield look much the same – all of which suggests that the singulands that go into the making of *løma*$^{\bullet}$ stand, in a sense, halfway between that bimould and its swappee *lømi*$^{\bullet}$.

Singulation-desingulation[86]

In keeping with the above remarks, we may (and shall), without loss of generality, limit ourselves to singulands $S^{x_1,\dots,x_l}_{[r_1,\dots,r_l]}$ and $S^{x_1,\dots,x_l}_{r_1,\dots,r_l}$ that are *even* (respectively *odd*) in each x_i if the corresponding index r_i is *odd* (respectively *even*). We may also (and shall), again without loss of generality, impose divisibility by $x_1 \dots x_l$.[87]

The construction of $l\o ma^{\boldsymbol{w}}$ is by induction, and goes like this.

Fix any odd integer r_* and assume we have already found singulates $\Sigma^{\bullet}_{[\boldsymbol{r}]}$ or $\Sigma^{\bullet}_{\boldsymbol{r}}$ of total index $|\boldsymbol{r}| := \sum r_i$ *odd* and $\leq r_*$, such that the truncated expansion:

$$\Theta^{\bullet}_{r_*} := \sum_{|\boldsymbol{r}|\leq r_*} \Sigma^{\bullet}_{[\boldsymbol{r}]} = \sum_{|\boldsymbol{r}|\leq r_*} \Sigma^{\bullet}_{\boldsymbol{r}} \tag{6.11}$$

has only *entire* components for all lengths $r \leq r_*$. One can then show the following:

(i) the component of $\Theta^{\bullet}_{r_*}$ of (even) length $1+r_*$ is automatically entire;

(ii) the component of $\Theta^{\bullet}_{r_*}$ of (odd) length $2+r_*$ is not entire, but possesses mulipoles of order r_* at the origin;

(iii) it is always possible to pick singulands $S^{\bullet}_{[\boldsymbol{r}]}$ or $S^{\bullet}_{\boldsymbol{r}}$ of total index $|\boldsymbol{r}| = 2+r_*$ and such that the corresponding singulates $\Sigma^{\bullet}_{[\boldsymbol{r}]}$ or $\Sigma^{\bullet}_{\boldsymbol{r}}$ exactly compensate the multipoles mentioned in (ii), so that the truncated sum $\Theta^{\bullet}_{2+\boldsymbol{r}}$ will coincide with $\Theta^{\bullet}_{\boldsymbol{r}}$ for all its components of length $r \leq 1+r_*$ but will have a singularity-free component of length $r = 2+r_*$;

(iv) the constraints on the newly added singulates are found by writing down, successively, the conditions for multipoles of order r_*, r_*-1, r_*-2 etc to be absent from the component $\Theta^{w_1,\dots,w_{2+r_*}}_{2+r_*}$;

(v) these constraints do not exactly determine the new singulates, but *very nearly so*[88], and in any case there exist two (closely related) privileged choices, leading to two closely related, canonical choices $lama^{\bullet}$, $loma^{\bullet}$ for $l\o ma^{\boldsymbol{w}}$;

(vi) there is also a third choice, $luma^{\bullet}$, whose components aren't sought in the ring of power series in $\boldsymbol{u}$ but rather in the space of meromorphic functions of $\boldsymbol{u}$, with multipoles located at the multiintegers $\boldsymbol{n}$, and with

[86] We prefer this pair to the unwieldy *singularisation-desingularisation* not just for reasons of euphony, but also to keep close to the coinages *singulator, singuland, singulate*.

[87] The reason being that to a constant singuland $S^{w_1}_{r_1} \equiv 1$ there always answers a vanishing singulate $\Sigma^{\bullet}_{r_1} \equiv 0$.

[88] In the sense that the *wandering bialternals*, which are ultimately responsible for this indeterminacy, are "few and far between". See Section 6.9 and the concluding comments in Section 9.1.

essentially bounded behaviour at infinity:[89]

$$S^{x_1,\dots,x_l}_{[r_1,\dots,r_l]} \overset{\text{ess}^{\text{lly}}}{=} \sum_{n_i \in \mathbb{Z}^*} R^{n_1,\dots,n_l}_{[r_1,\dots,r_l]}\, P(x_1+n_1)\dots P(x_l+n_l) \tag{6.12}$$

$$S^{x_1,\dots,x_l}_{r_1,\dots,r_l} \overset{\text{ess}^{\text{lly}}}{=} \sum_{n_i \in \mathbb{Z}^*} R^{n_1,\dots,n_l}_{r_1,\dots,r_l}\, P(x_1+n_1)\dots P(x_l+n_l). \tag{6.13}$$

Here, the solution $luma^{\bullet}$ turns out to be unique, its search essentially reducing to that of the multiresidues $R^{\boldsymbol{n}}_{[\boldsymbol{r}]}$ or $R^{\boldsymbol{n}}_{\boldsymbol{r}}$ carried by the multipoles of the singulands.[90] These multiresidues are uniquely determined rational numbers, and *perinomal functions*[91] of their argument $\boldsymbol{n}$. So the difficulty here is not the search for a canonical solution, but the elucidation of the arithmetical nature of the Taylor coefficients at the origin of the various components $luma^{\boldsymbol{w}}$, at least for lengths $r(\boldsymbol{w}) \geq 5$, since for lesser lengths the answer is elementary.

6.2 Singulation-desingulation up to length 2

As usual, we set $1/t =: P(t) =: P^t$ throughout, and favour the third variant inside mould equations, for greater visual coherence. At lengths $r \leq 2$, one singuland only contributes to $l\o ma^{\bullet}$. At length 1, both singuland and singulate coincide. At length 2, the formula for the singulate involves poles of order 1, but these cancel out, duly yielding an entire $l\o ma^{w_1,w_2}$.

$$\text{l}\o\text{ma}^{w_1} = \text{l}\o\text{ma}_1^{w_1} = \Sigma^{w_1}_{[1]} = S^{u_1}_{[1]} = u_1^2 + u_1^4 + u_1^6 + u_1^8 + \dots$$

$$\text{l}\o\text{ma}^{w_1,w_2} = \text{l}\o\text{ma}_1^{w_1,w_2} = \Sigma^{w_1,w_2}_{[1]} =$$

$$\frac{1}{2}\,P^{u_1}\,(S^{u_{12}}_{[1]} - S^{u_2}_{[1]}) + \frac{1}{2}\,P^{u_2}\,(S^{u_1}_{[1]} - S^{u_{12}}_{[1]}) + \frac{1}{2}\,P^{u_{12}}\,(S^{u_2}_{[1]} - S^{u_1}_{[1]})$$

89 Away from the multipoles, of course. Exactly what this means shall become clear in in the sequel: see Section 6.7 and Section 9. As for the warning *essentially* stacked over the = sign in the identities (6.12), (6.13), it means that we neglect simple corrective terms (with lower order multiples) that ensure convergence on the right-hand side.

90 These multiresidues $R^{n_1,\dots,n_l}_{[r_1,\dots,r_l]}$ have to be even (respectively odd) in n_i when r_i is even (respectively odd) to ensure that the singulate $S^{x_1,\dots,x_l}_{[r_1,\dots,r_l]}$ be odd (respectively even) when r_i is even (respectively odd).

91 See Section 6.7 and Section 9.

6.3 Singulation-desingulation up to length 4

The condition expressing that $l\phi ma^{w_1,w_2,w_3}$ has no poles of order 1 at the origin involves only the singulands and singulates of indices [1] and [1, 2]. For power series singulands, it reads:

$$\begin{aligned}0 = &+\frac{1}{12}\left(P^{x_2}\, S^{x_{12}}_{[1]} - P^{x_{12}}\, S^{x_2}_{[1]} - P^{x_2}\, S^{x_1}_{[1]} + P^{x_{12}}\, S^{x_1}_{[1]}\right)\\ &+ S^{x_1,x_2}_{[1,2]} + S^{x_2,x_{12}}_{[1,2]} - S^{x_1,x_{12}}_{[1,2]} - S^{x_{12},x_2}_{[1,2]}.\end{aligned} \tag{6.14}$$

For meromorphic singulands (of type (6.12)), it translates into a condition on the multiresidues $R_{[\bullet]}^{\bullet}$, which reads:

$$0=1/12\big(\delta^{n_{12}}\, R^{n_1}_{[1]} - \delta^{n_2}\, R^{n_1}_{[1]}\big) + R^{n_1,n_2}_{[1,2]} - R^{n_1,n_{12}}_{[1,2]} \tag{6.15}$$

$$0=1/12\big(\delta^{n_2}\, R^{n_{12}}_{[1]} - \delta^{n_2}\, R^{n_1}_{[1]} - \delta^{n_{12}}\, R^{n_2}_{[1]}\big) + R^{n_1,n_2}_{[1,2]} + R^{n_2,n_{12}}_{[1,2]} - R^{n_{12},n_2}_{[1,2]}. \tag{6.16}$$

When fulfilled, the above conditions ensure the entireness not just of $l\phi ma^{w_1,\dots,w_3}$ but also of $l\phi ma^{w_1,\dots,w_4}$.

6.4 Singulation-desingulation up to length 6

At this stage of the construction, we are dealing with a component $l\phi ma^{w_1,\dots,w_5}$ that may have multipoles of order 3, 2, 1 at the origin. Expressing that there are no such multipoles of order 3 leads to a single equation:

$$\mathcal{S}_{[1]} + \mathcal{S}_{[1,4]} + \mathcal{S}_{[1,4]} = 0 \tag{6.17}$$

with contributions:

$$\mathcal{S}_{[1]} := +\frac{1}{120}\left(P^{x_2}\, S^{x_{12}}_{[1]} - P^{x_2}\, S^{x_1}_{[1]}\right)$$

$$\mathcal{S}_{[1,4]} := -S^{x_1,x_2}_{[1,4]} + S^{x_{12},x_2}_{[1,4]}$$

$$\mathcal{S}_{[2,3]} := +2\, S^{x_{12},x_2}_{[2,3]} + S^{x_1,x_2}_{[2,3]} - S^{x_1,x_{12}}_{[2,3]} - S^{x_2,x_{12}}_{[2,3]}.$$

We may note that the singulate $S_{[1,2]}$ remains, somewhat surprisingly, uninvolved at this stage.

Next, we must write down the condition for $l\phi ma^{w_1,\dots,w_5}$ to have no multipoles of order 2 at the origin. This again leads to a single equation[92]

[92] This new condition, of course, makes sense, only *modulo* the earlier one, *i.e.* assuming the removal of order 3 multipoles.

that involves all singulands save the last one (*i.e.* $S_{[1,1,1,2]}$):

$$\mathcal{S}^*_{[1]} + \mathcal{S}^*_{[1,2]} + \mathcal{S}^*_{[1,4]} + \mathcal{S}^*_{[2,3]} + \mathcal{S}^*_{[1,1,3]} + \mathcal{S}^*_{[2,1,2]} = 0 \tag{6.18}$$

with contributions:

$$\begin{aligned}
720\,\mathcal{S}^*_{[1]} := & -P^{x_2}\,P^{x_3}\,S^{x_{123}}_{[1]} - P^{x_1}\,P^{x_{23}}\,S^{x_2}_{[1]} + P^{x_1}\,P^{x_{23}}\,S^{x_3}_{[1]} \\
& +4\,P^{x_2}\,P^{x_{23}}\,S^{x_{123}}_{[1]} - 4\,P^{x_2}\,P^{x_{23}}\,S^{x_1}_{[1]} - 4\,P^{x_1}\,P^{x_{123}}\,S^{x_3}_{[1]} \\
& +11\,P^{x_{12}}\,P^{x_{123}}\,S^{x_2}_{[1]} - 11\,P^{x_{12}}\,P^{x_{123}}\,S^{x_1}_{[1]} - 11\,P^{x_1}\,P^{x_{123}}\,S^{x_2}_{[1]} \\
& +14\,P^{x_{12}}\,P^{x_3}\,S^{x_1}_{[1]} - 14\,P^{x_2}\,P^{x_3}\,S^{x_1}_{[1]} - 14\,P^{x_3}\,P^{x_{12}}\,S^{x_2}_{[1]} \\
& -15\,P^{x_1}\,P^{x_3}\,S^{x_{23}}_{[1]} - 15\,P^{x_2}\,P^{x_{123}}\,S^{x_{12}}_{[1]} + 15\,P^{x_2}\,P^{x_{123}}\,S^{x_3}_{[1]} \\
& +15\,P^{x_1}\,P^{x_3}\,S^{x_2}_{[1]} - 15\,P^{x_1}\,P^{x_3}\,S^{x_{12}}_{[1]} + 15\,P^{x_2}\,P^{x_{123}}\,S^{x_1}_{[1]} \\
& +15\,P^{x_1}\,P^{x_{123}}\,S^{x_{12}}_{[1]} + 15\,P^{x_1}\,P^{x_3}\,S^{x_{123}}_{[1]} + 15\,P^{x_2}\,P^{x_3}\,S^{x_{12}}_{[1]} \\
& -15\,P^{x_2}\,P^{x_{123}}\,S^{x_{23}}_{[1]} + 25\,P^{x_3}\,P^{x_{123}}\,S^{x_{23}}_{[1]} - 25\,P^{x_3}\,P^{x_{23}}\,S^{x_{123}}_{[1]} \\
& -25\,P^{x_3}\,P^{x_{123}}\,S^{x_1}_{[1]} + 25\,P^{x_3}\,P^{x_{23}}\,S^{x_1}_{[1]}
\end{aligned}$$

$$\begin{aligned}
12\,\mathcal{S}^*_{[1,2]} := & +2\,P^{x_{123}}\,S^{x_3,x_{23}}_{[1,2]} - 2\,P^{x_{123}}\,S^{x_{23},x_3}_{[1,2]} + 2\,P^{x_{123}}\,S^{x_2,x_3}_{[1,2]} - 2\,P^{x_{123}}\,S^{x_2,x_{23}}_{[1,2]} \\
& -2\,P^{x_3}\,S^{x_{23},x_{123}}_{[1,2]} + 2\,P^{x_3}\,S^{x_{123},x_{23}}_{[1,2]} + 2\,P^{x_3}\,S^{x_1,x_{123}}_{[1,2]} - 2\,P^{x_3}\,S^{x_1,x_{23}}_{[1,2]} \\
& -3\,P^{x_1}\,S^{x_2,x_3}_{[1,2]} + 3\,P^{x_1}\,S^{x_{23},x_3}_{[1,2]} + 3\,P^{x_1}\,S^{x_2,x_{23}}_{[1,2]} - 3\,P^{x_1}\,S^{x_3,x_{23}}_{[1,2]} \\
& +3\,P^{x_1}\,S^{x_3,x_{123}}_{[1,2]} - 3\,P^{x_1}\,S^{x_{123},x_3}_{[1,2]} - 3\,P^{x_1}\,S^{x_{12},x_{123}}_{[1,2]} + 3\,P^{x_1}\,S^{x_{12},x_3}_{[1,2]} \\
& +3\,P^{x_2}\,S^{x_1,x_{23}}_{[1,2]} + 3\,P^{x_2}\,S^{x_{123},x_3}_{[1,2]} + 3\,P^{x_2}\,S^{x_{23},x_{123}}_{[1,2]} - 3\,P^{x_2}\,S^{x_1,x_{123}}_{[1,2]} \\
& -3\,P^{x_2}\,S^{x_{12},x_3}_{[1,2]} - 3\,P^{x_2}\,S^{x_3,x_{123}}_{[1,2]} + 3\,P^{x_2}\,S^{x_{12},x_{123}}_{[1,2]} - 3\,P^{x_2}\,S^{x_{123},x_{23}}_{[1,2]}
\end{aligned}$$

$$\begin{aligned}
12\,\mathcal{S}^*_{[1,4]} := & -2\,P^{x_{123}}\,S^{x_3,x_{23}}_{[1,4]} + 2\,P^{x_{123}}\,S^{x_1,x_3}_{[1,4]} - 2\,P^{x_{123}}\,S^{x_2,x_3}_{[1,4]} + 2\,P^{x_{123}}\,S^{x_2,x_{23}}_{[1,4]} \\
& -2\,P^{x_{23}}\,S^{x_1,x_3}_{[1,4]} - 2\,P^{x_{23}}\,S^{x_2,x_{123}}_{[1,4]} + 2\,P^{x_{23}}\,S^{x_3,x_{123}}_{[1,4]} + 2\,P^{x_{23}}\,S^{x_{123},x_3}_{[1,4]} \\
& +2\,P^{x_3}\,S^{x_1,x_{23}}_{[1,4]} - 2\,P^{x_3}\,S^{x_1,x_{123}}_{[1,4]} + 2\,P^{x_3}\,S^{x_2,x_{123}}_{[1,4]} - 2\,P^{x_3}\,S^{x_{123},x_{23}}_{[1,4]} \\
& +3\,P^{x_{12}}\,S^{x_1,x_3}_{[1,4]} - 3\,P^{x_{12}}\,S^{x_2,x_3}_{[1,4]} + 3\,P^{x_{12}}\,S^{x_1,x_{123}}_{[1,4]} - 3\,P^{x_{12}}\,S^{x_2,x_{123}}_{[1,4]} \\
& -3\,P^{x_1}\,S^{x_3,x_{23}}_{[1,4]} + 3\,P^{x_1}\,S^{x_2,x_{23}}_{[1,4]} - 3\,P^{x_1}\,S^{x_3,x_{123}}_{[1,4]} + 3\,P^{x_1}\,S^{x_2,x_{123}}_{[1,4]} \\
& -3\,P^{x_2}\,S^{x_1,x_3}_{[1,4]} - 3\,P^{x_2}\,S^{x_1,x_{23}}_{[1,4]} + 3\,P^{x_2}\,S^{x_{123},x_3}_{[1,4]} + 3\,P^{x_2}\,S^{x_{123},x_{23}}_{[1,4]}
\end{aligned}$$

$$
\begin{aligned}
12\,\mathcal{S}^{*}_{[2,3]}:=&-P^{x_{123}}\,S^{x_3,x_1}_{[2,3]}+P^{x_{23}}\,S^{x_3,x_1}_{[2,3]}+P^{x_{23}}\,S^{x_{123},x_2}_{[2,3]}-P^{x_3}\,S^{x_{123},x_2}_{[2,3]}\\
&+2\,P^{x_{23}}\,S^{x_3,x_{123}}_{[2,3]}+2\,P^{x_{23}}\,S^{x_{123},x_3}_{[2,3]}-3\,P^{x_{123}}\,S^{x_1,x_3}_{[2,3]}+3\,P^{x_{123}}\,S^{x_{23},x_3}_{[2,3]}\\
&+3\,P^{x_{12}}\,S^{x_1,x_3}_{[2,3]}-3\,P^{x_{12}}\,S^{x_2,x_3}_{[2,3]}+3\,P^{x_{12}}\,S^{x_2,x_{123}}_{[2,3]}-3\,P^{x_{12}}\,S^{x_1,x_{123}}_{[2,3]}\\
&-3\,P^{x_1}\,S^{x_2,x_3}_{[2,3]}-3\,P^{x_1}\,S^{x_{12},x_3}_{[2,3]}+3\,P^{x_1}\,S^{x_{12},x_{123}}_{[2,3]}+3\,P^{x_1}\,S^{x_2,x_{123}}_{[2,3]}\\
&+3\,P^{x_1}\,S^{x_{123},x_3}_{[2,3]}+3\,P^{x_1}\,S^{x_{23},x_3}_{[2,3]}+3\,P^{x_2}\,S^{x_1,x_3}_{[2,3]}-3\,P^{x_2}\,S^{x_1,x_{123}}_{[2,3]}\\
&-3\,P^{x_2}\,S^{x_3,x_{123}}_{[2,3]}-3\,P^{x_2}\,S^{x_{12},x_{123}}_{[2,3]}+3\,P^{x_2}\,S^{x_{12},x_3}_{[2,3]}-3\,P^{x_2}\,S^{x_{23},x_{123}}_{[2,3]}\\
&+3\,P^{x_3}\,S^{x_2,x_{123}}_{[2,3]}+3\,P^{x_3}\,S^{x_{23},x_{123}}_{[2,3]}+3\,P^{x_{23}}\,S^{x_2,x_{123}}_{[2,3]}-3\,P^{x_{23}}\,S^{x_1,x_3}_{[2,3]}\\
&-5\,P^{x_{123}}\,S^{x_3,x_{23}}_{[2,3]}-5\,P^{x_3}\,S^{x_{123},x_{23}}_{[2,3]}-6\,P^{x_1}\,S^{x_3,x_{23}}_{[2,3]}-6\,P^{x_1}\,S^{x_3,x_{123}}_{[2,3]}\\
&+6\,P^{x_2}\,S^{x_{123},x_3}_{[2,3]}+6\,P^{x_2}\,S^{x_{123},x_{23}}_{[2,3]}
\end{aligned}
$$

$$
\begin{aligned}
2\,\mathcal{S}^{*}_{[1,1,3]}:=&+S^{x_1,x_2,x_3}_{[1,1,3]}-S^{x_1,x_2,x_{23}}_{[1,1,3]}-S^{x_1,x_{23},x_3}_{[1,1,3]}+S^{x_3,x_1,x_{23}}_{[1,1,3]}-S^{x_1,x_{12},x_3}_{[1,1,3]}\\
&+S^{x_2,x_{12},x_3}_{[1,1,3]}+S^{x_1,x_3,x_{23}}_{[1,1,3]}-S^{x_2,x_1,x_{23}}_{[1,1,3]}-S^{x_{123},x_2,x_3}_{[1,1,3]}+S^{x_1,x_{123},x_3}_{[1,1,3]}\\
&+S^{x_2,x_3,x_{123}}_{[1,1,3]}+S^{x_2,x_1,x_{123}}_{[1,1,3]}-S^{x_1,x_3,x_{123}}_{[1,1,3]}-S^{x_3,x_1,x_{123}}_{[1,1,3]}-S^{x_{123},x_3,x_{23}}_{[1,1,3]}\\
&-S^{x_3,x_{123},x_{23}}_{[1,1,3]}+S^{x_2,x_{123},x_{23}}_{[1,1,3]}+S^{x_{123},x_{23},x_3}_{[1,1,3]}+S^{x_{123},x_2,x_{23}}_{[1,1,3]}-S^{x_2,x_{23},x_{123}}_{[1,1,3]}\\
&-S^{x_2,x_{12},x_{123}}_{[1,1,3]}+S^{x_3,x_{23},x_{123}}_{[1,1,3]}+S^{x_1,x_{12},x_{123}}_{[1,1,3]}-S^{x_2,x_{123},x_3}_{[1,1,3]}
\end{aligned}
$$

$$
\begin{aligned}
\mathcal{S}^{*}_{[2,1,2]}:=&+S^{x_3,x_1,x_{23}}_{[2,1,2]}-S^{x_{123},x_2,x_3}_{[2,1,2]}-S^{x_3,x_1,x_{123}}_{[2,1,2]}+S^{x_{123},x_2,x_{23}}_{[2,1,2]}+S^{x_{123},x_{23},x_3}_{[2,1,2]}\\
&-S^{x_{123},x_3,x_{23}}_{[2,1,2]}-S^{x_3,x_{123},x_{23}}_{[2,1,2]}+S^{x_3,x_{23},x_{123}}_{[2,1,2]}.
\end{aligned}
$$

Lastly, we must write down the condition for $l\phi ma^{w_1,\ldots,w_5}$ to have no poles of order 1 at the origin. This once again leads to a single equation, but one that now involves all seven relevant singulands:

$$
\mathcal{S}^{**}_{[1]}+\mathcal{S}^{**}_{[1,2]}+\mathcal{S}^{**}_{[1,4]}+\mathcal{S}^{**}_{[2,3]}+\mathcal{S}^{**}_{[1,1,3]}+\mathcal{S}^{**}_{[2,1,2]}+\mathcal{S}^{**}_{[1,1,1,2]}=0. \quad (6.19)
$$

Though easy to compute, the various contributions $\mathcal{S}^{**}_{[r]}$ are too unwieldy for us to write down. So we simply mention their number $\#(\mathcal{S}^{**}_{[r]})$ of summands. Here is the list:

$\#(\mathcal{S}^{**}_{[1]})=126,\quad \#(\mathcal{S}^{**}_{[1,2]})=299,\quad \#(\mathcal{S}^{**}_{[1,4]})=176,\quad \#(\mathcal{S}^{**}_{[2,3]})=314$
$\#(\mathcal{S}^{**}_{[1,1,3]})=288,\ \#(\mathcal{S}^{**}_{[2,1,2]})=324,\ \#(\mathcal{S}^{**}_{[1,1,1,2]})=192.$

If we now look for *meromorphic* singulands of type (6.12), the absence of multipoles of order 3 at the origin is equivalent to a system of two independent identities of the form $\mathcal{R}_{[1]}+\mathcal{R}_{[1,4]}+\mathcal{R}_{[2,3]}=0$, namely:

$$
0=-\frac{1}{120}\,\delta^{n_2}R^{n_1}_{[1]}-R^{n_1,n_2}_{[1,4]}+R^{n_1,n_2}_{[2,3]}-R^{n_1,n_{12}}_{[2,3]} \quad (6.20)
$$

$$
0=\frac{1}{120}\left(\delta^{n_2}R^{n_{12}}_{[1]}-\delta^{n_2}R^{n_1}_{[1]}\right)-R^{n_1,n_2}_{[1,4]}+R^{n_{12},n_2}_{[1,4]}+R^{n_1,n_2}_{[2,3]}-R^{n_2,n_{12}}_{[2,3]}+2R^{n_{12},n_2}_{[2,3]}.
$$

The absence of multipoles of order 2 at the origin is also equivalent to a system of two independent identities, with effective involvement of all singulands except the last one:

$$\mathcal{R}^{*}_{[1]}+\mathcal{R}^{*}_{[1,2]}+\mathcal{R}^{*}_{[1,4]}+\mathcal{R}^{*}_{[2,3]}+\mathcal{R}^{*}_{[1,1,3]}+\mathcal{R}^{*}_{[2,1,2]}=0 \tag{6.21}$$

$$\mathcal{R}^{\dagger}_{[1]}+\mathcal{R}^{\dagger}_{[1,2]}+\mathcal{R}^{\dagger}_{[1,4]}+\mathcal{R}^{\dagger}_{[2,3]}+\mathcal{R}^{\dagger}_{[1,1,3]}+\mathcal{R}^{\dagger}_{[2,1,2]}=0 \tag{6.22}$$

$$\begin{aligned}360\,\mathcal{R}^{*}_{[1]}=&-\delta^{n_1}\delta^{n_{23}}R^{n_2}_{[1]}-4\delta^{n_2}\delta^{n_{23}}R^{n_1}_{[1]}-11\delta^{n_1}\delta^{n_{123}}R^{n_2}_{[1]}-11\delta^{n_{12}}\delta^{n_{123}}R^{n_1}_{[1]}\\&+14\delta^{n_{12}}\delta^{n_3}R^{n_1}_{[1]}-14\delta^{n_2}\delta^{n_3}R^{n_1}_{[1]}+15\delta^{n_1}\delta^{n_3}R^{n_2}_{[1]}+15\delta^{n_2}\delta^{n_{123}}R^{n_1}_{[1]}\\&-15\delta^{n_1}\delta^{n_3}R^{n_{12}}_{[1]}+15\delta^{n_1}\delta^{n_{123}}R^{n_{12}}_{[1]}\end{aligned}$$

$$\begin{aligned}360\,\mathcal{R}^{\dagger}_{[1]}=&+\delta^{n_1}\delta^{n_{23}}R^{n_3}_{[1]}-\delta^{n_2}\delta^{n_3}R^{n_{123}}_{[1]}-14\delta^{n_2}\delta^{n_3}R^{n_1}_{[1]}-14\delta^{n_{12}}\delta^{n_3}R^{n_2}_{[1]}\\&+15\delta^{n_1}\delta^{n_3}R^{n_2}_{[1]}+15\delta^{n_2}\delta^{n_3}R^{n_{12}}_{[1]}-15\delta^{n_1}\delta^{n_3}R^{n_{23}}_{[1]}+15\delta^{n_2}\delta^{n_{123}}R^{n_3}_{[1]}\\&+25\delta^{n_{23}}\delta^{n_3}R^{n_1}_{[1]}+25\delta^{n_3}\delta^{n_{123}}R^{n_{23}}_{[1]}-25\delta^{n_{23}}\delta^{n_3}R^{n_{123}}_{[1]}\end{aligned}$$

$$\begin{aligned}2\,\mathcal{R}^{*}_{[1,2]}=&+\delta^{n_1}R^{n_2,n_3}_{[1,2]}-\delta^{n_1}R^{n_2,n_{23}}_{[1,2]}-\delta^{n_2}R^{n_1,n_{23}}_{[1,2]}+\delta^{n_2}R^{n_1,n_{123}}_{[1,2]}\\&-\delta^{n_1}R^{n_{12},n_3}_{[1,2]}+\delta^{n_1}R^{n_{12},n_{123}}_{[1,2]}\end{aligned}$$

$$\begin{aligned}6\,\mathcal{R}^{\dagger}_{[1,2]}=&+2\,\delta^{n_3}R^{n_1,n_{23}}_{[1,2]}-2\,\delta^{n_{123}}R^{n_2,n_3}_{[1,2]}-2\,\delta^{n_{123}}R^{n_3,n_{23}}_{[1,2]}+2\,\delta^{n_{123}}R^{n_{23},n_3}_{[1,2]}\\&-2\,\delta^{n_3}R^{n_{123},n_{23}}_{[1,2]}+2\,\delta^{n_3}R^{n_{23},n_{123}}_{[1,2]}+3\,\delta^{n_1}R^{n_2,n_3}_{[1,2]}+3\,\delta^{n_2}R^{n_{12},n_3}_{[1,2]}\\&-3\,\delta^{n_1}R^{n_{23},n_3}_{[1,2]}+3\,\delta^{n_1}R^{n_3,n_{23}}_{[1,2]}+3\,\delta^{n_2}R^{n_3,n_{123}}_{[1,2]}-3\,\delta^{n_2}R^{n_{123},n_3}_{[1,2]}\end{aligned}$$

$$\begin{aligned}2\,\mathcal{R}^{*}_{[1,4]}=&\,\delta^{n_{12}}R^{n_1,n_3}_{[1,4]}-\delta^{n_2}R^{n_1,n_3}_{[1,4]}-\delta^{n_2}R^{n_1,n_{23}}_{[1,4]}+\delta^{n_1}R^{n_2,n_{23}}_{[1,4]}\\&+\delta^{n_{12}}R^{n_1,n_{123}}_{[1,4]}+\delta^{n_1}R^{n_2,n_{123}}_{[1,4]}\end{aligned}$$

$$\begin{aligned}6\,\mathcal{R}^{\dagger}_{[1,4]}=&\,2\,\delta^{n_3}R^{n_2,n_{123}}_{[1,4]}-2\,\delta^{n_{23}}R^{n_1,n_3}_{[1,4]}+2\,\delta^{n_3}R^{n_1,n_{23}}_{[1,4]}-2\,\delta^{n_{123}}R^{n_2,n_3}_{[1,4]}\\&+2\,\delta^{n_{23}}R^{n_3,n_{123}}_{[1,4]}+2\,\delta^{n_{23}}R^{n_{123},n_3}_{[1,4]}-2\,\delta^{n_{123}}R^{n_3,n_{23}}_{[1,4]}-2\,\delta^{n_3}R^{n_{123},n_{23}}_{[1,4]}\\&-3\,\delta^{n_2}R^{n_1,n_3}_{[1,4]}-3\,\delta^{n_{12}}R^{n_2,n_3}_{[1,4]}-3\,\delta^{n_1}R^{n_3,n_{23}}_{[1,4]}+3\,\delta^{n_2}R^{n_{123},n_3}_{[1,4]}\end{aligned}$$

$$\begin{aligned}2\,\mathcal{R}^{*}_{[2,3]}=&\,\delta^{n_2}R^{n_1,n_3}_{[2,3]}-\delta^{n_1}R^{n_2,n_3}_{[2,3]}+\delta^{n_{12}}R^{n_1,n_3}_{[2,3]}-\delta^{n_2}R^{n_1,n_{123}}_{[2,3]}\\&+\delta^{n_1}R^{n_2,n_{123}}_{[2,3]}-\delta^{n_1}R^{n_{12},n_3}_{[2,3]}-\delta^{n_{12}}R^{n_1,n_{123}}_{[2,3]}+\delta^{n_1}R^{n_{12},n_{123}}_{[2,3]}\end{aligned}$$

$$\begin{aligned}6\,\mathcal{R}^{\dagger}_{[2,3]}=&\,\delta^{n_{23}}R^{n_3,n_1}_{[2,3]}-\delta^{n_3}R^{n_{123},n_2}_{[2,3]}+2\,\delta^{n_{23}}R^{n_{123},n_3}_{[2,3]}+2\,\delta^{n_{23}}R^{n_3,n_{123}}_{[2,3]}\\&+3\,\delta^{n_2}R^{n_1,n_3}_{[2,3]}-3\,\delta^{n_{23}}R^{n_1,n_3}_{[2,3]}-3\,\delta^{n_1}R^{n_2,n_3}_{[2,3]}-3\,\delta^{n_{12}}R^{n_2,n_3}_{[2,3]}\\&+3\,\delta^{n_2}R^{n_{12},n_3}_{[2,3]}+3\,\delta^{n_1}R^{n_{23},n_3}_{[2,3]}+3\,\delta^{n_3}R^{n_2,n_{123}}_{[2,3]}-3\,\delta^{n_2}R^{n_3,n_{123}}_{[2,3]}\\&+3\,\delta^{n_3}R^{n_{23},n_{123}}_{[2,3]}+3\,\delta^{n_{123}}R^{n_{23},n_3}_{[2,3]}-5\,\delta^{n_{123}}R^{n_3,n_{23}}_{[2,3]}\\&-5\,\delta^{n_3}R^{n_{123},n_{23}}_{[2,3]}-6\,\delta^{n_1}R^{n_3,n_{23}}_{[2,3]}+6\,\delta^{n_2}R^{n_{123},n_3}_{[2,3]}\end{aligned}$$

$$\mathcal{R}^{*}_{[1,1,3]}=R^{n_1,n_2,n_3}_{[1,1,3]}-R^{n_1,n_2,n_{23}}_{[1,1,3]}+R^{n_1,n_{12},n_{123}}_{[1,1,3]}-R^{n_1,n_{12},n_3}_{[1,1,3]}+R^{n_2,n_1,n_{123}}_{[1,1,3]}-R^{n_2,n_1,n_{23}}_{[1,1,3]}$$

$$\begin{aligned}\mathcal{R}^{\dagger}_{[1,1,3]}=&R^{n_1,n_2,n_3}_{[1,1,3]}+R^{n_2,n_{12},n_3}_{[1,1,3]}-R^{n_1,n_{23},n_3}_{[1,1,3]}+R^{n_3,n_1,n_{23}}_{[1,1,3]}+R^{n_1,n_3,n_{23}}_{[1,1,3]}\\&-R^{n_{123},n_2,n_3}_{[1,1,3]}-R^{n_2,n_{123},n_3}_{[1,1,3]}+R^{n_2,n_3,n_{123}}_{[1,1,3]}-R^{n_{123},n_3,n_{23}}_{[1,1,3]}\\&-R^{n_3,n_{123},n_{23}}_{[1,1,3]}+R^{n_{123},n_{23},n_3}_{[1,1,3]}+R^{n_3,n_{23},n_{123}}_{[1,1,3]}\end{aligned}$$

$$\mathcal{R}^{*}_{[2,1,2]}=0$$

$$\mathcal{R}^{\dagger}_{[2,1,2]}=2\big(R^{n_3,n_1,n_{23}}_{[2,1,2]}-R^{n_{123},n_2,n_3}_{[2,1,2]}+R^{n_3,n_{23},n_{123}}_{[2,1,2]}-R^{n_{123},n_3,n_{23}}_{[2,1,2]}-R^{n_3,n_{123},n_{23}}_{[2,1,2]}+R^{n_{123},n_{23},n_3}_{[2,1,2]}\big).$$

Lastly, the condition for $l\emptyset ma^{w_1,\dots,w_5}$ to have no poles of order 1 at the origin can be expressed by a single equation, that involves all seven relevant singulands:

$$\mathcal{R}^{**}_{[1]}+\mathcal{R}^{**}_{[1,2]}+\mathcal{R}^{**}_{[1,4]}+\mathcal{R}^{**}_{[2,3]}+\mathcal{R}^{**}_{[1,1,3]}+\mathcal{R}^{**}_{[2,1,2]}+\mathcal{R}^{**}_{[1,1,1,2]}=0. \quad (6.23)$$

Once again the $\mathcal{R}^{**}_{[r]}$ are too unwieldy for us to write down, and we merely mention their number $\#(\mathcal{R}^{**}_{[r]})$ of summands:

$$\begin{aligned}&\#(\mathcal{R}^{**}_{[1]})=34,\quad \#(\mathcal{R}^{**}_{[1,2]})=58,\quad \#(\mathcal{R}^{**}_{[1,4]})=40,\quad \#(\mathcal{R}^{**}_{[2,3]})=74\\&\#(\mathcal{R}^{**}_{[1,1,3]})=48,\ \#(\mathcal{R}^{**}_{[2,1,2]})=64,\ \#(\mathcal{R}^{**}_{[1,1,1,2]})=24\end{aligned}$$

6.5 The basis *lama*$^\bullet$/*lami*$^\bullet$.

As already pointed out, the desingulation conditions listed above admit multiple solutions when the singulands are sought in the space of power series, even after imposing the proper parity in each variable. To ensure uniqueness, many additional constraints are theoretically possible, but two stand out as clearly privileged, in the sense that they, and they alone, guarantee coefficients with arithmetically simple denominators.

We mention here the first constraint, leading to the bimould *lama*$^\bullet$, for the first non-trivial singulands $S^{\bullet}_{[1,2]}=S^{\bullet}_{1,2}$. For the coefficients of weight s, the equation (6.14) admits exactly *one* solution of the form:

$$\mathrm{Sa}^{x_1,x_2}_{1,2}=\sum_{1\le\delta\le\mathrm{ent}(\frac{s-1}{2})-\mathrm{ent}(\frac{s+1}{6})} a_{2\delta}\; x_1^{2\delta}\, x_2^{s-2-2\delta}. \qquad (6.24)$$

This is, moreover, the choice for which the prime factors in the denominator admit the best universal bound $p\le Cst\, s$. In fact, for this choice, the bound is $p\le\frac{s}{3}$.

6.6 The basis *loma*$^\bullet$/*lomi*$^\bullet$

Now, let us move on to the second type of constraints, leading to the bimould *loma*$^\bullet$, again for the first non-trivial singulands $S^{\bullet}_{[1,2]}=S^{\bullet}_{1,2}$.

For the coefficients of weight s, the equation (6.14) admits exactly *one* solution of the form:

$$\mathrm{So}^{x_1,x_2}_{1,2} = x_1^2\, x_2 \sum_{0\le\delta\le \mathrm{ent}(\frac{s-3}{6})} a_{2\,\delta}\,\left(x_1^{2\,\delta}\, x_2^{s-5-2\,\delta} + x_1^{s-5-2\,\delta}\, x_2^{2\,\delta}\right) \qquad (6.25)$$

which entails far fewer coefficients. This is basically the only other choice[93] for which the prime factors in the denominator admit a universal bound $p \le Cst\, s$. In this case the bound is $p \le \frac{2\,s-5}{3}$.

6.7 The basis *luma*$^\bullet$/*lumi*$^\bullet$

Here, we may deal at once with all length-2 singulands:

$$S^{x_1,x_2}_{[r_1,r_2]} = S^{x_1,x_2}_{r_1,r_2} \overset{\mathrm{ess^{ly}}}{=} \sum_{n_i\in\mathbb{Z}^*} R^{n_1,n_2}_{r_1,r_2}\, P(x_1{+}n_1)\, P(x_2{+}n_2). \qquad (6.26)$$

The multiresidues are simple enough:[94]

$$R^{n_1,n_2}_{[r_1,r_2]} = R^{n_1,n_2}_{r_1,r_2} = \gamma_{r_1,r_2}\, \mu(n_1,n_2)\, n_1^{n_2-1}\, n_2^{n_1-1} \qquad (6.27)$$

with γ_{r_1,r_2} a simple rational constant, and with $\mu(n_1,n_2)$ being 1 (respectively 0) if n_1, n_2 are co-prime (respectively otherwise). The Taylor coefficients of the singulates, however, are less simple: they carry Bernoulli numbers in their denominators, and sometimes very large prime factors, that can exceed any given bound of the form $Cst\, s$:

$$\mathrm{Su}^{x_1,x_2}_{r_1,r_2}(s) = (-1)^{r_1}\,\frac{B_{r_1+r_2-1}}{r_1+r_2-1} \sum_{\substack{\delta_1\ge r_1\\ \delta_2\ge r_2}}^{\delta_1+\delta_2=s+2} \frac{B^*_{\delta_1-r_1}\, B^*_{\delta_2-r_2}}{B^*_{\delta_1+\delta_2-r_1-r_2}}\, u_1^{\delta_2-2} u_2^{\delta_1-2} \qquad (6.28)$$

with $B^*_n = \dfrac{B_n}{n!}$, $B_{2n} :=$ Bernoulli number, $B_n := 0$ for n odd or <0.

Pay attention to the exponents: it is δ_2-2 on top of u_1 and δ_1-2 on top of u_2. In fact, since both s and $r_1{+}r_2$ are always odd, the summation rule produces only *positive* powers of u_1, u_2 (one *even*, the other *odd*), except for the pairs $(r_1,r_2) = (1,2)$ respectively $(2,1)$ where constant monomials in u_1 respectively u_2 do appear – but these may be neglected, since they contribute nothing to the singulate. Of course, the usual identity $Su^{x_1,x_2}_{r_1,r_2} + Su^{x_2,x_1}_{r_2,r_1} = 0$ holds.

93 Leaving aside, of course, simple *averages* of the first and second choice.

94 They cease to be simple for singulands of length $l \ge 3$. Here, we get full-blown 'perinomalness'. See Section 9.5.

6.8 Arithmetical vs analytic smoothness

To show how the three choices compare, arithmetically speaking, we list the weight-s component $S^{\bullet}_{1,2}(s)$ of the first non-trivial singuland in all three variants $Sa^{\bullet}_{1,2}(s)$, $So^{\bullet}_{1,2}(s)$, $Su^{\bullet}_{1,2}(s)$, up to the weight $s = 17$:

$$\mathrm{Sa}^{x_1,x_2}_{1,2}(5) = \mathrm{So}^{x_1,x_2}_{1,2}(5) = \mathrm{Su}^{x_1,x_2}_{1,2}(5) = -\frac{5}{12}\,x_1^2\,x_2$$

$$\mathrm{Sa}^{x_1,x_2}_{1,2}(7) = \mathrm{So}^{x_1,x_2}_{1,2}(7) = \mathrm{Su}^{x_1,x_2}_{1,2}(7) = -\frac{7}{24}\,x_1^2\,x_2^3 - \frac{7}{24}\,x_1^4\,x_2$$

$$\mathrm{Sa}^{x_1,x_2}_{1,2}(9) = \mathrm{So}^{x_1,x_2}_{1,2}(9) = \mathrm{Su}^{x_1,x_2}_{1,2}(9) = -\frac{5}{18}\,x_1^2\,x_2^5 - \frac{7}{36}\,x_1^4\,x_2^3 - \frac{5}{18}\,x_1^6\,x_2$$

$$\mathrm{Sa}^{x_1,x_2}_{1,2}(11) = -\frac{11}{8}\,x_1^2\,x_2^7 + \frac{55}{24}\,x_1^4\,x_2^5 - \frac{11}{6}\,x_1^6\,x_2^3$$

$$\mathrm{So}^{x_1,x_2}_{1,2}(11) = -\frac{11}{40}\,x_1^2\,x_2^7 - \frac{11}{60}\,x_1^4\,x_2^5 - \frac{11}{60}\,x_1^6\,x_2^3 - \frac{11}{40}\,x_1^8\,x_2$$

$$\mathrm{Su}^{x_1,x_2}_{1,2}(11) = \mathrm{So}^{x_1,x_2}_{1,2}(11)$$

$$\mathrm{Sa}^{x_1,x_2}_{1,2}(13) = -\frac{91}{48}x_1^4x_2^7 + \frac{65}{24}x_1^6x_2^5 - \frac{91}{48}x_1^8x_2^3,$$

$$\mathrm{So}^{x_1,x_2}_{1,2}(13) = -\frac{65}{252}x_1^2x_2^9 - \frac{143}{504}x_1^4x_2^7 - \frac{143}{504}x_1^8x_2^3 - \frac{65}{252}x_1^{10}x_2$$

$$\mathrm{Su}^{x_1,x_2}_{1,2}(13) = -\frac{2275}{8292}x_1^2x_2^9 - \frac{1001}{5528}x_1^4x_2^7 - \frac{715}{4146}x_1^6x_2^5 - \frac{1001}{5528}x_1^8x_2^3 - \frac{2275}{8292}x_1^{10}x_2$$

$$\mathrm{Sa}^{x_1,x_2}_{1,2}(15) = -\frac{691}{360}x_1^2x_2^{11} + \frac{665}{144}x_1^4x_2^9 - \frac{2233}{360}x_1^6x_2^7 + \frac{209}{48}x_1^8x_2^5 - \frac{21}{10}x_1^{10}x_2^3$$

$$\mathrm{So}^{x_1,x_2}_{1,2}(15) = -\frac{691}{2520}x_1^2x_2^{11} - \frac{13}{72}x_1^4x_2^9 - \frac{143}{840}x_1^6x_2^7 - \frac{143}{840}x_1^8x_2^5 - \frac{13}{72}x_1^{10}x_2^3 - \frac{691}{2520}x_1^{12}x_2$$

$$\mathrm{Su}^{x_1,x_2}_{1,2}(15) = \mathrm{So}^{x_1,x_2}_{1,2}(15)$$

$$\mathrm{Sa}^{x_1,x_2}_{1,2}(17) = -\frac{442}{15}x_1^2x_2^{13} + \frac{1105}{12}x_1^4x_2^{11} - \frac{1666}{15}x_1^6x_2^9 + \frac{187}{3}x_1^8x_2^7 - \frac{153}{10}x_1^{10}x_2^5$$

$$\mathrm{So}^{x_1,x_2}_{1,2}(17) = -\frac{17}{60}x_1^2x_2^{13} - \frac{17}{144}x_1^4x_2^{11} - \frac{221}{720}x_1^6x_2^9 - \frac{221}{720}x_1^{10}x_2^5 - \frac{17}{144}x_1^{12}x_2^3 - \frac{17}{60}x_1^{14}x_2$$

$$\mathrm{Su}^{x_1,x_2}_{1,2}(17) = -\frac{2975}{10851}x_1^2x_2^{13} - \frac{11747}{65106}x_1^4x_2^{11} - \frac{5525}{32553}x_1^6x_2^9 - \frac{2431}{14468}x_1^8x_2^7 - \frac{5525}{32553}x_1^{10}x_2^5 - \frac{11747}{65106}x_1^{12}x_2^3 - \frac{2975}{10851}x_1^{14}x_2.$$

6.9 Singulator kernels and "wandering" bialternals

Let $BIMU_l^s$ be the space of all bimoulds $M^\bullet$ whose only non-vanishing component $M^{w_1,\dots,w_l}$ is constant in the v_i-variables, and homogeneous polynomial of total degree $d = s - l$ in the u_i-variables.[95]

Likewise, let $BIMU^s_{r_1,\dots,r_l}$ be the subspace of $BIMU_l^s$ consisting of all bimoulds $M^\bullet$ whose only non-vanishing component $M^{w_1,\dots,w_l}$:
– is divisible by each u_i;
– is *even* in u_i if r_i is *odd*, and *vice versa*.

For each pair r and s large enough ($s \geq s_r$), there always exist non-trivial collections of special *singulands* $S_r^\bullet$:

$$\{S^\bullet_{r_1,\dots,r_l} \in \mathrm{BIMU}^s_{r_1,\dots,r_l} \ ; \ 1 < l < r \ , \ r_1 + \cdots + r_l = r\} \tag{6.29}$$

such that the corresponding bialternal *singulates* $\Sigma_r^\bullet$ combine to form a $\Theta_r^\bullet$ that is singularity-free, *i.e. polynomial*, with the predictable total degree $s-r$ and an unchanged 'weight' s:

$$\Theta_r^\bullet := \sum_{1<l<r} \sum_{r_1+\ldots+r_l=r} \mathrm{slank}_{r_1,\dots,r_l}.S^\bullet_{r_1,\dots,r_l} \in \mathrm{ALAL} \cap \mathrm{BIMU}_r^s \tag{6.30}$$

instead of presenting at the origin multipoles of order τ:

$$\tau := r - l_{\min} \quad \text{with} \quad 2 \leq l_{\min} := \inf(l) \quad \text{for} \quad S^\bullet_{r_1,\dots,r_l} \neq 0^\bullet \tag{6.31}$$

as would be the case for randomly chosen *singulands* $S_r^\bullet$. The result holds even if we impose that there be a least one nonzero singuland $S^\bullet_{r_1,r_2}$ of minimal length $l = 2$.

These paradoxical *non-singular singulates* $\Theta_r^\bullet$ are known as *wandering bialternals*. They span a subspace of *BIMU* which is in fact a (small) subalgebra $ALAL_{wander}$ of $ALAL \subset ARI^{\underline{al}/\underline{al}}$. On top of the natural gradation by r (the *length*), $ALAL_{wander}$ admits a natural filtration by τ (the *'avoided polar order'*).

The presence of these *wandering bialternals* is responsible for the *very slight* indeterminacy that exists in the construction by singulation-desingulation of a basis of $ALIL \subset ARI^{\underline{al}/\underline{il}}$. As we saw, to remove that indeterminacy, additional criteria (arithmetical or functional) are called for, leading to the three (distinct yet closely related) bases of Section 6.5, Section 6.6, Section 6.7.

---

95 So that s may be called the 'weight' of $M^\bullet$.

7 A conjectural basis for $ALAL \subset ARI^{\underline{al}/\underline{al}}$. The three series of bialternals

7.1 Basic bialternals: the enumeration problem

We shall have to handle three series of bialternals, each with a single non-zero component, of length 1, 2, 4 respectively. Here they are, with their names and natural indexation:

$$\begin{aligned} \text{ekma}_d^\bullet/\text{ekmi}_d^\bullet &\in \text{BIMU}_1, \quad d \text{ even } \geq 2 \\ \text{doma}_{d,b}^\bullet/\text{domi}_{d,b}^\bullet &\in \text{BIMU}_2, \quad d \text{ even } \geq 6,\ 1 \leq b \leq \beta(d) \\ \text{carma}_{d,c}^\bullet/\text{carmi}_{d,c}^\bullet &\in \text{BIMU}_4, \quad d \text{ even } \geq 8,\ 1 \leq c \leq \gamma(d). \end{aligned}$$

As usual, the vocalic alternation $a \leftrightarrow i$ is indicative of the basic involution *swap*. The integers $\alpha(d)$, $\beta(d)$, $\gamma(d)$ are given by the generating functions:

$$\sum \alpha(d)t^d := t^6(1-t^2)^{-1}(1-t^4)^{-1} = t^6+t^8+2\,t^{10}+2\,t^{12}+3\,t^{14}\ldots \qquad (7.1)$$

$$\sum \beta(d)t^d := t^6(1-t^2)^{-1}(1-t^6)^{-1} = t^6+t^8+t^{10}+2\,t^{12}+2\,t^{14}\ldots \qquad (7.2)$$

$$\sum \gamma(d)t^d := t^8(1-t^4)^{-1}(1-t^6)^{-1} = t^8+t^{12}+t^{14}+t^{16}+t^{18}+2\,t^{20}\ldots \qquad (7.3)$$

and clearly verify $\alpha(d) \equiv \beta(d) + \gamma(d-2)$. Mark the absence of t^{10} in (7.3).

7.2 The regular bialternals: *ekma, doma*

The *ekma* bialternals are utterly elementary

$$\text{ekma}_d^{w_1} := u_1^d;\ \text{ekmi}_d^{w_1} := v_1^d \qquad (7.4)$$

since, for length 1, bialternality reduces to *neg*-invariance. If the *ekmas* freely generated a subalgebra *EKMA* of *ALAL*, the dimension of $EKMA_{2,d}$ (length 2, degree d) would be exactly $\alpha(d)$. This, however, is not the case. Indeed, since the bialternality constraints for length 2 are *finitary*[96], Hilbert's invariant theory applies, and it is a simple matter to verify that $ALAL_2$:
(i) is spanned by *ekma* brackets;
(ii) admits the following *domas* as a canonical (in the sense of 'simplest') basis:

$$\text{doma}_{d,b}^{w_1,w_2} := \text{fa}(u_1,u_2)\ (\text{ga}(u_1,u_2))^{b-1}\ (\text{ha}(u_1,u_2))^{d/2-3b} \qquad (7.5)$$

$$\text{domi}_{d,b}^{w_1,w_2} := \text{fi}(v_1,v_2)\ (\text{gi}(v_1,v_2))^{b-1}\ (\text{hi}(v_1,v_2))^{d/2-3b} \qquad (7.6)$$

96 *I.e.* correspond to invariance under a finite subgroup of $Gl_2(\mathbb{C})$, which in the present instance is isomorphic to $\mathfrak{S}_3$. Finitariness ceases from length 3 onwards.

with

$$\mathrm{fa}(u_1, u_2) := u_1 u_2 (u_1 - u_2)(u_1 + u_2)(2u_1 + u_2)(2u_2 + u_1) \quad (7.7)$$

$$\mathrm{ga}(u_1, u_2) := (u_1 + u_2)^2 u_1^2 u_2^2; \quad \mathrm{ha}(u_1, u_2) := u_1^2 + u_1 u_2 + u_2^2 \quad (7.8)$$

$$\mathrm{fi}(v_1, v_2) := v_1 v_2 (v_1 - v_2)(v_1 + v_2)(2v_1 - v_2)(2v_2 - v_1) \quad (7.9)$$

$$\mathrm{gi}(v_1, v_2) := (v_1 - v_2)^2 v_1^2 v_2^2; \quad \mathrm{hi}(v_1, v_2) := v_1^2 - v_1 v_2 + v_2^2. \quad (7.10)$$

Therefore $dim(EKMA_{2,d}) = dim(ALAL_{2,d}) = \beta(d) \leq \alpha(d)$ and, for each even degree $d+2$, the *ekma*-brackets verify exactly $\gamma(d)$ independent relations of the form:

$$\sum_{d_1+d_2=d+2} Q_c^{d_1,d_2} \mathrm{ari}(\mathrm{ekma}^{\bullet}_{d_1}, \mathrm{ekma}^{\bullet}_{d_2}) = 0^{\bullet} \quad (1 \leq c \leq \gamma(d),\ Q_c^{d_1,d_2} \in \mathbb{Q}) \quad (7.11)$$

easily derivable from the decompositions:

$$\mathrm{ari}(\mathrm{ekma}^{\bullet}_{d_1}, \mathrm{ekma}^{\bullet}_{d_2}) = \sum_{1 \leq b \leq \beta(d_1+d_2)} K^b_{d_1,d_2}\, \mathrm{doma}^{\bullet}_{d_1+d_2,b} \quad (K^b_{d_1,d_2} \in \mathbb{Q}). \quad (7.12)$$

7.3 The irregular bialternals: *carma*

Not all bialternals of length $r = 4$ may be obtained as superpositions of *ekma* brackets. Thus, there exists (up to scalar multiplication) exactly *one* bialternal of length $r = 4$ and degree $d = 8$, which clearly cannot be generated by *ekmas*, since the first *ekma* has degree 2, and self-bracketting it four times yields nothing.

One of our conjectures (for which there is compelling theoretical and numerical evidence[97]) is that the number of these independent *exceptional* or *irregular* bialternals – we call them *carma* bialternals – is exactly $\gamma(d)$ as given by (7.3), and that these bialternals $carma_{d,c}$ $(1 \leq c \leq \gamma(d))$ are in one-to-one, constructive correspondence (see Section 7.7) with the elements (7.11) of length 2 and degree $d+2$ in the *ekma* ideal, under a transparent and quite universal *restoration mechanism* (see Section 7.9).

7.4 Main differences between regular and irregular bialternals

For one thing, the algebra $EKMA \subset ALAL$ generated by the *ekmas* is intrinsical, while the algebra $CARMA \subset ALAL$ generated by the *carmas* depends, as we shall see in Section 8.5, on the choice of a basis for

[97] See Section 7.9, Section 8.5, Section 8.10.

ALIL. (That said, there exist clearly canonical bases of *ALIL*, and therefore canonical choices for *CARMA* as well.)

Then, the definition of the $ekma^{\bullet}_{d}$ is as elementary as the construction of the $carma^{\bullet}_{d,c}$ is complex. Unsurprisingly, this difference finds its reflection in the arithmetical properties (divisibility etc) and above all in the sheer size of their coefficients.[98] For instance, if we consider the first 'cells' $ALAL_{r,d}$ where elements of *EKMA* and *DOMA* coexist with *unique* elements of *CARMA*, namely the 'cells' $r = 4$ and $d \leq 18$, and then compare typical elements of $EKMA_{r,d}$ and $DOMA_{r,d}$ with those of $CARMA_{r,d}$, we find that the latter are strikingly more complex.

For illustration, here is, with self-explanatory labels, a list of representatives chosen in the three algebras, with *red* signalling that our polynomials are taken in their simplest form, *i.e.* with coprime coefficients:

$$\mathrm{cara}_d := \mathrm{red}(\mathrm{carma}_{d,1}) \qquad (d=8,12,14,16,18)$$
$$\mathrm{eka}_d := \mathrm{red}(\mathrm{ari}(\mathrm{ekma}_{d-6}, \mathrm{ekma}_2, \mathrm{ekma}_2, \mathrm{ekma}_2)) \qquad (d=10,12,14,16,18)$$
$$\mathrm{doa}_{14} := \mathrm{red}(\mathrm{ari}(\mathrm{doma}_{6,1}, \mathrm{doma}_{8,1}))$$
$$\mathrm{doa}_{16} := \mathrm{red}(\mathrm{ari}(\mathrm{doma}_{6,1}, \mathrm{doma}_{10,1}))$$
$$\mathrm{doa}_{18} := \mathrm{red}(\mathrm{ari}(\mathrm{doma}_{6,1}, \mathrm{doma}_{12,2})).$$

The first table (below) mentions the exact number of monomials effectively present in each polynomial. That number is always larger in the $\boldsymbol{u}$-variables (vowel *a*) than in the $\boldsymbol{v}$-variables (vowel *i*), and the figures in boldface represent the difference. For comparison, the first column *FULL* mentions the maximum number of monomials in general homogeneous polynomials of the corresponding degree.

d	FULL	CARMA	CARMI		EKMA	EKMI		DOMA	DOMI	
8	165	142	118	**24**						
10	286				254	254	**0**			
12	455	434	420	**14**	422	408	**14**			
14	680	658	640	**18**	650	586	**64**	498	420	**78**
16	969	946	924	**22**	940	752	**188**	778	616	**162**
18	1330	1306	1280	**26**	1300	922	**378**	930	798	**132** .

The next table (below) mentions the approximate norms of our (reduced!) polynomials, *i.e.* the sum of the absolute values of all their co-prime coefficients. Here again, the norms are much larger for the $\boldsymbol{u}$- than for the $\boldsymbol{v}$-variables, and the numbers in boldface represent the approximate

[98] This applies equally to the $ekma^{\bullet}_{d}$, $carma^{\bullet}_{d,c}$ and their swappees $ekmi^{\bullet}_{d}$, $carmi^{\bullet}_{d,c}$.

ratios of the two.

d	CARMA	CARMI		EKMA	EKMI		DOMA	DOMI	
8	$8.6\ 10^6$	$2.6\ 10^6$	**3**						
10				$1.9\ 10^4$	$1.4\ 10^4$	**1.3**			
12	$1.2\ 10^{11}$	$6.0\ 10^9$	**19**	$1.8\ 10^5$	$3.3\ 10^4$	**5**			
14	$6.8\ 10^{12}$	$7.9\ 10^{10}$	**87**	$2.0\ 10^6$	$8.5\ 10^4$	**23**	$3.6\ 10^5$	$2.5\ 10^4$	**14**
16	$7.6\ 10^{13}$	$3.8\ 10^{11}$	**200**	$9.5\ 10^7$	$1.0\ 10^6$	**95**	$5.2\ 10^6$	$8.8\ 10^4$	**59**
18	$1.3\ 10^{17}$	$1.6\ 10^{14}$	**845**	$1.5\ 10^9$	$3.9 10^6$	**379**	$4.9\ 10^6$	$2.3\ 10^5$	**21** .

Thus, while the polynomials in *CARMA* are only marginally *fuller* (*i.e.* less lacunary) than those in *DOMA* and *EKMA*, the main difference lies in their dramatically *larger* coefficients. Arithmetically, too, their coefficients are more complex, as borne out by their various reductions *mod p*.

7.5 The *pre-doma* potentials

Rectifying $\sigma_{1,1}$ to $\sigma^*_{1,1}$

The mapping $(A^\bullet, B^\bullet) \in \mathrm{BIMU}_1 \times \mathrm{BIMU}_1 \mapsto C^\bullet := \mathrm{ari}(A^\bullet, B^\bullet) \in \mathrm{BIMU}_2$ induces by bilinearity a mapping $\sigma_{1,1} : S^\bullet \in \mathrm{BIMU}_2 \mapsto \Sigma^\bullet \in \mathrm{BIMU}_2$ with:

$$\begin{aligned}\Sigma^{\left(\begin{smallmatrix}u_1,&u_2\\ v_1,&v_2\end{smallmatrix}\right)} &= +S^{\left(\begin{smallmatrix}u_1,&u_2\\ v_1,&v_2\end{smallmatrix}\right)} + S^{\left(\begin{smallmatrix}u_2,&u_{12}\\ v_{2:1},&v_1\end{smallmatrix}\right)} + S^{\left(\begin{smallmatrix}u_{12},&u_1\\ v_2,&v_{1:2}\end{smallmatrix}\right)}\\ &\quad - S^{\left(\begin{smallmatrix}u_2,&u_1\\ v_2,&v_1\end{smallmatrix}\right)} - S^{\left(\begin{smallmatrix}u_{12},&u_2\\ v_1,&v_{2:1}\end{smallmatrix}\right)} - S^{\left(\begin{smallmatrix}u_1,&u_{12}\\ v_{1:2},&v_2\end{smallmatrix}\right)}.\end{aligned}$$

For arguments S^{w_1,w_2} that are *even* in both w_1 and w_2, $\sigma_{1,1}$ coincides with the simpler mapping $\sigma^*_{1,1} : S^\bullet \in \mathrm{BIMU}_2 \mapsto \Sigma^\bullet_* \in \mathrm{BIMU}_2$ with:

$$\begin{aligned}\Sigma_*^{\left(\begin{smallmatrix}u_1,&u_2\\ v_1,&v_2\end{smallmatrix}\right)} &= +S^{\left(\begin{smallmatrix}u_1,&u_2\\ v_1,&v_2\end{smallmatrix}\right)} + S^{\left(\begin{smallmatrix}u_2,&-u_{12}\\ v_{2:1},&-v_1\end{smallmatrix}\right)} + S^{\left(\begin{smallmatrix}-u_{12},&u_1\\ -v_2,&v_{1:2}\end{smallmatrix}\right)}\\ &\quad - S^{\left(\begin{smallmatrix}u_2,&u_1\\ v_2,&v_1\end{smallmatrix}\right)} - S^{\left(\begin{smallmatrix}-u_{12},&u_2\\ -v_1,&v_{2:1}\end{smallmatrix}\right)} - S^{\left(\begin{smallmatrix}u_1,&-u_{12}\\ v_{1:2},&-v_2\end{smallmatrix}\right)}\\ &= +S^{\left(\begin{smallmatrix}u_1,&u_2\\ v_{1:0},&v_{2:0}\end{smallmatrix}\right)} + S^{\left(\begin{smallmatrix}u_2,&u_0\\ v_{2:1},&v_{0:1}\end{smallmatrix}\right)} + S^{\left(\begin{smallmatrix}u_0,&u_1\\ v_{0:2},&v_{1:2}\end{smallmatrix}\right)}\\ &\quad - S^{\left(\begin{smallmatrix}u_2,&u_1\\ v_{2:0},&v_{1:0}\end{smallmatrix}\right)} - S^{\left(\begin{smallmatrix}u_0,&u_2\\ v_{0:1},&v_{2:1}\end{smallmatrix}\right)} - S^{\left(\begin{smallmatrix}u_1,&u_0\\ v_{1:2},&v_{0:2}\end{smallmatrix}\right)}\end{aligned}$$

which in the "long notation" (*i.e.* under adjunction of $u_0 := -u_{1,2}$ and $v_0 :=$ '*free*') takes on the pleasant form:

$$\begin{aligned}\Sigma_*^{[w_0],w_1,w_2} &= +S^{[w_0],w_1,w_2} + S^{[w_1],w_2,w_0} + S^{[w_2],w_0,w_1}\\ &\quad - S^{[w_0],w_2,w_1} - S^{[w_1],w_0,w_2} - S^{[w_2],w_1,w_0}.\end{aligned}$$

In this form, the 'finitariness' of $\sigma^*_{1,1}$ is conspicuous, since the right-hand side involves exactly all six permutations of the sequence (w_0, w_1, w_2).

But $\sigma^*_{1,1}$ has another merit: it turns not just all *bi-even*, but also all *bi-odd* alternals S^{w_1,w_2} into bialternals Σ^{w_1,w_2}_* (whereas $\sigma_{1,1}$ only turns *bi-even* alternals into bialternals). When acting on bi-even (respectively bi-odd) alternals, $\sigma^*_{1,1}$ bilinearly extends the action of *ari* (respectively that of *oddari*: see (2.80)). The mappings $\sigma_{1,1}$ and $\sigma^*_{1,1}$ are of course reminiscent of the mappings from *singulands* $S^\bullet$ to *singulates* $\Sigma^\bullet$ which we studied at length in Section 5, except that now neither $\Sigma^\bullet$ nor $\Sigma^\bullet_*$ carries poles.

The bi-even *pre-doma* potentials
Before turning to our proper object – the kernel of $\sigma^*_{1,1}$ – let us look for *pre-doma*-potentials, *i.e.* for (alternal, bi-even) pre-images of the $doma^\bullet_{d,b}$ under $\sigma^*_{1,1}$. If we impose the a priori form:

$$\text{predoma}^{x_1,x_2}_{d,b} = \sum_{1\leq\delta\,\text{ent}(\frac{d}{6})} c_{d,b;\delta}\,(x_1^{2\,\delta}\,x_2^{d-2\,\delta} - x_2^{2\,\delta}\,x_1^{d-2\,\delta})$$

$$\left(d\ even\ ,\ 1\leq b\leq \text{ent}\left(\frac{d}{6}\right)\right)$$

the solution is unique, and this is essentially the only choice that yields *arithmetical smoothness*, *i.e.* that ensures for the prime factors p in the denominators of the coefficients $c_{d,b;\delta}$ universal bounds of type $p \leq C\,d$. In fact, the bound here is $p \leq d-3$.

The bi-odd *pre-doma* potentials
Here again, there is only one (alternal, bi-odd) a priori form (analogous to the above) that ensures arithmetical smoothness.

Arithmetical smoothness
So, even in the case of the atypical, *singularity-free singulator* $\sigma^*_{1,1}$ we encounter anew the phenomenon which, in the preceding section, led us to the privileged bases $lama^\bullet_s$ and $loma^\bullet_s$, namely the existence of very specific conditions on the singulates that ensure uniqueness and simple 'factorial' bounds for the coefficients' denominators.

7.6 The *pre-carma* potentials

Natural basis for $\ker(\sigma^*_{1,1})$
On the space of alternal bimoulds that are independent of (v_1, v_2) and polynomial in (u_1, u_2) of even (total) degree d, the dimension of $\ker(\sigma^*_{1,1})$ is $s_d := ent(\frac{d-1}{3})$. Let us look for a convenient basis. Reverting to the (x_1, x_2) variables favoured for singulands, we see that the alternal bimoulds

$$H_{d,s}^{\binom{x_1}{0}\,,\,\binom{x_2}{0}} := (x_1{+}x_2)^s\,(x_1^s\,x_2^{d-2\,s} - x_2^s\,x_1^{d-2\,s}) \tag{7.13}$$

clearly belong to $\ker(\sigma^*_{1,1})$. Consider now the sequences:

$$\mathcal{H}_{d;s_1,s_2} = \{H^\bullet_{d,s_1}, H^\bullet_{d,s_1+1}, \ldots, H^\bullet_{d,s_2-1}, H^\bullet_{d,s_2}\}. \tag{7.14}$$

The main facts here are these:
(i) The elements of $\mathcal{H}_{d;1,s_d}$ constitute a basis of $\ker(\sigma^*_{1,1})$.
(ii) The same holds for the shifted sets $\mathcal{H}_{d;1+k,s_d+k}$.
(iii) But it is only the first basis $\mathcal{H}_{d;1,s_d}$ that leads to arithmetically smooth expansions.

Natural basis for the *pre-carma* space
The *pre-carmas* (so-called because they are the raw material from which the *carmas* shall be built) are the elements of $\ker(\sigma^*_{1,1})$ which are bi-even (*i.e.* even separately in x_1 and x_2) and divisible by $x_1^2 x_2^2$.[99] The main result here[100] is that there exists a complete system of arithmetically smooth *pre-carmas* of the form:

$$\text{precarma}_{d,k}^{x_1,x_2} = Q_{\tau(d)}(x_1,x_2) R_8(x_1,x_2)^k S_4(x_1,x_2)^{\kappa(d)-k} T_{d,k}(x_1,x_2) \tag{7.15}$$

with $1 \le k \le \kappa(d) = \gamma(d-2)$ and γ as in (7.3) or, equivalently:

$$\kappa(d) = \text{ent}\left(\frac{d-2}{12}\right) \quad \text{if } d \neq 10 \mod 12 \qquad (\text{ent} = \textit{entire part})$$

$$\kappa(d) = \text{ent}\left(\frac{d-2}{12}\right) + 1 \text{ if } d = 10 \mod 12.$$

The first factor depends on $\tau(d) := \gcd(d, 12)$. It is of degree $\tau^*(d) := \tau(d)$ except when $12|d$, in which case $\tau^*(d) := 8$. It is given for the four possible values of $\tau(d)$ by:

$$Q_2 := x_1^2 - x_2^2,\quad Q_4 := x_1^4 - x_2^4,\quad Q_6 := x_1^6 - x_2^6,\quad Q_{12} := \frac{Q_4\, Q_6}{Q_2}.$$

The second factor, of degree 8, is given by:

$$R_8 := x_1^2 x_2^2 (x_1^2 - x_2^2)^2$$

and the reason for its spontaneous occurrence is that the six following polynomials are divisible by $x_1^2 x_2^2 (x_1+x_2)^2$:

$$R_8(x_i, x_j), \qquad R_8(x_i, x_i+x_j), \qquad R_8(x_i+x_j, x_j) \qquad \text{with } i, j \subset \{1,2\}.$$

[99] We add this last condition for the reason that one-variable elements of $\ker(\sigma^*_{1,1})$ would contribute no *carmas*: see the construction in Section 7.7.

[100] Arrived at by expanding the bi-even solutions of $\sigma^*_{1,1}.S^\bullet_{1,1} = 0$ in the 'good' basis $\mathcal{H}_{d;1,s_d}$.

The third factor, of degree 4, can be chosen arbitrarily, provided it is symmetric in (x_1, x_2), *even* in each variable, and co-prime with R_8. The following choices:

$$S_4 := \frac{Q_4^2}{Q_2^2} = (x_1^2 + x_2^2)^2; \qquad S_4 := \frac{Q_6}{Q_2} = x_1^4 + x_1^2 x_2^2 + x_2^4$$

are natural candidates to the extent that they introduce no *new* factors, but there seems to exist no really privileged choice, *i.e.* no choice that would render the last factor $T_{d,k}$ indisputably *simplest*.

That last factor, symmetric in x_1,x_2 and with the right degree $\delta(d,k)$,[101] is then fully determined by the condition $\sigma^*_{1,1}.precarma_{d,k} = 0$. It is thus simplest for k maximal, *i.e.* $k = \kappa$. The corresponding $precarma_{d,\kappa}$ is also the only fully canonical $precarma_{d,k}$, since it does not depend on the choice of S_4.

7.7 Construction of the *carma* bialternals

The idea behind the construction

Fix a polynomial basis $\{l\varnothing ma^\bullet_s, s = 3, 5, 7\ldots\}$ of $ALIL \subset ARI^{\underline{\mathrm{al}}/\underline{\mathrm{il}}}$ [102] and consider a pre-carma polynomial *precar* of total degree d+2 (recall that d has to be *even* and either $= 8$ or ≥ 12) with alternal coefficients $c_{2\delta_1,2\delta_2}$:

$$\mathrm{precar}^{x_1,x_2} = \sum_{2(\delta_1+\delta_2)=d+2}^{\delta_i\geq 1} c_{2\delta_1,2\delta_2}\; x_1^{2\delta_1}\, x_2^{2\delta_2}. \tag{7.16}$$

Next, form the bimould $c\varnothing r^\bullet$ by bracketting the $l\varnothing ma^\bullet_s$ with the coefficients $c_{2\delta_1,2\delta_2}$ as weights:

$$\mathrm{c\varnothing r}^\bullet := \sum_{2(\delta_1+\delta_2)=d+2}^{\delta_i\geq 1} c_{2\delta_1,2\delta_2}\; \mathrm{preari}(l\varnothing ma^\bullet_{1+2\delta_1}, l\varnothing ma^\bullet_{1+2\delta_2}) \in \mathrm{ALIL} \tag{7.17}$$

$$= \sum_{2(\delta_1+\delta_2)=d+2}^{\delta_i\geq 1} c_{2\delta_1,2\delta_2}\frac{1}{2}\; \mathrm{ari}(l\varnothing ma^\bullet_{1+2\delta_1}, l\varnothing ma^\bullet_{1+2\delta_2}) \in \mathrm{ALIL} \tag{7.18}$$

and consider the projection $c\varnothing rma^\bullet$ of $c\varnothing r^\bullet$ on $BIMU_4$. By construction, $c\varnothing r^\bullet$ is of type $\underline{\mathrm{al}}/\underline{\mathrm{il}}$ and its *first* non-vanishing component is therefore, *on*

---

[101] *I.e.* to ensure degree d for $precarma_{d,k}$. Thus $\delta(d,k) = d - \tau^*(d) - 4k - 4\kappa(d)$.

[102] As usual, *ALIL* and *ALAL* are short-hand for $ARI^{\underline{\mathrm{al}}/\underline{\mathrm{il}}}_{ent/cst}$ and $ARI^{\underline{\mathrm{al}}/\underline{\mathrm{al}}}_{ent/cst}$. Constructing a basis of *ALIL* is of course no easy matter, as we saw in Section 6, but what we require here is only a basis up to length 3, which is quite simple to construct: see Section 6.3.

its own, of type $\underline{\text{al}}/\underline{\text{al}}$, *i.e.* bialternal. That first component cannot have length $r = 2$, because *precar* is a pre-carma. It cannot have length $r = 3$ either, because the component of length 3 is a polynomial of *odd* degree $1 + d$ and for that reason cannot possibly be bialternal. This implies, therefore, that $c\phi rma^\bullet$, *i.e.* the component of length 4, is either $\equiv 0$ or a *non-trivial bialternal* of degree d. Based on extensive computational and theoretical evidence, we conjecture that *the latter is always the case*, and more precisely, that when *precar* runs through a basis of the precarma space, the corresponding $c\phi rma^\bullet$ span a space $C\O RMA_4$ such that

$$\mathrm{C\O RMA}_4 \oplus \mathrm{EKMA}_4 = \mathrm{ALAL}_4 \subset \mathrm{ARI}_4^{\underline{\mathrm{al}}/\underline{\mathrm{al}}}. \tag{7.19}$$

In simpler words: the $c\phi rma^\bullet$ provide *all the missing bialternals* of length $r = 4$ and put them in one-to-one correspondance with the *precarma*$^\bullet$, *i.e.* with the “unproductive” brackets of *ekma*$^\bullet$.

The construction works for any basis $\{l\phi ma_s^\bullet\}$ of *ALIL*. Specialising it to the three canonical bases $\{lama_s^\bullet\}$, $\{loma_s^\bullet\}$, $\{luma_s^\bullet\}$, we get three series of ‘exceptional’ bialternals $\{carma_s^\bullet\}$, $\{corma_s^\bullet\}$, $\{curma_s^\bullet\}$, spanning spaces $CARMA_4$, $CORMA_4$, $CURMA_4$ which, though distinct, each verify the complementarity relation (7.19).

7.8 Alternative approach

In the expansion (6.5) for $\{l\phi ma_s^\bullet\}$, let us retain only the first two singulates (those namely that contribute to the components of length $r \leq 4$) and then let us restrict everything to the homogeneous parts of weight s. We get:

$$l\phi ma_s^\bullet = \Sigma^\bullet_{[1],s} + \Sigma^\bullet_{[1,2],s} \qquad (\mathrm{mod}\ \mathrm{BIMU}_{5\leq}). \tag{7.20}$$

If we now plug this into (7.17) for pairs$(s_1,s_2) = (1+2\delta_1, 1+2\delta_2)$, we get four contributions $\mathcal{P}_{[r^1],[r^2]}$, consisting of the terms linear in $\Sigma^\bullet_{[r^1],1+2\delta_1}$ and $\Sigma^\bullet_{[r^2],1+2\delta_2}$. The contribution $\mathcal{P}_{[1,2],[1,2]}$ begins with a non-zero component of length 6 and therefore vanishes *modulo* $BIMU_{5\leq}$. The contribution $\mathcal{P}_{[1],[1]}$ vanishes *exactly*, for the reason that, *adari*(*pal*$^\bullet$) being an algebra isomorphism, $\mathcal{P}_{[1],[1]}$ is necessarily of the form *adari*(*pal*$^\bullet$)$.leng_4.\mathcal{P}_{[1],[1]}$, with $leng_r$ denoting as usual the projector of *BIMU* onto $BIMU_r$. But $leng_4.\mathcal{P}_{[1],[1]} \equiv 0$ since we have assumed *precar*$^\bullet$ of (7.16) to be a *precarma*. Thus $\mathcal{P}_{[1],[1]} \equiv 0$. That leaves only the two contributions $\mathcal{P}_{[1],[1,2]} + \mathcal{P}_{[1,2],[1]}$, whose component of length 4 is clearly a singulate of type

$\Sigma\Sigma^{\bullet}_{[1,1,2]} := \mathit{slank}_{[1,1,2]}.S^{\bullet}_{[1,1,2]}$. It can in fact be shown to be of the form:

$$\begin{aligned}
\Sigma\Sigma^{w_1,w_2,w_3,w_4}_{[1,1,2]} &= (\mathrm{slank}_{[1,1,2]}.S_{[1,1,2]})^{w_1,w_2,w_3,w_4} \\
&= +X^{u_1,u_2,u_3,u_4}\, P(u_0) + Y^{u_1,u_2,u_3,u_4}\, P(u_2+u_3) \\
&\quad + X^{u_2,u_3,u_4,u_0}\, P(u_1) + Y^{u_2,u_3,u_4,u_0}\, P(u_3+u_4) \\
&\quad + X^{u_3,u_4,u_0,u_1}\, P(u_2) + Y^{u_3,u_4,u_0,u_1}\, P(u_4+u_0) \\
&\quad + X^{u_4,u_0,u_1,u_2}\, P(u_3) + Y^{u_4,u_0,u_1,u_2}\, P(u_0+u_1) \\
&\quad + X^{u_0,u_1,u_2,u_3}\, P(u_4) + Y^{u_0,u_1,u_2,u_3}\, P(u_1+u_2)
\end{aligned} \tag{7.21}$$

with polynomials $X^{\bullet}$ and $Y^{\bullet}$ given by:

$$\begin{aligned}
2\,X^{u_1,u_2,u_3,u_4} &= S^{u_3,u_2,u_1}_{[1,1,2]} + S^{u_1,u_3,u_4}_{[1,1,2]} + S^{u_3,u_4,u_1}_{[1,1,2]} + S^{u_3,u_1,u_4}_{[1,1,2]} - S^{u_2,u_1,u_4}_{[1,1,2]} \\
&- S^{u_2,u_3,u_4}_{[1,1,2]} - S^{u_2,u_4,u_1}_{[1,1,2]} - S^{u_4,u_2,u_1}_{[1,1,2]} + S^{u_3,u_1,u_{12}}_{[1,1,2]} + S^{u_4,u_2,u_{12}}_{[1,1,2]} + S^{u_2,u_4,u_{12}}_{[1,1,2]} + S^{u_1,u_3,u_{12}}_{[1,1,2]} \\
&+ S^{u_2,u_3,u_{34}}_{[1,1,2]} + S^{u_4,u_1,u_{34}}_{[1,1,2]} + S^{u_1,u_4,u_{34}}_{[1,1,2]} + S^{u_3,u_2,u_{34}}_{[1,1,2]} + S^{u_2,u_{12},u_4}_{[1,1,2]} + S^{u_2,u_{23},u_4}_{[1,1,2]} \\
&+ S^{u_2,u_{23},u_1}_{[1,1,2]} + S^{u_4,u_{34},u_1}_{[1,1,2]} - S^{u_2,u_3,u_{12}}_{[1,1,2]} - S^{u_3,u_2,u_{12}}_{[1,1,2]} - S^{u_4,u_1,u_{12}}_{[1,1,2]} - S^{u_1,u_4,u_{12}}_{[1,1,2]} - S^{u_1,u_3,u_{34}}_{[1,1,2]} \\
&- S^{u_4,u_2,u_{34}}_{[1,1,2]} - S^{u_3,u_1,u_{34}}_{[1,1,2]} - S^{u_2,u_4,u_{34}}_{[1,1,2]} - S^{u_1,u_{12},u_4}_{[1,1,2]} - S^{u_3,u_{23},u_4}_{[1,1,2]} - S^{u_3,u_{23},u_1}_{[1,1,2]} - S^{u_3,u_{34},u_1}_{[1,1,2]} \\
&+ S^{u_2,u_3,u_{123}}_{[1,1,2]} + S^{u_2,u_1,u_{123}}_{[1,1,2]} + S^{u_2,u_4,u_{234}}_{[1,1,2]} + S^{u_4,u_2,u_{234}}_{[1,1,2]} - S^{u_1,u_3,u_{123}}_{[1,1,2]} - S^{u_3,u_1,u_{123}}_{[1,1,2]} \\
&- S^{u_3,u_4,u_{234}}_{[1,1,2]} - S^{u_3,u_2,u_{234}}_{[1,1,2]} + S^{u_1,u_{12},u_{123}}_{[1,1,2]} + S^{u_3,u_{23},u_{123}}_{[1,1,2]} + S^{u_3,u_{23},u_{234}}_{[1,1,2]} + S^{u_3,u_{34},u_{234}}_{[1,1,2]} \\
&- S^{u_2,u_{23},u_{123}}_{[1,1,2]} - S^{u_2,u_{12},u_{123}}_{[1,1,2]} - S^{u_2,u_{23},u_{234}}_{[1,1,2]} - S^{u_4,u_{34},u_{234}}_{[1,1,2]}
\end{aligned}$$

$$\begin{aligned}
2\,Y^{u_1,u_2,u_3,u_4} &= S^{u_4,u_3,u_1}_{[1,1,2]} + S^{u_1,u_4,u_2}_{[1,1,2]} + S^{u_4,u_1,u_2}_{[1,1,2]} + S^{u_1,u_3,u_4}_{[1,1,2]} + S^{u_3,u_4,u_1}_{[1,1,2]} + S^{u_3,u_1,u_4}_{[1,1,2]} \\
&- S^{u_2,u_1,u_4}_{[1,1,2]} - S^{u_1,u_4,u_3}_{[1,1,2]} - S^{u_4,u_2,u_1}_{[1,1,2]} - S^{u_1,u_2,u_4}_{[1,1,2]} - S^{u_2,u_4,u_1}_{[1,1,2]} - S^{u_4,u_1,u_3}_{[1,1,2]} + S^{u_3,u_4,u_{123}}_{[1,1,2]} \\
&+ S^{u_4,u_3,u_{123}}_{[1,1,2]} + S^{u_1,u_3,u_{234}}_{[1,1,2]} + S^{u_3,u_1,u_{234}}_{[1,1,2]} + S^{u_4,u_{123},u_3}_{[1,1,2]} + S^{u_1,u_{234},u_3}_{[1,1,2]} - S^{u_4,u_2,u_{123}}_{[1,1,2]} \\
&- S^{u_2,u_4,u_{123}}_{[1,1,2]} - S^{u_2,u_1,u_{234}}_{[1,1,2]} - S^{u_1,u_2,u_{234}}_{[1,1,2]} - S^{u_4,u_{123},u_2}_{[1,1,2]} - S^{u_1,u_{234},u_2}_{[1,1,2]} + S^{u_2,u_{1234},u_4}_{[1,1,2]} \\
&+ S^{u_2,u_{1234},u_1}_{[1,1,2]} + S^{u_{1234},u_4,u_3}_{[1,1,2]} + S^{u_{1234},u_2,u_4}_{[1,1,2]} + S^{u_{1234},u_2,u_1}_{[1,1,2]} + S^{u_{1234},u_1,u_3}_{[1,1,2]} - S^{u_3,u_{1234},u_1}_{[1,1,2]} \\
&- S^{u_3,u_{1234},u_4}_{[1,1,2]} - S^{u_{1234},u_3,u_4}_{[1,1,2]} - S^{u_{1234},u_4,u_2}_{[1,1,2]} - S^{u_{1234},u_3,u_1}_{[1,1,2]} - S^{u_{1234},u_1,u_2}_{[1,1,2]} \\
&+ S^{u_2,u_{1234},u_{123}}_{[1,1,2]} + S^{u_2,u_{1234},u_{234}}_{[1,1,2]} + S^{u_{1234},u_2,u_{123}}_{[1,1,2]} + S^{u_{1234},u_2,u_{234}}_{[1,1,2]} + S^{u_{1234},u_{123},u_2}_{[1,1,2]} \\
&+ S^{u_{1234},u_{234},u_2}_{[1,1,2]} - S^{u_3,u_{1234},u_{123}}_{[1,1,2]} - S^{u_3,u_{1234},u_{234}}_{[1,1,2]} - S^{u_{1234},u_3,u_{123}}_{[1,1,2]} - S^{u_{1234},u_3,u_{234}}_{[1,1,2]} \\
&- S^{u_{1234},u_{123},u_3}_{[1,1,2]} - S^{u_{1234},u_{234},u_3}_{[1,1,2]}
\end{aligned}$$

and with a singuland $S^{\bullet}_{[1,1,2]}$ that has to be a homogeneous polynomial of total degree $1+d$ subject to three types of constraints.

First, it must be *even* in x_1, x_2, *odd* in x_3, and divisible by $x_1\,x_2\,x_3$.

Second, it must verify the identity:

$$\begin{aligned}0 = &-S_{[1,1,2]}^{x_1,x_2,x_3} + S_{[1,1,2]}^{x_1,x_2,x_{23}} + S_{[1,1,2]}^{x_2,x_1,x_{23}} + S_{[1,1,2]}^{x_1,x_{23},x_3} + S_{[1,1,2]}^{x_1,x_{12},x_3} - S_{[1,1,2]}^{x_1,x_3,x_{23}}\\ &- S_{[1,1,2]}^{x_3,x_1,x_{23}} - S_{[1,1,2]}^{x_2,x_{12},x_3} + S_{[1,1,2]}^{x_1,x_3,x_{123}} + S_{[1,1,2]}^{x_3,x_1,x_{123}} + S_{[1,1,2]}^{x_2,x_{123},x_3} + S_{[1,1,2]}^{x_{123},x_2,x_3}\\ &- S_{[1,1,2]}^{x_2,x_1,x_{123}} - S_{[1,1,2]}^{x_2,x_3,x_{123}} - S_{[1,1,2]}^{x_1,x_{123},x_3} + S_{[1,1,2]}^{x_2,x_{23},x_{123}} + S_{[1,1,2]}^{x_2,x_{12},x_{123}} + S_{[1,1,2]}^{x_3,x_{123},x_{23}}\\ &+ S_{[1,1,2]}^{x_{123},x_3,x_{23}} - S_{[1,1,2]}^{x_1,x_{12},x_{123}} - S_{[1,1,2]}^{x_3,x_{23},x_{123}} - S_{[1,1,2]}^{x_2,x_{123},x_{23}} - S_{[1,1,2]}^{x_{123},x_2,x_{23}} - S_{[1,1,2]}^{x_{123},x_{23},x_3}\end{aligned}$$

which ensures the absence of poles at the origin and therefore, in the terminology of Section 5.9, makes $\Sigma\Sigma^{\bullet}_{[1,1,2]}$ into a *wandering bialternal.*

Lastly, it must verify a third, similar-looking identity, which reflects the fact that *precar*$^\bullet$ is a *pre-carma* and, by so doing, guarantees that the bialternal $\Sigma\Sigma^{\bullet}_{[1,1,2]}$ won't be in *EKMA*.

Caveat: for each d, there is exactly one *carma* bialternal that is not captured by the above formula (7.21) but by a slight modification of the same.[103] This, however, is a minor technicality.

7.9 The global bialternal ideal and the universal 'restoration' mechanism

Suppose that, contrary to all evidence (see Section 8.5) the ideal *IDEKMA* is not generated by *IDEKMA*$_2$, *i.e.* by the sole *pre-carmas*. There would then exist at least one $r > 2$ and one identity of the form:

$$\sum_{2(\delta_1+\ldots+\delta_r)=2\delta}^{\delta_i\geq 1} c_{2\delta_1,\ldots,2\delta_r}\, x_1^{2\delta_1}\ldots x_r^{2\delta_r}\,\mathrm{ari}(\mathrm{ekma}^{\bullet}_{2\delta_1},\ldots,\mathrm{ekma}^{\bullet}_{2\delta_r})\equiv 0 \quad (7.22)$$

corresponding to a *'prime'* (*i.e.* non-derivative) element of *IDEKMA*$_r$. We might then form the polynomial *prehar*:

$$\mathrm{prehar}^{x_1,\ldots,x_r} = \sum_{2(\delta_1+\ldots+\delta_r)=2\delta}^{\delta_i\geq 1} c_{2\delta_1,\ldots,2\delta_r}\, x_1^{2\delta_1}\ldots x_r^{2\delta_r} \quad (7.23)$$

as an analogue of *precar* (see Section 7.7) and then use the alternal coefficients of *prehar* to construct a bimould *hør*$^\bullet$:

$$\mathrm{hør}^{\bullet} := \sum_{2(\delta_1+\ldots+\delta_r)=2\delta}^{\delta_i\geq 1} c_{2\delta_1,\ldots,2\delta_r}\,\mathrm{preari}(\mathit{l\o ma}^{\bullet}_{1+2\delta_1},\ldots,\mathit{l\o ma}^{\bullet}_{1+2\delta_r})\in \mathit{ALIL} \quad (7.24)$$

$$= \sum_{2(\delta_1+\ldots+\delta_r)=2\delta}^{\delta_i\geq 1} c_{2\delta_1,\ldots,2\delta_r}\,\frac{1}{r}\,\mathrm{ari}(\mathit{l\o ma}^{\bullet}_{1+2\delta_1},\ldots,\mathit{l\o ma}^{\bullet}_{1+2\delta_r})\in \mathit{ALIL} \quad (7.25)$$

[103] Due to the presence of the corrective term $Ca^{\bullet}_3/Ci^{\bullet}_3$ in the formula linking the components of length 3 and weight 3 of *løma*$^\bullet$/*lømi*$^\bullet$. See (6.3), (6.4).

exactly analogous to $c\phi r^\bullet$. By arguing on the same lines as in Section 7.7, we would see that the first non-vanishing component $h\phi rma^\bullet$ of $h\phi r^\bullet$, necessarily of *even* degree $2\,\delta-2\,k$ and therefore of length $r+2\,k$ with $k \geq 1$, would automatically provide an 'exceptional' bialternal that would 'make up' for the missing element of *EKMA* corresponding to (7.22). Although, in keeping with our general conjectures, the existence of *prime* relations (7.22) is most unlikely, it is reasonable to speculate that, *if perchance they exist*, the corresponding $h\phi rma^\bullet$ must then have length $r+2$ and degree $2\,\delta-2$, although they might conceivably have length $r+2\,k$ and degree $2\,\delta-2\,k$ for some $k \geq 2$. In any case, we have here a transparent *stop-gap mechanism* which automatically associates one *exceptional* bialternal to any 'missing' *regular* bialternal.

8 The enumeration of bialternals. Conjectures and computational evidence

8.1 Primary, sesquary, secondary algebras

Before addressing the enumeration of bialternals, let us return to the main subalgebras $\mathcal{A}$ of *ARI* listed in Section 2.5, but in the special case of bimoulds that are polynomial in $\boldsymbol{u}$ and constant in $\boldsymbol{v}$. For each such subalgebra $\mathcal{A}$, we tabulate the dimension $dim(\mathcal{A}_{r,d})$ of the cells of length $r \geq 3$ and of total $\boldsymbol{u}$-degree d. The reason for neglecting the length $r = 1$ respectively 2 is that the results there are trivial respectively elementary.[104] As in Section 2.5, we reserve bold-face for the *secondary* subalgebras.

$r=3 \quad \backslash\backslash \quad d =$	*1*	*2*	*3*	*4*	*5*	*6*	*7*	*8*	*9*	*10*	*11*	*12*	*13*	*14*
$\mathbf{ARI}^{al/*}$	**1**	**2**	**3**	**5**	**7**	**9**	**12**	**15**	**18**	**22**	**26**	**30**	**35**	**40**
$\mathrm{ARI}^{mantar/*}$	2	4	6	9	12	16	20	25	30	36	42	49	56	64
$\mathrm{ARI}^{pusnu/*}$	2	4	6	10	14	18	24	30	36	44	52	60	70	80
$\mathrm{ARI}^{pusnu/*}_{mantar/*}$	1	2	3	5	7	9	12	15	18	22	26	30	35	40
$\mathbf{ARI}^{\underline{al}/\underline{al}}$	**0**	**0**	**0**	**0**	**0**	**0**	**0**	**1**	**0**	**2**	**0**	**2**	**0**	**4**
$\mathrm{ARI}^{al/push}$	0	1	0	2	1	3	2	5	3	7	5	9	7	12
ARI^{push}	0	2	2	5	4	8	8	13	12	18	18	25	24	32
$\mathrm{ARI}^{\overline{pusnu}/\overline{pusnu}}_{\underline{mantar}/.}$	1	0	1	0	2	0	3	1	4	2	6	2	8	4
$\mathrm{ARI}^{\overline{pusnu}/\overline{pusnu}}$	1	2	2	5	7	8	12	15	17	22	26	29	35	40

[104] Since for $r = 2$ the constraints that define $\mathcal{A}$ are always *finitary*.

$r=4 \quad \backslash\backslash \quad d=$	*1*	*2*	*3*	*4*	*5*	*6*	*7*	*8*	*9*	*10*	*11*	*12*	*13*	*14*
$\mathbf{ARI}^{\mathbf{al}/*}$	**1**	**2**	**5**	**8**	**14**	**20**	**30**	**40**	**55**	**70**	**91**	**112**	**140**	**168**
$\mathrm{ARI}^{\mathrm{mantar}/*}$	2	5	10	16	28	40	60	80	110	140	182	224	280	336
$\mathrm{ARI}^{\mathrm{pusnu}/*}$	3	7	15	25	42	62	90	122	165	213	273	339	420	508
$\mathrm{ARI}^{\mathrm{pusnu}/*}_{\mathrm{mantar}/*}$	2	4	9	14	24	34	50	66	90	114	147	180	224	268
$\mathbf{ARI}^{\underline{\mathbf{al}}/\underline{\mathbf{al}}}$	**0**	**0**	**0**	**0**	**0**	**0**	**0**	**1**	**0**	**1**	**0**	**3**	**0**	**5**
$\mathrm{ARI}^{\mathrm{al}/\mathrm{push}}$	0	0	1	1	3	3	6	7	11	13	18	21	28	32
$\mathrm{ARI}^{\mathrm{push}}$	1	4	5	7	12	16	24	33	44	58	72	91	112	136
$\mathrm{ARI}^{\overline{\mathrm{pusnu}}/\overline{\mathrm{pusnu}}}_{\mathrm{mantar}/*}$	1	2	4	6	10	15	20	28	35	48	56	74	84	109
$\mathrm{ARI}^{\overline{\mathrm{pusnu}}/\overline{\mathrm{pusnu}}}$	2	4	10	15	28	40	60	79	110	140	182	223	280	336

$r=5 \quad \backslash\backslash \quad d=$	*1*	*2*	*3*	*4*	*5*	*6*	*7*	*8*	*9*	*10*	*11*	*12*	*13*	*14*
$\mathbf{ARI}^{\mathbf{al}/*}$	**1**	**3**	**7**	**14**	**25**	**42**	**66**	**99**	**143**	**200**	**273**	**364**	**476**	**612**
$\mathrm{ARI}^{\mathrm{mantar}/*}$	3	9	19	38	66	110	170	255	365	511	693	924	1204	1548
$\mathrm{ARI}^{\mathrm{pusnu}/*}$	4	12	28	56	100	168	264	396	572	800	1092	1456	1904	2448
$\mathrm{ARI}^{\mathrm{pusnu}/*}_{\mathrm{mantar}/*}$	2	6	14	28	50	84	132	198	286	400	546	728	952	1224
$\mathbf{ARI}^{\underline{\mathbf{al}}/\underline{\mathbf{al}}}$	**0**	**0**	**0**	**0**	**0**	**0**	**0**	**0**	**0**	**1**	**0**	**2**	**0**	**5**
$\mathrm{ARI}^{\mathrm{al}/\mathrm{push}}$	0	1	1	3	3	9	9	19	22	36	42	66	74	108
$\mathrm{ARI}^{\mathrm{push}}$	1	5	6	12	20	38	52	85	118	169	224	310		
$\mathrm{ARI}^{\overline{\mathrm{pusnu}}/\overline{\mathrm{pusnu}}}_{\mathrm{mantar}/*}$	2	3	8	14	26	42	69	99		200		364		612
$\mathrm{ARI}^{\overline{\mathrm{pusnu}}/\overline{\mathrm{pusnu}}}$	2	6	14	28	50	84	132	198	286	400	546	728	952	1224.

Let us now tabulate the corresponding generating functions. These are always *rational*. For brevity, we set $X_m^n := (1-x^m)^{-n}$.

$r=3$	*generating function* $\sum \dim(d)\, x^d$
$\mathbf{ARI}^{\mathbf{al}/*}$	$x\, X_1^2\, X_3^1$
$\mathrm{ARI}^{\mathrm{mantar}/*}$	$x\,(2-2x^2+x^3)\, X_1^2\, X_2^1$
$\mathrm{ARI}^{\mathrm{pusnu}/*}$	$2x\, X_1^2\, X_3^1$
$\mathrm{ARI}^{\mathrm{pusnu}/*}_{\mathrm{mantar}/.}$	$x\, X_1^2\, X_3^1$
$\mathbf{ARI}^{\underline{\mathbf{al}}/\underline{\mathbf{al}}}$	$x^8\,(1+x^2-x^4)\, X_2^1\, X_4^1\, X_6^1$
$\mathrm{ARI}^{\mathrm{al}/\mathrm{push}}$	$x^2\, X_2^2\, X_3^1$
$\mathrm{ARI}^{\mathrm{push}}$	$x^2\,(2+x^2-x^3-x^4+x^5)\, X_1^1\, X_2^1\, X_4^1$
$\mathrm{ARI}^{\overline{\mathrm{pusnu}}/\overline{\mathrm{pusnu}}}_{\mathrm{mantar}/*}$	$x\,(1+x^7+x^9+x^{10}-x^{11})\, X_2^1\, X_4^1\, X_6^1$
$\mathrm{ARI}^{\overline{\mathrm{pusnu}}/\overline{\mathrm{pusnu}}}$	$x\,(1-x+x^2)\,(1+x-x^2)\, X_1^2\, X_3^1$

$r=4$		*generating function* $\sum \dim(d)\, x^d$
$\mathbf{ARI}^{\mathrm{al}/*}$	$\Vert$	$x\, X_1^2\, X_2^2$
$\mathrm{ARI}^{\mathrm{mantar}/*}$	$\Vert$	$2\,x\, X_1^2\, X_2^2$
$\mathrm{ARI}^{\mathrm{pusnu}/*}$	$\Vert$	$x\,(3+x+x^2+x^3)\, X_1^2\, X_2^1\, X_4^1$
$\mathrm{ARI}^{\mathrm{pusnu}/*}_{\mathrm{mantar}/*}$	$\Vert$	$x\,(2+x^2)\, X_1^2\, X_2^1\, X_4^1$
$\mathbf{ARI}^{\underline{\mathrm{al}}/\underline{\mathrm{al}}}$	$\Vert$	$x^8\,(1+2\,x^4+x^6+x^8+2\,x^{10}+x^{14}-x^{16})\, X_2^1\, X_6^1\, X_8^1\, X_{12}^1$
$\mathrm{ARI}^{\mathrm{al}/\mathrm{push}}$	$\Vert$	$x^3\, X_1^1\, X_2^2\, X_5^1$
$\mathrm{ARI}^{\mathrm{push}}$	$\Vert$	$x\,(1+x-4x^2+3x^3+2x^4-5x^5+4x^6+x^7-2x^8-x^9+x^{10})\,X_1^3 X_5^1$
$\mathrm{ARI}^{\overline{\mathrm{pusnu}}/\overline{\mathrm{pusnu}}}_{\mathrm{mantar}/*}$	$\Vert$	$x\,(1+x+x^5-x^6)\, X_1^1\, X_2^2\, X_4^1$
$\mathrm{ARI}^{\overline{\mathrm{pusnu}}/\overline{\mathrm{pusnu}}}$	$\Vert$	$x\,(2+2\,x^2-x^3+2\,x^4-2\,x^6+x^7)\, X_1^2\, X_2^1\, X_4^1$

$r=5$		*generating function* $\sum \dim(d)\, x^d$
$\mathbf{ARI}^{\mathrm{al}/*}$	$\Vert$	$x\,(1+x^3)\, X_1^3\, X_2^1\, X_5^1$
$\mathrm{ARI}^{\mathrm{mantar}/*}$	$\Vert$	$x\,(3-5\,x^2+5\,x^3+x^4-3\,x^5+x^6)\, X_1^3\, X_2^2$
$\mathrm{ARI}^{\mathrm{pusnu}/*}$	$\Vert$	$4\,x\,(1+x^3)\, X_1^3\, X_2^1\, X_5^1$
$\mathrm{ARI}^{\mathrm{pusnu}/*}_{\mathrm{mantar}/*}$	$\Vert$	$2\,x\,(1+x^3)\, X_1^3\, X_2^1\, X_5^1$
$\mathbf{ARI}^{\underline{\mathrm{al}}/\underline{\mathrm{al}}}$	$\Vert$	$x^{10}\,(1+2\,x^2+3\,x^4+3\,x^6+2\,x^8)\, X_4^2\, X_6^2\, X_{10}^1$
$\mathrm{ARI}^{\mathrm{al}/\mathrm{push}}$	$\Vert$	$x^2\,(1+x+x^2+3x^4+2x^5+x^6+x^7+2x^8)\, X_2^2\, X_3^1\, X_5^1\, X_6^1$
$\mathrm{ARI}^{\mathrm{push}}$	$\Vert$	???
$\mathrm{ARI}^{\overline{\mathrm{pusnu}}/\overline{\mathrm{pusnu}}}_{\mathrm{mantar}/*}$	$\Vert$	???
$\mathrm{ARI}^{\overline{\mathrm{pusnu}}/\overline{\mathrm{pusnu}}}$	$\Vert$	$2\,x\,(1+x^3)\, X_1^3\, X_2^1\, X_5^1$.

8.2 The 'factor' algebra *EKMA* and its subalgebra *DOMA*

Of these two subalgebras of *ALAL*, generated respectively by the *ekmas* and *domas*, the first is obviously far from free (though all relations between the *ekmas* are conjectured to be generated by the sole *bilinear* relations) but the second is conjectured to be free, with the $doma^{\bullet}_{d,b}$, of length 2, as canonical generators.

The main unresolved point, even at the conjectural level, is this: how much of *EKMA* must one 'add' to *DOMA* to recover (ideally, with unique decomposition) the whole of *EKMA*? While the inclusion

$$\mathrm{DOMA} \oplus \mathrm{ari}(\mathrm{DOMA}, \mathrm{EKMA}_1) \subset \mathrm{EKMA}$$

is strict, the (rather small) gap between the two spaces would seem to be bridgeable, but exactly how is unclear at the moment.

8.3 The 'factor' algebra *CARMA*

Like *DOMA*, *CARMA* is conjectured to be free (the theoretical case as well as the computational evidence here are even more overwhelming) but, unlike *DOMA*, it is not intrinsically defined: it exists in various isomorphic realisations (some canonical), all of which are conjectured to verify:

$$\mathrm{EKMA}\,\widehat{\otimes}\,\mathrm{CARMA} = \mathrm{ALAL}$$

with the notation $E\,\widehat{\otimes}\,C = A$ (not a tensor product!) signalling that A is freely generated by E and C, *i.e.* without constraints other than those internal to E and C: see Section 8.5, $\boldsymbol{C_1}$ *infra*.

8.4 The total algebra of bialternals *ALAL* and the original BK-conjecture

How many multizeta irreducibles of weight s and length r must one retain to freely generate the $\mathbb{Q}$-ring $\mathbb{Z}eta$ of formal (uncoloured) multizetas? How many independent bialternals of weight s and length r are there in *ALAL*? It is easy to show that the answer to both questions is the same number $\mathcal{D}_{s,r}$, but harder to find these numbers. Based on their numerical investigation of *genuine* rather that *formal* multizetas, and on the assumptions that both rings are actually "the same", Broadhurst and Kreimer conjectured in [1] that the $\mathcal{D}_{s,r}$ are deducible, after Möbius inversion, from the formula:

$$\prod_{2\leq d,1\leq r} \left(1 - z^s\, y^r\right)^{\mathcal{D}_{s,r}} = 1 - \frac{z^3\, y}{1-z^2} + \frac{z^{12}y^2(1-y^2)}{(1-z^4)(1-z^6)}. \tag{8.1}$$

8.5 The factor algebras and our sharper conjectures

$\boldsymbol{C_1}$ **:** Under the *ari*-bracket, the factor algebras *EKMA* and *CARMA* *freely* generate the total algebra *ALAL* of all polynomial bialternals. *Freely* means: without other relations than those internal to each factor algebra.

$\boldsymbol{C_2}$ **:** Only the factor *EKMA* has internal relations, and all of these are generated by the bilinear relations between the $\{ekma^{\bullet}_{d}\,;\ d=2,4,6\ldots\}$. We recall[105] that for each even degree d there are exactly $[[\frac{d-2}{4}]]-[[\frac{d}{6}]]$ such relations.[106]

$\boldsymbol{C_3}$ **:** The $\{doma^{\bullet}_{d,\delta}\,;\ d=6,8\ldots,\delta\leq[[\frac{d}{6}]]\}$ freely generate *DOMA*.

$\boldsymbol{C_4}$ **:** The $\{carma^{\bullet}_{d,\delta};\ d=8,12,14\ldots,\delta\leq[[\frac{d}{4}]]-[[\frac{d+2}{6}]]\}$ freely generate *CARMA*.

If we now denote $D_{d,r}$, $D^{ek}_{d,r}$, $D^{do}_{d,r}$, $D^{car}_{d,r}$ the dimensions of the cells of *ALAL, EKMA, DOMA, CARMA* of degree d and length r, the above conjectures translate into the following formulas:

$$\boldsymbol{C^*_1}:\prod_{2\leq d,1\leq r}\left(1-x^d\,y^r\right)^{D_{d,r}}=1-\frac{x^2\,y}{1-x^2}+\frac{x^8y^2(x^2-y^2)}{(1-x^4)(1-x^6)}\tag{8.2}$$

$$\boldsymbol{C^*_2}:\prod_{2\leq d,1\leq r}\left(1-x^d\,y^r\right)^{D^{ek}_{d,r}}=1-\frac{x^2\,y}{1-x^2}+\frac{x^{10}y^2}{(1-x^4)(1-x^6)}\tag{8.3}$$

$$\boldsymbol{C^*_3}:\prod_{6\leq d,1\leq r}\left(1-x^d\,y^r\right)^{D^{do}_{d,r}}=1-\frac{x^6\,y^2}{(1-x^2)\,(1-x^6)}\tag{8.4}$$

$$\boldsymbol{C^*_4}:\prod_{8\leq d,1\leq r}\left(1-x^d\,y^r\right)^{D^{car}_{d,r}}=1-\frac{x^8\,y^4}{(1-x^4)(1-x^6)}.\tag{8.5}$$

Formula $\boldsymbol{C^*_1}$ merely restates the classical BK-conjecture in the (d,r)-parameters, but $\boldsymbol{C^*_2}$, $\boldsymbol{C^*_3}$, $\boldsymbol{C^*_4}$ are sharp improvements. Above all, these formulas, together with the compellingly natural *restoration mechanism*[107] that underpins them, provide a convincing explanation for the complicated y^4-term in $\boldsymbol{C^*_1}$ and completely divest it of its mysterious character.

For explicitness, we shall now list the partial generating functions $D^*_r(x)=\sum D^*_{d,r}x^d$ for each algebra and the first lengths r.

---

[105] See Section 7.2.

[106] With $[[x]]:=$ *entire part of x*.

[107] See Section 7.7 and Section 7.9.

8.6 Cell dimensions for *ALAL*

$$D_1 = \frac{x^2}{(1-x^2)}$$

$$D_2 = \frac{x^6}{(1-x^2)\,(1-x^6)}$$

$$D_3 = \frac{x^8\,(1+x^2-x^4)}{(1-x^2)\,(1-x^4)\,(1-x^6)}$$

$$D_4 = \frac{x^8\,(1+2\,x^4+x^6+x^8+2\,x^{10}+x^{14}-x^{16})}{(1-x^2)\,(1-x^6)\,(1-x^8)\,(1-x^{12})}$$

$$D_5 = \frac{x^{10}\,(1+2\,x^2+3\,x^4+3\,x^6+2\,x^8)}{(1-x^4)^2\,(1-x^6)^2\,(1-x^{10})}$$

$$\begin{aligned} D_6 = {} & x^{12}(1+x^8-2x^{10}+x^{14}-4x^{16}+4x^{18}-2x^{20}-x^{22}+2x^{24}-2x^{26} \\ & +2\,x^{28}-x^{32}+3\,x^{34}-3\,x^{36}+x^{38})(1-x^2)^{-3}(1-x^4)^{-1} \\ & \times (1-x^6)^{-1}(1-x^8)^{-1}(1-x^{12})^{-1}(1-x^{18})^{-1} \end{aligned}$$

$$\begin{aligned} D_7 = {} & x^{14}(1+4\,x^2+8x^4+8x^6+6x^8+4x^{10}+5x^{12}+6x^{14}+3x^{16}-2x^{18} \\ & -3\,x^{20}-x^{22}+x^{24}+x^{26})(1-x^4)^{-3}(1-x^6)^{-3}(1-x^{14})^{-1} \end{aligned}$$

$$\begin{aligned} D_8 = {} & x^{16}\,(1+3\,x^2+7\,x^4+8\,x^6+13\,x^8+14\,x^{10}+15\,x^{12}+16\,x^{14}+8\,x^{16} \\ & +10\,x^{18}+4\,x^{22}-3\,x^{24}+x^{26}-2\,x^{28}+x^{30}+x^{34}) \\ & \times (1-x^2)^{-2}(1-x^6)^{-2}(1-x^8)^{-2}(1-x^{12})^{-2} \end{aligned}$$

8.7 Cell dimensions for *EKMA*

$$D_1^{ek} = \frac{x^2}{(1-x^2)}$$

$$D_2^{ek} = \frac{x^6}{(1-x^2)\,(1-x^6)}$$

$$D_3^{ek} = \frac{x^8\,(1+x^2-x^4)}{(1-x^2)\,(1-x^4)\,(1-x^6)}$$

$$D_4^{ek} = \frac{x^{10}\,(1+x^2+2\,x^4+x^6+2\,x^8+x^{10}-x^{16})}{(1-x^2)\,(1-x^6)\,(1-x^8)\,(1-x^{12})}$$

$$D_5^{ek} = \frac{x^{12}\,(1+3\,x^2+4\,x^4+3\,x^6+x^8+x^{10}-x^{14}-x^{16})}{(1-x^4)^2\,(1-x^6)^2\,(1-x^{10})}$$

$$\begin{aligned} D_6^{ek} = {} & x^{14}\,(1+x^4-x^6+x^8-2\,x^{10}-x^{14}+x^{16}-x^{18}-x^{20}+x^{22} \\ & -x^{26}+2\,x^{28}+x^{34}-x^{36})(1-x^2)^{-3}(1-x^4)^{-1}(1-x^8)^{-1} \\ & \times (1-x^6)^{-1}(1-x^{12})^{-1}(1-x^{18})^{-1} \end{aligned}$$

$$\begin{aligned} D_7^{ek} = {} & x^{16}\,(1+4\,x^2+8\,x^4+10\,x^6+8\,x^8+6\,x^{10}+6\,x^{12}+6\,x^{14}+2\,x^{16} \\ & -3\,x^{18}-5\,x^{20}-3\,x^{22}+x^{26})(1-x^4)^{-3}(1-x^6)^{-3}(1-x^{14})^{-1} \end{aligned}$$

$$\begin{aligned} D_8^{ek} = {} & x^{18}\,(1+2x^2+7x^4+8x^6+17x^8+14x^{10}+23x^{12}+13x^{14}+17x^{16} \\ & +6\,x^{18}+3\,x^{20}-x^{22}-5\,x^{24}-2\,x^{26}-5\,x^{28}-x^{30}-x^{32}+x^{36}) \\ & \times (1-x^2)^{-2}(1-x^6)^{-2}(1-x^8)^{-2}(1-x^{12})^{-2} \end{aligned}$$

8.8 Cell dimensions for *DOMA*.

$$D_1^{do}=D_3^{do} = D_5^{do} \dots = 0$$

$$D_2^{do}=\frac{x^6}{(1-x^2)\,(1-x^6)}$$

$$D_4^{do}= \frac{x^{14}\,(1+x^4)}{(1-x^2)\,(1-x^4)\,(1-x^6)\,(1-x^{12})}$$

$$D_6^{do} =\frac{x^{20}\,(1+x^{10})}{(1-x^2)^2\,(1-x^4)\,(1-x^6)^2\,(1-x^{18})}$$

$$D_8^{do}=\frac{x^{26}\,(1+x^2)\,(1+x^4)\,(1+x^8)}{(1-x^2)\,(1-x^4)^3\,(1-x^6)^2\,(1-x^{12})^2}$$

$$D_{10}^{do}=\frac{x^{32}(1+x^4+2x^8+2x^{10}+x^{12}+x^{14}+4x^{18}+x^{22}+x^{24}+2x^{26}+2x^{28}+x^{32}+x^{36})}{(1-x^2)^3\,(1-x^4)\,(1-x^6)^3\,(1-x^{10})\,(1-x^{12})\,(1-x^{30})}.$$

8.9 Cell dimensions for *CARMA*

$$D_1^{car} = D_2^{car} = D_3^{car} = 0$$

$$D_4^{car} = \frac{x^8}{(1-x^4)\,(1-x^6)}$$

$$D_5^{car} = D_6^{car} = D_7^{car} = 0$$

$$D_8^{car} = \frac{x^{20}}{(1-x^2)\,(1-x^6)\,(1-x^8)\,(1-x^{12})}$$

$$D_9^{car} = D_{10}^{car} = D_{11}^{car} = 0$$

$$D_{12}^{car} = \frac{x^{28}\,(1+x^{12})}{(1-x^2)\,(1-x^4)\,(1-x^6)\,(1-x^8)\,(1-x^{12})\,(1-x^{18})}.$$

Predictably, for the two *free* subalgebras of *ALAL*, *i.e.* *DOMA* and *CARMA*, the generating functions verify self-symmetry relations:

$$(x)^{-2n}\, D_{2n}^{do}(x) = \left(\frac{1}{x}\right)^{-2n}\, D_{2n}^{do}\left(\frac{1}{x}\right) \tag{8.6}$$

$$(x)^{-3n}\, D_{4n}^{car}(x) = \left(\frac{1}{x}\right)^{-3n}\, D_{4n}^{car}\left(\frac{1}{x}\right). \tag{8.7}$$

8.10 Computational checks (Sarah Carr)

We checked conjecture $\boldsymbol{C_3^*}$ (which of course is not independent of $\boldsymbol{C_2^*}$) for $r \leq 8$ and $d \leq 100$, by using the following, highly efficient method: (i) form the *domi*-generating functions (see notations *fi*, *hi*, *gi* in Section 7.2):

$$\text{gedomi}_{t;a,b}^{w_1,w_2} := \frac{t^6\, fi(v_1, v_2)}{(1 - t^2\, a\, hi(v_1, v_2))(1 - t^6\, b\, gi(v_1, v_2))}; \tag{8.8}$$

(ii) form the *ari*-brackets of several copies of $gedomi_{t;a_i,b_i}^{\boldsymbol{w}}$; keep the variables v_1, v_2 and parameters a_i, b_i provisionally unassigned; and studiously refrain from simplifying the rational functions obtained in the process;
(iii) assign random entire values to the v_1, v_2 and a_i, b_i and reduce everything modulo some moderately large prime number p (8 or 9 digits);
(iv) expand everything into power series of t and, for each d, study the dimensions of the spaces generated by the coefficient in front of t^d.

We then requested Sarah Carr, during her 2010 stay at Orsay, to computationally check the other conjectures $\boldsymbol{C_1^*}$, $\boldsymbol{C_2^*}$, $\boldsymbol{C_4^*}$ for lengths r up to 8 and degrees d up to 100. To that end, we supplied her with a complete system of independent *carma*/*carmi*-polynomials[108] of degree $d \leq 40$ (there are exactly 44 such polynomials). Here is her own account of the method she used and the scope of her verifications.

Checking the conjectures $\boldsymbol{C_2^*}$ about *EKMA*

Checking $\mathbf{C_2^*}$ *is equivalent to checking conjecture* $\mathbf{C_2^{**}}$*, according to which all* ari*-relations between the* $\text{ekma}_d^\bullet$ *are generated by the sole bilinear relations (whose exact number is known from the theory). To test* $\mathbf{C_2^{**}}$*, I created the generators in the lengths and degrees given in Table* A *infra. To slightly reduce the complexity of the calculations, I opted for working with the* $\text{ekmi}_d^\bullet$ *rather than the* $\text{ekma}_d^\bullet$*, so as to deal with pair-wise differences of* v_i*'s rather than multiple sums of* u_i*'s.*

For each length r *and degree* d*, I calculated and stored all elements of the form* $\text{ari}(\text{f}_{\text{d}',\text{r}'}, \text{f}_{\text{d}-\text{d}',\text{r}-\text{r}'})$ *where* $1 \leq \text{r}' \leq [\text{r}/2]$ *and* $2r'+2 \leq d' \leq d$*, and where* $\text{f}_{\text{d}',\text{r}'}$ *(respectively* $\text{f}_{\text{d}-\text{d}',\text{r}-\text{r}'}$*) is a basis element of the length* r' *(respectively* $\text{r}-\text{r}'$*) degree* d' *(respectively* $\text{d}-\text{d}'$*) graded part of the Lie algebra. Let the number of such generators be denoted by* $\text{G}_{\text{d,r}}^{\text{ek}}$ *and let the elements in the set of generators be denoted by* $(\text{g}^{\text{ek}})_{\text{d,r}}^{\text{i}}$*;* $1 \leq \text{i} \leq \text{G}_{\text{d,r}}^{\text{ek}}$*.*

[108] They are those constructed from the *lama*/*lami*-basis of *ALIL* (see Section 6 and Section 7).

Since we know that the integers $\mathrm{D}^{\mathrm{ek}}_{\mathrm{d,r}}$ *are upper bounds for the dimensions, we need to verify that we have at least* $\mathrm{D}^{\mathrm{ek}}_{\mathrm{d,r}}$ *linearly independent elements. To check this, I created the generating series* $\sum_{1\le i\le G^{\mathrm{ek}}_{d,r}} \alpha_i\,(g^{ek})^i_{d,r}$. *The polynomials have many terms with large coefficients. I first zeroed out some of the terms in this series by setting a number of variables (between none and* 5, *depending on the length and degree) equal to zero. Then I defined* $\mathrm{G}^{\mathrm{ek}}_{\mathrm{d,r}}$ *randomly generated vectors from the series, by substituting a randomly chosen number (using the* Linear Algebra [Random Vector] Maple function) *between* 1 *and* 20 *for each of the variables, and repeating the process* $\mathrm{G}^{\mathrm{ek}}_{\mathrm{d,r}}$ *times. Lastly, I reduced these vectors modulo either* 101 *or* 100003. *Now, given the linear system defined by these matrices, there are a number of options for solving it. Since we expect this system to have some relations coming from the universal Jacobi identity and from the bilinear relations special to our problem, I tested the efficiency of the Maple commands* linalg[rank],linalg[ker] *and* solve. *The* solve *command proved to be the most efficient. I then used the solution of the linear system to find a basis for the length* r, *degree* d *part.*

The tests confirmed the conjecture $\boldsymbol{C_2^{**}}$ *to lengths and degrees given in Table* A. *More precisely, the dimensions of all degrees between* $2+2\times$ length *and the highest degree entered in the table were verified.*

Table A.

Length	Highest degree generators	Dimension highest degree
1	100	100
2	100	100
3	100	100
4	100	58
5	50	40
6	38	32
7	32	28
8	26	24

Checking the conjectures $\boldsymbol{C_4^*}$ **about** *CARMA*
The calculations were done with the same method as for EKMA. *The scope of the verification is indicated in the following Table.*

Table B.

Length	Highest degree generators	Range of degrees verifying $\boldsymbol{C_4^*}$
4	46	8 – 46
8	54	20 – 54
12	58	28 – 58

Checking the conjectures C_1^* about $EKMA \,\widehat{\otimes}\, CARMA$[109]

Here again I used the same method as for the two previous conjectures. The results are consigned in the following Table.

Table C.

Length	Highest degree generators	Range of degrees verifying C_1^*
4	46	8 – 46
8	54	20 – 54
12	58	28 – 58

Acknowledgments. *My computations were done on the calculation servers at the Max Planck Institut für Math. in Bonn, the Medicis servers at the Ecole Polytechnique and the calculation servers at the Math. Dept. of Orsay University. I would like to thank these institutions for their permission and trust, and warmly thank the system administrators for their indispensable and patient guidance. (Sarah Carr).*

9 Canonical irreducibles and perinomal algebra

9.1 The general scheme

The trifactorisation of $Zag^\bullet$

Let $Zag^\bullet$ denote the generating functions of the (uncoloured) multizetas, defined as in (1.9), but with all $\epsilon_i = 0$ and all $e_i = 1$. This generating function $Zag^\bullet$ admits a remarkable trifactorisation in *GARI*, with a first factor $\mathrm{Zag}_{\mathrm{I}}^\bullet$ which in turn splits into three subfactors:

$$\mathrm{Zag}^\bullet := \mathrm{gari}\big(\mathrm{Zag}_{\mathrm{I}}^\bullet, \mathrm{Zag}_{\mathrm{II}}^\bullet, \mathrm{Zag}_{\mathrm{III}}^\bullet\big) \tag{9.1}$$

$$\mathrm{Zag}_{\mathrm{I}}^\bullet := \mathrm{gari}\big(\mathrm{tal}^\bullet\,,\ \mathrm{invgari.pal}^\bullet\,,\ \mathrm{R\text{ø}ma}^\bullet\big) \tag{9.2}$$

$$\mathrm{Zag}_{\mathrm{I}}^\bullet := \mathrm{gari}\big(\mathrm{tal}^\bullet\,,\ \mathrm{invgari.pal}^\bullet\,,\ \mathrm{expari.r\text{ø}ma}^\bullet\big). \tag{9.3}$$

Here is where the three factors or sub-factors belong:

$$\mathrm{tal}^\bullet\,,\ \mathrm{pal}^\bullet \in \mathrm{GARI}^{\mathrm{as/as}} \tag{9.4}$$

$$\mathrm{invgari.pal}^\bullet\,,\ \mathrm{Zag}^\bullet\,,\ \mathrm{Zag}_{\mathrm{I}}^\bullet \in \mathrm{GARI}^{\mathrm{as/is}} \tag{9.5}$$

$$\mathrm{R\text{ø}ma}^\bullet\,,\ \mathrm{Zag}_{\mathrm{II}}^\bullet\,,\ \mathrm{Zag}_{\mathrm{III}}^\bullet \ \in \mathrm{GARI}^{\underline{\mathrm{as}}/\underline{\mathrm{is}}} \tag{9.6}$$

$$\mathrm{r\text{ø}ma}^\bullet\,,\ \mathrm{logari.Zag}_{\mathrm{II}}^\bullet\,,\ \mathrm{logari.Zag}_{\mathrm{III}}^\bullet \ \in \mathrm{ARI}^{\underline{\mathrm{al}}/\underline{\mathrm{il}}} \tag{9.7}$$

[109] For the meaning of $\widehat{\otimes}$, see Section 8.3.

and here is their real meaning in terms of multizeta irreducibles:
(i) the factor $\mathrm{Zag}_{\mathrm{I}}^{\bullet}$ carries only powers of the special irreducibe $\zeta(2) = \pi^2/6$, of weight 2;
(ii) the factor $\mathrm{Zag}_{\mathrm{II}}^{\bullet}$ carries only irreducibles of even weight $s \geq 4$ and their products;
(iii) the factor $\mathrm{Zag}_{\mathrm{III}}^{\bullet}$ carries only irreducibles of odd weight $s \geq 3$ and their products.

Now, since *weight, length, and degree* are related by $s = r + d$, it is obvious that under the involution *neg.pari*:
(j) elements of *ARI* or *GARI* that carry only *even* weights remain unchanged;
(jj) elements of *ARI* that carry only *odd* weights change sign, and their exponentials in *GARI* change into their *gari*-inverses.

With respect to our three factors, this yields:

$$\mathrm{neg.pari.Zag}_{\mathrm{I}}^{\bullet} = \mathrm{Zag}_{\mathrm{I}}^{\bullet} \tag{9.8}$$

$$\mathrm{neg.pari.Zag}_{\mathrm{II}}^{\bullet} = \mathrm{Zag}_{\mathrm{II}}^{\bullet} \tag{9.9}$$

$$\mathrm{neg.pari.Zag}_{\mathrm{III}}^{\bullet} = \mathrm{invgari.Zag}_{\mathrm{III}}^{\bullet} \tag{9.10}$$

$$\mathrm{gari}(\mathrm{Zag}_{\mathrm{III}}^{\bullet}, \mathrm{Zag}_{\mathrm{III}}^{\bullet}) = \mathrm{gari}(\mathrm{neg.pari.invgari.Zag}^{\bullet}, \mathrm{Zag}^{\bullet}). \tag{9.11}$$

Since all elements of *GARI* have one well-defined square-root,[110] the last identity (9.11) readily yields $Zag_{\mathrm{III}}^{\bullet}$. Separating the last factor from the first two is thus an easy matter (assuming the flexion machinery). But separating $Zag_{\mathrm{I}}^{\bullet}$ from $Zag_{\mathrm{II}}^{\bullet}$ is much trickier, and requires the construction of a bimould $røma^{\bullet}$ rather analogous to $løma^{\bullet}$ but not quite. More precisely, the sought-after $røma^{\bullet}$
– must (like $løma^{\bullet}$) be of type $\underline{al}/\underline{il}$
– must (unlike $løma^{\bullet}$) carry multipoles at the origin that are so chosen as to cancel those of $tal^{\bullet}$ and $pal^{\bullet}$ in the trifactorisation (9.3).

The auxiliary bimoulds $løma^{\bullet}$, $røma^{\bullet}$
The building blocks are the elementary singulands $sa_{s_1}^{\bullet} \in BIMU_1$ and the corresponding elementary singulates $sa_{\binom{s_1}{r_1}}^{\bullet} \in ARI^{\underline{al}/\underline{il}}$:

$$\mathrm{sa}_{s_1}^{w_1} := u_1^{s_1-1}; \qquad \mathrm{sa}_{\binom{s_1}{r_1}}^{\bullet} := \mathrm{slang}_{r_1}.\mathrm{sa}_{s_1}^{\bullet}. \tag{9.12}$$

The singulates $sa_{\binom{s_1}{r_1}}^{\bullet}$ are $\neq 0$ *iff* s_1+r_1 is even and $s_1 \geq 2$.

110 Apply $expari.\frac{1}{2}.logari$.

We then define $l\phi ma^\bullet$ and $r\phi ma^\bullet$ as sums of their homogeneous components of weight s:

$$l\phi ma^\bullet := \sum_{s\ \mathrm{odd}\geq 3} l\phi ma^\bullet_s; \qquad r\phi ma^\bullet := \sum_{s\ \mathrm{odd}\geq 2} r\phi ma^\bullet_s \tag{9.13}$$

and proceed to construct these homogeneous components by bracketting the singulates, in *PREARI* rather than *ARI* (– because that is by far the theoretically cleaner way –), with the multibrackets always defined from *left to right*, as in (2.49).

$$\mathrm{l\phi ma}^\bullet_s = \sum_{1\leq l} \sum_{\substack{\{ {s_i+\cdots+s_l=s \atop r_1+\cdots+r_l\ odd} \} \\ \{ {1\leq s_i,1\leq r_i \atop s_i+r_i\ even} \}}} \mathrm{l\phi m}^{\binom{s_1\ ,\dots,\ s_l}{r_1\ ,\dots,\ r_l}}\ \mathrm{preari}(\mathrm{sa}^\bullet_{\binom{s_1}{r_1}},\dots,\mathrm{sa}^\bullet_{\binom{s_l}{r_l}})\ (\forall s\ odd) \tag{9.14}$$

$$\mathrm{r\phi ma}^\bullet_s = \sum_{1\leq l} \sum_{\substack{\{ {s_i+\cdots+s_l=s \atop r_1+\cdots+r_l\ even} \} \\ \{ {1\leq s_i,1\leq r_i \atop s_i+r_i\ even} \}}} \mathrm{r\phi m}^{\binom{s_1\ ,\dots,\ s_l}{r_1\ ,\dots,\ r_l}}\ \mathrm{preari}(\mathrm{sa}^\bullet_{\binom{s_1}{r_1}},\dots,\mathrm{sa}^\bullet_{\binom{s_l}{r_l}})\ (\forall s\ even). \tag{9.15}$$

As for $R\phi ma^\bullet$, it may be sought either in the form $expari.r\phi ma^\bullet$ or, equivalently but more directly, in the form:

$$\mathrm{R\phi ma}^\bullet = 1^\bullet + \sum_{1\leq l} \sum_{\substack{\{ {s_i+\cdots+s_l\ even \atop r_1+\cdots+r_l\ even} \} \\ \{ {1\leq s_i,1\leq r_i \atop s_i+r_i\ even} \}}} \mathrm{R\phi m}^{\binom{s_1\ ,\dots,\ s_l}{r_1\ ,\dots,\ r_l}}\ \mathrm{preari}(\mathrm{sa}^\bullet_{\binom{s_1}{r_1}},\dots,\mathrm{sa}^\bullet_{\binom{s_l}{r_l}})\quad (\forall s\ even).$$

Of course, in the above expansions, all summands must be true singulates,[111] with a least a pole of order 1 at the origin, so that at least one of their indices r_i must be ≥ 2.

Due to the condition $\sum s_i = s$, the right-hand sides of (9.14) and (9.15) carry only finitely many summands. Each summand that goes into the making of $l\phi ma^\bullet_s$ or $r\phi ma^\bullet_s$ is of type $\underline{al}/\underline{il}$ and its *shortest component* is of *even* degree $d = \sum(s_i - r_i)$, which is compatible with its being of type $\underline{al}/\underline{al}$.

The moulds $l\phi m^\bullet$ or $r\phi m^\bullet$ (respectively $R\phi m^\bullet$) must be alternal (respectively symmetral) and one goes from $r\phi m^\bullet$ to $R\phi m^\bullet = expmu(r\phi m^\bullet)$ by the straightforward mould exponential.

At this stage (*i.e.* provisionally setting aside all considerations of canonicity) the only additional constraints on the alternal moulds $l\phi m^\bullet$,

[111] with the sole exception of the first summand in the expansion (9.14) for $l\phi ma^\bullet_s$, which is of the form $l\phi m^{\binom{s}{1}}\ sa^\bullet_{\binom{s}{1}}$ with $l\phi m^{\binom{s}{1}} = 1$.

$r\o m^\bullet$, and the symmetral mould $R\o m^\bullet$ are these:
(k) $l\o m^\bullet$ must make $l\o ma^\bullet_s$ singularity-free;
(kk) $r\o m^\bullet$ (or $R\o m^\bullet$) must, within the *gari*-product:

$$\mathrm{Zag}^\bullet_{\mathrm{I}} := \mathrm{gari}\big(\mathrm{tal}^\bullet,\ \mathrm{invgari.pal}^\bullet,\ \mathrm{Roma}^\bullet\big) \tag{9.16}$$

$$:= \mathrm{gari}\big(\mathrm{tal}^\bullet,\ \mathrm{invgari.pal}^\bullet,\ \mathrm{expari.}\sum_s \mathrm{roma}^\bullet_s\big) \tag{9.17}$$

eliminate all the singularities present in *gari*($tal^\bullet$, $invgari.pal^\bullet$);
(kkk) the moulds $l\o m^\bullet$ or $r\o m^\bullet$ must be rational-valued.

Explicit decomposition of multizetas into irreducibles

Anticipating on the construction of $l\o ma^\bullet$ and its iso-weight parts $l\o ma^\bullet_s$, the *preari*-product gives us an extremely elegant and explicit representation of the multizetas in terms of irreducibles:

$$\mathrm{Zag}^\bullet_{\mathrm{II}} := 1^\bullet + \sum_{\substack{1\le l\\ l\ \mathrm{even}}} \sum_{\substack{3\le s_i\\ s_i\ \mathrm{odd}}} \mathrm{Irr\o}^{s_1,\dots,s_l}_{\mathrm{II}}\,\mathrm{preari}(\mathrm{l\o ma}^\bullet_{s_1},\dots,\mathrm{l\o ma}^\bullet_{s_l}) \tag{9.18}$$

$$\mathrm{Zag}^\bullet_{\mathrm{III}} = 1^\bullet + \sum_{\substack{1\le l\\ l\ \mathrm{free}}} \sum_{\substack{3\le s_i\\ s_i\ \mathrm{odd}}} \mathrm{Irr\o}^{s_1,\dots,s_l}_{\mathrm{III}}\,\mathrm{preari}(\mathrm{l\o ma}^\bullet_{s_1},\dots,\mathrm{l\o ma}^\bullet_{s_l}) \tag{9.19}$$

$$\mathrm{logari.Zag}^\bullet_{\mathrm{II}} = +\sum_{\substack{1\le l\\ l\ \mathrm{even}}} \sum_{\substack{3\le s_i\\ s_i\ \mathrm{odd}}} \mathrm{irr\o}^{s_1,\dots,s_l}_{\mathrm{II}}\,\mathrm{preari}(\mathrm{l\o ma}^\bullet_{s_1},\dots,\mathrm{l\o ma}^\bullet_{s_l}) \tag{9.20}$$

$$\mathrm{logari.Zag}^\bullet_{\mathrm{III}} := +\sum_{\substack{1\le l\\ l\ \mathrm{odd}}} \sum_{\substack{3\le s_i\\ s_i\ \mathrm{odd}}} \mathrm{irr\o}^{s_1,\dots,s_l}_{\mathrm{III}}\,\mathrm{preari}(\mathrm{l\o ma}^\bullet_{s_1},\dots,\mathrm{l\o ma}^\bullet_{s_l}). \tag{9.21}$$

The irreducible carriers $Irr\o^\bullet_{\mathrm{III}}$, $Irr\o^\bullet_{\mathrm{III}}$ (respectively $irr\o^\bullet_{\mathrm{II}}$, $irr\o^\bullet_{\mathrm{II}}$) are scalar moulds of symmetral (respectively alternal) type. They are related under ordinary mould exponentiation:

$$\mathrm{Irr\o}^\bullet_{\mathrm{II}} = \mathrm{expmu.irr\o}^\bullet_{\mathrm{II}} \tag{9.22}$$

$$\mathrm{Irr\o}^\bullet_{\mathrm{III}} = \mathrm{expmu.irr\o}^\bullet_{\mathrm{III}}. \tag{9.23}$$

The pair $irr\o^\bullet_{\mathrm{II}}$, $Irr\o^\bullet_{\mathrm{II}}$ has only (non-vanishing) components of even length. In the pair $irr\o^\bullet_{\mathrm{III}}$, $Irr\o^\bullet_{\mathrm{III}}$, however, $irr\o^\bullet_{\mathrm{III}}$ has only (non-vanishing) components of odd length, but $Irr\o^\bullet_{\mathrm{III}}$ has of course components of any length, even or odd.

There are two ways of looking at the expansions (9.18)-(9.21).

If we are dealing with *formal multizetas*, then our four moulds (9.22)-(9.23) are subject to no other constraints than the above, *i.e.* symmetrality or alternality, and a definite length parity. They subsume all multizeta irreducibles other than π^2 in the theoretically most satisfactory manner, *i.e.* without introducing any artificial dissymmetry.[112]

In practice, to decompose *formal* multizetas into irreducibles, one may:
– calculate $Zag^\bullet_{\mathrm{I}}$ according to (9.2) or (9.3);
– calculate $Zag^\bullet_{\mathrm{II}}$ and $Zag^\bullet_{\mathrm{III}}$ according to (9.18) and (9.19);
– calculate $Zag^\bullet$ according to the trifactorisation (9.1);
– calculate the *swappee* $Zig^\bullet$ of $Zag^\bullet$;
– harvest the Taylor coefficients of $Zig^\bullet$.

Since any given multizeta appears once and only once as Taylor coefficient of $Zig^\bullet$, it can thus be expressed in purely algorithmic manner, via the flexion machinery, in terms of $irr\phi^\bullet_{\mathrm{II}}$ and $irr\phi^\bullet_{\mathrm{III}}$, or $Irr\phi^\bullet_{\mathrm{II}}$ and $Irr\phi^\bullet_{\mathrm{III}}$.

When dealing with the *genuine* multizetas, on the other hand, the irreducibles are well-defined *numbers* and the five-step procedure works in both directions: it also enables one to express $irr\phi^\bullet_{\mathrm{II}}$, $irr\phi^\bullet_{\mathrm{III}}$ and $Irr\phi^\bullet_{\mathrm{II}}$, $Irr\phi^\bullet_{\mathrm{III}}$ in terms of the multizetas. This 'reverse expression', however, is not unique. To get a unique, privileged expression of the irreducibles – not in terms of multizetas, but of *perinomal* numbers – there is no (known) alternative to the approach sketched in Section 9.4 *infra*.

Explicit decomposition of multizetas into canonical irreducibles

To qualify as *canonical*, the irreducible carriers $irr\phi^\bullet_{II}$, $irr\phi^\bullet_{III}$ or $Irr\phi^\bullet_{II}$, $Irr\phi^\bullet_{III}$ just defined must correspond to a *compellingly natural* solution ($l\phi ma^\bullet_s$, $r\phi ma^\bullet_s$). The constraints (*k*), (*kk*), (*kkk*), however, do not quite suffice to uniquely determine the solution – due to the existence of *wandering bialternals*, which was pointed out in Section 6.9.

One cannot stress enough that this *residual indeterminacy*, compared with the huge *a priori indeterminacy* inherent in all other approaches, is quite negligible, and that too in a precise and measurable sense. Indeed, let $\mathcal{I}rr(r, s)$ be the space of prime irreducibles of length r and total weight s. Next, let $Wander(r, s)$ be the indeterminacy (*i.e.* number of free parameters) in the definition of the irreducibles in $\mathcal{I}rr(r, s)$ that comes from the existence of *wandering bialternals*. Lastly, let $Naive(r, s)$ be the indeterminacy that we would be stuck with in the *naive approach*, *i.e.* if we had no criteria for privileging any given irreducible $\rho_{r,s}$ in $\mathcal{I}rr(r, s)$ over

[112] If one wishes for a basis of scalar irreducibles totally free of constraints, one can readily produce one by picking any minimal system of components of, say, $irr\phi^\bullet_{\mathrm{II}}$ and $irr\phi^\bullet_{\mathrm{III}}$, that is large enough to determine all other components by alternality. That essentially amounts to selecting a basis in the Lie algebra freely generated by the symbols $\epsilon_3, \epsilon_5, \epsilon_7 \ldots$. Many such bases exist (Lyndon's etc) but none is truly canonical. Thus, while in calculations it may often be convenient to opt for *free i.e. unconstrained* systems of irreducibles, from a theoretical viewpoint it is far preferable to stick with the *constrained* systems implicit in $irr\phi^\bullet_{\mathrm{II}}$ and $irr\phi^\bullet_{\mathrm{III}}$ or their symmetral counterparts $Irr\phi^\bullet_{\mathrm{II}}$ and $Irr\phi^\bullet_{\mathrm{III}}$.

all its variants of the form:

$$\rho_{r,s} + \sum_{s_1+\ldots+s_l=s}^{l\geq 2} c^{r_1,\ldots,r_l}_{s_1,\ldots,s_l} \prod_{1\leq i\leq l} \rho_{r_i,s_i} \text{ with } c^{r_1,\ldots,r_l}_{s_1,\ldots,s_l} \in \mathbb{Q};\ \rho_{r_i,s_i} \in \mathcal{I}rr(r_i,s_i). \quad (9.24)$$

One shows that, for each r fixed and $s \to \infty$, we have:

$$Wander(r,s)/Naive(r,s) = \mathcal{O}(s^{-1}). \quad (9.25)$$

So this small residual indeterminacy due to the wandering bialternals is something we could live with. We can remove it, however, and ensure both uniqueness and canonicity, by imposing additional conditions – of arithmetical or function-theoretical nature. As we shall see, there are three basic choices (two arithmetical options and a function-theoretical one) but we go with relative ease from the one to the others, so that we are still justified in speaking, in the singular, of *the* canonical choice.

9.2 Arithmetical criteria

One way of lifting the residual indeterminacy in the construction of the pair ($l\phi ma^{\bullet}_{s}$, $r\phi ma^{\bullet}_{s}$) is to impose additional linear constraints on the Taylor coefficients of the singulates $S^{\bullet}_{r}$ being used in the successive[113] inductive steps. As it happens, there are two natural systems of linear constraints that do the trick. We mentioned them in Section 6.5 and Section 6.6 in the case of $l\phi ma^{\bullet}_{s}$ and only at the first occurence (*i.e.* for $r=3$) but they extend to all lengths, and have their exact counterparts for $r\phi ma^{\bullet}_{s}$. They lead to two distinct pairs ($lama^{\bullet}_{s}$, $rama^{\bullet}_{s}$) and ($loma^{\bullet}_{s}$, $roma^{\bullet}_{s}$), which stand out on account of their arithmetical properties. Very roughly speaking: with the first pair, both singulators and singulates possess "more" independent Taylor coefficients but these have "smaller" denominators, whereas with the second pair the position is exactly reversed. In both cases, however, the denominators of the Taylor coefficients are always divisors of simple *factorials* that depend only on *length* and *degree*. That changes completely with the third pair ($luma^{\bullet}_{s}$, $ruma^{\bullet}_{s}$), which we shall examine next and which is characterised by its functional properties.

9.3 Functional criteria

To transport entire multipoles, we require dilation operators δ^n:
(i) that define a group action: $\delta^{n_1}\,\delta^{n_2} \equiv \delta^{n_1 n_2}$, $\forall n_i \in \mathbb{Q}^+$;
(ii) that act as flexion automorphisms;

[113] For *odd* lengths r in the case of $l\phi ma^{\bullet}_{s}$ and *even* lengths in the case of $r\phi ma^{\bullet}_{s}$.

(iii) that commute with the singulators (simple or composite);
(iv) that conserve multiresidues.

This imposes the definition:

$$(\delta^n.A)^{\binom{u_1\ ,\dots,\ u_r}{v_1\ ,\dots,\ v_r}} := n^{-r}\ A^{\binom{u_1/n\ ,\dots,\ u_r/n}{v_1.n\ ,\dots,\ v_r.n}} \qquad (\forall n \in \mathbb{Z}) \tag{9.26}$$

which ensures the required properties:

$$\delta^n : ARI^{\underline{\mathrm{al}}/\underline{\mathrm{al}}} \overset{\mathrm{isom.}}{\to} ARI^{\underline{\mathrm{al}}/\underline{\mathrm{al}}},\quad ARI^{\underline{\mathrm{al}}/\underline{\mathrm{il}}} \overset{\mathrm{isom.}}{\to} ARI^{\underline{\mathrm{al}}/\underline{\mathrm{il}}} \tag{9.27}$$

$$\delta^n\ \mathrm{slank}_{r_1,\dots,r_l}\ S^\bullet \equiv \mathrm{slank}_{r_1,\dots,r_l}\ \delta^n\ S^\bullet \tag{9.28}$$

$$\delta^n\ \mathrm{slang}_{r_1,\dots,r_l}\ S^\bullet \equiv \mathrm{slang}_{r_1,\dots,r_l}\ \delta^n\ S^\bullet. \tag{9.29}$$

Next, to reflect the change from *power series* to *meromorphic functions*, we must replace:
– the monomial singulands $sa^\bullet_{s_1} \in BIMU_1$ of singulates $sa^\bullet_{\binom{s_1}{r_1}} \in ARI^{\underline{\mathrm{al}}/\underline{\mathrm{il}}}_{r_1\leq}$;
– by monopolar singulands $ta^\bullet_{n_1} \in BIMU_1$ of singulates $ta^\bullet_{\binom{n_1}{r_1}} \in ARI^{\underline{\mathrm{al}}/\underline{\mathrm{il}}}_{r_1\leq}$.

Concretely, we set:

$$\mathrm{ta}^{w_1} := (1-u_1)^{-1},\quad \mathrm{ta}^{w_1,\dots,w_r} := 0 \ \text{ if } \ r \neq 1 \tag{9.30}$$

$$\mathrm{ta}^\bullet_{n_1} := \delta^{n_1}.\mathrm{ta}^\bullet,\quad ta^\bullet_{\binom{n_1}{r_1}} := \mathrm{slang}_{r_1}.\delta^{n_1}.\mathrm{ta}^\bullet = \delta^{n_1}.\mathrm{slang}_{r_1}.\mathrm{ta}^\bullet. \tag{9.31}$$

We may now look for bimoulds *luma*$^\bullet$ and *ruma*$^\bullet$ given by expansions of the form:

$$\mathrm{luma}^\bullet = \sum_{1\leq l}\ \sum_{1\leq n_i, 1\leq r_i}^{\{\substack{n_i\ coprime \\ r_1+\dots+r_l\ odd}\}} \mathrm{lum}^{\binom{n_1\ ,\dots,\ n_l}{r_1\ ,\dots,\ r_l}}\ \mathrm{preari}(\mathrm{ta}^\bullet_{\binom{n_1}{r_1}},\dots,\mathrm{ta}^\bullet_{\binom{n_l}{r_l}}) \tag{9.32}$$

$$\mathrm{ruma}^\bullet = \sum_{1\leq l}\ \sum_{1\leq n_i, 1\leq r_i}^{\{\substack{n_i\ anything \\ r_1+\dots+r_l\ even}\}} \mathrm{rum}^{\binom{n_1\ ,\dots,\ n_l}{r_1\ ,\dots,\ r_l}}\ \mathrm{preari}(\mathrm{ta}^\bullet_{\binom{n_1}{r_1}},\dots,\mathrm{ta}^\bullet_{\binom{n_l}{r_l}}) \tag{9.33}$$

that run exactly parallel to (9.14) and (9.15), and may also be rewritten as:

$$\mathrm{luma}^\bullet = \sum_{1\leq l}\ \sum_{1\leq n_i\ ,\ 1\leq r_i}^{\{\substack{n_i\ coprime \\ r_1+\dots+r_l\ odd}\}} \mathrm{lum}^{\binom{n_1\ ,\dots,\ n_l}{r_1\ ,\dots,\ r_l}}\ \mathrm{slang}_{r_1,\dots,r_l}.\mathrm{mu}(\delta^{n_1}\mathrm{ta}^\bullet,\dots,\delta^{n_l}\mathrm{ta}^\bullet) \tag{9.34}$$

$$\mathrm{ruma}^\bullet = \sum_{1\leq l}\ \sum_{1\leq n_i\ ,\ 1\leq r_i}^{\{\substack{n_i\ anything \\ r_1+\dots+r_l\ even}\}} \mathrm{rum}^{\binom{n_1\ ,\dots,\ n_l}{r_1\ ,\dots,\ r_l}}\ \mathrm{slang}_{r_1,\dots,r_l}.\mathrm{mu}(\delta^{n_1}\mathrm{ta}^\bullet,\dots,\delta^{n_l}\mathrm{ta}^\bullet). \tag{9.35}$$

The remarkable fact is that if we impose:[114]

$$\mathrm{lum}^{\binom{1}{1}} = 1, \quad \mathrm{lum}^{\binom{n_1}{1}} = 0 \qquad \forall n_1 \geq 2$$

and

$$\mathrm{lum}^{\binom{n_1\ ,\dots,\ n_l}{1\ ,\dots,\ 1}} = 0 \qquad \forall l \geq 2,\ \forall n_i \tag{9.36}$$

$$\mathrm{rum}^{\binom{n_1\ ,\dots,\ n_l}{1\ ,\dots,\ 1}} = 0 \qquad \forall l \geq 2,\ \forall n_i \tag{9.37}$$

then there is only one mould *lum*$^\bullet$ (respectively *rum*$^\bullet$) such that *luma*$^\bullet$ be *free* of singularities at the origin (resp that *ruma*$^\bullet$ carry exactly the *right* singularities[115] there). So the problem now is no longer that of determining a canonical solution, but of ascertaining the arithmetical nature of the Taylor coefficients at the origin of the unique *luma*$^\bullet$ and the unique *ruma*$^\bullet$. With *luma*$^\bullet$ the problem arises only for lengths $r \geq 5$, and with *ruma*$^\bullet$ only for lengths $r \geq 4$. This, however, is not a matter for this Survey.

But even without addressing this question, we may note that the pair *luma*$^\bullet$, *ruma*$^\bullet$ leads to a trifactorisation (9.1) of *Zag*$^\bullet$ exactly as the pair *løma*$^\bullet$, *røma*$^\bullet$ did at the end of Section 9.1. Explicitly:

$$\mathrm{Zag}_{\mathrm{I}}^{\bullet} := \mathrm{gari}\left(\mathrm{tal}^\bullet, \mathrm{invgari}(\mathrm{pal}^\bullet), \mathrm{expari}\left(\sum_{1\leq n} \delta^n \mathrm{ruma}^\bullet\right)\right) \tag{9.38}$$

$$\mathrm{Zag}_{\mathrm{II}}^{\bullet} := 1^\bullet + \sum_{1\leq r}\sum_{1\leq n_i} \mathrm{Urr}_{\mathrm{II}}^{n_1,\dots,n_l}\mathrm{preari}(\delta^{n_1}\mathrm{luma}^\bullet, \dots, \delta^{n_l}\mathrm{luma}^\bullet) \tag{9.39}$$

$$\mathrm{Zag}_{\mathrm{III}}^{\bullet} = 1^\bullet + \sum_{1\leq r}\sum_{1\leq n_i} \mathrm{Urr}_{\mathrm{III}}^{n_1,\dots,n_l}\mathrm{preari}(\delta^{n_1}\mathrm{luma}^\bullet, \dots, \delta^{n_l}\mathrm{luma}^\bullet) \tag{9.40}$$

$$\mathrm{logari.Zag}_{\mathrm{II}}^{\bullet} = +\sum_{1\leq r}\sum_{1\leq n_i} \mathrm{urr}_{\mathrm{II}}^{n_1,\dots,n_l}\mathrm{preari}(\delta^{n_1}\mathrm{luma}^\bullet, \dots, \delta^{n_l}\mathrm{luma}^\bullet). \tag{9.41}$$

$$\mathrm{logari.Zag}_{\mathrm{III}}^{\bullet} := +\sum_{1\leq r}\sum_{1\leq n_i} \mathrm{urr}_{\mathrm{III}}^{n_1,\dots,n_l}\mathrm{preari}(\delta^{n_1}\mathrm{luma}^\bullet, \dots, \delta^{n_l}\mathrm{luma}^\bullet). \tag{9.42}$$

Instead of the symmetral pair of irreducible carriers $Irr\phi^\bullet_{II}$, $Irr\phi^\bullet_{III}$ and the alternal pair $irr\phi^\bullet_{II}$, $irr\phi^\bullet_{III}$, we now have the symmetral pair $Irru^\bullet_{II}$, $Irru^\bullet_{III}$ and the alternal pair $irru^\bullet_{II}$, $irru^\bullet_{III}$, with indices no longer running through $\{3, 5, 7\dots\}$ but through $\mathbb{N}^*$. Moreover, when dealing with

[114] No such condition is requires for *rum*$^\bullet$ since it automatically vanishes when the sum $r_1+\ldots+r_l$ is odd, and in particular when it reduces to $r_1 = 1$.

[115] *I.e.* singularities capable of compensating those of *tal*$^\bullet$ and *pal*$^\bullet$ and of ensuring the regularity of $Zag^\bullet_I$ at the origin.

the genuine (rather than formal) multizetas, these four new moulds are well-determined, *rational-valued*, and, for any given length r, *perinomal* functions of their indices n_i. So it is about time to explain what perinomal functions are, and what they can accomplish.

9.4 Notions of perinomal algebra

A function $\rho \in C(\mathbb{Z}^r, \mathbb{C})$ is said to be *perinomal* (of arity r and rank r^*) iff:
(i) there exist $S_1, \ldots, S_{r^*} \in Sl_r(\mathbb{Z})$ such that the functions $\rho \circ S_1, \ldots, \rho \circ S_{r^*}$ be linearly independent;
(ii) for any $r^{**} > r^*$ and any $S_1, \ldots, S_{r^{**}} \in Sl_r(\mathbb{Z})$, the $\rho\circ S_1, \ldots, \rho\circ S_{r^{**}}$ are linearly dependent.

We set $S\rho := \rho\circ S$, which defines an anti-action of $Sl_r(\mathbb{Z})$ on $\mathbb{C}(\mathbb{Z}^r, \mathbb{C})$. If $T \in Sl_r(\mathbb{Z})$, $\boldsymbol{S} := [S_1, \ldots, S_{r^*}] \in (Sl_r(\mathbb{Z}))^{r^*}$ and $\boldsymbol{n} \in \mathbb{Z}^r$, we also set:

$$\begin{aligned}
\boldsymbol{S}\rho &:= [S_1\rho, \ldots, S_{r^*}\rho] = [\rho\circ S_1, \ldots, \rho\circ S_{r^*}]\\
T\boldsymbol{S}\rho &:= [TS_1\rho, \ldots, TS_d\rho] = [\rho\circ S_1\circ T, \ldots, \rho\circ S_{r^*}\circ T]\\
T\boldsymbol{S}\rho(\boldsymbol{n}) &:= [TS_1\rho(\boldsymbol{n}), \ldots, TS_{r^*}\rho(\boldsymbol{n})] = [(\rho\circ S_1\circ T)(\boldsymbol{n}), \ldots, (\rho\circ S_{r^*}\circ T)(\boldsymbol{n})].
\end{aligned}$$

If the S_i are now chosen so as to make $S_1\rho, \ldots, S_{r^*}\rho$ linearly independent, for each T there must exist scalars $M_i^j(\boldsymbol{S}\rho\,; T)$ such that

$$TS_i\rho(\boldsymbol{n}) \equiv \sum_{1\le j\le d} S_j\rho(\boldsymbol{n})\, M_i^j(\boldsymbol{S}\rho\,; T) \quad (\forall\, i, \forall \boldsymbol{n}) \quad \textit{i.e. in matrix notation:}$$

$$T\boldsymbol{S}\rho(\boldsymbol{n}) \equiv (\boldsymbol{S}\rho(\boldsymbol{n}))\,.\, M(\boldsymbol{S}.\rho\,; T). \tag{9.43}$$

But changing $\boldsymbol{S}$ into another choice $\boldsymbol{S}'$ would simply subject M to some T*-independent* matrix conjugation $M \to M'$:

$$M'(\boldsymbol{S}'\rho\,; T) = C(\boldsymbol{S}'\rho\,; \boldsymbol{S}\rho)\, M(\boldsymbol{S}\rho\,; T)\, C(\boldsymbol{S}\rho\,; \boldsymbol{S}'\rho). \tag{9.44}$$

Moreover, we clearly have:

$$M(\boldsymbol{S}\rho\,; T_1T_2) \equiv M(\boldsymbol{S}\rho\,; T_1)M(\boldsymbol{S}\rho\,; T_2). \tag{9.45}$$

The upshot is that the identity (9.43) defines a linear representation of $Sl_r(\mathbb{Z})$ into $Gl_{r^*}(\mathbb{Z})$ or rather $Sl_{r^*}(\mathbb{Z})$:

$$\mathrm{Sl}_r(\mathbb{Z}) \to \mathrm{Sl}_{r^*}(\mathbb{Z}) \tag{9.46}$$

$$T \mapsto M(\boldsymbol{S}\rho;\, T) \sim M_\rho(T). \tag{9.47}$$

This representation M_ρ in turn splits into irreducible factor representations M_{ρ,r_i^*}:

$$M_\rho = M_{\rho,r_1^*} \otimes \cdots \otimes M_{\rho,r_s^*} \quad \text{with} \quad r_1^* + \ldots r_s^* = r^*. \tag{9.48}$$

Analogy with polynomials and action of $sl_r(\mathbb{Z})$

Let ρ be perinomal of type (r, r^*). For $T \in Sl_r(\mathbb{Z})$ of the form *id* + *nilpotent* and with logarithm $t = \log(T) \in sl_r(\mathbb{Z})$, the image $M_\rho(T)$ of T in $Sl_{r^*}(\mathbb{Z})$ is also of the form *id* + *nilpotent*. For any $\boldsymbol{n}$ fixed in $\mathbb{Z}^r$ the sequence $\{T^k \rho(\boldsymbol{n}), k = \mathbb{Z}\}$ is therefore polynomial in k and it makes sense to set:

$$t\rho(\boldsymbol{n}) := \left[\partial_k \, T^k \, \rho(\boldsymbol{n})\right]_{k=0} \qquad (\forall \boldsymbol{n},\ T = \exp(t)) \tag{9.49}$$

as if k were a continuous variable. This defines a *coherent* anti-action on $Peri_r$ (the ring of perinomal functions of arity r), first of the nilpotent part of $sl_r(\mathbb{Z})$, and then, by composition, of $sl_r(\mathbb{Z})$ in its entirety. This applies in particular for the elementary operators:

$$e_{i,j} \in \mathrm{sl}_r(\mathbb{Z}) \text{ ``=''} n_j \partial_{n_i}$$

$$E_{i,j} \in \mathrm{Sl}_r(\mathbb{Z}) \ E_{i,j} : \boldsymbol{n} \mapsto \boldsymbol{n}' \ \text{ with } \ n'_i := n_i + n_j \ \text{ and } \ n'_k = n_k \ \text{ if } \ k \neq i.$$

But despite this analogy with polynomial functions, perinomal functions as a rule do not admit sensible extensions beyond $\mathbb{Z}^r$: they are essentially discrete creatures.

Perinomal continuation

Even for functions ρ defined only on a "full-measure" cone of $\mathbb{Z}^r$, *e.g.* on $\mathbb{N}^r$, the above definitions of perinomalness still applies, but under restriction to the sub-semigroup of $\Gamma \subset Sl_r(\mathbb{Z})$ that sends that cone into itself. When these conditions of "partial perinomalness" are fulfilled, on can then pick in Γ elements of the form *id*+*nilpotent* and take advantage of the polynomial dependence of $T^k\rho(\boldsymbol{n})$ in k for $k \in \mathbb{N}$ to extend, in unique and coherent manner, the function ρ to the whole of $\mathbb{Z}^r$, and then define, on this extended function, the anti-action not just of Γ but of the whole of $Sl_r(\mathbb{Z}) \supset \Gamma$.

Stability properties of perinomal functions

Perinomal functions are stable under most common operations, such as:
(i) ordinary addition and multiplication (assuming a common arity r);
(ii) concatenation or, what amounts to the same, mould mutiplication;
(iii) the whole range of flexion operations, and notably *ari*/*gari*.

The latter means that bimoulds $A^{\boldsymbol{w}}$ whose indices $w_i = \binom{u_i}{v_i}$ assume only entire values and whose dependence on the sequences $\boldsymbol{u}$ and/or $\boldsymbol{v}$ is *perinomal*, are stable under *ari, gari* etc.

Basic transforms $\rho \leftrightarrow \rho^* \leftrightarrow \rho^\#$
The definitions read:

$$\rho^*(s_1,\dots,s_r) := \sum_{n_i \in \mathbb{N}^*} \rho(n_1,\dots,n_r)\, n_1^{-s_1} \dots n_r^{-s_r} \quad (s_i \in \mathbb{C} \text{ or } \mathbb{N}) \tag{9.50}$$

$$\rho^\#(x_1,\dots,x_r) \overset{\text{dom.}}{:=} \sum_{n_i \in \mathbb{N}^*} \frac{\rho(n_1,\dots,n_r)}{(n_1 - x_1)\dots(n_r - x_r)} \qquad (x_i \in \mathbb{C}). \tag{9.51}$$

For $s_i \in \mathbb{C}$ and $\Re(s_i) > C_i$ with C_i large enough, the sum (9.50) converges to an analytic function ρ^* which may or may not possess a meromorphic continuation to the whole of $\mathbb{C}^r$. But one usually considers entire arguments s_i. The corresponding *perinomal numbers* $\rho^*(\boldsymbol{s})$ constitute a remarkable $\mathbb{Q}$-ring that not only extends the $\mathbb{Q}$-ring of multizetas, but is also the proper framework for studying the *"impartial"* multizeta irreducibles.

As for the sum (9.51), it usually converges only if we subtract from the generic summand suitable corrective monomials (of bounded degrees) in the x_i. Hence the caveat *"dom." i.e. "dominant"* over the sign = . The resulting meromorphic function $\rho^\#(\boldsymbol{x})$ is known as a *perinomal carrier*. Its Taylor coefficients are clearly related to the *perinomal numbers* $\rho^*(\boldsymbol{s})$ and its multiresidues $\rho(\boldsymbol{n})$ are *perinomal functions* of $\boldsymbol{n}$.

9.5 The all-encoding perinomal mould *peri*•

Definition of *peri*•

For any $l \geq 1$ and any integers n_i , $r_i \geq 1$ we set:

$$\begin{aligned}
\mathrm{peri}^{\binom{n_1\ ,\dots,\ n_l}{r_1\ ,\dots,\ r_l}} &:= \mathrm{urr}_{\mathrm{III}}^{n_1,\dots,n_l} && \text{if } r_i \equiv 1\ \forall i && \text{and } \textstyle\sum r_i = l \ \textit{ is odd}\\
&:= \mathrm{urr}_{\mathrm{II}}^{n_1,\dots,n_l} && \text{if } r_i \equiv 1\ \forall i && \text{and } \textstyle\sum r_i = l \ \textit{ is even}\\
&:= \mathrm{lum}^{\binom{n_1\ ,\dots,\ n_l}{r_1\ ,\dots,\ r_l}} && \text{if } \max_i(r_i) > 1 && \text{and } \textstyle\sum r_i \ \textit{ is odd}\\
&:= \mathrm{rum}^{\binom{n_1\ ,\dots,\ n_l}{r_1\ ,\dots,\ r_l}} && \text{if } \max_i(r_i) > 1 && \text{and } \textstyle\sum r_i \ \textit{ is even.}
\end{aligned}$$

The following table recalls the origin and role of the four *parts* of *peri*•, depending on l and $\boldsymbol{r}$:

peri•	$\sum r_i$ *odd*	$\sum r_i$ *even*
$r_i = 1 \quad \forall i$	*constructs* $\mathrm{Zag}^\bullet_{\mathrm{III}}$ from luma•	*constructs* $\mathrm{Zag}^\bullet_{\mathrm{II}}$ from luma•
$\max(r_i) > 1$	*constructs* luma• from ta•	*constructs* ruma• from ta•.

In view of its definition, this holds-all mould *peri*• may seem a hopelessly heterogeneous and ramshackle construct. However, upon closer examination, its four *parts* turn out to be so closely interrelated that they cannot be described or understood in isolation. This amply justifies our welding them together into a unique mould $peri^\bullet$ which, far from being composite, is almost "seamless" .

Properties of $peri^\bullet$

(i) As a mould[116] with indices $\binom{n_i}{r_i}$, $peri^\bullet$ is *alternal.*[117]

(ii) For any fixed sequence $(r_1,\ldots,r_l)$, $\mathrm{peri}^{\binom{n_1,\ldots,n_l}{r_1,\ldots,r_l}}$ is a *perinomal* function of $(n_1,\ldots,n_l)$.

(iii) Although the above formulas define $peri^{\binom{\boldsymbol{n}}{\boldsymbol{r}}}$ only for an upper sequence $\boldsymbol{n}$ in $\mathbb{N}^l$, *perinomal continuation* ensures a unique extension to $\mathbb{Z}^l$.

(iv) There is another natural way of extending $peri^\bullet$ for $\boldsymbol{n}\in\mathbb{Z}^l$, namely by parity continuation, according to the formula:

$$\mathrm{peri}^{\binom{n_1,\ldots,n_l}{r_1,\ldots,r_l}} := (\mathrm{sign}(n_1))^{r_1}\ldots(\mathrm{sign}(n_l))^{r_l}\;\mathrm{peri}^{\binom{|n_1|,\ldots,|n_l|}{r_1,\ldots,r_l}} \tag{9.52}$$

(v) Whether the *perinomal* and *parity* continuations coincide – wholly, partially, or not at all – depends on the sequence $\boldsymbol{r}$ via simple criteria.

(vi) The *perinomal numbers* associated with $urr^\bullet_{II}$ and $urr^\bullet_{III}$ generate a $\mathbb{Q}$-ring that contains the $\mathbb{Q}$-ring of multizetas.

(vii) The *perinomal numbers* associated with $lum^\bullet$ respectively $rum^\bullet$ "tend" to be in $\mathbb{Q}$ respectively $\mathbb{Q}[\pi^2]$ (they are definitely there for very small sequence lengths l) but it is still a moot point whether this holds true for all l.

9.6 A glimpse of perinomal splendour

As an illustration, we shall mention the remarkable perinomal equations involving the elementary transformations $E_{i,j}$ and $e_{i,j}$ relative to neighbouring indices i, j. So let us set:

$$\begin{array}{l} E_i^+ := E_{i,i+1}\in \mathrm{Sl}_l(\mathbb{Z});\; e_i^+ \text{ ":=" } n_{i+1}\partial_{n_i}\in \mathrm{sl}_l(\mathbb{Z}) \\ E_i^- := E_{i,i-1}\in \mathrm{Sl}_l(\mathbb{Z});\; e_i^- \text{ ":=" } n_{i-1}\partial_{n_i}\in \mathrm{sl}_l(\mathbb{Z}) \end{array}$$

E_i^+ and E_i^- clearly commute, and so do e_i^+ and e_i^-.

[116] Despite having two-layered indices $\binom{n_i}{r_i}$, $peri^\bullet$ should be viewed as a mould rather than a bimould, since it would be meaningless to subject the r_i-part (as opposed to the n_i-part) to the flexion operations.

[117] As a consequence, is is enough to know a rather small subset of all numbers $peri^{\binom{n_1,\ldots,n_l}{r_1,\ldots,r_l}}$, *e.g.* those with $r_1=\min(r_i)$, to know them all.

Given a sequence $\boldsymbol{r} = (r_1, \ldots, r_l)$ and $1 \le i \le l$, we set

$$r_i^+ := \sum_{i<j\le l} r_j; \qquad r_i^- := \sum_{1\le j<i} r_j \tag{9.53}$$

$$(E_i^+ - id)^{1+r_i^+}\,(E_i^- - id)^{r_i^-}\ \mathrm{peri}^{\binom{n_1\ ,\ldots,\ n_l}{r_1\ ,\ldots,\ r_l}} \equiv 0 \quad (\forall \boldsymbol{r},\ \forall i) \tag{9.54}$$

$$(E_i^+ - id)^{r_i^+}\,(E_i^- - id)^{1+r_i^-}\ \mathrm{peri}^{\binom{n_1\ ,\ldots,\ n_l}{r_1\ ,\ldots,\ r_l}} \equiv 0 \quad (\forall \boldsymbol{r},\ \forall i). \tag{9.55}$$

In particular, for extreme values of i:

$$(E_1^+ - id)^{1+r_2+\cdots+r_l}\ \mathrm{peri}^{\binom{n_1\ ,\ldots,\ n_l}{r_1\ ,\ldots,\ r_l}} \equiv 0 \quad (\forall \boldsymbol{r}) \tag{9.56}$$

$$(E_l^- - id)^{1+r_1+\cdots+r_{l-1}}\ \mathrm{peri}^{\binom{n_1\ ,\ldots,\ n_l}{r_1\ ,\ldots,\ r_l}} \equiv 0 \quad (\forall \boldsymbol{r}). \tag{9.57}$$

In the above identities, the discrete difference operators $E_i^{\pm} - id$ may of course be replaced by the derivations $e_i^{\pm}$. But the most interesting identities are these:

$$(E_1^+ - id)^{r_2+\cdots+r_l}\ \mathrm{peri}^{\binom{n_1\ ,\ldots,\ n_l}{r_1\ ,\ldots,\ r_l}} \equiv \mathrm{peri}_L^{\binom{n_2\ ,\ldots,\ n_l}{r_2\ ,\ldots,\ r_l}} \quad (\forall \boldsymbol{r}) \tag{9.58}$$

$$(E_l^- - id)^{r_1+\cdots+r_{l-1}}\ \mathrm{peri}^{\binom{n_1\ ,\ldots,\ n_l}{r_1\ ,\ldots,\ r_l}} \equiv \mathrm{peri}_R^{\binom{n_1\ ,\ldots,\ n_{l-1}}{r_1\ ,\ldots,\ r_{l-1}}} \quad (\forall \boldsymbol{r}) \tag{9.59}$$

$$(e_1^+)^{r_2+\cdots+r_l}\ \mathrm{peri}^{\binom{n_1\ ,\ldots,\ n_l}{r_1\ ,\ldots,\ r_l}} \equiv \mathrm{peri}_{L*}^{\binom{n_2\ ,\ldots,\ n_l}{r_2\ ,\ldots,\ r_l}} \quad (\forall \boldsymbol{r}) \tag{9.60}$$

$$(e_l^-)^{r_1+\cdots+r_{l-1}}\ \mathrm{peri}^{\binom{n_1\ ,\ldots,\ n_l}{r_1\ ,\ldots,\ r_l}} \equiv \mathrm{peri}_{R*}^{\binom{n_1\ ,\ldots,\ n_{l-1}}{r_1\ ,\ldots,\ r_{l-1}}} \quad (\forall \boldsymbol{r}) \tag{9.61}$$

because they yield new, simpler perinomal fonctions $peri_L^\bullet, peri_R^\bullet$ (or their infinitesimal variants $peri_{L*}^\bullet$, $peri_{R*}^\bullet$) that are themselves closely related to the *jump functions* that measure the differences between the 2^l perinomal continuations $C^{\epsilon_1,\ldots,\epsilon_l} peri^{\binom{n_1\ ,\ldots,\ n_l}{r_1\ ,\ldots,\ r_l}}$ of $peri^{\binom{n_1\ ,\ldots,\ n_l}{r_1\ ,\ldots,\ r_l}}$ starting from the 'multioctant':

$$\mathcal{O}^{\epsilon_1,\ldots,\epsilon_l} := \{(n_1, \ldots, n_l) \in \mathbb{Z}^l \quad \text{with} \quad \epsilon_i n_i \in \mathbb{N}^*, \quad \epsilon_i \in \{+, -\}\}. \tag{9.62}$$

They are also related to the shorter components of $peri^\bullet$.

It is probably no exaggeration to say that this wondrous, double-layered mould $peri^\bullet$ is some sort of *algebraic Mandelbrot set* – its equal in terms of complexity and richness of sub-structure at all scales, but much tidier, because here the structure is algebraic in nature, consisting as it does of:

– the infinite series of perinomal fonctions encoded in $peri^\bullet$;
– their seemingly inexhaustible properties and relations;
– the degrees of the induced representations of $Sl_l(\mathbb{Z})$ for all l;
– the irreducible factor representations of these induced representations;
– the arithmetic properties of the corresponding perinomal numbers;
etc etc...

10 Provisional conclusion

10.1 Arithmetical and functional dimorphy

The word ‘dimorphy’ points to the parallel existence of two distinct multiplication rules, but the interpretation differs for *functions* and for *numbers*. For functions, the two multiplication rules define *distinct and independent products*. For numbers, they are merely *distinct and independent expressions* of one and the same product.

• Dimorphy for functions rings
A function space $\mathbb{F}$ is said to be dimorphic if it is endowed with, and stable under, two distinct (bilinear) products – usually, *pointwise multiplication* and some form or other of *convolution*. One often adds the requirement that both products should have the same unit – usually, the constant function 1. Moreover, dimorphic function rings often possess two sets of *exotic derivations*, *i.e.* linear operators irreducible to ordinary differentiation but acting as abstract derivations respective to the first or second product. (It would be tempting to attach to these dimorphic function rings the label “bialgebra”, had it not long ago acquired a different connotation – namely, stability under a product and a coproduct.)

• Dimorphy for numbers rings
A countable $\mathbb{Q}$-ring $\mathbb{D} \subset \mathbb{C}$ is dimorphic if it has two countable prebases[118] $\{\alpha_m\}$ and $\{\beta_n\}$, with a simple conversion rule linking the two, and a multiplication rule[119] attached to each prebasis:

$$\alpha_m = \sum{}^{*} H_m^n \, \beta_n \quad , \quad \beta_n = \sum{}^{*} K_n^m \, \alpha_m \qquad (H_m^n,\ K_n^m \in \mathbb{Q})$$
$$\alpha_{m_1}\alpha_{m_2} = \sum{}^{*} A_{m_1,m_2}^{m_3} \, \alpha_{m_3}, \quad \beta_{n_1}\beta_{n_2} = \sum{}^{*} B_{n_1,n_2}^{n_3} \, \beta_{n_3} \quad (A_{n_1,n_2}^{n_3},\ B_{n_1,n_2}^{n_3} \in \mathbb{Q}).$$

All sums Σ^* have to be finite. Moreover, the two multiplication rules must be “independent”, in the precise sense that neither should follow *algebraically* from the other under the conversion rule. This in turn implies that neither $\{\alpha_m\}$ nor $\{\beta_n\}$ can be a $\mathbb{Q}$-basis of $\mathbb{D}$: there have to be non-trivial, linear $\mathbb{Q}$-relations between the α_m, and others between the β_n. The main challenges, when studying a dimorphic $\mathbb{Q}$-ring $\mathbb{D} \subset \mathbb{C}$, are therefore:
(i) ascertaining whether $\mathbb{D}$ is a *polynomial algebra* (generated by a countable set of *irreducibles*) or the quotient of a polynomial algebra by some ideal;

---

[118] See definition at the beginning of Section 1.1.

[119] Compatible with $\mathbb{D}$’s natural product, which is induced by that of $\mathbb{C}$.

(ii) pruning each prebasis $\{\alpha_m\}$ and $\{\beta_n\}$ of redundant elements, so as to turn them into true bases;
(iii) whenever possible, constructing an *impartial* or *'non-aligned'* basis $\{\gamma_p\}$, positioned 'halfway' between $\{\alpha_m\}$ and $\{\beta_n\}$;
(iv) whenever possible, finding for the impartial γ_p's a *direct* expression that is itself *impartial* and leans neither towards the α_m's nor the β_n's.

• Kinship and difference between the two types of dimorphy: functional and numerical
The two notions have much in common: indeed, most dimorphic number rings are derived from dimorphic function rings either via *function evaluation* at some special points, or via *function integration*, or again via the application of *exotic derivations* to the functions and the harvesting of the constants produced in the process. And yet there is this striking difference: whereas the notion of dimorphic ring is entirely objective (– the two products are just *there* –), that of numerical dimorphy is embarrassingly subjective: on any countable $\mathbb{Q}$-ring $\mathbb{D} \subset \mathbb{C}$, one may always construct two prebases $\{\alpha_m\}$ and $\{\beta_n\}$ with the required properties. So what makes a $\mathbb{Q}$-ring $\mathbb{D}$ truly dimorphic is the existence of genuinely *natural* prebases, and the – often considerable difficulty – of solving the four problems (i), (ii), (iii), (iv) listed above. The irony, withal, is that the notion of numerical dimorphy, despite its conceptual shakiness, is much more interesting and basic than that of functional dimorphy, and throws up much harder problems.

• Hyperlogarithmic functions: the dimorphic ring $\mathcal{H}$
An interesting dimorphic space is the space $\mathcal{H}$ of hyperlogarithmic functions, which is spanned by the $\overline{\mathcal{H}}^{\boldsymbol{\alpha}}$ thus defined:[120]

$$\overline{\mathcal{H}}^{\alpha_1,\ldots,\alpha_r}(\zeta) = \int_0^\zeta \overline{\mathcal{H}}^{\alpha_1,\ldots,\alpha_{r-1}}(\zeta_r)\,\frac{d\zeta_r}{\zeta_r-\alpha_r} \quad \text{with} \quad \mathcal{H}^{\emptyset}(\zeta)\equiv 1 \quad (10.1)$$

$\mathcal{H}$ is stable under pointwise multiplication and under the unit-preserving convolution $\star$:

$$(\underline{\mathcal{H}}_1\star\underline{\mathcal{H}}_2)(\zeta)=\int_0^\zeta d\underline{\mathcal{H}}_1(\zeta_1)\,\underline{\mathcal{H}}_2(\zeta-\zeta_1)=\int_0^\zeta \underline{\mathcal{H}}_1(\zeta-\zeta_2)\,d\underline{\mathcal{H}}_2(\zeta_2). \quad (10.2)$$

Side by side with the $\boldsymbol{\alpha}$-encoding, it is convenient to consider an $\boldsymbol{\omega}$-encoding via the correspondence:

$$\underline{\mathcal{H}}^{\omega_1,\ldots,\omega_r} := \overline{\mathcal{H}}^{\omega_1,\,\omega_1+\omega_2,\ldots,\,\omega_1+\ldots+\omega_r} \quad (10.3)$$

120 First for ζ small, and then in the large by analytic continuation.

if all $\alpha_i := \omega_1 + \ldots + \omega_i$ are $\neq 0$, and by a slightly modified formula otherwise.

For this function ring $\mathcal{H}$, the basic dimorphic stability follows from the fact that the moulds $\overline{\mathcal{H}}^{\bullet}$ and $\underline{\mathcal{H}}^{\bullet}$ are both *symmetral*, the former under pointwise multiplication, the latter under convolution. Moreover, there exist on $\mathcal{H}$ two rich arrays of exotic derivations: the *foreign derivations* ∇_{α_0} and the *alien derivations* Δ_{ω_0}. These are linear operators that basically 'analyse' the singularities 'over'[121] the points α_0 or ω_0, but in such a way as to make the ∇_{α_0} and Δ_{ω_0} act as *derivations* on $\mathcal{H}$ relative to, respectively, multiplication and convolution.

• Hyperlogarithmic numbers: the dimorphic ring $\mathbb{H}$

If we now restrict ourselves to rational-complex sequences $\boldsymbol{\alpha}$ or $\boldsymbol{\omega}$ (*i.e.* with all indices α_i or ω_i in $\mathbb{Q}+i\,\mathbb{Q}$) and evaluate the corresponding $\overline{\mathcal{H}}^{\boldsymbol{\alpha}}$ or $\underline{\mathcal{H}}^{\boldsymbol{\omega}}$ *at* or *over* rational-complex points ζ, the space $\mathbb{Q}$-spanned by these numbers is in fact a $\mathbb{Q}$-ring: the $\mathbb{Q}$-ring $\mathbb{H}$ of so-called *hyperlogarithmic numbers*, which is in fact dimorphic, since it possesses two natural prebases $\{\overline{H}^{\boldsymbol{\alpha}}\}$ and $\{\underline{H}^{\boldsymbol{\omega}}\}$, each with its own, independent multiplication rule.[122]

Clearly, $\mathcal{H}$ contains the space of polylogarithms with singularities over the unit roots. Likewise, $\mathbb{H}$ contains the dimorphic $\mathbb{Q}$-ring of all (colourless or coloured) multizetas, but it also contains much more: in fact, the structure of $\mathbb{H}$ is still farther from a complete elucidation than that of the ring of multizetas.

10.2 Moulds and bimoulds. The flexion structure

• Moulds have their origin in alien calculus

Alien calculus deals with the totally non-commutative derivations Δ_ω and with the Hopf algebra $\boldsymbol{\Delta}$ freely generated by them. Let $\mathbb{A}$ be any commutative algebra. Multiplying several elements $B_i \in \mathbb{A} \otimes \boldsymbol{\Delta}$:

$$B_i = \sum_{\bullet} A_i^{\bullet}\, \Delta_{\bullet} = \sum_{r\geq 0} \sum_{\omega_i} A_i^{\omega_1,\ldots,\omega_r}\; \Delta_{\omega_1} \ldots \Delta_{\omega_r} \tag{10.4}$$

reduces to multiplying the corresponding moulds $A_i^{\bullet}$, which in many contexts (*e.g.* in formal computation) is much more convenient.

---

121 Since we are dealing here with highly ramified functions, we have to consider various leaves *over* any given point.

122 Their origin, roughly, is as follows: when we subject some 'monomial' $\overline{\mathcal{H}}^{\boldsymbol{\alpha}}$ (respectively $\underline{\mathcal{H}}^{\boldsymbol{\omega}}$) to an exotic derivation ∇_{α_0} (respectively Δ_{ω_0}), what we get is a linear combination of simpler monomials $\overline{\mathcal{H}}^{\boldsymbol{\alpha}'}$ (respectively $\underline{\mathcal{H}}^{\boldsymbol{\omega}'}$) with constant coefficients $\overline{H}^{\boldsymbol{\alpha}''}$ (respectively $\underline{H}^{\boldsymbol{\omega}''}$), which are precisely the elements of our two prebases.

Alien calculus led straightaway to the four hyperlogarithmic moulds:

$$\mathcal{U}^\bullet(z),\ \mathcal{V}^\bullet(z)\ \text{(resurgent-valued)};\qquad U^\bullet,\ V^\bullet\ \text{(scalar-valued)} \tag{10.5}$$

with their many properties and symmetries, and this is what really got the whole subject of mould calculus started.

One way of looking at moulds is to think of them as permitting the handling of non-commutative objects by means of commutative operations.

Another is to view them as permitting the explicit calculation of objects (like the Taylor coefficients of the power series expansions of the solutions of very complex, non-linear equations) that would otherwise resist explicitation.

• Moulds have found their second largest application in local differential geometry

Expansions of type (10.4) but with scalar- or function-valued moulds $A_i^\bullet$ and with homogeneous (ordinary) differential operators in place of the Δ_ω, are very useful in local differential geometry (especially when all data are analytic) for expressing and investigating normal forms, normalising transformations, fractional iterates etc. Here again, moulds make it possible to render explicit the seemingly inexplicitable – with all the advantages that accrue from transparency.

• Mould operations and mould symmetries

Moulds of natural origin usually come with a definite symmetry type – *symmetral* or *symmetrel*, *alternal* or *alternel* – and most mould operations either preserve these symmetries or transmute them in a predictable manner.

• Moulds and arborification

When natural mould-comould expansions such as (10.4) display normal divergence and yet "ought to converge" (because they stand for really existing function germs or 'local' geometric objects), a general and very effective remedy is at hand: the transform known as *arborification-coarborification* nearly always suffices to restore normal convergence. Roughly speaking, the transform in question replaces, dually in $A^\bullet$ and $\Delta_\bullet$, the totally ordered sequences $\boldsymbol{\omega}$ by sequences carrying a weaker, arborescent order, and it does so in such a way as to leave the global series formally unchanged, while effecting the proper internal reordering that restores convergence.

• Bimoulds

There is much more to being a bimould than just carrying double-layered indices $w_i := \binom{u_i}{v_i}$. On top of being subject to the usual mould operations, like *mu* and *lu*, and being eligible for the four basic mould symmetries (see above), bimoulds can also display new symmetries *sui generis*, and can be subjected to numerous (unary or binary) operations without 'classical' equivalents. These are the so-called *flexion operations*, under which the u_i get added bunch-wise, and the v_i subtracted pair-wise, in such a way as to preserve $\sum u_i v_i$ and $\sum du_i \wedge dv_i$.

• **The flexion structure**
A non-pedantic, if slightly cavalier, way of defining the *flexion structure* is to characterise it as the collection of all interesting objects (unary or binary operations, symmetry types, algebras, groups etc) that may be constructed on bimoulds from the sole *flexions*. It turns out that, up to isomorphy, the flexion structure consists of exactly:
(i) seven algebras, notably *ARI* and *ALI*;
(ii) seven groups, notably *GARI* and *GALI*;
(iii) five super-algebras, notably *SUARI* and *SUALI*.

• **Recovering most classical moulds from bimoulds**
Many classical moulds (especially when, as is often the case, their analytical expression involves partial sums or pairwise differences of their indices ω_i) can be recovered, and their properties better understood, when viewed as special bimoulds with one vanishing row of indices (either $\boldsymbol{v} = \mathbf{0}$ or $\boldsymbol{u} = \mathbf{0}$).

• **Monogenous substructures**
These are the spaces $Flex(\mathfrak{E}) = \oplus_{0 \leq r} Flex_r(\mathfrak{E})$ generated by a single length-one bimould $\mathfrak{E}^\bullet$ under *all* flexion operations. The most natural monogenous structures correspond to the case when $\mathfrak{E}^{w_1}$ is totally 'random' (*i.e.* when there are no unexpected relations in its flexion offspring) or possesses a given parity in u_1 and v_1 (four possibilities).

• **Flexion units and their offspring**
In terms of applications the most important monogenous structures $Flex(\mathfrak{E})$ correspond to special generators $\mathfrak{E}^\bullet$ that verify the so-called *tripartite identity* (3.9). These $\mathfrak{E}^\bullet$ are known as *flexion units* and admit various realisations as concrete functions of w_1: polar, trigonometric, 'flat' etc.

• **Algebraisation of the substructures**
Each type of abstract generator $\mathfrak{E}^\bullet$ subject to a given set of constraints[123] may admit several realisations (as a function or distribution etc.), or just

[123] Like (3.9) or (3.28) or (3.29) etc.

one, or none at all. But in all cases the flexion structure $Flex(\mathfrak{E}) = \oplus_{0\leq r} Flex_r(\mathfrak{E})$ generated by $\mathfrak{E}^\bullet$ is a well-defined algebraic object, with an integer sequence $d_r = \dim(Flex_r(\mathfrak{E}))$ that reflects the strength of the constraints on $\mathfrak{E}^\bullet$. Moreover, in most cases, the length-r component $Flex_r(\mathfrak{E})$ of $Flex(\mathfrak{E})$ possesses one (or several) *natural* bases $\{\mathfrak{e}_t^\bullet\} = \{\mathfrak{e}_t^{w_1,\dots,w_r}\}$, with basis elements *naturally* indexed by r-node *trees* $\boldsymbol{t}$ of a well-defined sort – like for instance *binary trees* if $\mathfrak{E}^\bullet$ is a *flexion unit* or *ternary trees* if $\mathfrak{E}^\bullet$ is 'random'.[124] This automatically endows the abstract space spanned by those trees with the full flexion structure and all its wealth of operations, opening the way for fascinating (and as yet largely unexplored) developments in combinatorics.[125]

• Origins of the flexion structure

The flexion structure arose in the early 1990s in an analysis context, as a tool for describing a very specific type of resurgence, variously known as *quantum resurgence*[126] or *parametric resurgence*[127] or *co-equational resurgence*.[128]

• Present and future of the flexion structure

In the early 2000s, the flexion structure began to be used, to great effect, in the investigation of multizeta arithmetics and numerical dimorphy, and this is likely to remain the theory's main area of application for quite some time to come. However, the *algebraisation* of monogenous (respectively polygenous) structures like $Flex(\mathfrak{E})$ (respectively $Flex(\mathfrak{E}_1,\dots,\mathfrak{E}_n)$) also suggests promising applications in algebra and combinatorics. We can even discern the outlines of a future 'flexion Galois theory' that would concern itself with the way in which a given type of constraints on $\mathfrak{E}^\bullet$ or on the $\mathfrak{E}_i^\bullet$ impacts the structure, dimensions, etc, of such objects as $Flex_r(\mathfrak{E})$ or $Flex_r(\mathfrak{E}_1,\dots,\mathfrak{E}_n)$.

---

124 But with a given parity in u_1 and v_1.

125 There exists of course an abundant botanical literature on trees of various descriptions, their enumeration, generation, classification etc. But so far these trees have not been studied, generated, classified etc from the angle of the flexion operations, for the obvious reason that these operations are new.

126 Because often encountered in the 'semi-classical' mechanics – *i.e.* when expanding formal solutions of the Schrödinger equation in power series of the Planck constant $\hbar$. See Section 11.1, Section 11.2, Section 11.3 *infra*.

127 Since it is typically encountered in power series of a (singular perturbation) parameter.

128 Because it is loosely dual to 'equational resurgence', that is to say, to the type of resurgence encountered in power series of the equation's proper variable.

10.3 *ARI/GARI* and the handling of double symmetries

• Simple symmetries or subsymmetries at home in *LU/MU*
The uninflected mould bracket *lu* preserves *alternality* and its two subsymmetries: *mantar*-invariance and *pus*-neutrality.[129] Similarly, the uninflected mould product *mu* preserves *symmetrality* and its two subsymmetries: *gantar*-invariance and *gus*-neutrality.[130] And that's about all. Even when *lu* or *mu* are made to act on bimoulds, they preserve none of the double symmetries[131] and none of the induced subsymmetries[132] – not even the so crucial *push*- or *gush*-invariance.

• Double symmetries or subsymmetries at home in *ARI/GARI*
Things change when we go over to the inflected operations, or rather to the right ones, since of all seven pairs consisting of a flexion Lie algebra and its group, only *ARI//GARI* and *ALI//GALI* are capable of preserving double symmetries and subsymmetries. In the case of *ARI* (respectively *GARI*) the full picture has been summarised on the table of Section 2.5 (respectively Section 2.6). Things differ slightly with *ALI* (respectively *GALI*), but we need not bother with these differences since, when restricted to bimoulds of type $\underline{al}/\underline{al}$ (respectively $\underline{as}/\underline{as}$), the Lie brackets *ari* and *ali* (respectively the group laws *gari* and *gali*) exactly coincide.

All the above, it should be noted, applies to *straight* (*i.e.* uninflected) double symmetries, but similar results hold for the *twisted*[133] double symmetries that really matter, beginning with $\underline{al}/\underline{il}$ and $\underline{as}/\underline{is}$.

• Ubiquity of poles at the origin: associator
In the canonical trifactorisation of $Zag^\bullet$, the leftmost factor $Zag_I^\bullet$ which, we recall, encodes all the information about the canonical-rational associator, admits in its turn a trifactorisation of the form

$$\mathrm{Zag}_{\mathrm{I}}^\bullet = \mathrm{gari}(\mathrm{tal}^\bullet, \mathrm{invgari.pal}^\bullet, \mathrm{Roma}^\bullet) \tag{10.6}$$

[129] As defined in Section 2.4.

[130] As defined in Section 2.4.

[131] *I.e.* symmetries affecting simultaneously a bimould $M^\bullet$ and its swappee $swap.M^\bullet$.

[132] Meaning of course the *strictly double* subsymmetries – *i.e.* those that don't follow from a *single* symmetry.

[133] Or, should we say, *half-twisted*, since it is not the bimould $M^\bullet$ itself, but only its swappee $swap.M^\bullet$, that may display a twisted symmetry. No other combination would be stable under the flexion operations.

and the strange thing is that, although $Zag_I^\bullet$, as a function of the u_i variables, is of course free of poles at the origin, all three factors are replete with them.
(i) The (polar) mid-factor $pal^\bullet$ contains nothing but multipoles at the origin, and so does its *gari*-inverse.
(ii) The (trigonometric) first factor $tal^\bullet$, which is a *periodised* variant of $pal^\bullet$, carries multipoles *at* and *off* the origin, and those *at* the origin are roughly the same as those of $pal^\bullet$.
(iii) Since the multipoles of $pal^\bullet$ and $tal^\bullet$ very nearly, but not exactly, cancel out at the origin, a (highly transcendental) third factor $Roma^\bullet$ is called for to remove the remaining singularities, and the construction of that third factor involves at every step special operators, the so-called *singulators*, whose function it is to introduce, in a systematic and controlled way, all the required corrective singularities at the origin.

• Ubiquity of poles at the origin: singulators and generation of $ALIL \subset ARI^{\underline{al}/\underline{il}}$

To construct any of the three alternative bases $\{luma_s^\bullet\}$, $\{loma_s^\bullet\}$, $\{lama_s^\bullet\}$ of *ALIL*, we start from the arch-elementary bimoulds *ekmas*, purely of length-1 and trivially of type $\underline{al}/\underline{al}$, and then apply $adari(pal^\bullet)$ to produce new bimoulds, this time of the right type $\underline{al}/\underline{il}$ but ridden with unwanted singularities at the origin. To remove these without losing the property $\underline{al}/\underline{il}$, we must then engage in a double process of *singularity destruction* and *singularity re-introduction* (at higher lengths), which is painstakingly described in Section 6. The operators behind the construction, the so-called *singulators*, are themselves built from the *purely singular*, polar bimould $pal^\bullet$. Poles, therefore, completely dominate the process – first as obstacles, then as remedies.

• Ubiquity of poles at the origin: singulators and generation of $ALAL \subset ARI^{\underline{al}/\underline{al}}$**. The exceptional bialternals**

That the construction of pole-free bases for *ALIL* should involve poles at all intermediary steps, is surprising enough, but still halfway understandable, since the very definition of *alternility* involves (mutually cancelling) polar terms. But the really weird thing is that poles should also be required to construct bases of *ALAL*, since the double symmetry here is completely straight. Nevertheless, such is the case: to the elementary *ekma* bialternals, one must adjoin the exceptional and very complex *carma* bialternals, whose construction cannot bypass the introduction of poles, since it requires the prior knowledge of an *ALIL*-basis up to length $r = 3$ (but, thankfully, no farther), as shown in Section 7.

• **Ubiquity of poles away from the origin: perinomal analysis**
Perinomal analysis deals with meromorphic functions that possess multipoles all over the place: their location admits a natural indexation over $\mathbb{Z}^r$, their multiresidues are also defined on $\mathbb{Z}^r$ and are of *perinomal* nature. So, here again, multipoles have a way of inviting themselves into all calculations.

• ***ARI* and the Ihara algebra**
The fact that the Ihara algebra is isomorphic to a *twee tiny little* subalgebra of *ARI*[134] – namely, the subalgebra of bimoulds of type $\underline{al}/\underline{il}$, polynomial in $\boldsymbol{u}$ and constant in $\boldsymbol{v}$ – is no reason for 'equating' the two structures, or even their Lie bracket. But since there still reigns much confusion around this fraught issue, a short clarification is in order.
(i) To begin with, none of the dozens of pole-carrying bimoulds such as $pal^\bullet$ or $tal^\bullet$ or $r\o ma^\bullet$, which are key to the understanding of $Zag^\bullet$, possess any counterpart in the Ihara algebra. As a consequence, neither can the *carma* bialternals be constructed in that framework, nor can the reason behind their presence be understood, nor can anything even remotely resembling $l\o ma^\bullet$ be constructed.
(ii) Second, unlike the Ihara algebra, the *ARI* approach puts both symmetries – alternal and alternil – on exactly the same footing and does full justice to the duality that underpins multizeta (and general arithmetical) dimorphy. Indeed, with its involution *swap*, its built-in duality between upper and lower indices, and all the main bimoulds like $pal^\bullet/pil^\bullet$, $tal^\bullet/til^\bullet$ etc that always occur in pairs, *ARI* is itself 'dimorphic' to the marrow.
(iii) Third, the whole subject of perinomal algebra and of canonical irreducibles is beyond not just the computational reach of the Ihara algebra, but even its means of conception.
(iv) Fourth, unlike the Ihara algebra, *ARI*, with its double row of indices, lends itself effortlessly to the passage from uncoloured to coloured multizetas.
(v) Lastly, *ARI* arose independently of the Ihara algebra, in direct answer to a problem of analysis and resurgence. In fact, unlike the Ihara algebra, *ARI* is serviceable in analysis no less than in algebra.

10.4 What has already been achieved

Finding the proper setting was the first and arguably main step. The rest followed rather naturally.

---

[134] Though it houses the multizetas themselves (in a formalised version), the subalgebra in question is too cramped a framework for their complete elucidation, since most auxiliary constructions required in the process lie *outside*.

• Correction formula

Moving from the scalar multizetas $Wa^\bullet/Ze^\bullet$ to the generating functions $Zag^\bullet/Zig^\bullet$ makes it much easier to understand the reason for the corrective terms $Mana^\bullet/Mini^\bullet$ in (1.27), (1.23). As meromorphic functions, $Zag^\bullet$ and $Zig^\bullet$ are both given by *semi-convergent* series of multipoles. Formally, the involution *swap* exchanges both series *exactly*, but alters their *summation order*, leading to simple corrective terms constructed from monozetas.

• Meromorphic continuation of multizetas and arithmetical nature at negative points

When taken in the *Ze* encoding, the scalar multizetas $Ze^{\binom{\epsilon_1\ ,\ \ldots,\ \epsilon_r}{s_1\ ,\ \ldots,\ s_r}}$ possess a meromorphic extension to the whole of $\mathbb{C}^r$, with all their multipoles on $\mathbb{Z}^r$.

(i) The density of multipoles decreases with the 'coloration' of the multizetas, *i.e.* with the number of non-vanishing ϵ_i's.

(ii) The values (respectively residues) found at the regular (respectively irregular) places $s \in \mathbb{Z}^r \dot{-} \mathbb{N}^r$ are themselves rational combinations of *simpler* multizetas.[135]

(iii) The symmetrelity relations verified by $Ze^\bullet$, which hold for positive s_i's, extend by meromorphic continuation to the whole of $\mathbb{C}^r$, including to the points of $\mathbb{Z}^r$ where, in view of (ii), they *might* – but in fact *do not* – generate new multizeta relations.[136]

• Unit-cleansing

Any 'uncoloured' multizeta $Ze^{\binom{0\ ,\ \ldots,\ 0}{s_1\ ,\ \ldots,\ s_r}}$ with $s_i \in (\mathbb{N}^*)^r$ can in fact be expressed (in non-unique manner) as a rational-linear combinations of analogous but *unit-free* multizetas (*i.e.* with $s_i \geq 2$). The proof rests on a reformulation of the problem in terms of bialternals, and then on the so-called *redistribution* identities (of rich combinatorial content) which make it possible to recover any bialternal polynomial $Mi^{\binom{0\ ,\ \ldots,\ 0}{v_1\ ,\ \ldots,\ v_r}}$ from its *essential part*, *i.e.* from the collection of its constituant monomials that are divisible by $v_1 \ldots v_r$.

• Parity reduction

Any 'uncoloured' multizeta $Ze^{\binom{0\ ,\ \ldots,\ 0}{s_1\ ,\ \ldots,\ s_r}}$ with $s_i \in (\mathbb{N}^*)^r$ can in fact be expressed as a rational-linear combinations of analogous multizetas of

---

135 *I.e.* of multizetas of length $r' < r$. The more 'negative' s_i's there are, the smaller the number r' becomes.

136 For details, see [4].

even degree.[137] While this follows from the general result on the decomposition of multizetas into *irreducibles* (these correspond here to uncoloured bialternal polynomials, which are necessarily of *even* degree d), there exists a more elementary derivation, based on the properties of the symmetrel bimould $Tig^{\bullet}(z)$, or "multitangent bimould", thus defined:[138]

$$Tig^{\binom{\epsilon_1\ ,\dots,\ \epsilon_r}{v_1\ ,\dots,\ v_r}}(z) := \sum_{s_i\geq 1} Te^{\binom{\epsilon_1\ ,\dots,\ \epsilon_r}{s_1\ ,\dots,\ s_r}}(z)\, v_1^{s_1-1}\dots v_r^{s_r-1} \tag{10.7}$$

$$Te^{\binom{\epsilon_1\ ,\dots,\ \epsilon_r}{s_1\ ,\dots,\ s_r}}(z) := \sum_{+\infty>n_1>\dots>n_r>-\infty} e_1^{-n_1}\dots e_r^{-n_r}(n_1+z)^{-s_1}\dots(n_r+z)^{-s_r} \tag{10.8}$$

and on the *two different ways* of expressing each uncoloured multitangent $Tig^{\binom{\mathbf{0}}{s}}(z)$ as sums of uncoloured monotangents $Tig^{\binom{0}{s_1}}(z)$ with uncoloured multizeta coefficients. See Section 11.7.

• The senary relation and palindromy formula

The *senary relations* on bimoulds of type $\underline{al}/\underline{il}$ are the only double subsymmetries of finite arity – they involve exactly six terms. In polar (respectively universal) mode, they assume the form (3.64) (respectively (3.58)). They result from the double symmetry $\underline{al}/\underline{il}$ of a bimould $M^{\bullet}$, more precisely from the *mantar*-invariance of $M^{\bullet}$ (consequence of its alternality) and the *mantir*-invariance of *swap*.$M^{\bullet}$ (consequence of its alternility).

The *palindromy relations*, on the other hand, apply to homogeneous elements $C \in IHARA \subset \mathbb{Q}[x_0, x_1]$ of the Ihara algebra (x_0 and x_1 don't commute), or more precisely to their left or right decompositions:

$$C = A_0\, x_0 + A_1\, x_1 = x_0\, B_0 + x_1\, B_1 \quad (A_i, B_i \in \mathbb{Q}[x_0, x_1]) \tag{10.9}$$

and state that the sums $A_0 + A_1$ and $B_0 + B_1$ are invariant under the palindromic involution:

$$x_{\epsilon_1}\, x_{\epsilon_2} \dots x_{\epsilon_s} \mapsto (-1)^s\, x_{\epsilon_s} \dots x_{\epsilon_2}\, x_{\epsilon_1}. \tag{10.10}$$

[137] Recall that $d := s - r$: the degree d of a scalar multizeta (in the *Ze* encoding) is equal to its total weight s minus its length r.

[138] In the second sum, $e_j := \exp(2\pi i \epsilon_j)$ as usual, and we apply standard symmetrel renormalisation to get a finite result when either s_1 or s_r is $= 1$.

The palindromy relations[139], which according to the above involve *four clusters of terms*, can easily be shown to be equivalent to a special case of the senary relations[140], which involve *six*.

• Coloured multizetas. Bicolours and tricolours

The statement about the eliminability of unit weights[141] in uncoloured multizetas still applies in the coloured case, but here another, almost opposite results holds: every bicoloured or tricoloured multizeta[142] with arbitrary weights can be (in non-unique manner) expressed as a rational-linear combination of multizetas with unit-weights only.

• Canonical-rational associator and explicit decomposition into canonical irreducibles

We would rate this as the second-most encouraging result obtained so far with the flexion apparatus. The existence of a truly canonical decomposition[143] was by no means a foregone conclusion – in fact, it had gone completely unsuspected. Moreover, since everything rests on the construction of an explicit basis of $ALIL \subset ARI^{\underline{al}/\underline{il}}$, which in turn requires the repeated introduction and elimination of *singularities at the origin*,[144] the construction cannot be duplicated in any other framework than the flexion structure.

• The impartial expression of irreducibles as perinomal numbers

We would, in all humility, regard this as the crowning achievement of the flexion method so far. The two circumstances which made it possible are: the exact adequation of *ARI//GARI* to dimorphy; and the 'vastness' of the structure, which accommodates not just polynomials in the $\boldsymbol{u}$ or $\boldsymbol{v}$ variables, but also meromorphic functions (and much else).

• The first forays into perinomal territory

Though we only stand at the beginning of what looks like an open-ended exploration, we can already rely on two firm facts to guide the search: one

---

139 They were empirically observed by followers of the Ihara approach, and pointed out to me, as conjectures, by L. Schneps in March 2010.

140 Namely for $\boldsymbol{u}$-polynomial and $\boldsymbol{v}$-constant bimoulds. The senary relations first appeared, among many similar consequences of *double symmetries*, in a 2002 paper by us and were mentioned, the next year, during a series of Orsay lectures.

141 *I.e.* of all indices s_i that are equal to 1.

142 *I.e.* with $\epsilon_i \in \frac{1}{2}\mathbb{Z}/\mathbb{Z}$ or $\epsilon_i \in \frac{1}{3}\mathbb{Z}/\mathbb{Z}$

143 The existence of three closely related variants (see Section 9.1 and Section 9.3) in no way detracts from the canonicity.

144 The infinite process is described in Section 6.

is the *perinomal nature* of the multiresidues 'hidden' in the constituent parts of $Zag^\bullet/Zig^\bullet$; the other is the existence, attached to each integer sequence $\boldsymbol{r} := (r_1, \ldots, r_l)$, of a *specific linear representation* of $Sl_l(\mathbb{Z})$.

10.5 Looking ahead: what is within reach and what beckons from afar

• **Arithmetical and analytic properties of** $lama^\bullet/lami^\bullet$
Of all three 'co-canonical' pairs, this is the simplest, arithmetically speaking. As power series of $\boldsymbol{u}$ or $\boldsymbol{v}$, these bimoulds carry Taylor coefficients that have, globally, the smallest possible denominators. But the series themselves are divergent-resurgent – with a resurgence pattern that is still poorly understood.[145]

• **Arithmetical and analytic properties of** $loma^\bullet/lomi^\bullet$
Arithmetically, this second pair is less simple (the Taylor coefficients have slightly larger denominators) but the associated power series are convergent, with a finite multiradius of convergence. At the moment, however, it is unclear whether the corresponding functions admit endless analytic continuation and, if so, what the exact nature of their isolated singularities might be.

• **Arithmetical and analytic properties of** $luma^\bullet/lumi^\bullet$
This last pair, being defined by semi-convergent series of multipoles, has a completely transparent meromorphic structure. The difficulty, here, is with the arithmetics of the Taylor coefficients: up to length $r = 4$, they are all rational, but (for $3 \leq r \leq 4$) with very irregular denominators.[146] Beyond that (for $5 \geq r$), it is not even known whether the coefficients are rational.[147]

Needless to say, analogous questions arise for the three parallel pairs $rama^\bullet/rami^\bullet$, $roma^\bullet/romi^\bullet$, and $ruma^\bullet/rumi^\bullet$.

• **Perinomal algebra. Ranks of** $Sl_r(\mathbb{Z})$ **representations**
As repeatedly noted, to each integer sequence $\boldsymbol{r} := (r_1, \ldots, r_l)$, our approach to multizeta algebra attaches a perinomal function $\boldsymbol{n} \mapsto peri^{\binom{\boldsymbol{n}}{\boldsymbol{r}}}$, which in turn induces a linear representation $\mathcal{R}_{\boldsymbol{r}}$ of $Sl_l(\mathbb{Z})$. The (clearly fast increasing) ranks of these $\mathcal{R}_{\boldsymbol{r}}$ are unknown except in a few special cases, and their structure (*e.g.* their decomposition into irreducible representations) is equally unknown.

---

[145] For any given length r, the resulting resurgence algebra is probably finite dimensional, which would be an additional incentive for unravelling its structure.

[146] See Section 6.7.

[147] See Section 9.3.

• Links between the four series of perinomal functions
To each perinomal function carried by $peri^\bullet$, identities such as (9.58) or (9.59) attach simpler but related perinomal functions, but a clear overall picture is probably still a long way off. For aught we know, the two-layered mould $peri^\bullet$ may turn out to be as complex (though more tidy) than the Mandelbrot set, with algebraic (rather than fractal-geometric) detail "as far as the sight reaches".

• Arithmetical nature of all perinomal numbers
The $\mathbb{Q}$-ring *PERI* of all perinomal numbers (see Section 8.4) exceeds the $\mathbb{Q}$-ring $\mathbb{Z}eta$ of multizetas (even if we allow colour) but the range and structure of the *difference* remains unexplored.

• The quest for numerical derivations.
Does there exist on *PERI* an algebra *DERI* of *direct* numerical derivations, that is to say, of linear operators D verifying:

$$D(x.y) \equiv Dx.y + x.Dy \qquad (\forall x, y \in \mathrm{PERI}, \ \ \forall D \in \mathrm{DERI}) \qquad (10.11)$$

$$D.\mathbb{Q} = \{0\}, \ \ \{0\} \neq D.\mathrm{PERI} \subset \mathrm{PERI} \qquad (\forall D \in \mathrm{DERI}). \qquad (10.12)$$

The emphasis here is on *direct*, meaning that the action of D on any $x \in PERI$ ought to be defined in universal terms, *i.e.* based on a universal expansion (decimal, continued fraction, etc) of x, and not on its *mode of construction*. This at the moment is little more than a dream, but if it came true, it would give us a key – possibly, the only workable key – to unlock the *exact*, as opposed to *formal*, arithmetics[148] of *PERI* and its subring $\mathbb{Z}eta$. But this is purest *terra incognita* and, as it said on ancient maps where unchartered territory began, *ibi sunt leones*...

11 Complements

11.1 Origin of the flexion structure

The flexion structure has its origin (ca 1990) in the investigation of *parametric resurgence* – typically, the sort of resurgence associated with formal expansions in series of a *singular perturbation parameter* ϵ.[149] Set

[148] No one would seriously expect the two arithmetics – exact and formal – to differ, but proving their identity is another matter.

[149] Think for definiteness of a differential equation with a small ϵ sitting in front of the highest order derivative.

$x = \epsilon^{-1}$ (x large, ϵ small) and consider the standard system:

$$((u_1 + \ldots + u_r)\, x + \partial_z) \mathcal{W}^{\binom{u_1\ ,\ldots,\ u_r}{v_1\ ,\ldots,\ v_r}}(z, x) = \mathcal{W}^{\binom{u_1\ ,\ldots,\ u_{r-1}}{v_1\ ,\ldots,\ v_{r-1}}}(z, x) \frac{1}{z - v_r} \quad (11.1)$$

with $\mathcal{W}^{\binom{\emptyset}{\emptyset}}(z, x) := 1$ to start the induction.

We may fix x and expand the solutions as formal power series of z^{-1}. These turn out to be divergent, Borel-summable, and resurgent, with a simple resurgence locus[150] consisting of the sums of u_i indices.

We may also fix z and expand the solutions as formal power series of x^{-1}. These are again divergent, Borel-summable, and resurgent, but with a much more intricate resurgence locus generated (bi-linearly) by the two sets of indices, the u_i and v_i, under 'flexion operations'.

As functions of z, the $\mathcal{W}^{\binom{u}{v}}(z, x)$ do not differ significantly[151] from the standard resurgence monomials $\mathcal{V}^{\omega}(z) := \mathcal{W}^{\binom{\omega}{0}}(z, 1)$ defined by the induction:

$$(\omega_1 + \ldots + \omega_r + \partial_z) \mathcal{V}^{\omega_1,\ldots,\omega_r}(z) = \mathcal{V}^{\omega_1,\ldots,\omega_{r-1}}(z) \frac{1}{z} \quad \text{with} \quad \mathcal{V}^{\emptyset}(z) := 1. \quad (11.2)$$

As functions of x, on the other hand, the $\mathcal{W}^{\binom{u}{v}}(z, x)$ can be expressed as linear combinations[152] of standard resurgence monomials $\mathcal{V}^{\omega}(x) = \mathcal{V}^{\omega_1,\ldots,\omega_r}(x)$, with indices ω_j that depend bilinearly on the indices u_i and v_i (to which one must add z itself). Formally, the u_j's and v_j's contribute in much the same way to the ω_j's, although the natural way of expressing the ω_j's is via *sums* of (several consecutive) u_j's and *differences* of (two non-necessarily consecutive) v_j's or of v_j's and z.

As to their origin, however, the u_j's and v_j's could not differ more. In all natural problems, the u_j's depend only on the *principal part* of the differential equation or system and tend to be generated by a finite number of scalars (such as the system's *multipliers*, *i.e.* the eigenvalues of its linear part). There is thus considerable *rigidity* about the u_j's. With the v_j's, on the other hand, we have complete *flexibility*: they reflect pre-existing singularities in the (multiplicative) z-plane and can be *anything*.

---

150 The *resurgence locus* of a resurgent function f is the set $\Omega \subset \mathbb{C}_\bullet := \widetilde{(\mathbb{C} \dot{-} 0)}$ of all ω_0 that give rise to non-vanishing alien derivatives $\Delta_{\omega_0} f$ or $\Delta_{\omega_0} \Delta_{\omega_1} \ldots \Delta_{\omega_r} f$.

151 In terms of their resurgence properties.

152 The number of summands is exactly $r!! := 1.3.5 \ldots (2r - 1)$ and all coefficients are of the form ± 1.

11.2 From simple to double symmetries. The *scramble* transform

Originally, the *scramble* transform arose during the search for a systematic expression of the complex $\mathcal{W}^\bullet$ of (11.1) in terms of the simpler $\mathcal{V}^\bullet$ of (11.2). Our reason for mentioning it here is because the transform in question led:
(i) to the first systematic use of flexions;
(ii) to the first systematic production of double symmetries.

The *scramble* is a linear transform on *BIMU*:

$$M^\bullet \to S^\bullet = \mathrm{scram}.M^\bullet \qquad \text{with} \qquad S^{\boldsymbol{w}} := \sum_{\boldsymbol{w}^*} \epsilon(\boldsymbol{w}, \boldsymbol{w}^*)\, M^{\boldsymbol{w}^*} \quad (11.3)$$

which not only preserves simple symmetries (*alternal* or *symmetral*) but, in the case of *all-even* bimoulds[153] $M^\bullet$, turns simple into double symmetries (*alternal* into *bialternal* and *symmetral* into *bisymmetral*).

$$\begin{array}{llll}
\mathrm{scramble} : M^\bullet \mapsto S^\bullet & & & \\
\mathrm{scramble} : \mathrm{LU}^{\mathrm{al}} & \to \mathrm{ARI}^{\mathrm{al}} & \| \ \mathrm{LU}^{\mathrm{al}}_{\text{all-even}} & \to \mathrm{ARI}^{\underline{\mathrm{al}}/\underline{\mathrm{al}}} \\
\mathrm{scramble} : \mathrm{MU}^{\mathrm{as}} & \to \mathrm{GARI}^{\mathrm{as}} & \| \ \mathrm{MU}^{\mathrm{as}}_{\text{all-even}} & \to \mathrm{GARI}^{\underline{\mathrm{as}}/\underline{\mathrm{as}}}.
\end{array}$$

To define the sums $S^{\boldsymbol{w}}$ in (11.3) we have the choice between a *forward* and *backward* induction, quite dissimilar in outward form but equivalent nonetheless. They involve respectively the 'mutilation' operators *cut* and *drop*:

$$\begin{array}{llll}
(\mathrm{cut}_{w_0} M)^{w_1,\ldots,w_r} & := M^{w_2,\ldots,w_r} & \text{if } w_0 = w_1 \\
& := 0 & \text{if } w_0 \neq w_1 \\
(\mathrm{drop}_{w_0} M)^{w_1,\ldots,w_r} & := M^{w_1,\ldots,w_{r-1}} & \text{if } w_0 = w_r \\
& := 0 & \text{if } w_0 \neq w_r.
\end{array}$$

We get each induction started by setting $S^{w_1} := M^{w_1}$ and then apply the following rules.

Forward induction rule
We set $(\mathrm{cut}_{w_0}.S)^{\mathbf{w}} := 0$ unless w_0 be of the form $\lceil w_i \rceil$ with respect to some sequence factorisation $\mathbf{w} = \mathbf{a} w_i \mathbf{bc}$, in which case we set:

$$(\mathrm{cut}_{\lceil w_i \rceil} S)^{\mathbf{w}} := (-1)^{r(\mathbf{b})} \sum_{\mathbf{w}' \in \mathrm{sha}\left(\mathbf{a}\rfloor,\, \lfloor\tilde{\mathbf{b}}, \mathbf{c}\right)} S^{\mathbf{w}'} \qquad (\text{if } \mathbf{w} = \mathbf{a} w_i \mathbf{bc}) \quad (11.4)$$

with $\tilde{\boldsymbol{b}}$ denoting the sequence $\boldsymbol{b}$ in reverse order. If $M^\bullet$ is symmetral, so is $S^\bullet$ (see below). In that important case the forward induction rules

153 *I.e.* in the case of bimoulds $M^{\boldsymbol{w}}$ that are *even* separately in each double index w_i.

assumes the much simpler form:

$$(\mathrm{cut}_{\lceil w_i \rceil} S)^{\mathbf{w}} := S^{\mathbf{a}\rfloor}(\mathrm{invmu}.S)^{\lfloor\mathbf{b}} S^{\mathbf{c}} \qquad (\text{if } \mathbf{w} = \mathbf{a}w_i\mathbf{bc}) \qquad (11.5)$$

Backward induction rule
We set $(\mathrm{drop}_{w_0}.S)^{\mathbf{w}} := 0$ unless w_0 be of the form $\lfloor w_i$ or $w_i\rfloor$ with respect to some sequence factorisation $\mathbf{w} = \mathbf{a}w_i\mathbf{b}$, in which case we set:

$$(\mathrm{drop}_{\lfloor w_i}.S)^{\mathbf{w}} := -S^{\mathbf{a}\rceil\mathbf{b}} \qquad (\text{if } \mathbf{w} = \mathbf{a}w_i\mathbf{b}) \qquad (11.6)$$

$$(\mathrm{drop}_{w_i\rfloor}.S)^{\mathbf{w}} := +S^{\mathbf{a}\lceil\mathbf{b}} \qquad (\text{if } \mathbf{w} = \mathbf{a}w_i\mathbf{b}). \qquad (11.7)$$

Remark 1: *mu* is bilinear whereas *gari* is heavily non-linear in its second argument. So how can the scramble inject $\mathrm{MU}^{\mathrm{as}}$ into $\mathrm{GARI}^{\mathrm{as}}$? The answer is that under the above algebra morphism, the non-linearity of *gari* gets "absorbed" by the bimoulds' symmetrality. This is easy to check up to length 3, on the formulas:

$$\begin{aligned}
S^{\binom{u1}{v1}} &= +M^{\binom{u1}{v1}}\\
S^{\binom{u_1\,,\,u_2}{v_1\,,\,v_2}} &= +M^{\binom{u_1\,,\,u_2}{v_1\,,\,v_2}} + M^{\binom{u_{12}\,,\,u_1}{v_2\,,\,v_{1:2}}} - M^{\binom{u_{12}\,,\,u_2}{v_1\,,\,v_{2:1}}}\\
S^{\binom{u_1\,,\,u_2\,,\,u_3}{v_1\,,\,v_2\,,\,v_3}} &= +M^{\binom{u_1\,,\,u_2\,,\,u_3}{v_1\,,\,v_2\,,\,v_3}} + M^{\binom{u_1\,,\,u_{23}\,,\,u_2}{v_1\,,\,v_3\,,\,v_{2:3}}} - M^{\binom{u_1\,,\,u_{23}\,,\,u_3}{v_1\,,\,v_2\,,\,v_{3:2}}}\\
&\quad + M^{\binom{u_{12}\,,\,u_1\,,\,u_3}{v_2\,,\,v_{1:2}\,,\,v_3}} - M^{\binom{u_{12}\,,\,u_2\,,\,u_3}{v_1\,,\,v_{2:1}\,,\,v_3}}\\
&\quad + M^{\binom{u_{12}\,,\,u_3\,,\,u_1}{v_2\,,\,v_3\,,\,v_{1:2}}} - M^{\binom{u_{12}\,,\,u_3\,,\,u_2}{v_1\,,\,v_3\,,\,v_{2:1}}}\\
&\quad + M^{\binom{u_{123}\,,\,u_{23}\,,\,u_3}{v_1\,,\,v_{2:1}\,,\,v_{3:2}}} - M^{\binom{u_{123}\,,\,u_{23}\,,\,u_2}{v_1\,,\,v_{3:1}\,,\,v_{2:3}}} + M^{\binom{u_{123}\,,\,u_3\,,\,u_2}{v_1\,,\,v_{3:1}\,,\,v_{2:1}}}\\
&\quad - M^{\binom{u_{123}\,,\,u_1\,,\,u_3}{v_2\,,\,v_{1:2}\,,\,v_{3:2}}} - M^{\binom{u_{123}\,,\,u_3\,,\,u_1}{v_2\,,\,v_{3:2}\,,\,v_{1:2}}}\\
&\quad + M^{\binom{u_{123}\,,\,u_1\,,\,u_2}{v_3\,,\,v_{1:3}\,,\,v_{2:3}}} - M^{\binom{u_{123}\,,\,u_{12}\,,\,u_2}{v_3\,,\,v_{1:3}\,,\,v_{2:3}}} + M^{\binom{u_{123}\,,\,u_{12}\,,\,u_1}{v_3\,,\,v_{2:3}\,,\,v_{1:2}}}.
\end{aligned}$$

The number of summands M^{w^*} in the expression of $S^{w_1,\ldots,w_r}$ is exactly $r!! := 1.3.5\ldots(2r-1)$

Remark 2: Extending the *scramble* to ordinary moulds.
We must often let the *scramble* act on moulds $M^\bullet$ by first 'lifting' these into bimoulds $\underline{M}^\bullet$ according to the rule: $\underline{M}^{\binom{u_1\,,\ldots,\,u_r}{v_1\,,\ldots,\,v_r}} = M^{u_1 v_1+\ldots+u_r v_r}$. Of course, the *scramble* of a mould is a bimould – not a mould. Thus, the bimould $\mathcal{W}^\bullet$ of (11.1) is essentially the *scramble* of the mould $\mathcal{V}^\bullet$ of (11.2).

11.3 The bialternal tesselation bimould

Let $V^\bullet$ be the classical scalar mould produced under alien derivation from the equally classical resurgent mould $\mathcal{V}^\bullet(z)$:

$$\Delta_{\omega_0}\mathcal{V}^{\boldsymbol{\omega}}(z) = \sum_{\boldsymbol{\omega}=\boldsymbol{\omega}'\boldsymbol{\omega}''}^{\|\boldsymbol{\omega}'\|=\omega_0} V^{\boldsymbol{\omega}'}\,\mathcal{V}^{\boldsymbol{\omega}''}(z) \qquad (11.8)$$

$\mathcal{V}^\bullet(z)$ is symmetral; $V^\bullet$ is alternal.

If we now apply the *scramble* transform to the alternal mould $V^\bullet$ (see Remark 2 *supra* about the lift $V^\bullet \mapsto \underline{V}^\bullet$), we get a bialternal bimould $tes^\bullet$:[154]

$$\mathrm{tes}^\bullet = \mathrm{scram}.V^\bullet \qquad \text{with} \qquad \mathrm{tes}^{w} := \sum_{w^*} \epsilon(w, w^*)\, \underline{V}^{w^*} \tag{11.9}$$

which (surprisingly) turns out to be piecewise constant in each u_i and v_i, despite being a sum of hyperlogarithmic summands $\underline{V}^{w^*}$. This begs for an alternative, simpler expression of $tes^\bullet$. The following induction formula provides such an elementary alternative:

$$\mathrm{tes}^{\mathbf{w}} = \sum_{0 \le n \le r(\mathbf{w})} \mathrm{push}^n \sum_{\mathbf{w}'\mathbf{w}''=\mathbf{w}} \mathrm{sig}^{\mathbf{w}',\mathbf{w}''}\, \mathrm{tes}^{\mathbf{w}^*}\, \mathrm{tes}^{\mathbf{w}^{**}}. \tag{11.10}$$

The notations are as follows.

We fix $\theta \in \mathbb{R}/2\pi\mathbb{Z}$ and set $\Re_\theta : z \in \mathbb{C} \mapsto \Re(e^{i\theta} z) \in \mathbb{R}$. Then we define:

$$f_{\mathbf{w}}^{\mathbf{w}'} := <\mathbf{u}', \mathbf{v}'><\mathbf{u}, \mathbf{v}>^{-1}, \qquad g_{\mathbf{w}}^{\mathbf{w}'} := <\mathbf{u}', \Re_\theta \mathbf{v}'><\mathbf{u}, \Re_\theta \mathbf{v}>^{-1} \tag{11.11}$$

$$f_{\mathbf{w}}^{\mathbf{w}''} := <\mathbf{u}'', \mathbf{v}''><\mathbf{u}, \mathbf{v}>^{-1}, \qquad g_{\mathbf{w}}^{\mathbf{w}''} := <\mathbf{u}'', \Re_\theta \mathbf{v}''><\mathbf{u}, \Re_\theta \mathbf{v}>^{-1}. \tag{11.12}$$

From these scalars we construct the crucial sign factor *sig* which takes its values in $\{-1, 0, 1\}$. Here, the abbreviation *si*(.) stands for sign($\Im$(.)).

$$\begin{aligned} \mathrm{sig}^{\mathbf{w}',\mathbf{w}''} = \mathrm{sig}_\theta^{\mathbf{w}',\mathbf{w}''} := \frac{1}{8} & \left(\mathrm{si}(f_{\mathbf{w}}^{\mathbf{w}'} - f_{\mathbf{w}}^{\mathbf{w}''}) - \mathrm{si}(g_{\mathbf{w}}^{\mathbf{w}'} - g_{\mathbf{w}}^{\mathbf{w}''})\right) \\ & \times \left(1 + \mathrm{si}(f_{\mathbf{w}}^{\mathbf{w}'}/g_{\mathbf{w}}^{\mathbf{w}'})\ \mathrm{si}(f_{\mathbf{w}}^{\mathbf{w}'} - g_{\mathbf{w}}^{\mathbf{w}'})\right) \\ & \times \left(1 + \mathrm{si}(f_{\mathbf{w}}^{\mathbf{w}''}/g_{\mathbf{w}}^{\mathbf{w}''})\ \mathrm{si}(f_{\mathbf{w}}^{\mathbf{w}''} - g_{\mathbf{w}}^{\mathbf{w}''})\right). \end{aligned} \tag{11.13}$$

Lastly, the pair $(\mathbf{w}^*, \mathbf{w}^{**})$ is constructed from the pair $(\mathbf{w}', \mathbf{w}'')$ according to:

$$\mathbf{u}^* := \mathbf{u}', \quad \mathbf{v}^* := \mathbf{v}'<\mathbf{u}, \mathbf{v}>^{-1} \Im g_{\mathbf{w}}^{\mathbf{w}'} - \Re_\theta \mathbf{v}'<\mathbf{u}, \Re_\theta \mathbf{v}>^{-1} \Im g_{\mathbf{w}}^{\mathbf{w}'} \tag{11.14}$$

$$\mathbf{u}^{**} := \mathbf{u}'', \quad \mathbf{v}^{**} := \mathbf{v}''<\mathbf{u}, \mathbf{v}>^{-1} \Im g_{\mathbf{w}}^{\mathbf{w}''} - \Re_\theta \mathbf{v}''<\mathbf{u}, \Re_\theta \mathbf{v}>^{-1} \Im g_{\mathbf{w}}^{\mathbf{w}''}. \tag{11.15}$$

Remark 1: The above induction for $tes^\bullet$ is elementary in the sense of being non-transcendental: it depends only on the *sign function*. But on the face of it, it looks non-intrinsical. Indeed, the partial sum:

$$\mathrm{urtes}_\theta^{\mathbf{w}} := \sum_{\mathbf{w}'\mathbf{w}''=\mathbf{w}} \mathrm{sig}^{\mathbf{w}',\mathbf{w}''} \mathrm{tes}^{\mathbf{w}^*} \mathrm{tes}^{\mathbf{w}^{**}} = \sum_{\mathbf{w}'\mathbf{w}''=\mathbf{w}} \mathrm{sig}_\theta^{\mathbf{w}',\mathbf{w}''} \mathrm{tes}^{\mathbf{w}_\theta^*} \mathrm{tes}^{\mathbf{w}_\theta^{**}} \tag{11.16}$$

---

[154] Its proper place is in resurgence theory – in the description of the "geometry" of *co-equational resurgence*.

is *polarised*, *i.e.* θ-dependent. However, its *push*-invariant offshoot:

$$\text{tes}^\bullet := \sum_{0\leq n\leq r(\mathbf{w})} \text{push}^n \ \text{urtes}^\bullet_\theta \tag{11.17}$$

is duly *unpolarised.* We might of course remove the polarisation in $urtes^\bullet_\theta$ itself by replacing it by this isotropic variant:

$$\text{urtes}^\bullet_{\text{iso}} := \frac{1}{2\pi}\int_0^{2\pi} \text{urtes}^\bullet_\theta \ d\theta \tag{11.18}$$

but at the cost of rendering it less elementary, since $urtes^\bullet_{iso}$ would assume its value in $\mathbb{R}$ rather than $\{-1, 0, 1\}$. It would also depend hyperlogarithmically on its indices, and thus take us back to something rather like formula (11.9), which we wanted to get away from. So the alternative for $tes^\bullet$ is: *either* an intrinsical but heavily transcendental expression *or* an elementary but heavily polarised one!

Remark 2: In the induction (11.10) we might exchange everywhere the role of $\boldsymbol{u}$ and $\boldsymbol{v}$ and still get the correct answer $tes^\bullet$, but via a different polarised intermediary $urtes^\bullet_\theta$. The natural setting for studying $tes^\bullet$ is the *biprojective* space $\mathbb{P}^{r,r}$ equal to $\mathbb{C}^{2r}$ quotiented by the relation $\{\boldsymbol{w}^1 \sim \boldsymbol{w}^2\} \Leftrightarrow \{\boldsymbol{u}^1 = \lambda \boldsymbol{u}^2\ ,\ \boldsymbol{v}^1 = \mu\, \boldsymbol{v}^2\ (\lambda, \mu \in \mathbb{C}^*)\}$. But rather than using biprojectivity to get rid of two coordinates (u_i, v_i), it is often useful, on the contrary, to resort to the *augmented* or *long* notation, by *adding* two redundant coordinates (u_0, v_0). The *long* coordinates (u_i^*, v_i^*) relate to the short ones (u_i, v_i) under the rules:

$$u_i = u_i^*, \quad v_i = v_i^* - v_0^* \qquad (1 \leq i \leq r). \tag{11.19}$$

The *long* u_i^* are constrained by $u_0^* + \cdots + u_r^* = 0$ while the *long* v_i^* are, dually, regarded as defined up to a common additive constant. Thus we have $<u^*, v^*>=<u, v>$.The indices i of the *long* coordinates are viewed as elements of $\mathbb{Z}_{r+1} = \mathbb{Z}/(r+1)\mathbb{Z}$ with the natural circular ordering on triplets $circ(i_1 < i_2 < i_3)$ that goes with it. Lastly, we require r^2-1 basic "homographies" $H_{i,j}$ on $\mathbb{P}^{r,r}$, defined by:

$$H_{i,j}(\mathbf{w}) := Q_{i,j}(\mathbf{w})/Q^*_{i,j}(\mathbf{w}) \qquad (i - j \neq 0;\, i, j \in \mathbb{Z}_{r+1}) \tag{11.20}$$

$$Q_{i,j}(\mathbf{w}) := \sum_{\text{circ}(j\leq q<i)} u_q^*\,(v_q^* - v_j^*) \tag{11.21}$$

$$Q^*_{i,j}(\mathbf{w}) := \sum_{\text{circ}(i\leq q<j)} u_q^*\,(v_q^* - v_j^*) \ \neq\ Q_{j,i}(\mathbf{w}). \tag{11.22}$$

Main properties of $tes^\bullet$

$\boldsymbol{P_1}$: the bimould $tes^\bullet$ is bialternal, *i.e.* alternal and of alternal *swappee*;
$\boldsymbol{P_2}$: in fact $swap\, tes^\bullet = tes^\bullet$;
$\boldsymbol{P_3}$: $tes^\bullet$ is *push*-invariant;
$\boldsymbol{P_4}$: $tes^\bullet$ is *pus*-variant, *i.e.* of zero *pus*-average;
$\boldsymbol{P_5}$: $tes^\bullet$ assumes the sole values -1,0,1;
$\boldsymbol{P_6}$: for r fixed but large, the sets $S_\pm \subset \mathbb{P}^{r,r}$ where $tes^{\mathbf{w}}$ is ± 1, have positive but incredibly small Lebesgue measure;
$\boldsymbol{P_7}$: for r fixed, all three sets S_-, S_0, S_+ are path-connected;
$\boldsymbol{P_8}$: for r fixed, the hypersurfaces $\Im(H_{i,j}(\mathbf{w})) = 0$ *limit*[155] but do not *separate*[156] the sets S_-, S_0, S_+;
$\boldsymbol{P_9}$: $tes^{\mathbf{w}} = 0$ whenever $\mathbf{w}$ is semi-real, *i.e.* whenever one of its two components $\mathbf{u}$ or $\mathbf{v}$ is real.[157]

11.4 Polar, trigonometric, bitrigonometric symmetries

The trigonometric symmetries *iil* and *uul* coincide *modulo c* with the polar symmetries *il* and *ul*, but their exact expression is much more complex. So let us first restate the polar symmetries in terms that lend themselves to the extension to the trigonometric case.

Polar symmetries: symmetril/alternil

A bimould $M^\bullet$ is symmetril (respectively alternil) iff for all pairs $\mathbf{w}', \mathbf{w}'' \neq \emptyset$ the identity holds:

$$\sum_{\mathbf{w} \in \mathrm{shi}(\mathbf{w}',\mathbf{w}'')} M^{\mathbf{w}} \prod_{1 \le k \le r(\mathbf{w})} \mathrm{li}^{w_k} \equiv M^{\mathbf{w}'} M^{\mathbf{w}''} \quad (\text{respectively} \equiv 0) \qquad (11.23)$$

with a sum ranging over all sequences $\mathbf{w}$ that are order-compatible with $(\mathbf{w}', \mathbf{w}'')$ and whose indices w_k are of the form:
(i) either w'_i or w''_j, in which case $\mathrm{li}^{w_k} := 1$;
(ii) or $\binom{u'_i+u''_j}{v'_i}$, in which case $\mathrm{li}^{w_k} := -P(v''_j - v'_i)$;
(iii) or $\binom{u'_i+u''_j}{v''_j}$, in which case $\mathrm{li}^{w_k} := -P(v'_i - v''_j)$.

---

[155] That is to say, the boundaries of these sets lie on the hypersurfaces.

[156] That is to say, none of the three sets can be defined in terms of the sole signs $si(H_{i,j}(\mathbf{w})) := \mathrm{sign}(\Im(H_{i,j}(\mathbf{w})))$, at least for $r \geq 3$. For $r = 1$, $tes^\bullet \equiv 1$ and for $r = 2$, $tes^\bullet = \pm 1$ iff $si(H_{0,1}(\mathbf{w})) = si(H_{1,2}(\mathbf{w})) = si(H_{2,0}(\mathbf{w})) = \pm$ and 0 otherwise.

[157] Or purely imaginary, since under biprojectivity this amounts to the same. Of course, $tes^{\mathbf{w}}$ vanishes in many more cases. In fact it vanishes most of the time: see $\boldsymbol{P_6}$ above.

Polar symmetries: symmetrul/alternul

A bimould $M^\bullet$ is symmetrul (respectively alternul) iff for all pairs $\mathbf{w}', \mathbf{w}'' \neq \emptyset$ the identity holds:

$$\sum_{\mathbf{w} \in \mathrm{shu}(\mathbf{w}', \mathbf{w}'')} M^{\mathbf{w}} \prod_{1 \le k \le r(\mathbf{w})} \mathrm{lu}^{w_k} \equiv M^{\mathbf{w}'} M^{\mathbf{w}''} \quad (\text{respectively} \equiv 0) \qquad (11.24)$$

with a sum ranging over all sequences $\mathbf{w}$ that are order-compatible with $(\mathbf{w}', \mathbf{w}'')$ and whose indices w_k are of the form:
(i) either w_i' or w_j'', in which case $\mathrm{lu}^{w_k} := 1$;
(ii) or $\binom{u_i'+u_j''}{v_i'}$, in which case $\mathrm{lu}^{w_k} := -P(u_j'')$;
(iii) or $\binom{u_i'+u_j''}{v_j''}$, in which case $\mathrm{lu}^{w_k} := -P(u_i')$.

Trigonometric symmetries: auxiliary functions
To handle the trigonometric case, we require four series of rational coefficients:
(*) $xii_{p,q}$, $zii_{p,q}$, $xuu_{p,q}$, $zuu_{p,q}$;
which are best defined as Taylor coefficients of the following functions:
(**) $Xii(x, y)$, $Zii(x, y)$, $Xuu(x, y)$, $Zuu(x, y)$.
Here are the definitions:

$$Q(t) := \frac{1}{\tan(t)} \,\|\, R(t) := \frac{1}{\arctan(t)} \qquad (11.25)$$

$$\mathrm{Xii}(x, y) := \frac{x^{-1} + y^{-1}}{Q(x) + Q(y)} \,\|\mathrm{Xuu}(x, y) := \frac{x^{-1} + y^{-1}}{R(x) + R(y)} \qquad (11.26)$$

$$\mathrm{Zii}(x, y) := \frac{x^{-1}Q(x) - y^{-1}Q(y)}{Q(x) + Q(y)} \,\|\mathrm{Zuu}(x, y) := \frac{x^{-1}R(x) - y^{-1}R(y)}{R(x) + R(y)}. \qquad (11.27)$$

Thus:

$$\mathrm{Xii}(x, y) = 1 + \frac{1}{3}xy + \frac{1}{45}y^3 + \frac{4}{45}x^2y^2 + \frac{1}{45}x^3y$$
$$+ \frac{2}{945}xy^5 + \frac{4}{315}x^2y^4 + \frac{23}{945}x^3y^3 + \frac{4}{315}x^4y^2 + \frac{2}{945}x^5y + \ldots$$

$$\mathrm{Xuu}(x, y) = 1 - \frac{1}{3}xy + \frac{4}{45}xy^3 + \frac{1}{45}x^2y^2 + \frac{4}{45}x^3y$$
$$- \frac{44}{945}xy^5 - \frac{4}{315}x^2y^4 - \frac{23}{945}x^3y^3 - \frac{4}{315}x^4y^2 - \frac{44}{945}x^5y + \ldots$$

$$\mathrm{Zii}(x,y)=x^{-1}-y^{-1}-\frac{1}{3}x+\frac{1}{3}y-\frac{1}{45}x^3-\frac{4}{45}x^2y+\frac{4}{45}xy^2+\frac{1}{45}y^3$$
$$-\frac{2}{945}x^5-\frac{4}{315}x^4y-\frac{16}{945}x^3y^2+\frac{16}{945}x^2y^3$$
$$+\frac{4}{315}xy^4+\frac{2}{945}y^5+\dots$$
$$\mathrm{Zuu}(x,y)=x^{-1}-y^{-1}+\frac{1}{3}x-\frac{1}{3}y-\frac{4}{45}x^3-\frac{1}{45}x^2y+\frac{1}{45}xy^2+\frac{4}{45}y^3$$
$$+\frac{44}{945}x^5+\frac{4}{315}x^4y-\frac{1}{189}x^3y^2+\frac{1}{189}x^2y^3$$
$$-\frac{4}{315}xy^4-\frac{44}{945}y^5+\dots$$

Trigonometric symmetries: symmetriil/alterniil

A bimould $M^\bullet$ is symmetriil (respectively alterniil) iff for all pairs $\mathbf{w}',\mathbf{w}''\neq\emptyset$ the identity holds:

$$\sum_{\mathbf{w}\in\mathrm{shii}(\mathbf{w}',\mathbf{w}'')} M^{\mathbf{w}} \prod_{1\le k\le r(\mathbf{w})} \mathrm{lii}^{w_k} \equiv M^{\mathbf{w}'} M^{\mathbf{w}''} \quad (\text{respectively} \equiv 0) \quad (11.28)$$

with a sum ranging over all sequences $\mathbf{w}$ that are order-compatible with $(\mathbf{w}',\mathbf{w}'')$ and whose indices w_k are of the form:
(i) either w'_i or w''_j, in which case $\mathrm{lii}^{w_k} := 1$;
(ii) or $\left({}^{u'_i+\dots u'_{i+p}+u''_j+\dots u''_{j+q}}_{v'_i}\right)$ with $p,q\geq 0$, in which case

$$\mathrm{lii}^{w_k} := -c^{p+q}\,\mathrm{xii}_{p,q}\,Q_c(v''_j-v'_i) - c^{p+q+1}\,\mathrm{zii}_{p,q}; \quad (11.29)$$

(iii) or $\left({}^{u'_i+\dots u'_{i+p}+u''_j+\dots u''_{j+q}}_{v''_j}\right)$ with $p,q\geq 0$, in which case

$$\mathrm{lii}^{w_k} := +c^{p+q}\,\mathrm{xii}_{p,q}\,Q_c(v'_i-v''_j) + c^{p+q+1}\,\mathrm{zii}_{p,q}. \quad (11.30)$$

Trigonometric symmetries: symmetruul/alternuul

A bimould $M^\bullet$ is symmetruul (respectively alternuul) iff for all pairs $\mathbf{w}',\mathbf{w}''\neq\emptyset$ the identity holds:

$$\sum_{\mathbf{w}\in\mathrm{shuu}(\mathbf{w}',\mathbf{w}'')} M^{\mathbf{w}} \prod_{1\le k\le r(\mathbf{w})} \mathrm{luu}^{w_k} \equiv M^{\mathbf{w}'} M^{\mathbf{w}''} \quad (\text{respectively} \equiv 0) \quad (11.31)$$

with a sum ranging over all sequences $\mathbf{w}$ that are order-compatible with $(\mathbf{w}', \mathbf{w}'')$ and whose indices w_k are of the form:
(i) either w_i' or w_j'', in which case $\text{luu}^{w_k} := 1$;
(ii) or $\binom{u_i'+u_j''}{v_i'}$, in which case $\text{luu}^{w_k} := -Q_c(u_j'')$;
(iii) or $\binom{u_i'+u_j''}{v_j''}$, in which case $\text{luu}^{w_k} := -Q_c(u_i')$;
(iv) or $\binom{u_i'+\ldots u_{i+p}'+u_j''+\ldots u_{j+q}''}{v_i'}$ with $p, q \geq 0$ and $p+q \geq 1$, in which case

$$\text{luu}^{w_k} := -\sum_{\substack{0\leq p_1\leq p\\ 0\leq q_1\leq q}} c^{p+q+1}\,\text{zuu}_{p_1,q_1}\,\text{Sym}_{p-p_1}\Bigg(\bigcup_{i<s<i+p} Q_c(u_s')\Bigg)\text{Sym}_{q-q_1}\Bigg(\bigcup_{j<s<j+q} Q_c(u_s'')\Bigg);$$

(v) or $\binom{u_i'+\ldots u_{i+p}'+u_j''+\ldots u_{j+q}''}{v_j''}$ with $p, q \geq 0$ and $p+q \geq 1$, in which case

$$\text{luu}^{w_k} := +\sum_{\substack{0\leq p_1\leq p\\ 0\leq q_1\leq q}} c^{p+q+1}\,\text{zuu}_{p_1,q_1}\,\text{Sym}_{p-p_1}\Bigg(\bigcup_{i<s<i+p} Q_c(u_s')\Bigg)\text{Sym}_{q-q_1}\Bigg(\bigcup_{j<s<j+q} Q_c(u_s'')\Bigg)$$

with $Sym_s(x_1, \ldots, x_r)$ standing for the s-th symmetric function of the x_i:

$$\text{Sym}_s(x_1, \ldots, x_r) := \sum_{1\leq i_1<\cdots<i_s\leq r} x_{i_1} \ldots x_{i_s}. \tag{11.32}$$

However, to get the formula for luu^{w_k} right, we must observe the following convention:

$$\begin{aligned} &\text{Sym}_0(x_1, \ldots, x_r) := 1 && (\textit{even}\ \text{if}\ r = 0)\\ &\text{Sym}_s(x_1, \ldots, x_r) := 0\ \ \text{if}\ 1 \leq r < s && (\text{but}\ \text{Sym}_0(\emptyset) := 1). \end{aligned}$$

We may also note the complete absence, from the expression of luu^{w_k}, of the four extreme terms $Q_c(u_i')$, $Q_c(u_{i+p}')$, $Q_c(u_j'')$, $Q_c(u_{j+q}'')$.

Dimorphic transport

As in the polar case, the adjoint action of the bisymmetrals $tal_c^\bullet$ and $til_c^\bullet$ exchanges double symmetries, but without respecting entireness.

$$\begin{array}{ccc} \text{GARI}^{\underline{\text{as}}/\underline{\text{as}}} & \xrightarrow{\text{adgari}(\text{tal}_c^\bullet)} & \text{GARI}^{\underline{\text{as}}/\underline{\text{iis}}}\\ \text{logari} \downarrow\uparrow \text{expari} & & \text{logari} \downarrow\uparrow \text{expari}\\ \text{ARI}^{\underline{\text{al}}/\underline{\text{al}}} & \xrightarrow{\text{adari}(\text{tal}_c^\bullet)} & \text{ARI}^{\underline{\text{al}}/\underline{\text{iil}}} \end{array}$$

$$\begin{array}{ccc} \text{GARI}^{\underline{\text{as}}/\underline{\text{as}}} & \xrightarrow{\text{adgari}(\text{til}_c^\bullet)} & \text{GARI}^{\underline{\text{as}}/\underline{\text{uus}}}\\ \text{logari} \downarrow\uparrow \text{expari} & & \text{logari} \downarrow\uparrow \text{expari}\\ \text{ARI}^{\underline{\text{al}}/\underline{\text{al}}} & \xrightarrow{\text{adari}(\text{til}_c^\bullet)} & \text{ARI}^{\underline{\text{al}}/\underline{\text{uul}}}. \end{array}$$

Bitrigonometric symmetries

As usual, the trigonometric case fully determines the bitrigonmetric extension.

Symmetries associated with other approximate units $\mathfrak{E}^\bullet$

$\mathfrak{E}^\bullet$ of course replaces Q_c in the expressions of lii^{w_k} and luu^{w_k} but the structure coefficients $xii_{p,q}, zii_{p,q}, xuu_{p,q}, zuu_{p,q}$ do not change, and must still be calculated from Q and the related R (see (11.25)) even in the case of the *flat* approximate units $Sa^\bullet$ or $Si^\bullet$ of (3.25).

Remark 1. While bimoulds polynomial or entire in the u_i and v_i variables may be alternil or symmetril or alterniil or symmetriil, they *can never* be alternul nor symmetrul nor alternuul nor symmetruul.

Remark 2. Of course, just as with the *straight* symmetries (see Section 2.4), when expressing the new, *twisted* symmetries, one should take care to allow only sequences $\mathbf{w}$ that are order-compatible with $\mathbf{w}'$ and $\mathbf{w}''$, *i.e.* that never carry pairs u_i', v_i' or u_j'', v_j'' (whether in isolation or within sums or differences) in an order that clashes with their relative position within the parent sequences $\mathbf{w}'$ or $\mathbf{w}''$.

11.5 The separative algebras *Inter*(Qi_c) and *Exter*(Qi_c)

Introduction

The subalgebra *Exter*(Qi_c) of *Flex*(Qi_c) is the trigonometric equivalent of the polar subalgebra $ARI_{<pi>}$ of *Flex*(Pi) which itself is but the specialisation, for $\mathfrak{E} = Pi$, of the subalgebra $ARI_{<\mathfrak{re}>}$ of *Flex*$(\mathfrak{E})$ which was investigated in Section 3.6. Both *Exter*(Qi_c) and $ARI_{<pi>}$ consist of $\boldsymbol{u}$-constant, $\boldsymbol{v}$-dependent, alternal bimoulds, and both are indispensable to an in-depth understanding of the fundamental bialternals $pil^\bullet$ and $til_c^\bullet$ since they house their *ari*-logarithms $logari.pil^\bullet$ and $logari.til_c^\bullet$.

However, due to $Pi^\bullet$ being an *exact* flexion unit, the algebra $ARI_{<pi>}$ has a very simple structure: it is spanned by the bimoulds $pi_r^\bullet$ $(1 \leq r)$, which self-reproduce under the *ari*-bracket: $ari(pi_{r_1}^\bullet, pi_{r_2}^\bullet) \equiv (r_1 - r_2) pi_{r_1+r_2}^\bullet$.

Its trigonometric counterpart *Exter*(Qi_c), on the other hand, is vaster and much more complex: it does indeed contain a series of bimoulds $qi_r^\bullet$ defined in the same way as the $pi_r^\bullet$ or the $\mathfrak{re}_r^\bullet$ of (4.5), but these $qi_r^\bullet$ no longer self-reproduce under the *ari*-bracket: they do so only modulo c^2.

Nonetheless, the structure of *Exter*(Qi_c) is highly interesting, and can be exhaustively described by decomposing *Exter*(Qi_c) into a direct sum of subspaces $\mathfrak{g}^n.Inter(Qi_c)$ $(0 \leq n)$ which are all derived from a subalgebra $Inter(Qi_c) \subset Exter(Qi_c)$ consisting of all alternals in $Flex(Qi_c)$ that depend only on the differences $v_i - v_j$.[158] The algebra $Inter(Qi_c)$ and its elements shall be called *internal*, whereas elements of $Exter(Qi_c) \dot{-} Inter(Qi_c)$ shall be called *external*. The internal algebra is quite elementary: on it, most flexion operations reduce to non-inflected operations. Thus, the *ari*-bracket of two internals coincides (up to a sign change) with their *lu*-bracket.

The external and internal algebras are also called *separative*, since under the action of the operator *separ*, which is to *ARI* what the operator *gepar* of Section 4.1. was to *GARI*:

$$\mathrm{separ}.M^\bullet := \mathrm{anti.swap}.M^\bullet + \mathrm{swap}.M^\bullet \tag{11.33}$$

$$\mathrm{gepar}.M^\bullet := \mathrm{mu}(\mathrm{anti.swap}.M^\bullet, \mathrm{swap}.M^\bullet) \tag{11.34}$$

their bimoulds experience a *separation* of their variables[159] and assume the elementary form:

$$\{M^\bullet \in \mathrm{Exter(Qi_c)}\} \Rightarrow \{(\mathrm{separ}.M)^{w_1,\ldots,w_r} \in \mathbb{C}[c^2, Q_c(u_1), \ldots, Q_c(u_r)]\}$$

Remark: strictly speaking, elements of $Flex(Qi_c)$ can involve only *even* powers of c, but it is convenient to enlarge $Exter(Qi_c)$ and $Inter(Qi_c)$ with *odd* powers of c, so as to make room for the bimoulds $qin_r^\bullet$ and the operators $\mathfrak{h}_n$ (defined *infra*). Ultimately, however, we shall end up with structure formulas where these $qin_r^\bullet$ and $\mathfrak{h}_n$ appear only in pairs, thus ensuring that there is no violation of c-parity.

The external $qi_r^\bullet$ and the internal $qin_r^\bullet$

They are the first ingredients of the 'separative' structure. These alternal bimoulds of $BIMU_r$ are defined by the induction:

$$\mathrm{qi}_1^{w_1} := \mathrm{Qi}_c^{w_1} = Q_c(v_1) = \frac{c}{\tan(c\,v1)} \quad \| \quad \mathrm{qi}_r^\bullet := \mathrm{arit}(\mathrm{qi}_{r-1}^\bullet).\mathrm{qi}_1^\bullet \quad \forall (r \geq 2)$$

$$\mathrm{qin}_1^{w_1} := c \quad \| \quad \mathrm{qin}_r^\bullet := \mathrm{ari}(\mathrm{qin}_1^\bullet, \mathrm{qi}_{r-1}^\bullet) \quad \forall (r \geq 2)$$

---

[158] *In the short notation*, of course. In the long notation (with the additional variable v_0), this is automatic and implies no constraint at all.

[159] Due to the *swap* which is implicit in the definition of *separ* and *gepar*, the new variables are no longer v_i's but u_i's.

The auxiliary mould *har*$^\bullet$
Our second ingredient is a scalar mould whose only non-vanishing components have *odd* length. Here again, the definition is by induction:

$$\mathrm{har}^{n_1,\dots,n_r} := 0 \qquad \forall r \textit{ even} \geq 0 \tag{11.35}$$

$$\mathrm{har}^{n_1} := \frac{1}{n_1} \tag{11.36}$$

$$\mathrm{har}^{n_1,\dots,n_r} := \frac{1}{n_1+\ldots+n_r} \sum_{1<i<r} \mathrm{har}^{n_1,\dots,n_{i-1}} \mathrm{har}^{n_{i+1},\dots,n_r} \quad \forall r \textit{ odd} \geq 3. \tag{11.37}$$

Thus:

$$\mathrm{har}^{n_1,n_2,n_3} = \frac{1}{n_1 n_3 n_{123}} \tag{11.38}$$

$$\mathrm{har}^{n_1,n_2,n_3,n_4,n_5} = \frac{1}{n_1 n_3 n_5 n_{12345}} \left(\frac{1}{n_{123}} + \frac{1}{n_{345}}\right). \tag{11.39}$$

The operators $\mathfrak{g}^n$, $\mathfrak{h}_n$
These linear operators of $BIMU_r$ into $BIMU_{r+n}$ are our third ingredient. The first are mere powers of a single operator $\mathfrak{g}$ defined by:

$$\mathfrak{g} : \mathcal{A}^\bullet \to \mathcal{B}^\bullet := \mathrm{arit}(\mathcal{A}^\bullet)\, \mathrm{qi}^\bullet_1 = \mathrm{arit}(\mathcal{A}^\bullet)\, \mathrm{Qi}^\bullet_c \tag{11.40}$$

which, since $Qi^\bullet_c \in BIMU_1$, may be rewritten as:

$$\mathcal{B}^{\binom{u_1,\dots,u_r}{v_1,\dots,v_r}} = \mathcal{A}^{\binom{u_1,\dots,u_{r-1}}{v_{1:r},\dots,v_{r-1:r}}} Q_c(v_r) - \mathcal{A}^{\binom{u_2,\dots,u_r}{v_{2:1},\dots,v_{r:1}}} Q_c(v_1). \tag{11.41}$$

The operators $\mathfrak{h}_n$, on the other hand, must be defined singly:

$$\mathfrak{h}_n\, A^\bullet := \sum_{1\leq s} \sum_{\substack{1\leq n_i\\ n_1+\ldots n_s=n}} \mathrm{har}^{n_1,\dots,n_r}\, [\mathrm{qin}^\bullet_{n_1}, [\mathrm{qin}^\bullet_{n_2} \ldots [\mathrm{qin}^\bullet_{n_s}, A^\bullet]..]]_{\mathrm{lu}}. \tag{11.42}$$

Due to the imparity of *har*$^\bullet$, the $\mathfrak{h}_n$ too are strictly odd in c.

The operators of $\mathfrak{G}$ and $\mathfrak{H}$
If we set:

$$\mathfrak{G} := \mathrm{id} + \sum_{1\leq n} \mathfrak{g}^n; \qquad \mathfrak{H} := +\sum_{1\leq n} \mathfrak{h}_n \tag{11.43}$$

the operators of $\mathfrak{G}$ and $\mathfrak{H}$ so defined verify the identities:

$$\mathfrak{G}\mathrm{mu}(A^\bullet, B^\bullet) \equiv \mathrm{mu}(\mathfrak{G}A^\bullet, \mathfrak{G}B^\bullet) + \mathfrak{G}\mathrm{mu}(\mathfrak{H}A^\bullet, \mathfrak{H}B^\bullet) \tag{11.44}$$

$$\begin{aligned}\mathfrak{H}\mathrm{mu}(A^\bullet, B^\bullet) \equiv\ & \mathrm{mu}(\mathfrak{H}A^\bullet, B^\bullet)\\ & + \mathrm{mu}(A^\bullet, \mathfrak{H}B^\bullet) + \mathfrak{H}\mathrm{mu}(\mathfrak{H}A^\bullet, \mathfrak{H}B^\bullet)\end{aligned} \tag{11.45}$$

and of course analogous identities with *lu* in place of *mu*. The only restriction is that in (11.44) the inputs $A^\bullet$, $B^\bullet$ must be *internal*.

If we now 'iterate' these identites so as to rid their right-hand sides of all terms $\mathfrak{G}.mu(\ldots,\ldots)$ and $\mathfrak{H}.mu(\ldots,\ldots)$, we find that $\mathfrak{G}$ and $\mathfrak{H}$ verify the co-products:

$$\mathfrak{G} \to \mathfrak{G} \otimes \mathfrak{G} + \sum_{1\le s} \mathfrak{G}\mathfrak{H}^s \otimes \mathfrak{G}\mathfrak{H}^s \tag{11.46}$$

$$\mathfrak{H} \to \mathfrak{H} \otimes 1 + 1 \otimes \mathfrak{H} + \sum_{1\le s} (\mathfrak{H}^{s+1} \otimes \mathfrak{H}^s + \mathfrak{H}^s \otimes \mathfrak{H}^{s+1}). \tag{11.47}$$

Again, the coproducts (11.45), (11.47) for $\mathfrak{H}$ hold on the full algebra of bimoulds, whereas the coproducts (11.44), (11.46) for $\mathfrak{G}$ hold only on the algebra of *internals*.

The rectified operators $\mathfrak{G}_*$ and $\mathfrak{H}_*$.
As the above coproducts show, $\mathfrak{H}$ is an 'approximate' derivation and $\mathfrak{G}$ an 'approximate' automorphism. However, if we set:

$$\mathfrak{H}_* := \arctan(\mathfrak{H}) = \mathfrak{H} - \frac{1}{3}\mathfrak{H}^3 + \frac{1}{5}\mathfrak{H}^5 \ldots \tag{11.48}$$

$$\mathfrak{G}_* := \mathfrak{G}(\mathrm{id} + \mathfrak{H})^{-\frac{1}{2}} = \mathfrak{G} - \frac{1}{2}\mathfrak{G}\mathfrak{H}^2 + \frac{3}{8}\mathfrak{G}\mathfrak{H}^4 \ldots \tag{11.49}$$

$$= \mathfrak{G}\cos(\mathfrak{H}_*) = \mathfrak{G} - \frac{1}{2}\mathfrak{G}\mathfrak{H}_*^2 + \frac{1}{24}\mathfrak{G}\mathfrak{H}_*^4 \ldots \tag{11.50}$$

we get an operator $\mathfrak{H}_*$ that is an exact derivation and an operator $\mathfrak{G}_*$ that is an exact automorphism.

The chain $Inter(Qi_c) \subset Exter(Qi_c) \subset Flex(Qi_c)$**.**
The space $Inter(Qi_c)$ is separative, and so is the space $Exter(Qi_c)$ defined as the (direct) sum of all the $\mathfrak{g}$-translates of $Inter(Qi_c)$.

$$\mathrm{Exter}(\mathrm{Qi}_c^\bullet) := \bigoplus_{0\le n} \mathfrak{g}^n.\mathrm{Inter}(\mathrm{Qi}_c^\bullet). \tag{11.51}$$

In fact, both spaces are stable under the *ari*-bracket, and we shall now give a complete description of their structure with the help of our two series of operators $\mathfrak{g}^n$ and $\mathfrak{h}_n$.

Full structure of the *ari*-algebra $Inter(Qi_c)$**.**
The space $Inter(Qi_c^\bullet)$ is obviously stable under the *lu*-bracket, and also under the *ari*-bracket, due to the elementary identities:

$$\mathrm{ari}(A^\bullet, B^\bullet) = -\mathrm{lu}(A^\bullet, B^\bullet) \qquad \forall\, A^\bullet, B^\bullet \in \mathrm{Inter}(\mathrm{Qi_c}) \tag{11.52}$$

$$\mathrm{arit}(A^\bullet).B^\bullet = +\mathrm{lu}(A^\bullet, B^\bullet) \qquad \forall\, A^\bullet, B^\bullet \in \mathrm{Inter}(\mathrm{Qi_c}). \tag{11.53}$$

Full structure of the *ari*-algebra *Exter*(Qi_c).
The space *Exter*(Qi_c), though not closed under the *lu*-bracket, is stable under the *ari*-bracket and the *arit*-operation. Its full structure is given by the three following identities, where $A^\bullet$, $B^\bullet$ stand for arbitrary elements of *Inter*(Qi_c):

$$\begin{aligned}\mathrm{ari}(\mathfrak{g}^p A^\bullet, \mathfrak{g}^q B^\bullet) \equiv & -\mathfrak{g}^q\ \mathrm{arit}(\mathfrak{g}^p A^\bullet) B^\bullet + \mathfrak{g}^p\ \mathrm{arit}(\mathfrak{g}^q B^\bullet) A^\bullet \\ & + \mathfrak{g}^{p+q}\ \mathrm{lu}(A^\bullet, B^\bullet) \\ & - \sum_{\substack{1\le p_1\le p\\ 1\le q_1\le q}} \mathfrak{g}^{p+q-p_1-q_1}\ \mathrm{lu}(\mathfrak{h}_{p_1} A^\bullet, \mathfrak{h}_{q_1} B^\bullet)\end{aligned} \tag{11.54}$$

$$\begin{aligned}\mathrm{arit}(\mathfrak{g}^p A^\bullet)\ \mathfrak{g}^q B^\bullet \equiv & +\mathfrak{g}^q\ \mathrm{arit}(\mathfrak{g}^p A^\bullet) B^\bullet \\ & - \sum_{0\le q_1\le q-1} \mathrm{lu}(\mathfrak{g}^{p+q-q_1} A^\bullet, \mathfrak{g}^{q_1} B^\bullet) r \\ & - \sum_{\substack{1\le q_1\le q-1\\ p+1\le p_1\le p+q-q_1}} \mathfrak{g}^{p+q-p_1-q_1}\ \mathrm{lu}(\mathfrak{h}_{p_1} A^\bullet, \mathfrak{h}_{q_1} B^\bullet)\end{aligned} \tag{11.55}$$

$$\mathrm{arit}(\mathfrak{g}^p A^\bullet)\ \mathrm{tin}_q^\bullet \equiv \mathfrak{h}_{p,q} A^\bullet. \tag{11.56}$$

Since the above identities are linear in each *internal* argument $A^\bullet$ or $B^\bullet$ and since any *external* bimould $M^\bullet$ uniquely decomposes into a sum $\sum \mathfrak{g}^n.M^\bullet_{(n)}$ of $\mathfrak{g}$-tanslates of internal $M^\bullet_{(n)}$, one readily sees that the above identities do indeed encapsulate the whole structure of *Exter*(Qi_c), provided one adds to *Inter*(Qi_c) a symbolic bimould $\square^\bullet \in BIMU_0$ subject to the following rules:[160]

$$\mathrm{qi}_n^\bullet := +\mathfrak{g}^n\, \square^\bullet \tag{11.57}$$
$$\mathrm{qin}_n^\bullet := -\mathfrak{h}_n\, \square^\bullet \tag{11.58}$$

$$\mathrm{lu}(A^\bullet, \square^\bullet) := -r(\bullet)\ A^\bullet \tag{11.59}$$
$$\mathrm{ari}(A^\bullet, \square^\bullet) := +r(\bullet)\ A^\bullet \tag{11.60}$$
$$\mathrm{arit}(A^\bullet)\, \square^\bullet := -r(\bullet)\ A^\bullet \tag{11.61}$$
$$\mathrm{arit}(\square^\bullet)\ A^\bullet := +r(\bullet)\ A^\bullet \tag{11.62}$$

and of course

$$\mathrm{lu}(\square^\bullet, \square^\bullet) = \mathrm{ari}(\square^\bullet, \square^\bullet) = \mathrm{arit}(\square^\bullet)\, \square^\bullet = 0^\bullet. \tag{11.63}$$

[160] One should beware of applying to $\square^\bullet$ any other rules than these, and never forget than $\square^\bullet$ is just a convenient symbol rather than a true bimould. Indeed, the only *bona fide* bimould of $BIMU_0$ is (up to a scalar factor) the multiplication unit $1^\bullet$ with $1^\emptyset := 1$.

11.6 Multizeta cleansing: elimination of unit weights

Main statement

The present section is devoted to proving the following:
***P_0* : (*Unit cleansing*)**
Every uncoloured multizeta $\zeta(s_1, \ldots, s_r)$ can be expressed as a finite sum, with rational coefficients, of unit-free multizetas.[161] *The result extends to all coloured multizetas, but it is less relevant there.*[162]

We shall provide an effective algorithm for achieving the *unit-cleansing*. Along the way, we shall also come across some really fine combinatorics about bialternals, and construct a new *infinitary* subalgebra of *ARI* larger than *ALAL*.

Some heuristics

As is natural with heuristics, we proceed backwards:

Step 4: *restriction of the problem to bialternals.*
Since scalar irreducibles accompany homogeneous bialternals, it will be both necessary and sufficient to express the latter without recourse to unit weights.

Step 3: *the need for "reconstitution identities".*
Since in the v_i-encoding, unit weights correspond to monomials not divisible by $v_1 \ldots v_r$, the challenge it to reconstitute any homogeneous bialternal from its "essential part", *i.e.* the part that is divisible by $v_1 \ldots v_r$.

Step 2: *the need for "redistribution identities".*
To do this, it is more or less clear beforehand that we shall have to find a means of expressing any homogeneous bialternal $M^{\boldsymbol{w}}$ with one or several vanishing v_i's as a superpositions of $M^{\boldsymbol{w}^*}$, with new v_i^*'s formed from the sole non-vanishing v_i's.

Step 1: *the need for "pairing identities".*
To be able to extend the procedure to *coloured* multizetas (and also to respect the spirit of dimorphy), we must find a way of restating the redistribution identities for arbitrary bialternals that effectively depend on the u_i's as well as on the v_i's.

---

[161] *I.e.* of multizetas $\zeta(s'_1, \ldots, s'_{r*})$ with partial weights $s'_i \geq 2$.

[162] For two reasons: first, because the removal of the unit-weights necessitates a remixing of the colours; and second, because one may on the contrary play on the colours to express everything in terms of multizetas with *nothing but unit-weights!*

Step 1: The pairing identities
An *endoflexion* (of length r) is any self-mapping of $BIMU_r$ of the form

$$\text{flex}.M^{w_1,\ldots,w_r} = M^{w_1^*,\ldots,w_r^*} \qquad \left(w_i = \binom{u_i}{v_i},\ w_i^* = \binom{u_i^*}{v_i^*}\right) \tag{11.64}$$

with

$$u_i^* := \overbrace{u_{m_i} + \ldots + u_{n_i}}^{\text{circular}} = \sum^{(m_i \le k \le n_i)_{\mathbb{Z}_{r+1}}} u_k$$
$$v_i^* := v_{p_i} - v_{q_i} \qquad \text{and} \qquad p_i \in \mathcal{P}^+, q_i \in \mathcal{P}^-$$
$$\sum_{1 \le i \le r} u_i{}^* v_i^* \equiv \sum_{1 \le i \le r} u_i v_i .$$

Here, all indices m_i, n_i, p_i, q_i are in the set $\{0, 1, \ldots, r\} \sim \mathbb{Z}_{r+1}$ and $\mathcal{P} = (\mathcal{P}^+, \mathcal{P}^-)$ is any given (strict) partition of $\{0, 1, \ldots, r\}$. We say that *flex* is $\mathcal{P}$-compatible. Whereas *flex* determines $\mathcal{P}$ if we impose (as we shall do) that 0 be in $\mathcal{P}^-$, there are usually many endoflexions *flex* compatible with a given partition $\mathcal{P}$.

***P*$_1$:** *(Existence and uniqueness of the pairing identities.)*
For any strict partition $\mathcal{P}$ of $\{0, 1, \ldots, r\}$ into $\mathcal{P}^+$ ("white indices") and $\mathcal{P}^-$ ("black indices") there exists a self-mapping $\text{flex}_{\mathcal{P}}$ *of* BIMU_r *of the form:*

$$\text{flex}_{\mathcal{P}} = \sum_{\text{flex}_n\ \mathcal{P}\text{-compatible}} \epsilon_n\, \text{flex}_n \qquad (\epsilon_n \in \{0, 1, -1\}) \tag{11.65}$$

whose restriction to the bialternals is the identity:

$$\text{flex}_{\mathcal{P}}.M^\bullet \equiv M^\bullet \qquad \forall M^\bullet \in \text{ARI}_r^{\underline{\text{al}}/\underline{\text{al}}}. \tag{11.66}$$

Furthermore, $\text{flex}_{\mathcal{P}}$ *is unique modulo the alternality (not bialternality!) relations on* $\text{ARI}_r^{\underline{\text{al}}/\underline{\text{al}}}$.

Let us now return to the graph pairs $\boldsymbol{g} = (\boldsymbol{ga}, \boldsymbol{gi})$ defined in Section 3.1 (see also the examples and pictures *infra*). We say that such a pair $\boldsymbol{g}$ is $\mathcal{P}$-compatible if all edges of $\boldsymbol{gi}$ connect a "white" vertex S_{p_i} with a "black" vertex S_{q_i}. Now, both $\boldsymbol{gi}$ and $\boldsymbol{ga}$ have r edges each, and every edge of $\boldsymbol{gi}$ intersects exactly *one* edge of $\boldsymbol{ga}$ at exactly *one* point x, and there clearly exists a (topologically) unique graph $\boldsymbol{gai}$ with those r intersection points x as vertices, and with edges that intersect neither the unit circle nor the edges of $\boldsymbol{ga}$ nor those of $\boldsymbol{gi}$. To each vertex x_* of $\boldsymbol{gai}$ there corresponds one unique coherent orientation $\mathcal{O}_{g,x_*}$ of the edges of $\boldsymbol{gai}$ or, what amounts to the same, one coherent arborescent order, also noted $\mathcal{O}_{g,x_*}$, on the vertices x of $\boldsymbol{gai}$, with x_* as the lowest vertex.

Next, for any $\boldsymbol{g}$ that is $\mathcal{P}$-compatible and for any vertex x_* of the corresponding $\boldsymbol{gai}$, let $\boldsymbol{\gamma}$ be a total order on the vertices of $\boldsymbol{gai}$ that is compatible with the arborescent order $\mathcal{O}_{\boldsymbol{g},x_*}$. We write $\boldsymbol{\gamma} \in \mathcal{O}_{\boldsymbol{g},x_*}$ to denote this compatibility and associate with $\boldsymbol{\gamma}$ the following endoflexion:

$$\mathrm{flex}_{\boldsymbol{\gamma}}.M^{w_1,\dots,w_r} = M^{w_1^*(\boldsymbol{\gamma}),\dots,w_r^*(\boldsymbol{\gamma})} \quad \left(w_i = \begin{pmatrix} u_i \\ v_i \end{pmatrix}, w_i^*(\boldsymbol{\gamma}) = \begin{pmatrix} u_i^*(\boldsymbol{\gamma}) \\ v_i^*(\boldsymbol{\gamma}) \end{pmatrix}\right) \tag{11.67}$$

with

$$\begin{aligned} u_i^*(\boldsymbol{\gamma}) &:= \overbrace{u_{m_i} + \ldots + u_{n_i}}^{\text{circular}} = \sum^{(m_i \le k \le n_i)_{\mathbb{Z}_{r+1}}} u_k \\ v_i^*(\boldsymbol{\gamma}) &:= v_{p_i} - v_{q_i} \qquad \text{and} \qquad p_i \in \mathcal{P}^+, q_i \in \mathcal{P}^- \\ \sum_{1 \le i \le r} u_i{}^*(\boldsymbol{\gamma})\, v_i^*(\boldsymbol{\gamma}) &\equiv \sum_{1 \le i \le r} u_i\, v_i \end{aligned}$$

and with the following notations:
– $x_i(\boldsymbol{\gamma})$ is the i-th vertex of $\boldsymbol{gai}$ in the total order $\boldsymbol{\gamma}$;
– $ga_i(\boldsymbol{\gamma})$ is the unique edge of $\boldsymbol{ga}$ passing through $x_i(\boldsymbol{\gamma})$;
– $gi_i(\boldsymbol{\gamma})$ is the unique edge of $\boldsymbol{gi}$ passing through $x_i(\boldsymbol{\gamma})$;
– $u_i^*(\boldsymbol{\gamma})$ is the sum of all u_k with Si_k on the 'correct' side of $ga_i(\boldsymbol{\gamma})$, *i.e.* on the side that contains the "white" vertex Si_{p_i} of $gi_i(\boldsymbol{\gamma})$;
– $v_i^*(\boldsymbol{\gamma})$ is the difference $v_{p_i} - v_{q_i}$ with Si_{p_i} and Si_{q_i} being the "white" and "black" vertices joined by $gi_i(\boldsymbol{\gamma})$.

Next, we set:

$$\mathrm{flex}_{\boldsymbol{g},x_*} := \sum_{\boldsymbol{\gamma} \in \mathcal{O}_{\boldsymbol{g},x_*}} \epsilon_{\boldsymbol{\gamma}}\, \mathrm{flex}_{\boldsymbol{\gamma}} \qquad \text{with} \tag{11.68}$$

$$\epsilon_{\boldsymbol{\gamma}} := \prod_{gai_k \in \mathrm{Edge}(\boldsymbol{gai})} \epsilon(\mathrm{gai}_k) \in \{1, -1\} \tag{11.69}$$

with a product in (11.69) extending to all $r-1$ edges gai_k of $\boldsymbol{gai}$, and with factor signs $\epsilon(gai_k)$ defined as follows. Each edge gai_k of $\boldsymbol{gai}$ touches two edges $gi_{k'}$ and $gi_{k''}$ of $\boldsymbol{gi}$, which in turn meet at a vertex Si_{k^*} of $\boldsymbol{gi}$. What counts is the colour of that vertex Si_{k^*}, and the position of the triangle $\{gai_k, gi_{k'}, gi_{k''}\}$ respective to the oriented vertex $\vec{gai}_k$. Concretely, we set:
(i) $\epsilon(gai_k) := +1$ if $\vec{gai}_k$ sees a white Si_{k^*} to its right or a black Si_{k^*} to its left;
(ii) $\epsilon(gai_k) := -1$ if $\vec{gai}_k$ sees a white Si_{k^*} to its left or a black Si_{k^*} to its right.

It is readily seen that the operator flex_{g,x_*}, *when applied to alternal bimoulds*, is independent of the choice of the base vertex: indeed, replacing x_* by a neighbouring vertex x_{**} simultaneously changes the signs of $\epsilon(\boldsymbol{\gamma})$ and $flex_{\boldsymbol{\gamma}}.M^\bullet$, for any alternal $M^\bullet$. We shall therefore drop x_* and write simply flex_g whenever the operator flex_{g,x_*} is made to act on alternals (or *a fortiori* on bialternals).

Remark: each one of the graphs $\boldsymbol{ga}$ or $\boldsymbol{gi}$ completely determines the other as well as $\boldsymbol{gai}$. It also determines the only partition $\mathcal{P}$ of $\{0, 1, \ldots, r\}$ with which it is compatible, since 0 is automatically *black*, and so Si_0 is declared *black* too, and the colouring then extends to all Si_k by following $\boldsymbol{gi}$. On the other hand, the number of graph pairs $\boldsymbol{g} = \{\boldsymbol{ga}, \boldsymbol{gi}\}$ compatible with a given partition $\mathcal{P}$ is *on average* equal to $\frac{(3r)!}{(2r+1)!r!2^r}$ and therefore tends to be very large.

***P*$_2$:** (*Explicit formula for the pairing identities.*)
For each partition $\mathcal{P}$ of $\{0, 1, \ldots, r\}$, the pairing operator $\text{flex}_{\mathcal{P}}$ *of* (11.65) *is explicitely given by*:

$$\text{flex}_{\mathcal{P}} := \sum_{g\ \mathcal{P}\text{-compatible}} \text{flex}_g \qquad \text{with} \quad \epsilon_g \in \{1, -1\} \tag{11.70}$$

with a sum extending to all graph pairs $\boldsymbol{g} = (\boldsymbol{ga}, \boldsymbol{gi})$ compatible with the white-black partition $\mathcal{P}$.

***P*$_3$:** (*Unitary criterion for bialternality.*)
A bimould $M^\bullet \in \text{BIMU}_r$ is bialternal if and only if it verifies all pairing identities $\text{flex}_{\mathcal{P}}.M^\bullet \equiv M^\bullet$, for all partitions $\mathcal{P} = \mathcal{P}^+ \sqcup \mathcal{P}^-$ of $\{0, 1, \ldots, r\}$.

This is the only known characterisation of bialternality that is *unitary* – by which we mean that, unlike all the others, it does not split into two distinct sets of conditions, one bearing on $M^\bullet$ and another on $swap.M^\bullet$.

Step 2: The redistribution identities.
***P*$_4$:** (*Redistribution identity on* $\text{ARI}^{\underline{\text{al}}/\underline{\text{al}}}$ *and* swap.ALAL.)
If we take a bialternal $M^\bullet \in \text{ARI}_r^{\underline{\text{al}}/\underline{\text{al}}}$ and a partition $\mathcal{P} = \mathcal{P}^+ \sqcup \mathcal{P}^-$ of $\{0, 1, \ldots, r\}$ and then turn all u_i*'s into* 0 *and also turn all black v_i's (i.e. all v_i's with black indices) into* 0 *but leave all white v_i's unchanged, the* pairing identity *of Proposition P_2 becomes a* redistribution identity:

$$\{\text{flex}_P.M^\bullet = M^\bullet\} \Longrightarrow \{\text{redis}_P.M^\bullet = M^\bullet\} \tag{11.71}$$

so-called because it has the effect of 'spreading' or 'redistributing' the total multiplicity μ_0 of the vanishing black v_i's[163] *among the multiplicities μ_i of the remaining white v_i's, with $\mu_0 - 1 = \sum(\mu_i - 1)$. The redistribution identities apply in particular to all bimoulds of* swap.ALAL, *since they are bialternal, $\boldsymbol{u}$-constant and polynomial in $\boldsymbol{v}$.*

***P_5* :** *(The infinitary redistribution algebra.)*
The set of all "redistributive" bimoulds, i.e, of all bimoulds that are:
– *$\boldsymbol{u}$-constant;*
– *alternal;*
– *and verify all redistribution identities;*
constitutes a subalgebra of ARI that is:
– *much larger than that of the $\boldsymbol{u}$-constant bialternals;*
– *not subject to* neg-*invariance (unlike the bialternals);*
– *and yet defined by an infinitary group of constraints (like the bialternals).*

Although the *redistribution identities* have a more elementary appearance than the *pairing identities*, they are in fact:
– theoretically derivative;
– distinctly weaker (since they do not imply bialternality);
– and less transparent (since the terms on the right-hand side are *composite*[164] and preceded by general integers rather than by $\pm$ signs.)

Step 3: The reconstitution identities.
For any bimould $M^\bullet$, we denote by *essen.*$M^\bullet$ the "essential part" of $M^\bullet$, *i.e.* the "part" of $M^\bullet$ that is "divisible" by each v_i. In precise terms:

$$(\mathrm{essen}.M)^{\binom{u_1\ ,\ldots,\ u_r}{v_1\ ,\ldots,\ v_r}} = \sum_{\epsilon_i \in \{0,1\}} \left(\prod_{1 \le i \le r} (-1)^{1+\epsilon_i} \right) M^{\binom{u_1\ ,\ldots,\ u_r}{\epsilon_1 v_1\ ,\ldots,\ \epsilon_r v_r}}. \quad (11.72)$$

Likewise, to each partition $\mathcal{P}$ that makes 0 *black*, we associate the "slice" of $M^\bullet$ that is "divisible" by all *white* v_i's and constant in all *black* v_i's:[165]

$$(\mathrm{slice}_{\mathcal{P}}.M)^{\binom{u_1\ ,\ldots,\ u_r}{v_1\ ,\ldots,\ v_r}} = \sum_{\substack{0 \in \mathcal{P}^- \\ \left\{ \substack{\epsilon_i \in \{0,1\} \text{ if } i \in \mathcal{P}^+ \\ \epsilon_i = 0 \text{ if } i \in \mathcal{P}^-} \right\}}} \left(\prod_{i \in \mathcal{P}^+} (-1)^{1+\epsilon_i} \right) M^{\binom{u_1\ ,\ldots,\ u_r}{\epsilon_1 v_1\ ,\ldots,\ \epsilon_r v_r}} \quad (11.73)$$

---

[163] In the augmented notation, *i.e.* considering $\{v_0, v_1, \ldots, v_r\}$, with v_0 automatically regarded as black. When v_0 is the only black variable, *i.e.* when $\mathcal{P}^- = \{0\}$ and $\mathcal{P}^+ = \{1, \ldots, r\}$, then $\mu_0 = 1$ and both *flex*$_P$ and *redis*$_P$ reduce to the identity, so that in this case the pairing and redistribution identities become trivial.

[164] In the sense that they often conflate several contributions, which were clearly distinct in the pairing identities.

[165] If $\mathcal{P}^+ = \{1, \ldots, r\}$ and $\mathcal{P}^- := \{0\}$, the slice *slice*$_{\mathcal{P}}.M^\bullet$ coincides with *essen.*$M^\bullet$.

$M^\bullet$ is clearly the sum of all its slices:

$$M^\bullet = \sum_{\mathcal{P} \text{ with } 0 \in \mathcal{P}^-} \text{slice}_{\mathcal{P}}.M^\bullet \tag{11.74}$$

and if $M^\bullet$ happens to be bialternal, each slice may be separately recovered from $essen.M^\bullet$ by means of the redistribution identites, since:

$$\text{slice}_{\mathcal{P}}.M^\bullet \equiv \text{redis}_{\mathcal{P}}\,.\,\text{essen}.M^\bullet \tag{11.75}$$

as we can see by applying the $\mathcal{P}$-related redistribution identity separately to each summand on the right-hand side of (11.72). Therefore:

$\boldsymbol{P_6}$ **:** (*Reconstitution identity on* $ARI^{\underline{\mathrm{al}}/\underline{\mathrm{al}}}$.)
For each bialternal bimould $M^\bullet$ (*purely of length* r), *the identity holds*:

$$M^\bullet \equiv \text{induc.essen}.M^\bullet \tag{11.76}$$

with the linear operator

$$\text{induc} := \sum_{\mathcal{P} \text{ with } 0 \in \mathcal{P}^-} \text{redis}_{\mathcal{P}}. \tag{11.77}$$

This applies in particular to all elements of *swap.ALAL*, *i.e.* to all $\boldsymbol{u}$-constant, $\boldsymbol{v}$-polynomial, and bialternal bimoulds. For such bialternals, the possiblity of recovering $M^\bullet$ from $essen.M^\bullet$ was by no means a foregone conclusion, since for a not too large ratio $d/r := degree/length$[166] the essential part $essen.M^\bullet$ carries but a minute fraction of the total data of $M^\bullet$.

$\boldsymbol{P_7}$ **:** (*Involutive nature of* induc.)
While essen *is* (*trivially*) *a projector,* induc *becomes* (*non-trivially*) *an involution when restricted to the space of* $\boldsymbol{u}$*-constant bialternals.*

Step 4: The unit-cleansing algorithm.
The algorithm applies to all multizetas, coloured or uncoloured, but let us focus on the uncoloured case for simplicity.

Fix any basis $\{l\phi ma^\bullet_s\,;\ s = 3,5,7\ldots\}$ of *ALIL*. That automatically fixes a system of irreducibles $\{irr\phi^\bullet_{II}, irr\phi^\bullet_{III}\}$ and provides a way of expressing all multizetas in terms of these.

Now, reason inductively. Assume that all irreducibles of length $r < r_0$ have already been expressed in terms of unit-free multizetas $\zeta(s_1,\ldots,s_r)$.

166 Say, for $2 < d/r < 3$. (Recall that d/r can in no case be ≤ 2).

The machinery of Section 6 makes it possible to exactly determine the contribution that these "earlier" irreducibles (including π^2) are going to make to $Zig^\bullet := swap.Zag^\bullet$, at all higher lengths, including at length r_0. Next, subtract from $leng_{r_0}.Zig^\bullet$ (*i.e.* from the length-r_0 component of $Zig^\bullet$) all these contributions from the "earlier" irreducibles. What is left is a superposition $M^\bullet$ of independent bialternals $M_j^\bullet$ of length r_0:

$$M^\bullet = \sum \mathrm{irr}\phi_j\, M_j^\bullet \quad \text{with} \quad M_j^\bullet \in \mathrm{swap.ALAL}_r \text{ and } \mathrm{irr}\phi_j \in \mathbb{C} \qquad (11.78)$$

with scalar coefficients $irr\phi_j$ that are irreducibles of length r_0. But, as we just saw, $M^\bullet$, and therefore all $M_j^\bullet$ and all $irr\phi_j$, can be recovered from $essen.M^\bullet$, and as a consequence expressed in terms of *unit-free* multizetas $\zeta(s_1, \ldots, s_r)$. By induction, this applies to *all* irreducibles subsumed in the moulds $irr\phi^\bullet_{_{II}}$, $irr\phi^\bullet_{_{III}}$ and of course also to the exceptional irreducible $\pi^2 = 6\,\zeta(2)$.

But since every multizeta $\zeta(s_1, \ldots, s_r)$ can be (algorithmically) expressed in terms of irreducibles, this means that every multizeta can be expressed as a polynomial of *unit-free* multizetas $\zeta(s_1, \ldots, s_r)$, with rational coefficients. After *symmetrel linearisation*, this polynomial becomes a linear combination of multizetas, still unit-free and still with rational coefficients. □

Example of pairing identities

For $r = 5$, $\mathcal{P}^+ = \{1, 2, 4\}$, $\mathcal{P}^- = \{0, 3, 5\}$, the pairing identity $M^\bullet \equiv flex_{\mathcal{P}}.M^\bullet$ takes the form:

$$M^{\left(\begin{smallmatrix} u_1, & u_2, & u_3, & u_4, & u_5, \\ v_1, & v_2, & v_3, & v_4, & v_5, \end{smallmatrix}\right)} \equiv \qquad (***)$$

$$\begin{aligned}
&-M^{\left(\begin{smallmatrix} u_{5*0}, & u_{4*5}, & u_{1*4}, & u_{4*3}, & u_{2*1} \\ v_{1:0}, & v_{1:5}, & v_{1:3}, & v_{4:3}, & v_{2:3} \end{smallmatrix}\right)} - M^{\left(\begin{smallmatrix} u_{5*0}, & u_{4*5}, & u_{1*4}, & u_{2*1}, & u_{4*3} \\ v_{1:0}, & v_{1:5}, & v_{1:3}, & v_{2:3}, & v_{4:3} \end{smallmatrix}\right)} - M^{\left(\begin{smallmatrix} u_{1*0}, & u_{5*1}, & u_{4*5}, & u_{2*4}, & u_{4*3} \\ v_{1:0}, & v_{2:0}, & v_{2:5}, & v_{2:3}, & v_{4:3} \end{smallmatrix}\right)} \\
&-M^{\left(\begin{smallmatrix} u_{5*0}, & u_{1*5}, & u_{4*1}, & u_{2*4}, & u_{4*3} \\ v_{1:0}, & v_{1:5}, & v_{2:5}, & v_{2:3}, & v_{4:3} \end{smallmatrix}\right)} - M^{\left(\begin{smallmatrix} u_{4*5}, & u_{5*3}, & u_{3*1}, & u_{1*0}, & u_{2*3} \\ v_{4:5}, & v_{4:0}, & v_{2:0}, & v_{1:0}, & v_{2:3} \end{smallmatrix}\right)} - M^{\left(\begin{smallmatrix} u_{4*5}, & u_{5*3}, & u_{3*1}, & u_{2*3}, & u_{1*0} \\ v_{4:5}, & v_{4:0}, & v_{2:0}, & v_{2:3}, & v_{1:0} \end{smallmatrix}\right)} \\
&+M^{\left(\begin{smallmatrix} u_{2*1}, & u_{1*3}, & u_{3*0}, & u_{5*3}, & u_{4*5} \\ v_{2:3}, & v_{1:3}, & v_{4:0}, & v_{1:0}, & v_{4:5} \end{smallmatrix}\right)} - M^{\left(\begin{smallmatrix} u_{1*0}, & u_{2*1}, & u_{5*2}, & u_{4*5}, & u_{2*3} \\ v_{1:0}, & v_{2:0}, & v_{4:0}, & v_{4:5}, & v_{4:3} \end{smallmatrix}\right)} - M^{\left(\begin{smallmatrix} u_{1*0}, & u_{2*1}, & u_{5*2}, & u_{2*3}, & u_{4*5} \\ v_{1:0}, & v_{2:0}, & v_{4:0}, & v_{4:3}, & v_{4:5} \end{smallmatrix}\right)} \\
&+M^{\left(\begin{smallmatrix} u_{2*1}, & u_{1*3}, & u_{5*1}, & u_{1*0}, & u_{4*5} \\ v_{2:3}, & v_{4:3}, & v_{4:0}, & v_{1:0}, & v_{4:5} \end{smallmatrix}\right)} + M^{\left(\begin{smallmatrix} u_{2*1}, & u_{1*3}, & u_{5*1}, & u_{4*5}, & u_{1*0} \\ v_{2:3}, & v_{4:3}, & v_{4:0}, & v_{4:5}, & v_{1:0} \end{smallmatrix}\right)} + M^{\left(\begin{smallmatrix} u_{2*1}, & u_{1*0}, & u_{0*3}, & u_{5*0}, & u_{4*5} \\ v_{2:3}, & v_{1:3}, & v_{4:3}, & v_{4:0}, & v_{4:5} \end{smallmatrix}\right)} \\
&-M^{\left(\begin{smallmatrix} u_{1*0}, & u_{5*1}, & u_{2*5}, & u_{5*3}, & u_{4*5} \\ v_{1:0}, & v_{2:0}, & v_{2:3}, & v_{4:3}, & v_{4:5} \end{smallmatrix}\right)} + M^{\left(\begin{smallmatrix} u_{4*5}, & u_{5*3}, & u_{1*5}, & u_{5*0}, & u_{2*1} \\ v_{4:5}, & v_{4:3}, & v_{1:3}, & v_{1:0}, & v_{2:3} \end{smallmatrix}\right)} + M^{\left(\begin{smallmatrix} u_{4*5}, & u_{5*3}, & u_{1*5}, & u_{2*1}, & u_{5*0} \\ v_{4:5}, & v_{4:3}, & v_{1:3}, & v_{2:3}, & v_{1:0} \end{smallmatrix}\right)} \\
&-M^{\left(\begin{smallmatrix} u_{5*0}, & u_{1*5}, & u_{4*1}, & u_{1*3}, & u_{2*1} \\ v_{1:0}, & v_{1:5}, & v_{4:5}, & v_{4:3}, & v_{2:3} \end{smallmatrix}\right)} - M^{\left(\begin{smallmatrix} u_{1*0}, & u_{5*1}, & u_{2*5}, & u_{4*2}, & u_{2*3} \\ v_{1:0}, & v_{2:0}, & v_{2:5}, & v_{4:5}, & v_{4:3} \end{smallmatrix}\right)} - M^{\left(\begin{smallmatrix} u_{2*1}, & u_{1*3}, & u_{3*5}, & u_{5*0}, & u_{4*3} \\ v_{2:3}, & v_{1:3}, & v_{1:5}, & v_{1:0}, & v_{4:5} \end{smallmatrix}\right)} \\
&-M^{\left(\begin{smallmatrix} u_{2*1}, & u_{1*3}, & u_{3*5}, & u_{4*3}, & u_{5*0} \\ v_{2:3}, & v_{1:3}, & v_{1:5}, & v_{4:5}, & v_{1:0} \end{smallmatrix}\right)} - M^{\left(\begin{smallmatrix} u_{5*0}, & u_{1*5}, & u_{2*1}, & u_{4*2}, & u_{2*3} \\ v_{1:0}, & v_{1:5}, & v_{2:5}, & v_{4:5}, & v_{4:3} \end{smallmatrix}\right)} + M^{\left(\begin{smallmatrix} u_{1*0}, & u_{5*1}, & u_{3*5}, & u_{4*3}, & u_{2*3} \\ v_{1:0}, & v_{2:0}, & v_{2:5}, & v_{4:5}, & v_{2:3} \end{smallmatrix}\right)} \\
&+M^{\left(\begin{smallmatrix} u_{1*0}, & u_{5*1}, & u_{3*5}, & u_{2*3}, & u_{4*3} \\ v_{1:0}, & v_{2:0}, & v_{2:5}, & v_{2:3}, & v_{4:5} \end{smallmatrix}\right)} + M^{\left(\begin{smallmatrix} u_{5*0}, & u_{1*5}, & u_{3*1}, & u_{4*3}, & u_{2*3} \\ v_{1:0}, & v_{1:5}, & v_{2;5}, & v_{4:5}, & v_{2:3} \end{smallmatrix}\right)} + M^{\left(\begin{smallmatrix} u_{5*0}, & u_{1*5}, & u_{3*1}, & u_{2*3}, & u_{4*3} \\ v_{1:0}, & v_{1:5}, & v_{2:5}, & v_{2:3}, & v_{4:5} \end{smallmatrix}\right)}
\end{aligned}$$

with the usual convention $u_0 := -(u_1 + \cdots + u_r)$, $v_0 := 0$ and the convenient abbreviations:

$$u_{i*j} := su_i - su_j \quad \text{with} \quad su_k := u_0 + u_1 + \ldots + u_k = -u_{k+1} - u_{k+2} \ldots - u_r$$

$$v_{i:j} := v_i - v_j$$

To arrive at the pairing identity $(***)$, we form all graph triples $\boldsymbol{g} = \{\boldsymbol{ga}, \boldsymbol{gi}, \boldsymbol{gai}\}$ compatible with the partition $\mathcal{P}$. There exist exactly 16 such triples. They are pictured on Figure 11.1, with split lines for the edges of $\boldsymbol{ga}$, plain lines for those of $\boldsymbol{gi}$, and large plain lines for those of $\boldsymbol{gai}$. Next, on each $\boldsymbol{gai}$, we pick a vertex x_* so chosen as to minimise the number $\nu(\boldsymbol{gai}, x_*)$ of total orders $\boldsymbol{\gamma}$ on $\boldsymbol{gai}$ compatible with the partial order induced by x_*. In each case, x_* has to be at the extremity of the longest branch of $\boldsymbol{gai}$. For eight graphs $\boldsymbol{gai}$, this minimal number $\nu_{\min}(\boldsymbol{gai})$ is 1; for the remaining eight graphs, $\nu_{\min}(\boldsymbol{gai})$ is 2. Altogether, this yields the 24 elementary flexions flex_γ that contribute to the pairing identity $(* * *)$.

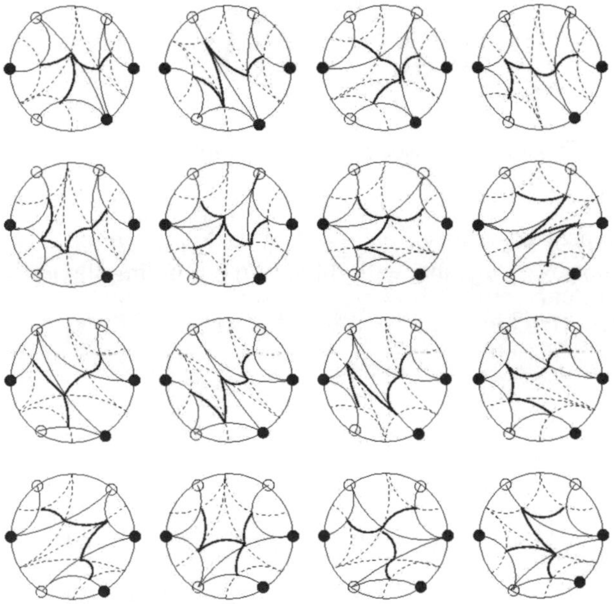

Figure 11.1. The 16 graph triads $\boldsymbol{g} = \{\boldsymbol{ga}, \boldsymbol{gi}, \boldsymbol{gai}\}$ compatible with the partition $\mathcal{P}$ of $\{0, 1, 2, 3, 4, 5\}$ defined by $\mathcal{P}^+ = \{1, 2, 4\}$, $\mathcal{P}^- = \{0, 3, 5\}$.

Lastly, to show how to calculate each $\mathit{flex}_{\boldsymbol{g}}$, we focus on the first graph triple (the one in top-left position on Figure 11.1) and reproduce it, enlarged, in Figure 11.2.

Applying the rules just after (11.67), we see that the flexion indices $w_i^* = \binom{u_i^*}{v_i^*}$ corresponding to the five vertices of $\boldsymbol{gai}$ are given by:

$$
\begin{array}{lll|l}
u_1^* = u_{1,2,3,4,5} & & & v_1^* = v_1 - v_0 = v_1 \\
u_2^* = u_{0,1} & = -u_{2,3,4,5} & & v_2^* = v_1 - v_5 \\
u_3^* = u_{2,3} & & & v_3^* = v_2 - v_5 \\
u_4^* = u_{4,5,0,1,2} = -u_3 & & & v_4^* = v_2 - v_3 \\
u_5^* = u_4 & & & v_5^* = v_4 - v_5
\end{array}
$$

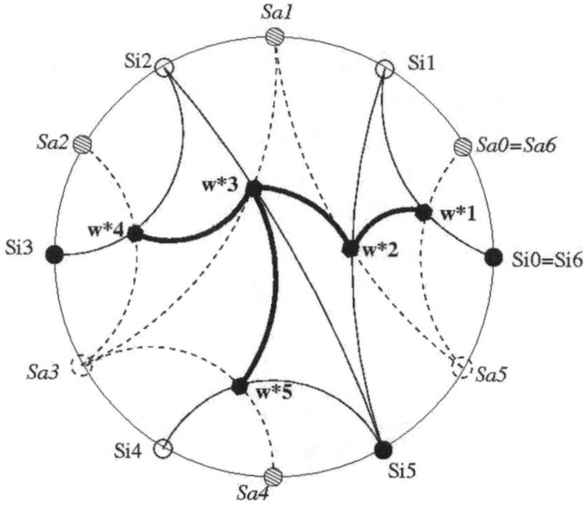

Figure 11.2. Flexion *flex*$_g$ associated with a graph triad $\boldsymbol{g} = \{\boldsymbol{ga}, \boldsymbol{gi}, \boldsymbol{gai}\}$.

with the expected identity $\sum_{1\le i\le 5} u_i^*\, v_i^* \equiv \sum_{1\le i\le 5} u_i\, v_i$. There are three possible roots, w_1^*, w_4^*, w_5^*, with three corresponding flexions:

$$(\mathrm{flex}_{\boldsymbol{g},w_1^*}.M)^{w_1,w_2,w_3,w_4,w_5} = +M^{w_1^*,w_2^*,w_3^*,w_4^*,w_5^*} + M^{w_1^*,w_2^*,w_3^*,w_5^*,w_4^*}$$

$$(\mathrm{flex}_{\boldsymbol{g},w_4^*}.M)^{w_1,w_2,w_3,w_4,w_5} = -M^{w_4^*,w_3^*,w_2^*,w_1^*,w_5^*} - M^{w_4^*,w_3^*,w_2^*,w_5^*,w_1^*} - M^{w_4^*,w_3^*,w_5^*,w_2^*,w_1^*}$$

$$(\mathrm{flex}_{\boldsymbol{g},w_5^*}.M)^{w_1,w_2,w_3,w_4,w_5} = -M^{w_5^*,w_3^*,w_2^*,w_1^*,w_4^*} - M^{w_5^*,w_3^*,w_2^*,w_4^*,w_1^*} - M^{w_5^*,w_3^*,w_4^*,w_2^*,w_1^*}$$

which coincide modulo the alternality relations:

$$\mathrm{flex}_{\boldsymbol{g},w_1^*}.M^\bullet \equiv \mathrm{flex}_{\boldsymbol{g},w_4^*}.M^\bullet \equiv \mathrm{flex}_{\boldsymbol{g},w_5^*}.M^\bullet \qquad \forall M^\bullet \ \mathit{alternal}.$$

One might also take *flex*$_{\boldsymbol{g},w_2^*}.M^\bullet$ and *flex*$_{\boldsymbol{g},w_3^*}.M^\bullet$, but here the number of summands would be much larger: 8 and 12 respectively.

Example of redistribution identity
For $r = 5$, $\mathcal{P}^+ = \{1, 2, 4\}$ and $\mathcal{P}^- = \{0, 3, 5\}$, we have a black multiplicity $\mu_0 = 3$, and the *redistribution identity* $M^\bullet \equiv redis_{\mathcal{P}}.M^\bullet$ follows from the preceding *pairing identity* $M^\bullet \equiv flex_{\mathcal{P}}.M^\bullet$ by setting all black v_i's equal to zero in $(***)$. For simplicity, we write the redistribution identity only for $\boldsymbol{u}$-constant bilaternals, and since for them the u_i's don't matter, we don't mention them.

$$\begin{aligned} M^{v_1,v_2,0,v_4,0} \equiv & -M^{v_1,v_1,v_1,v_4,v_2} - M^{v_1,v_1,v_1,v_2,v_4} - M^{v_4,v_4,v_2,v_1,v_2} - M^{v_4,v_4,v_2,v_2,v_1} \\ & + M^{v_2,v_1,v_1,v_4,v_4} - 2M^{v_1,v_2,v_4,v_4,v_4} + M^{v_2,v_4,v_4,v_1,v_4} + M^{v_2,v_4,v_4,v_4,v_1} \\ & + M^{v_2,v_1,v_4,v_4,v_4} - 2M^{v_1,v_2,v_2,v_4,v_4} + M^{v_4,v_4,v_1,v_1,v_2} + M^{v_4,v_4,v_1,v_2,v_1} \\ & - M^{v_1,v_1,v_4,v_4,v_2} - M^{v_2,v_1,v_1,v_1,v_4} - M^{v_2,v_1,v_1,v_4,v_1} - M^{v_1,v_1,v_2,v_4,v_4} \\ & + M^{v_1,v_2,v_2,v_4,v_2} + M^{v_1,v_1,v_2,v_4,v_2}. \end{aligned}$$

Examples of reconstitution identities

Up to length 2, the operator *induc* is trivial, but the number N_r of terms involved increases sharply thereafter. Thus:[167]

$$N_1=1, N_2=2, N_3\sim 7, N_4\sim 38, N_5\sim 273, N_6\sim 1837, N_7\sim 15199, \text{etc}\ldots$$

Here are the formulas up to length 4, for the case of $\boldsymbol{u}$-constant bimoulds (and after removal of the u_i's):

$$(\text{induc}.M)^{v_1} := M^{v_1}; \qquad (\text{induc}.M)^{v_1,v_2} := M^{v_1,v_2}$$

$$\begin{aligned}(\text{induc}.M)^{v_1,v_2,v_3} :=&+ M^{v_1,v_2,v_3} + M^{v_1,v_1,v_2} + M^{v_1,v_2,v_2} + M^{v_1,v_3,v_1}\\ &+ M^{v_3,v_1,v_3} + M^{v_2,v_2,v_3} + M^{v_2,v_3,v_3}\end{aligned}$$

$$\begin{aligned}(\text{induc}.M)^{v_1,v_2,v_3,v_4} :=&+ M^{v_1,v_2,v_3,v_4} + M^{v_1,v_1,v_2,v_3} + M^{v_1,v_2,v_2,v_3} + M^{v_1,v_2,v_3,v_3}\\
&+ M^{v_1,v_4,v_1,v_2} + M^{v_1,v_2,v_4,v_2} + M^{v_1,v_4,v_2,v_4} + M^{v_4,v_1,v_4,v_2}\\
&+ M^{v_4,v_1,v_2,v_4} + M^{v_3,v_4,v_1,v_4} + M^{v_3,v_1,v_3,v_4} + M^{v_1,v_3,v_1,v_4}\\
&+ M^{v_3,v_1,v_4,v_1} + M^{v_1,v_3,v_4,v_1} + M^{v_2,v_3,v_4,v_4} + M^{v_2,v_3,v_3,v_4}\\
&+ M^{v_2,v_2,v_3,v_4} + M^{v_1,v_1,v_1,v_2} + M^{v_1,v_1,v_3,v_1} + M^{v_1,v_4,v_1,v_1}\\
&+ M^{v_1,v_2,v_2,v_2} + M^{v_2,v_2,v_2,v_3} + M^{v_2,v_2,v_4,v_2} + M^{v_3,v_1,v_3,v_3}\\
&+ M^{v_2,v_3,v_3,v_3} + M^{v_3,v_3,v_3,v_4} + M^{v_4,v_4,v_1,v_4} + M^{v_4,v_2,v_4,v_4}\\
&+ M^{v_3,v_4,v_4,v_4} + M^{v_1,v_1,v_2,v_2} + M^{v_3,v_3,v_1,v_1} + M^{v_1,v_3,v_1,v_3}\\
&+ M^{v_1,v_1,v_4,v_4} + M^{v_4,v_1,v_4,v_1} + M^{v_2,v_2v,v_3,v_3} + M^{v_4,v_4,v_2v,v_2}\\
&+ M^{v_2,v_4,v_2,v_4} + M^{v_3,v_3,v_4,v_4}.\end{aligned}$$

11.7 Multizeta cleansing: elimination of odd degrees

We shall now construct a simple algorithm for *expressing every multizeta of odd degree as a finite sum, with rational coefficients, of multizetas of even degree.*[168]

We take as our starting point the symmetrel multitangent mould $Te^{\bullet}(z)$ and its generating function, the symmetril mould $Tig^{\bullet}(z)$, with definitions

---

[167] Recall that the expression of *induc* is unique only modulo the alternality relations. Hence the sign $\sim$ to caution that there is at least *one* expression of *induc* with the number N_r of summands mentioned. In any case, the minimal number N_r^{min} cannot be significantly less.

[168] Recall that the degree $d := s - r$ of a multizeta is defined as its total weight s minus its length (or depth) r.

transparently patterned on those of $Ze^\bullet$ and $Zig^\bullet$:

$$\mathrm{Te}^{\binom{\epsilon_1\ ,\dots,\ \epsilon_r}{s_1\ ,\dots,\ s_r}}(z) := \sum_{+\infty>n_1>\dots>n_r>-\infty} \prod_{i=1}^{i=r} \left(e_i^{-n_i}\ (n_i+z)^{-s_1} \right) \tag{11.79}$$

$$\mathrm{Tig}^{\binom{\epsilon_1\ ,\dots,\ \epsilon_r}{v_1\ ,\dots,\ v_1}}(z) := \sum_{s_i\geq 1} Te^{\binom{\epsilon_1\ ,\dots,\ \epsilon_r}{s_1\ ,\dots,\ s_r}}(z)\, v_1^{s_1-1} \dots v_r^{s_r-1}. \tag{11.80}$$

The next step is to express the multitangents in terms of multizetas. Here, we have the choice between an *uninflected* formula which leaves z spread over all terms, and an *inflected* formula which concentrates z in a few elementary central terms:

$$\mathrm{Tig}^{\boldsymbol{w}}(z) = \sum_{\boldsymbol{w}=\boldsymbol{w}^+\boldsymbol{w}^-} \mathrm{Zig}^{\boldsymbol{w}^+}(z)\mathrm{viZig}^{\boldsymbol{w}^-}(z) - \sum_{\boldsymbol{w}=\boldsymbol{w}^+w_0\boldsymbol{w}^-} \mathrm{Zig}^{\boldsymbol{w}^+}(z)\mathrm{Pi}^{w_0}(z)\mathrm{viZig}^{\boldsymbol{w}^-}(z)$$

$$\mathrm{Tig}^{\boldsymbol{w}}(z) = \mathrm{Rig}^{\boldsymbol{w}} - \sum_{\boldsymbol{w}=\boldsymbol{w}^+w_0\boldsymbol{w}^-} \mathrm{Zig}^{\boldsymbol{w}^+\rfloor}\ \mathrm{Qii}^{\lceil w_0 \rceil}(z)\ \mathrm{viZig}^{\lfloor \boldsymbol{w}^-}.$$

The ingredient $Rig^\bullet$ in the above formulas is defined as follows:

$$\mathrm{Rig}^{w_1,\dots,w_r} := 0 \quad \text{for } r = 0 \text{ or } r \ \textit{odd}$$

$$\mathrm{Rig}^{w_1,\dots,w_r} := \frac{(\pi i)^r}{r!}\ \delta(u_1)\dots\delta(u_r) \qquad \text{for } r \ \textit{even} > 0$$

with δ denoting as usual the discrete dirac.[169] The length-1 bimoulds $Pi^\bullet$ and $Qii^\bullet := Qii^\bullet_\pi$ denote the *polar* and *bitrigonometric* flexion units of Section 3.2, and $vi\,Zig^\bullet := neg.pari.anti.Zig^\bullet$. Lastly, the bimoulds $Pi^\bullet(z)$, $Qii^\bullet(z)$, $Zig^\bullet(z)$, $vi\,Zig^\bullet(z)$ are deduced from $Pi^\bullet$, $Qii^\bullet$, $Zig^\bullet$, $vi\,Zig^\bullet$ under the change $v_i \to v_i - z$ $(\forall i)$.

By equating our *uninflected* and *inflected* expressions of $Tig^\bullet(z)$ and then setting $z = 0$, we get the remarkable identity:

$$\begin{aligned} &\sum_{\boldsymbol{w}=\boldsymbol{w}^+\boldsymbol{w}^-} \mathrm{Zig}^{\boldsymbol{w}^+}\ \mathrm{viZig}^{\boldsymbol{w}^-} - \sum_{\boldsymbol{w}=\boldsymbol{w}^+w_0\boldsymbol{w}^-} \mathrm{Zig}^{\boldsymbol{w}^+}\ \mathrm{Pi}^{w_0}\ \mathrm{viZig}^{\boldsymbol{w}^-} \\ &= \mathrm{Rig}^{\boldsymbol{w}} - \sum_{\boldsymbol{w}=\boldsymbol{w}^+w_0\boldsymbol{w}^-} \mathrm{Zig}^{\boldsymbol{w}^+\rfloor}\ \mathrm{Qii}^{\lceil w_0 \rceil}\ \mathrm{viZig}^{\lfloor \boldsymbol{w}^-} \qquad (\forall \boldsymbol{w}) \end{aligned} \tag{11.81}$$

where the factor sequences $\boldsymbol{w}^\pm$ can be $\emptyset$. As a consequence, (11.81) is of the form:

$$\mathrm{Zig}^{w_1,\dots,w_r} + (-1)^r\ \mathrm{Zig}^{-w_r,\dots,-w_1} = \text{“\textit{shorter terms}”}. \tag{11.82}$$

[169] $\delta(0) := 1$ and $\delta(t) := 0$ for $t \neq 0$.

But $Zig^\bullet$ is symmetril and therefore *mantir*-invariant (see Section 3.4), which again yields an identity of the form:

$$\mathrm{Zig}^{-w_1,\dots,-w_r} + (-1)^r\,\mathrm{Zig}^{-w_r,\dots,-w_1} = \text{“\textit{shorter terms}”}. \qquad (11.83)$$

If we now take 'colourless' indices w_i, *i.e.* indices $w_i := \binom{0}{v_i}$, then subtract (11.83) from (11.82), and calculate therein the coefficient of $\prod v_i^{s_i-1}$, we find:

$$(1-(-1)^d)\mathrm{Ze}^{\binom{0\ ,\dots,\ 0}{s_1\ ,\dots,\ s_r}} = \text{“\textit{shorter terms}”} \quad \left(d := \sum s_i - r\right) \qquad (11.84)$$

with quite explicit 'shorter terms'.

We have here a very effective algorithm for the 'elimination' of all *uncoloured* multizetas $\zeta(s_1,\dots,s_r)$ of *odd* degree d. The argument extends to the case of *bicoloured* multizetas $Ze^{\binom{\epsilon_1\ ,\dots,\ \epsilon_r}{s_1\ ,\dots,\ s_r}}$ with $\epsilon_i \in \frac{1}{2}\mathbb{Z}/\mathbb{Z}$, since we then have $\epsilon_i \equiv -\epsilon_i$. In the case of more than two colours, however, equation (11.84) becomes a *singular* linear system, which allows the elimination of *most*, but not *all*, multizetas of odd degree.

Remark 1: elimination of irreducibles other than π^2.
A simple argument shows that identity (11.81) still holds if we neglect all irreducibles other than π^2, *i.e.* if we retain only the first factor $Zig_I{}^\bullet$ in the trifactorisation (9.1) of $Zig^\bullet$. But since $Zig_I^\bullet$ is invariant under *pari.neg*, we clearly have $viZig_I = anti.Zig_I{}^\bullet$, so that (11.81) becomes:

$$\mathrm{mu}(\mathrm{Zig}_{\mathrm{I}}{}^\bullet, 1^\bullet - \mathrm{Pi}^\bullet, \mathrm{anti.Zig}_{\mathrm{I}}{}^\bullet) = \mathrm{Rig}^\bullet - \mathrm{giwat}(\mathrm{Zig}_{\mathrm{I}}{}^\bullet).\mathrm{Qi}^\bullet. \qquad (11.85)$$

Remark 2: separation of π^2 from the rationals.
Actually, we may retain in (11.85) only the first two factors of $Zig_I{}^\bullet$ (see (9.2)) namely *gira*($til^\bullet$, $sripil^\bullet$) with $sripil^\bullet$:= *invgira*($pil^\bullet$). Furthermore, since in (11.85) the 'trigometric' part (which carries π^2) and the 'polar' part (which carries only rationals) do not mix, (11.85) leads to two distinct identities, to wit:

$$\mathrm{mu}(\mathrm{sripil}^\bullet,\ \mathrm{anti.sripil}^\bullet) = \mathrm{mu}(\mathrm{sripil}^\bullet,\ \mathrm{Pi}^\bullet,\ \mathrm{anti.sripil}^\bullet) \qquad (11.86)$$

$$\mathrm{mu}(\mathrm{til}^\bullet,\ \mathrm{anti.til}^\bullet) = \mathrm{Rig}^\bullet - \mathrm{giwat}(\mathrm{til}^\bullet).\mathrm{Qi}^\bullet. \qquad (11.87)$$

Remark 3: universalisation.
Identity (11.86) admits an automatic extension to all exact units, namely:

$$\mathrm{mu}(\mathfrak{ess}^\bullet,\ \mathrm{anti}.\mathfrak{ess}^\bullet) = \mathrm{mu}(\mathfrak{ess}^\bullet,\ \mathfrak{E}^\bullet,\ \mathrm{anti}.\mathfrak{ess}^\bullet). \qquad (11.88)$$

Identity (11.87), which involves the approximate unit $Qi^\bullet$, *does not* admit *extensions* to all approximate units,[170] but it does possess a *restriction* to the polar unit $Pi^\bullet$ [171] and hence an *extension* to all exact units:

$$\mathrm{mu}(\mathfrak{ess}^\bullet, \mathrm{anti}.\mathfrak{ess}^\bullet) = -\mathrm{giwat}(\mathfrak{ess}^\bullet).\mathfrak{E}^\bullet. \tag{11.89}$$

11.8 $GARI_{\mathfrak{se}}$ and the two separation lemmas

Let $\mathfrak{E}$ be an exact flexion unit and $\mathfrak{O}$ its conjugate unit. Reverting to the notations of Section 4.1, with any $f(x) := x + \sum_{1\leq r} x^{r+1}$ in the group *GIFF*, we associate its image $\mathfrak{Se}^\bullet_f$ in the group $GARI_{<\mathfrak{se}>} \subset GARI^{as}$. Being the exponential of an alternal bimould of *ARI*, $\mathfrak{Se}^\bullet_f$ is automatically symmetral but its swappee $\mathfrak{S\ddot{o}}^\bullet_f := swap.\mathfrak{Se}^\bullet_f$ is only exceptionnally so. It does possess, however, two remarkable *separation properties*, which may be viewed as weakened forms of symmetrality. Indeed, if we set

$$\mathrm{gepar}.\mathfrak{Se}^\bullet_f := \mathrm{mu}(\mathrm{anti.swap}.\mathfrak{Se}^\bullet_f, \mathrm{swap}.\mathfrak{Se}^\bullet_f) \tag{11.90}$$

$$\mathrm{hepar}.\mathfrak{Se}^\bullet_f := \sum_{1\leq r\leq r(\bullet)} \mathrm{pus}^k.\mathrm{logmu.swap}.\mathfrak{Se}^\bullet_f \tag{11.91}$$

then both $\mathrm{gepar}.\mathfrak{Se}^\bullet_f$ and $\mathrm{hepar}.\mathfrak{Se}^\bullet_f$ turn out to be expressible as simple, uninflected products of the conjugate unit $\mathfrak{O}$. More precisely:

$$\mathrm{gepar}.\mathfrak{Se}_f^{w_1,\dots,w_r} := a_r^*\, \mathfrak{O}^{w_1}\dots\mathfrak{O}^{w_r} \quad \text{with} \quad a_r^* := (r+1)a_r \tag{11.92}$$

$$\mathrm{hepar}.\mathfrak{Se}_f^{w_1,\dots,w_r} := a_r^{**}\mathfrak{O}^{w_1}\dots\mathfrak{O}^{w_r} \quad \text{with} \quad \sum_{1\leq r} a_r^{**} x^r := \frac{x}{2}\frac{f''(x)}{f'(x)}. \tag{11.93}$$

Remark 1: The definition of *hepar* involves *logmu*, which is of course the logarithm relative to the *mu*-product. It should be noted, however, that after simplification all rational coefficients disappear from the right-hand side of (11.91) and the only coefficients left are ± 1. In fact, the right-hand side of (11.91) is none other than the left-hand side of (2.75).

Remark 2: If $\mathfrak{S\ddot{o}}_f$ were exactly symmetral, it would verify the two subsymmetries implied by symmetrality, namely *gantar*-invariance (see (2.74)) and *gus*-neutrality (see (2.75)) and we would have

$$mu(pari.anti.\mathfrak{S\ddot{o}}^\bullet_f, \mathfrak{S\ddot{o}}^\bullet_f) \equiv 1^\bullet$$

$$\text{and} \quad \sum_{1\leq r\leq r(\bullet)} pus^k.logmu.\mathfrak{S\ddot{o}}^\bullet_f \equiv 0^\bullet \; mod\, BIMU_I.$$

[170] It has no simple counterpart with ($Qa^\bullet$, $tal^\bullet$) in place of ($Qi^\bullet$, $til^\bullet$).

[171] After automatic elimination of the $Rig^\bullet$ part.

As it is, we merely have the separation properties (11.92) and (11.93), with the addded twist that *separar* involves $mu(anti.\mathfrak{So}^{\bullet}_f, \ddot{\mathfrak{So}}^{\bullet}_f)$ rather than $mu(pari.anti.\ddot{\mathfrak{So}}^{\bullet}_f, \ddot{\mathfrak{So}}^{\bullet}_f)$.

Remark 3: The simplest way to prove the separation identities is to consider the infinitesimal dilator $f_{\#}(x) = \sum_{1\leq r} \eta_r\, x^{r+1}$ of f and to form its image $\mathfrak{Te}^{\bullet}_f = \sum_{1\leq r} \eta_r \mathfrak{te}^{\bullet}_r$ in *ARI*. One of the defining identities for $\mathfrak{Se}^{\bullet}_f$ then reads:

$$r(\bullet)\, \mathfrak{Se}^{\bullet}_f = \mathrm{preari}(\mathfrak{Se}^{\bullet}_f, \mathfrak{Te}^{\bullet}_f) = \mathrm{preawi}(\mathfrak{Se}^{\bullet}_f, \mathfrak{Te}^{\bullet}_f). \qquad (11.94)$$

Under the *swap* transform this becomes:[172]

$$r(\bullet)\, \ddot{\mathfrak{So}}^{\bullet}_f = \mathrm{preira}(\ddot{\mathfrak{So}}^{\bullet}_f, \ddot{\mathfrak{To}}^{\bullet}_f) = \mathrm{preiwa}(\ddot{\mathfrak{So}}^{\bullet}_f, \ddot{\mathfrak{To}}^{\bullet}_f). \qquad (11.95)$$

If we then set:

$$\ddot{\mathfrak{O}}_{*}^{w_1,\dots,w_r} := a^{*}_r\, \mathfrak{O}^{w_1} \dots \mathfrak{O}^{w_r}; \quad \ddot{\mathfrak{O}}_{**}^{w_1,\dots,w_r} := a^{**}_r\, \mathfrak{O}^{w_1} \dots \mathfrak{O}^{w_r} \qquad (11.96)$$

we readily sees that (11.92) is equivalent to the rather elementary identity:

$$r(\bullet)\, \ddot{\mathfrak{O}}^{\bullet}_{*} = \mathrm{iwat}(\ddot{\mathfrak{To}}^{\bullet}_f).\ddot{\mathfrak{O}}^{\bullet}_{*} + \mathrm{mu}(\ddot{\mathfrak{O}}^{\bullet}_{*}, \ddot{\mathfrak{To}}^{\bullet}_f) + \mathrm{mu}(\mathrm{anti}.\ddot{\mathfrak{To}}^{\bullet}_f, \ddot{\mathfrak{O}}^{\bullet}_{*}). \qquad (11.97)$$

The proof of the (11.93) follows the same pattern, with $\ddot{\mathfrak{O}}_{*}$ replaced by $\ddot{\mathfrak{O}}_{**}$, but is less direct.

Remark 4: In view of these two separation identities (11.92),(11.93), which involve respectively the coefficients a^{*}_r and a^{**}_r of f' and f''/f', *i.e.* of the differential operators of first and second order that give rise to simple *composition laws*, one may speculate about the existence of a third separation identity that would involve the coefficients a^{***}_r of the Schwarzian derivative of f. At the moment no such identity is known, but it may be pointed out that the formulas in Table 3 below also fall into the broad category of separation identities: see Remark 1 in Section 12.3.

11.9 Bisymmetrality of $\mathfrak{ess}^{\bullet}$: conceptual proof

The bimould $\mathfrak{ess}^{\bullet}$ of Section 4.2 is a special element $\mathfrak{Se}^{\bullet}_f$ of $GARI_{<\mathfrak{se}>}$ whose preimage f and dilator $f_{\#}$ are given by:

$$f(x) := 1 - e^{-x}, \quad f_{\#}(x) := 1 + x - e^{x}, \quad \frac{x}{2}\frac{f''(x)}{f'(x)} := -\frac{x}{2}. \qquad (11.98)$$

[172] The reasons why in this particular instance one may replace the pair *ari/ira* by the more convenient pair *awi/iwa* were explained in Section 4.1.

As a consequence, the two *separation lemmas* of Section 11.8 yield:

$$\text{mu}(\text{anti}.\ddot{\mathfrak{oss}}^\bullet,\, \ddot{\mathfrak{oss}}^\bullet) = \text{expmu}(-\mathfrak{O}^\bullet) \tag{11.99}$$

$$\sum_{1\le k\le r} \text{pus}^k.\text{logmu}.\ddot{\mathfrak{oss}}^\bullet = -\frac{1}{2}\,\mathfrak{O}^\bullet. \tag{11.100}$$

Both relations exhibit the only possibly form compatible with $\ddot{\mathfrak{oss}}^\bullet$ being symmetral, but we aren't quite there yet. To collect more information, let us harken back to the relation that defines $\ddot{\mathfrak{oss}}^\bullet$ in terms of its dilator $\ddot{\mathfrak{ott}}^\bullet$. It reads:

$$r(\bullet)\,\ddot{\mathfrak{oss}}^\bullet = \text{preiwa}(\ddot{\mathfrak{oss}}^\bullet,\, \ddot{\mathfrak{ott}}^\bullet) \tag{11.101}$$

with

$$\ddot{\mathfrak{ott}}^\bullet := -\sum_{1\le r} \frac{1}{(2r+1)!}\,\mathfrak{t}\ddot{\mathfrak{o}}^\bullet_{2r} = -\sum_{1\le r} \frac{1}{(2r+1)!}\,\text{swap}.\mathfrak{te}^\bullet_{2r}. \tag{11.102}$$

Let us further *mu*-factorise $\ddot{\mathfrak{oss}}^\bullet$ as in (4.46), with the same elementary right factor $\ddot{\mathfrak{oss}}^\bullet_\star$ but with a left factor $\ddot{\mathfrak{oss}}^\bullet_{\star\star}$ whose properties are a priori unknown:

$$\ddot{\mathfrak{oss}}^\bullet_\star = \text{mu}(\ddot{\mathfrak{oss}}^\bullet_{\star\star},\, \ddot{\mathfrak{oss}}^\bullet_\star) \quad \text{with} \quad \ddot{\mathfrak{oss}}^\bullet_\star := \text{expmu}\left(-\frac{1}{2}\mathfrak{O}^\bullet\right). \tag{11.103}$$

Elementary calculations show that (11.101) transforms into:

$$r(\bullet)\,\ddot{\mathfrak{oss}}^\bullet_{\star\star} = \text{preiwa}(\ddot{\mathfrak{oss}}^\bullet_{\star\star},\, \ddot{\mathfrak{ott}}^\bullet_{\star\star}) + \frac{1}{2}\,\text{mu}(\ddot{\mathfrak{oss}}^\bullet_{\star\star},\, \ddot{\mathfrak{ott}}^\bullet_\star) \tag{11.104}$$

with

$$\ddot{\mathfrak{ott}}^\bullet_\star := +\sum_{1\le r} \frac{1}{(2r)!}\,\text{mu}(\overbrace{\mathfrak{O}^\bullet,\ldots,\mathfrak{O}^\bullet}^{2\text{ r times}}) = \text{coshmu}(\mathfrak{O}^\bullet) \tag{11.105}$$

$$\ddot{\mathfrak{ott}}^\bullet_{\star\star} := -\sum_{1\le r} \frac{1}{(2r+1)!}\,\mathfrak{t}\ddot{\mathfrak{o}}^\bullet_{2r}. \tag{11.106}$$

But since $\ddot{\mathfrak{ott}}^\bullet_\star$ and $\ddot{\mathfrak{ott}}^\bullet_{\star\star}$ have only non-vanishing components of *even* length, (11.104) shows that the same must hold for $\ddot{\mathfrak{oss}}^\bullet_{\star\star}$. Reverting to the factorisation (11.101) and the separation identity (11.99) and using the invariance of $\ddot{\mathfrak{oss}}^\bullet$, we deduce from all this:

$$\text{mu}(\text{pari}.\text{anti}.\ddot{\mathfrak{oss}}^\bullet\,,\, \ddot{\mathfrak{oss}}^\bullet) = 1^\bullet \tag{11.107}$$

(11.107) expresses the *gantar*-invariance of $\ddot{\mathfrak{o}}\mathfrak{s}\mathfrak{s}^\bullet$ and (11.100) expresses its *gus*-neutrality. In other words, $\ddot{\mathfrak{o}}\mathfrak{s}\mathfrak{s}^\bullet$ possesses *the two fundamental subsymmetries implied by symmetrality.* Yet this still doesn't imply full symmetrality. Fortunately, two crucial facts save the situation:
(i) since $\ddot{\mathfrak{o}}\mathfrak{s}\mathfrak{s}^\bullet$ has a swappee $\mathfrak{e}\mathfrak{s}\mathfrak{s}^\bullet$ that is obviously symmetral, and therefore *gantar*-invariant, the *gantar*-invariance of $\ddot{\mathfrak{o}}\mathfrak{s}\mathfrak{s}^\bullet$, in view of the factorisation (4.46), also implies its invariance under *neg.gush* or, what here amounts to the same, *pari.gush*;
(ii) between themselves, the *neg.gush*-invariance and *gus*-neutrality of $\ddot{\mathfrak{o}}\mathfrak{s}\mathfrak{s}^\bullet$ ensure its symmetrality.[173]

This fact is akin to the analogous implication valid in the algebras:

$$\{\textit{pus}\text{-neutrality} + \textit{push}\text{- or } \textit{neg-push}\text{-invariance}\} \Rightarrow \{\text{alternality}\}.$$

Ultimately, it rests on the fact that *pus* and *push*, interpreted in the *short* and *long* notations,[174] amount to circular permutations of order r and $r+1$ respectively, which together generate the full symmetric group $\mathfrak{S}_{r+1}$. More precisely, each $\sigma \in \mathfrak{S}_{r+1}$ can be written as a product $\alpha^{m_1}\beta^{n_1}\dots\alpha^{m_{r-1}}\beta^{n_{r-1}}$ with $\alpha = \textit{pus}$ and $\beta = \textit{push}$.

11.10 Bisymmetrality of $\mathfrak{e}\mathfrak{s}\mathfrak{s}^\bullet$: combinatorial proof

This alternative proof uses the inductive expression of $\ddot{\mathfrak{o}}\mathfrak{s}\mathfrak{s}^\bullet$ in terms of its dilators $\ddot{\mathfrak{o}}\mathfrak{t}\mathfrak{t}^\bullet$ (direct) and $\ddot{\mathfrak{o}}\mathfrak{d}\mathfrak{d}^\bullet$ (inverse). Explicitely:

$$r(\bullet)\,\ddot{\mathfrak{o}}\mathfrak{s}\mathfrak{s}^\bullet = +\text{preiwa}(\ddot{\mathfrak{o}}\mathfrak{s}\mathfrak{s}^\bullet, \ddot{\mathfrak{o}}\mathfrak{t}\mathfrak{t}^\bullet) \tag{11.108}$$

$$r(\bullet)\,\ddot{\mathfrak{o}}\mathfrak{s}\mathfrak{s}^\bullet = -\text{giwa}(\ddot{\mathfrak{o}}\mathfrak{d}\mathfrak{d}^\bullet, \ddot{\mathfrak{o}}\mathfrak{s}\mathfrak{s}^\bullet) \tag{11.109}$$

with

$$\ddot{\mathfrak{o}}\mathfrak{t}\mathfrak{t}^\bullet := \text{swap}.\mathfrak{e}\mathfrak{t}\mathfrak{t}^\bullet \quad \text{and} \quad \mathfrak{e}\mathfrak{t}\mathfrak{t}^\bullet := -\sum_{1\le r}\frac{1}{(r+1)!}\,\mathfrak{t}\mathfrak{e}_r^\bullet \tag{11.110}$$

$$\ddot{\mathfrak{o}}\mathfrak{d}\mathfrak{d}^\bullet := \text{swap}.\mathfrak{e}\mathfrak{d}\mathfrak{d}^\bullet \quad \text{and} \quad \mathfrak{e}\mathfrak{d}\mathfrak{d}^\bullet := +\sum_{1\le r}\frac{1}{r\,(r+1)}\,\mathfrak{t}\mathfrak{e}_r^\bullet. \tag{11.111}$$

173 Which *gantar*-invariance + *gus*-neutrality do not!

174 See at the beginning of Section 5.1, right before (5.2).

These identities flow from the fact that the preimage of $\mathfrak{ess}^\bullet$ in *GIFF* is the diffeo $f(x) := 1 - e^{-x}$ with a reciprocal diffeo $f^{-1}(x) = -\log(1-x)$. The corresponding dilators therefore admit the expansions

$$f_{\#}(x) = 1 + x - e^x \qquad = -\sum_{1\le r} \frac{1}{(r+1)!}\, x^{r+1} \tag{11.112}$$

$$(f^{-1})_{\#}(x) = x + (1-x)\,\log(1-x) = +\sum_{1\le r} \frac{1}{r\,(r+1)}\, x^{r+1} \tag{11.113}$$

which provide us with the defining coefficients of $\mathfrak{ett}^\bullet$ and $\mathfrak{edd}^\bullet$.

On the face of it, relation (11.108), being linear in $\ddot{\mathfrak{o}}\mathfrak{ss}^\bullet$, would seem a more promising starting point than relation (11.109), whose right-hand side is heavily non-linear in $\ddot{\mathfrak{o}}\mathfrak{ss}^\bullet$. This appearance is deceptive, though, because the bimould $\mathfrak{ett}^\bullet$ possesses only a simple symmetry (alternal), unlike the bimould $\mathfrak{edd}^\bullet$, which possesses a double one: it is alternal, with an $\mathfrak{O}$-alternal swappee, as already observed in Section 4.1. Indeed, $\mathfrak{edd}^\bullet$ coincides with the bimould $\mathfrak{ste}^\bullet$ of (4.6). We shall therefore take our stand on (11.109) rather than (11.108). But first we require a general bimould identity.

For any two bimoulds $S^\bullet$, $D^\bullet$ in $BIMU^* \times BIMU_*$, *i.e.* such that $S^\emptyset = 1$ and $D^\emptyset = 0$, we introduce the following abbreviations

$$\boldsymbol{S}^{\{\{\boldsymbol{w}^1;\,\boldsymbol{w}^2\}\}} := -S^{\boldsymbol{w}^1} S^{\boldsymbol{w}^2} + \sum_{\boldsymbol{w}\in \mathrm{sha}(\boldsymbol{w}^1;\,\boldsymbol{w}^2)} S^{\boldsymbol{w}} \tag{11.114}$$

$$\boldsymbol{D}^{[[\boldsymbol{w}^1;\,\boldsymbol{w}^2]]} := \Big[\sum_{\boldsymbol{w}\in \mathrm{sh}\mathfrak{o}(\boldsymbol{w}^1;\,\boldsymbol{w}^2)} D^{\boldsymbol{w}}\Big]_{\mathfrak{O}^\bullet = -2\,S_1^\bullet} \tag{11.115}$$

$$\boldsymbol{S}^{\{\boldsymbol{w}\}} := \Big[\mathrm{mu}(S^\bullet, \mathrm{anti}.S^\bullet) + \mathrm{giwat}(S^\bullet).\mathfrak{O}^\bullet\Big]_{\mathfrak{O}^\bullet = -2\,S_1^\bullet}. \tag{11.116}$$

In all the above, $S_1^\bullet$ denotes the projection of $S^\bullet$ onto $BIMU_1$, and the interpretation of the three symbols is as follows:
(i) $\boldsymbol{S}^{\{\{\bullet\,;\,\bullet\}\}}$ measures the failure of $S^\bullet$ to be symmetral;
(ii) $\boldsymbol{D}^{[[\bullet\,;\,\bullet]]}$ measures the failure of $D^\bullet$ to be $\mathfrak{O}$-alternal, with $\mathfrak{O}$-alternality defined as in Section 3.4, but *after replacement of the flexion unit $\mathfrak{O}^\bullet$ by $-2S_1^\bullet$, which is not required to be a unit!*;
(iii) $\boldsymbol{S}^{\{\bullet\}}$ measure the failure of $S^\bullet$ to verify a property closely related to *gantar*-invariance, which is a subsymmetry of symmetrality.

Thus, for $r(\boldsymbol{w}^1) = r(\boldsymbol{w}^2) = 1$ and for any $\boldsymbol{w}$, we get (mark the signs and the position of *anti*):

$$\boldsymbol{S}^{\{\!\{(w_1)\,;\,(w_2)\}\!\}} = -S^{w_1}\,S^{w_2} + S^{w_1,w_2} + S^{w_2,w_1}$$

$$\boldsymbol{D}^{[[(w_1)\,;\,(w_2)]]} = D^{w_1,w_2} + D^{w_2,w_1} + 2\,D^{w_1\rceil}\,S^{\lfloor w_2} + 2\,D^{w_2\rceil}\,S^{\lfloor w_1}$$

$$\boldsymbol{S}^{\{\boldsymbol{w}\}} = \sum_{\boldsymbol{w}^1.\boldsymbol{w}^2=\boldsymbol{w}} S^{\boldsymbol{w}^1}(\text{anti}.S)^{\boldsymbol{w}^2} - 2\sum_{\boldsymbol{w}^1.w_0.\boldsymbol{w}^2=\boldsymbol{w}} S^{\boldsymbol{w}^1\rfloor}S^{\lceil w_0\rceil}(\text{anti}.S)^{\lfloor\boldsymbol{w}^2}.$$

We now require the following lemma:

If the bimoulds $S^\bullet$, $D^\bullet$ are related under the identity:[175]

$$-r(\bullet)\,S^\bullet = \text{giwa}(D^\bullet, S^\bullet) \tag{11.117}$$

then for any two $\mathbf{w}^1$, $\mathbf{w}^2$ *the identity holds*:

$$0 = (r_1 + r_2)\,\boldsymbol{S}^{\{\!\{\boldsymbol{w}^1;\,\boldsymbol{w}^2\}\!\}} + \boldsymbol{D}^{[[\boldsymbol{w}^1;\,\boldsymbol{w}^2]]} + \boldsymbol{\Sigma}_1 + \boldsymbol{\Sigma}_2 + \boldsymbol{\Sigma}_3 \tag{11.118}$$

(i) *with a sum* $\boldsymbol{\Sigma}_1$ *linear in earlier terms* $\boldsymbol{D}^{[[\boldsymbol{w}';\,\boldsymbol{w}'']]}$ *and multilinear in earlier terms* $S^{\boldsymbol{w}^*}$, *"earlier" meaning that* $r' + r''$ *and* r^* *are always* $< r_1 + r_2$;
(ii) *with a sum* $\boldsymbol{\Sigma}_2$ *bilinear in earlier terms* $\boldsymbol{S}^{\{\!\{\boldsymbol{w}';\,\boldsymbol{w}''\}\!\}}$, $D^{\boldsymbol{w}'''}$ *and multilinear in earlier terms* $S^{\boldsymbol{w}^*}$;
(iii) *with a sum* $\boldsymbol{\Sigma}_3$ *bilinear in earlier terms* $\boldsymbol{S}^{\{\boldsymbol{w}'\}}$, $D^{\boldsymbol{w}''}$ *and multilinear in earlier terms* $S^{\boldsymbol{w}^*}$.
Moreover, in all three sums, the coefficients in front of the monomials made up of 'earlier' terms are always equal to $+1$.

The way to prove (11.118) is:
(i) to start from the identity

$$(r_1{+}r_2)\,\boldsymbol{S}^{\{\!\{\boldsymbol{w}^1\,;\,\boldsymbol{w}^2\}\!\}} = -(r_1\,S^{\boldsymbol{w}^1})\,S^{\boldsymbol{w}^2} - S^{\boldsymbol{w}^1}\,(r_2\,S^{\boldsymbol{w}^2}) + \sum_{\boldsymbol{w}^1.\boldsymbol{w}^2=\boldsymbol{w}} (r_1{+}r_2)\,S^{\boldsymbol{w}};$$

(ii) to replace therein all terms of the form $r(\bullet)\,S^\bullet$ by $-giwa(D^\bullet, S^\bullet)$;
(iii) to replace (- this clearly is the crucial step -) the usual definition of *giwa* for totally ordered sequences $\boldsymbol{w}$ by an analogous expression valid

[175] We recall that *GIWA* is the unary subgroup of *GAXI* relative to the involution $\mathcal{M}_R = anti.\mathcal{M}_L$. Under normal circumstances, $giwa(A^\bullet, B^\bullet)$ has its two arguments $A^\bullet$, $B^\bullet$ in *BIMU**. Here, however, we have to consider giwa$(D^\bullet, S^\bullet)$, with a first argument in $BIMU_*$, but we can take recourse to the usual definition $giwa(D^\bullet, S^\bullet) := mu(giwat(S^\bullet).D^\bullet, S^\bullet)$, which still makes perfect sense.

for sequences $\boldsymbol{w}$ carrying a weaker, arborescent order[176] – in the present instance, for sequences $\boldsymbol{w}$ consisting of two totally ordered, but mutually non comparable branches $\boldsymbol{w}^{\mathbf{1}}$, $\boldsymbol{w}^{\mathbf{2}}$.

Thus, in the (very elementary) case $r_1 = 1, r_2 = 2$, we find

$$0 = (1+2)\, \boldsymbol{S}^{\{\!\{\binom{u_1}{v_1};\left(\begin{smallmatrix} u_2 & , & u_3 \\ v_2 & , & v_3\end{smallmatrix}\right)\}\!\}} + \boldsymbol{D}^{[[\binom{u_1}{v_1};\left(\begin{smallmatrix} u_2 & , & u_3 \\ v_2 & , & v_3\end{smallmatrix}\right)]]} + \boldsymbol{\Sigma_1} + \boldsymbol{\Sigma_2} + \boldsymbol{\Sigma_3}$$

with

$$\begin{aligned}
\boldsymbol{\Sigma_1} &= \boldsymbol{D}^{[[\binom{u_1}{v_1};\binom{u_2}{v_2}]]}\, S^{\binom{u_3}{v_3}} + \boldsymbol{D}^{[[\binom{u_1}{v_1};\binom{u_{23}}{v_3}]]}\, S^{\binom{u_2}{v_{2:3}}} + \boldsymbol{D}^{[[\binom{u_1}{v_1};\binom{u_{23}}{v_2}]]}\, S^{\binom{u_3}{v_{3:2}}}\\
\boldsymbol{\Sigma_2} &= \boldsymbol{S}^{\{\!\{\binom{u_1}{v_1};\binom{u_3}{v_3}\}\!\}}\, S^{\binom{u_2}{v_2}} + \boldsymbol{S}^{\{\!\{\binom{u_1}{v_{1:3}};\binom{u_2}{v_{2:3}}\}\!\}}\, S^{\binom{u_{123}}{v_3}} + \boldsymbol{S}^{\{\!\{\binom{u_1}{v_{1:2}};\binom{u_3}{v_{3:2}}\}\!\}}\, S^{\binom{u_{123}}{v_2}}\\
\boldsymbol{\Sigma_3} &= \boldsymbol{S}^{\left\{\begin{smallmatrix} u_2 & , & u_3 \\ v_{2:1} & , & v_{3:1}\end{smallmatrix}\right\}}\, \boldsymbol{T}^{\binom{u_{123}}{v_1}}.
\end{aligned}$$

At this point, all we have to do is:
(i) replace $S^\bullet$ by $\ddot{\mathfrak{o}}\mathfrak{ss}^\bullet$ and $D^\bullet$ by $\ddot{\mathfrak{o}}\mathfrak{dd}^\bullet$ in (11.118);
(ii) observe that since in this case $\mathfrak{D}^\bullet = -2\,\ddot{\mathfrak{o}}\mathfrak{dd}^\bullet_1$, all terms $\boldsymbol{D}^{[[\bullet\,,\,\bullet]]}$ automatically vanish, since $D^\bullet \equiv \ddot{\mathfrak{o}}\mathfrak{dd}^\bullet$ is indeed $\mathfrak{D}$-alternal;
(iii) observe that the identities $\boldsymbol{S}^{\{\bullet\}} = 0$ (up to length $r-1$) are an easy consequence of the symmetrality of $S^\bullet \equiv \ddot{\mathfrak{o}}\mathfrak{ss}^\bullet$ (up to length $r-1$) and of the factorisation (4.46). Besides, these identities $\boldsymbol{S}^{\{\bullet\}} = 0$ are also capable of an elementary, direct derivation, as we saw towards the end of Section 11.7: see (11.89).

Altogether, the identity (11.118) shows that if $\ddot{\mathfrak{o}}\mathfrak{ss}^\bullet$ is symmetral up to length $r-1$, it is automatically symmetral up to length r. □

There exist several other strategies for establishing the symmetrality of $\ddot{\mathfrak{o}}\mathfrak{ss}^\bullet$, all of more or less equal length,[177] but apparently no completely elementary proof.

12 Tables, index, references

12.1 Table 1: basis for *Flex*($\mathfrak{E}$)

Here are the bases of the first cells of the free *monogenous* structure $\oplus Flex_r(\mathfrak{E})$ generated by a general $\mathfrak{E}$ subject only to one of the four possible parity constraints (3.1): it doesn't matter which. By retaining only

[176] This, of course, does not apply for *giwa* alone: *all* flexion operations without exception extend to the case of arborescent sequences $\boldsymbol{w}$, provided we suitably redefine the product *mu* and the four flexions ⌋, ⌈, ⌉, ⌊ in accordance with the new order.

[177] Thus, there exists a heavily calculational proof based on formula (12.6) of Section 12.3.

the first $\frac{(2r)!}{(r+1)!r!}$ elements, one also obtains bases for the *eumonogenous* structure $\oplus Flex_r(\mathfrak{E})$ generated by an exact flexion unit $\mathfrak{E}$.

$$
\begin{array}{lll|lll}
\mathfrak{e}_{1,1}^{w_1} & := & \mathfrak{E}^{\binom{u_1}{v_1}} & & & \\
\mathfrak{e}_{2,1}^{w_1,w_2} & := & \mathfrak{E}^{\binom{u_{12}}{v_2}}\mathfrak{E}^{\binom{u_1}{v_{1:2}}} & \mathfrak{e}_{2,3}^{w_1,w_2} & := & \mathfrak{E}^{\binom{u_1}{v_1}}\mathfrak{E}^{\binom{u_2}{v_2}} \\
\mathfrak{e}_{2,2}^{w_1,w_2} & := & \mathfrak{E}^{\binom{u_{12}}{v_1}}\mathfrak{E}^{\binom{u_2}{v_{2:1}}} & & & \\
\mathfrak{e}_{3,1}^{w_1,w_2,w_3} & := & \mathfrak{E}^{\binom{u_{123}}{v_3}}\,\mathfrak{E}^{\binom{u_{12}}{v_{2:3}}}\,\mathfrak{E}^{\binom{u_1}{v_{1:2}}} & \mathfrak{e}_{3,6}^{w_1,w_2,w_3} & := & \mathfrak{E}^{\binom{u_{123}}{v_1}}\mathfrak{E}^{\binom{u_2}{v_{2:1}}}\mathfrak{E}^{\binom{u_3}{v_{3:1}}} \\
\mathfrak{e}_{3,2}^{w_1,w_2,w_3} & := & \mathfrak{E}^{\binom{u_{123}}{v_3}}\,\mathfrak{E}^{\binom{u_{12}}{v_{1:3}}}\,\mathfrak{E}^{\binom{u_2}{v_{2:1}}} & \mathfrak{e}_{3,7}^{w_1,w_2,w_3} & := & \mathfrak{E}^{\binom{u_{123}}{v_3}}\mathfrak{E}^{\binom{u_1}{v_{1:3}}}\mathfrak{E}^{\binom{u_2}{v_{2:3}}} \\
\mathfrak{e}_{3,3}^{w_1,w_2,w_3} & := & \mathfrak{E}^{\binom{u_{123}}{v_2}}\,\mathfrak{E}^{\binom{u_1}{v_{1:2}}}\,\mathfrak{E}^{\binom{u_3}{v_{3:2}}} & \mathfrak{e}_{3,8}^{w_1,w_2,w_3} & := & \mathfrak{E}^{\binom{u_{23}}{v_2}}\,\mathfrak{E}^{\binom{u_1}{v_1}}\,\mathfrak{E}^{\binom{u_3}{v_{3:2}}} \\
\mathfrak{e}_{3,4}^{w_1,w_2,w_3} & := & \mathfrak{E}^{\binom{u_{123}}{v_1}}\,\mathfrak{E}^{\binom{u_{23}}{v_{3:1}}}\,\mathfrak{E}^{\binom{u_2}{v_{2:3}}} & \mathfrak{e}_{3,9}^{w_1,w_2,w_3} & := & \mathfrak{E}^{\binom{u_{23}}{v_3}}\,\mathfrak{E}^{\binom{u_1}{v_1}}\,\mathfrak{E}^{\binom{u_2}{v_{2:3}}} \\
\mathfrak{e}_{3,5}^{w_1,w_2,w_3} & := & \mathfrak{E}^{\binom{u_{123}}{v_1}}\,\mathfrak{E}^{\binom{u_{23}}{v_{2:1}}}\,\mathfrak{E}^{\binom{u_3}{v_{3:2}}} & \mathfrak{e}_{3,10}^{w_1,w_2,w_3} & := & \mathfrak{E}^{\binom{u_{12}}{v_2}}\,\mathfrak{E}^{\binom{u_1}{v_{1:2}}}\,\mathfrak{E}^{\binom{u_3}{v_3}} \\
 & & & \mathfrak{e}_{3,11}^{w_1,w_2,w_3} & := & \mathfrak{E}^{\binom{u_{12}}{v_1}}\,\mathfrak{E}^{\binom{u_2}{v_{2:1}}}\,\mathfrak{E}^{\binom{u_3}{v_3}} \\
 & & & \mathfrak{e}_{3,12}^{w_1,w_2,w_3} & := & \mathfrak{E}^{\binom{u_1}{v_1}}\,\mathfrak{E}^{\binom{u_2}{v_2}}\,\mathfrak{E}^{\binom{u_3}{v_3}}.
\end{array}
$$

Here follows the graphic interpretation of the bases, with full lines for the graphs $\boldsymbol{gi}$ and broken lines for the graphs $\boldsymbol{ga}$ (see Section 3.3).

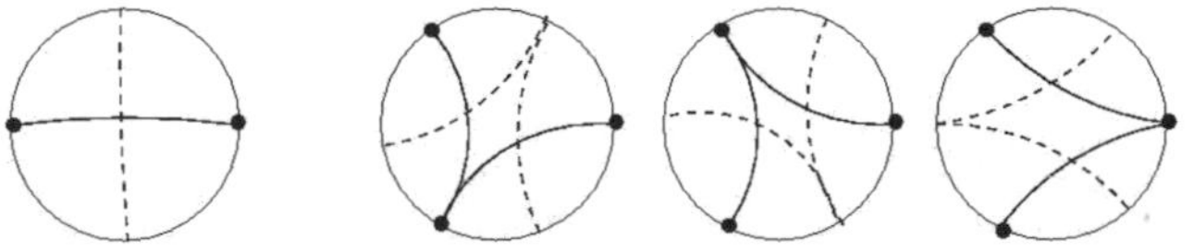

Figure 12.1. Length $r = 1, 2$. Basis vectors $\{\mathfrak{e}^{\bullet}_{1,1}\}$ and $\{\mathfrak{e}^{\bullet}_{2,1}, \mathfrak{e}^{\bullet}_{2,2}\} \cup \{\mathfrak{e}^{\bullet}_{2,3}\}$.

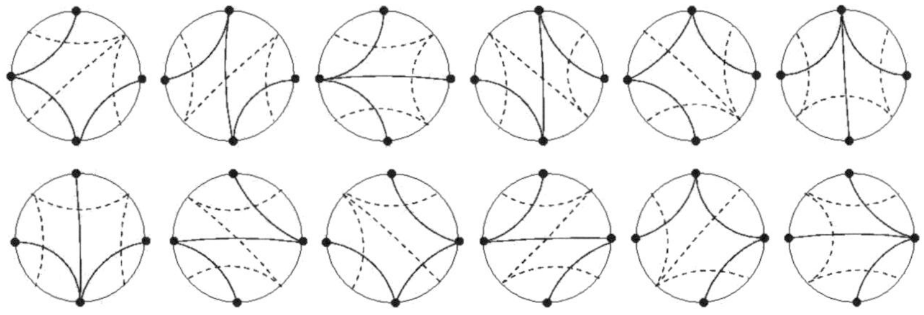

Figure 12.2. Length $r = 3$. Basis vectors $\{\mathfrak{e}^{\bullet}_{3,1}, \dots, \mathfrak{e}^{\bullet}_{3,5}\} \cup \{\mathfrak{e}^{\bullet}_{3,6}, \dots, \mathfrak{e}^{\bullet}_{3,12}\}$.

$$\mathfrak{e}^{w_1..w_4}_{4,1}=\mathfrak{E}\binom{u_{1}}{v_{1:2}}\mathfrak{E}\binom{u_{12}}{v_{2:3}}\mathfrak{E}\binom{u_{123}}{v_{3:4}}\mathfrak{E}\binom{u_{1234}}{v_{4}}\;\|\;\mathfrak{e}^{w_1..w_4}_{4,8}=\mathfrak{E}\binom{u_{1}}{v_{1:2}}\mathfrak{E}\binom{u_{1234}}{v_{2}}\mathfrak{E}\binom{u_{3}}{v_{3:4}}\mathfrak{E}\binom{u_{34}}{v_{4:2}}$$
$$\mathfrak{e}^{w_1..w_4}_{4,2}=\mathfrak{E}\binom{u_{12}}{v_{1:3}}\mathfrak{E}\binom{u_{2}}{v_{2:1}}\mathfrak{E}\binom{u_{123}}{v_{3:4}}\mathfrak{E}\binom{u_{1234}}{v_{4}}\;\|\;\mathfrak{e}^{w_1..w_4}_{4,9}=\mathfrak{E}\binom{u_{1}}{v_{1:2}}\mathfrak{E}\binom{u_{1234}}{v_{2}}\mathfrak{E}\binom{u_{34}}{v_{3:2}}\mathfrak{E}\binom{u_{4}}{v_{4:3}}$$
$$\mathfrak{e}^{w_1..w_4}_{4,3}=\mathfrak{E}\binom{u_{1}}{v_{1:2}}\mathfrak{E}\binom{u_{123}}{v_{2:4}}\mathfrak{E}\binom{u_{3}}{v_{3:2}}\mathfrak{E}\binom{u_{1234}}{v_{4}}\;\|\;\mathfrak{e}^{w_1..w_4}_{4,10}=\mathfrak{E}\binom{u_{1234}}{v_{1}}\mathfrak{E}\binom{u_{2}}{v_{2:3}}\mathfrak{E}\binom{u_{23}}{v_{3:4}}\mathfrak{E}\binom{u_{234}}{v_{4:1}}$$
$$\mathfrak{e}^{w_1..w_4}_{4,4}=\mathfrak{E}\binom{u_{123}}{v_{1:4}}\mathfrak{E}\binom{u_{2}}{v_{2:3}}\mathfrak{E}\binom{u_{23}}{v_{3:1}}\mathfrak{E}\binom{u_{1234}}{v_{4}}\;\|\;\mathfrak{e}^{w_1..w_4}_{4,11}=\mathfrak{E}\binom{u_{1234}}{v_{1}}\mathfrak{E}\binom{u_{23}}{v_{2:4}}\mathfrak{E}\binom{u_{3}}{v_{3:2}}\mathfrak{E}\binom{u_{234}}{v_{4:1}}$$
$$\mathfrak{e}^{w_1..w_4}_{4,5}=\mathfrak{E}\binom{u_{123}}{v_{1:4}}\mathfrak{E}\binom{u_{23}}{v_{2:1}}\mathfrak{E}\binom{u_{3}}{v_{3:2}}\mathfrak{E}\binom{u_{1234}}{v_{4}}\;\|\;\mathfrak{e}^{w_1..w_4}_{4,12}=\mathfrak{E}\binom{u_{1234}}{v_{1}}\mathfrak{E}\binom{u_{2}}{v_{2:3}}\mathfrak{E}\binom{u_{234}}{v_{3:1}}\mathfrak{E}\binom{u_{4}}{v_{4:3}}$$
$$\mathfrak{e}^{w_1..w_4}_{4,6}=\mathfrak{E}\binom{u_{1}}{v_{1:2}}\mathfrak{E}\binom{u_{12}}{v_{2:3}}\mathfrak{E}\binom{u_{1234}}{v_{3}}\mathfrak{E}\binom{u_{4}}{v_{4:3}}\;\|\;\mathfrak{e}^{w_1..w_4}_{4,13}=\mathfrak{E}\binom{u_{1234}}{v_{1}}\mathfrak{E}\binom{u_{234}}{v_{2:1}}\mathfrak{E}\binom{u_{3}}{v_{3:4}}\mathfrak{E}\binom{u_{34}}{v_{4:2}}$$
$$\mathfrak{e}^{w_1..w_4}_{4,7}=\mathfrak{E}\binom{u_{12}}{v_{1:3}}\mathfrak{E}\binom{u_{2}}{v_{2:1}}\mathfrak{E}\binom{u_{1234}}{v_{3}}\mathfrak{E}\binom{u_{4}}{v_{4:3}}\;\|\;\mathfrak{e}^{w_1..w_4}_{4,14}=\mathfrak{E}\binom{u_{1234}}{v_{1}}\mathfrak{E}\binom{u_{234}}{v_{2:1}}\mathfrak{E}\binom{u_{34}}{v_{3:2}}\mathfrak{E}\binom{u_{4}}{v_{4:3}}$$

$$\mathfrak{e}^{w_1..w_4}_{4,15}=\mathfrak{E}\binom{u_{1}}{v_{1:2}}\mathfrak{E}\binom{u_{12}}{v_{2}}\mathfrak{E}\binom{u_{3}}{v_{3}}\mathfrak{E}\binom{u_{4}}{v_{4}}\;\|\;\mathfrak{e}^{w_1..w_4}_{4,36}=\mathfrak{E}\binom{u_{12}}{v_{1:3}}\mathfrak{E}\binom{u_{2}}{v_{2:1}}\mathfrak{E}\binom{u_{123}}{v_{3}}\mathfrak{E}\binom{u_{4}}{v_{4}}$$
$$\mathfrak{e}^{w_1..w_4}_{4,16}=\mathfrak{E}\binom{u_{12}}{v_{1}}\mathfrak{E}\binom{u_{2}}{v_{2:1}}\mathfrak{E}\binom{u_{34}}{v_{3}}\mathfrak{E}\binom{u_{4}}{v_{4:3}}\;\|\;\mathfrak{e}^{w_1..w_4}_{4,37}=\mathfrak{E}\binom{u_{1}}{v_{1:2}}\mathfrak{E}\binom{u_{123}}{v_{2}}\mathfrak{E}\binom{u_{3}}{v_{3:2}}\mathfrak{E}\binom{u_{4}}{v_{4}}$$
$$\mathfrak{e}^{w_1..w_4}_{4,17}=\mathfrak{E}\binom{u_{12}}{v_{1}}\mathfrak{E}\binom{u_{2}}{v_{2:1}}\mathfrak{E}\binom{u_{3}}{v_{3}}\mathfrak{E}\binom{u_{4}}{v_{4}}\;\|\;\mathfrak{e}^{w_1..w_4}_{4,38}=\mathfrak{E}\binom{u_{123}}{v_{1}}\mathfrak{E}\binom{u_{2}}{v_{2:1}}\mathfrak{E}\binom{u_{3}}{v_{3:1}}\mathfrak{E}\binom{u_{4}}{v_{4}}$$
$$\mathfrak{e}^{w_1..w_4}_{4,18}=\mathfrak{E}\binom{u_{1}}{v_{1:2}}\mathfrak{E}\binom{u_{12}}{v_{2}}\mathfrak{E}\binom{u_{34}}{v_{3}}\mathfrak{E}\binom{u_{4}}{v_{4:3}}\;\|\;\mathfrak{e}^{w_1..w_4}_{4,39}=\mathfrak{E}\binom{u_{123}}{v_{1}}\mathfrak{E}\binom{u_{2}}{v_{2:3}}\mathfrak{E}\binom{u_{23}}{v_{3:1}}\mathfrak{E}\binom{u_{4}}{v_{4}}$$
$$\mathfrak{e}^{w_1..w_4}_{4,19}=\mathfrak{E}\binom{u_{1}}{v_{1:2}}\mathfrak{E}\binom{u_{12}}{v_{2}}\mathfrak{E}\binom{u_{3}}{v_{3:4}}\mathfrak{E}\binom{u_{34}}{v_{4}}\;\|\;\mathfrak{e}^{w_1..w_4}_{4,40}=\mathfrak{E}\binom{u_{1234}}{v_{1}}\mathfrak{E}\binom{u_{2}}{v_{2:1}}\mathfrak{E}\binom{u_{3}}{v_{3:1}}\mathfrak{E}\binom{u_{4}}{v_{4:1}}$$
$$\mathfrak{e}^{w_1..w_4}_{4,20}=\mathfrak{E}\binom{u_{1}}{v_{1}}\mathfrak{E}\binom{u_{234}}{v_{2}}\mathfrak{E}\binom{u_{3}}{v_{3:4}}\mathfrak{E}\binom{u_{34}}{v_{4:2}}\;\|\;\mathfrak{e}^{w_1..w_4}_{4,41}=\mathfrak{E}\binom{u_{1}}{v_{1:2}}\mathfrak{E}\binom{u_{1234}}{v_{2}}\mathfrak{E}\binom{u_{3}}{v_{3:2}}\mathfrak{E}\binom{u_{4}}{v_{4:2}}$$
$$\mathfrak{e}^{w_1..w_4}_{4,21}=\mathfrak{E}\binom{u_{123}}{v_{1}}\mathfrak{E}\binom{u_{23}}{v_{2:1}}\mathfrak{E}\binom{u_{3}}{v_{3:2}}\mathfrak{E}\binom{u_{4}}{v_{4}}\;\|\;\mathfrak{e}^{w_1..w_4}_{4,42}=\mathfrak{E}\binom{u_{1}}{v_{1:3}}\mathfrak{E}\binom{u_{2}}{v_{2:3}}\mathfrak{E}\binom{u_{1234}}{v_{3}}\mathfrak{E}\binom{u_{4}}{v_{4:3}}$$
$$\mathfrak{e}^{w_1..w_4}_{4,22}=\mathfrak{E}\binom{u_{12}}{v_{1}}\mathfrak{E}\binom{u_{2}}{v_{2:1}}\mathfrak{E}\binom{u_{3}}{v_{3:4}}\mathfrak{E}\binom{u_{34}}{v_{4}}\;\|\;\mathfrak{e}^{w_1..w_4}_{4,43}=\mathfrak{E}\binom{u_{1234}}{v_{1}}\mathfrak{E}\binom{u_{234}}{v_{2:1}}\mathfrak{E}\binom{u_{3}}{v_{3:2}}\mathfrak{E}\binom{u_{4}}{v_{4:2}}$$
$$\mathfrak{e}^{w_1..w_4}_{4,23}=\mathfrak{E}\binom{u_{1}}{v_{1}}\mathfrak{E}\binom{u_{2}}{v_{2}}\mathfrak{E}\binom{u_{3}}{v_{3:4}}\mathfrak{E}\binom{u_{34}}{v_{4}}\;\|\;\mathfrak{e}^{w_1..w_4}_{4,44}=\mathfrak{E}\binom{u_{1234}}{v_{1}}\mathfrak{E}\binom{u_{23}}{v_{2:1}}\mathfrak{E}\binom{u_{3}}{v_{3:2}}\mathfrak{E}\binom{u_{4}}{v_{4:1}}$$
$$\mathfrak{e}^{w_1..w_4}_{4,24}=\mathfrak{E}\binom{u_{1}}{v_{1}}\mathfrak{E}\binom{u_{2}}{v_{2}}\mathfrak{E}\binom{u_{3}}{v_{3}}\mathfrak{E}\binom{u_{4}}{v_{4}}\;\|\;\mathfrak{e}^{w_1..w_4}_{4,45}=\mathfrak{E}\binom{u_{1234}}{v_{1}}\mathfrak{E}\binom{u_{2}}{v_{2:3}}\mathfrak{E}\binom{u_{23}}{v_{3:1}}\mathfrak{E}\binom{u_{4}}{v_{4:1}}$$
$$\mathfrak{e}^{w_1..w_4}_{4,25}=\mathfrak{E}\binom{u_{1}}{v_{1}}\mathfrak{E}\binom{u_{2}}{v_{2}}\mathfrak{E}\binom{u_{34}}{v_{3}}\mathfrak{E}\binom{u_{4}}{v_{4:3}}\;\|\;\mathfrak{e}^{w_1..w_4}_{4,46}=\mathfrak{E}\binom{u_{1234}}{v_{1}}\mathfrak{E}\binom{u_{2}}{v_{2:4}}\mathfrak{E}\binom{u_{3}}{v_{3:4}}\mathfrak{E}\binom{u_{234}}{v_{4:1}}$$
$$\mathfrak{e}^{w_1..w_4}_{4,26}=\mathfrak{E}\binom{u_{1}}{v_{1}}\mathfrak{E}\binom{u_{2}}{v_{2:4}}\mathfrak{E}\binom{u_{3}}{v_{3:4}}\mathfrak{E}\binom{u_{234}}{v_{4}}\;\|\;\mathfrak{e}^{w_1..w_4}_{4,47}=\mathfrak{E}\binom{u_{1234}}{v_{1}}\mathfrak{E}\binom{u_{2}}{v_{2:1}}\mathfrak{E}\binom{u_{34}}{v_{3:1}}\mathfrak{E}\binom{u_{4}}{v_{4:3}}$$
$$\mathfrak{e}^{w_1..w_4}_{4,27}=\mathfrak{E}\binom{u_{1}}{v_{1}}\mathfrak{E}\binom{u_{2}}{v_{2:3}}\mathfrak{E}\binom{u_{23}}{v_{3}}\mathfrak{E}\binom{u_{4}}{v_{4}}\;\|\;\mathfrak{e}^{w_1..w_4}_{4,48}=\mathfrak{E}\binom{u_{1234}}{v_{1}}\mathfrak{E}\binom{u_{2}}{v_{2:1}}\mathfrak{E}\binom{u_{3}}{v_{3:4}}\mathfrak{E}\binom{u_{34}}{v_{4:1}}$$
$$\mathfrak{e}^{w_1..w_4}_{4,28}=\mathfrak{E}\binom{u_{1}}{v_{1}}\mathfrak{E}\binom{u_{23}}{v_{2}}\mathfrak{E}\binom{u_{3}}{v_{3:2}}\mathfrak{E}\binom{u_{4}}{v_{4}}\;\|\;\mathfrak{e}^{w_1..w_4}_{4,49}=\mathfrak{E}\binom{u_{1}}{v_{1:4}}\mathfrak{E}\binom{u_{2}}{v_{2:4}}\mathfrak{E}\binom{u_{3}}{v_{3:4}}\mathfrak{E}\binom{u_{1234}}{v_{4}}$$
$$\mathfrak{e}^{w_1..w_4}_{4,29}=\mathfrak{E}\binom{u_{1}}{v_{1}}\mathfrak{E}\binom{u_{2}}{v_{2:3}}\mathfrak{E}\binom{u_{23}}{v_{3:4}}\mathfrak{E}\binom{u_{234}}{v_{4}}\;\|\;\mathfrak{e}^{w_1..w_4}_{4,50}=\mathfrak{E}\binom{u_{123}}{v_{1:4}}\mathfrak{E}\binom{u_{2}}{v_{2:1}}\mathfrak{E}\binom{u_{3}}{v_{3:1}}\mathfrak{E}\binom{u_{1234}}{v_{4}}$$
$$\mathfrak{e}^{w_1..w_4}_{4,30}=\mathfrak{E}\binom{u_{1}}{v_{1}}\mathfrak{E}\binom{u_{23}}{v_{2:4}}\mathfrak{E}\binom{u_{3}}{v_{3:2}}\mathfrak{E}\binom{u_{234}}{v_{4}}\;\|\;\mathfrak{e}^{w_1..w_4}_{4,51}=\mathfrak{E}\binom{u_{1}}{v_{1:3}}\mathfrak{E}\binom{u_{2}}{v_{2:3}}\mathfrak{E}\binom{u_{123}}{v_{3:4}}\mathfrak{E}\binom{u_{1234}}{v_{4}}$$
$$\mathfrak{e}^{w_1..w_4}_{4,31}=\mathfrak{E}\binom{u_{1}}{v_{1}}\mathfrak{E}\binom{u_{2}}{v_{2:3}}\mathfrak{E}\binom{u_{234}}{v_{3}}\mathfrak{E}\binom{u_{4}}{v_{4:3}}\;\|\;\mathfrak{e}^{w_1..w_4}_{4,52}=\mathfrak{E}\binom{u_{1}}{v_{1:4}}\mathfrak{E}\binom{u_{23}}{v_{2:4}}\mathfrak{E}\binom{u_{3}}{v_{3:2}}\mathfrak{E}\binom{u_{1234}}{v_{4}}$$
$$\mathfrak{e}^{w_1..w_4}_{4,32}=\mathfrak{E}\binom{u_{1}}{v_{1}}\mathfrak{E}\binom{u_{234}}{v_{2}}\mathfrak{E}\binom{u_{3}}{v_{3:2}}\mathfrak{E}\binom{u_{4}}{v_{4:2}}\;\|\;\mathfrak{e}^{w_1..w_4}_{4,53}=\mathfrak{E}\binom{u_{1}}{v_{1:4}}\mathfrak{E}\binom{u_{2}}{v_{2:3}}\mathfrak{E}\binom{u_{23}}{v_{3:4}}\mathfrak{E}\binom{u_{1234}}{v_{4}}$$
$$\mathfrak{e}^{w_1..w_4}_{4,33}=\mathfrak{E}\binom{u_{1}}{v_{1}}\mathfrak{E}\binom{u_{234}}{v_{2}}\mathfrak{E}\binom{u_{34}}{v_{3:2}}\mathfrak{E}\binom{u_{4}}{v_{4:3}}\;\|\;\mathfrak{e}^{w_1..w_4}_{4,54}=\mathfrak{E}\binom{u_{12}}{v_{1:4}}\mathfrak{E}\binom{u_{2}}{v_{2:1}}\mathfrak{E}\binom{u_{3}}{v_{3:4}}\mathfrak{E}\binom{u_{1234}}{v_{4}}$$
$$\mathfrak{e}^{w_1..w_4}_{4,34}=\mathfrak{E}\binom{u_{1}}{v_{1:3}}\mathfrak{E}\binom{u_{2}}{v_{2:3}}\mathfrak{E}\binom{u_{123}}{v_{3}}\mathfrak{E}\binom{u_{4}}{v_{4}}\;\|\;\mathfrak{e}^{w_1..w_4}_{4,55}=\mathfrak{E}\binom{u_{1}}{v_{1:2}}\mathfrak{E}\binom{u_{12}}{v_{2:4}}\mathfrak{E}\binom{u_{3}}{v_{3:4}}\mathfrak{E}\binom{u_{1234}}{v_{4}}$$
$$\mathfrak{e}^{w_1..w_4}_{4,35}=\mathfrak{E}\binom{u_{1}}{v_{1:2}}\mathfrak{E}\binom{u_{12}}{v_{2:3}}\mathfrak{E}\binom{u_{123}}{v_{3}}\mathfrak{E}\binom{u_{4}}{v_{4}}\;\|.$$

We end with bases for the first cells of the structures $\oplus Flex_r(\mathfrak{E})$ and $\oplus Flex_r(\mathfrak{O})$ for an approximate flexion unit $\mathfrak{E}$ verifying the same tripartite equation (3.30) as Qaa_c and an approximate flexion unit $\mathfrak{O}$ verifying the same tripartite equation (3.31) as Qii_c. Here, $\mathfrak{d}$ denotes the discrete dirac multiplied by c. In other words: $\mathfrak{d}^t := c\,\delta(t)$.

$$
\begin{array}{lllll|lllll}
\mathfrak{ed}_1^{w_1} & =\mathfrak{E}^{\binom{u_1}{v_1}} & & & & \|\,\mathfrak{od}_1^{w_1} & =\mathfrak{O}^{\binom{u_1}{v_1}} & & &
\end{array}
$$

$$
\begin{array}{llll|llll}
\mathfrak{ed}_1^{w_1,w_2} & =\mathfrak{E}^{\binom{u_1}{v_{1:2}}} & \mathfrak{E}^{\binom{u_{12}}{v_2}} & & \|\,\mathfrak{od}_1^{w_1,w_2} & =\mathfrak{O}^{\binom{u_1}{v_{1:2}}} & \mathfrak{O}^{\binom{u_{12}}{v_2}} & \\
\mathfrak{ed}_2^{w_1,w_2} & =\mathfrak{E}^{\binom{u_{12}}{v_1}} & \mathfrak{E}^{\binom{u_2}{v_{2:1}}} & & \|\,\mathfrak{od}_2^{w_1,w_2} & =\mathfrak{O}^{\binom{u_{12}}{v_1}} & \mathfrak{O}^{\binom{u_2}{v_{2:1}}} & \\
\mathfrak{ed}_3^{w_1,w_2} & =\mathfrak{d}^{v_1} & \mathfrak{d}^{v_2} & & \|\,\mathfrak{od}_3^{w_1,w_2} & =\mathfrak{d}^{u_1} & \mathfrak{d}^{u_2} &
\end{array}
$$

$$
\begin{array}{llll|llll}
\mathfrak{ed}_1^{w_1..w_3} & =\mathfrak{E}^{\binom{u_1}{v_{1:2}}} & \mathfrak{E}^{\binom{u_{12}}{v_{2:3}}} & \mathfrak{E}^{\binom{u_{123}}{v_3}} & \|\,\mathfrak{od}_1^{w_1..w_3} & =\mathfrak{O}^{\binom{u_1}{v_{1:2}}} & \mathfrak{O}^{\binom{u_{12}}{v_{2:3}}} & \mathfrak{O}^{\binom{u_{123}}{v_3}} \\
\ldots\ldots & = \ldots\ldots & \ldots\ldots & \ldots\ldots & \|\,\ldots\ldots & = \ldots\ldots & \ldots\ldots & \ldots\ldots \\
\mathfrak{ed}_5^{w_1..w_3} & =\mathfrak{E}^{\binom{u_{123}}{v_1}} & \mathfrak{E}^{\binom{u_{23}}{v_{2:1}}} & \mathfrak{E}^{\binom{u_3}{v_{3:2}}} & \|\,\mathfrak{od}_5^{w_1..w_3} & =\mathfrak{O}^{\binom{u_{123}}{v_1}} & \mathfrak{O}^{\binom{u_{23}}{v_{2:1}}} & \mathfrak{O}^{\binom{u_3}{v_{3:2}}} \\
\mathfrak{ed}_6^{w_1..w_3} & =\mathfrak{E}^{\binom{u_1}{v_1}} & \mathfrak{d}^{v_2} & \mathfrak{d}^{v_3} & \|\,\mathfrak{od}_6^{w_1..w_3} & =\mathfrak{O}^{\binom{u_1}{v_1}} & \mathfrak{d}^{u_2} & \mathfrak{d}^{u_3} \\
\mathfrak{ed}_7^{w_1..w_3} & =\mathfrak{E}^{\binom{u_2}{v_2}} & \mathfrak{d}^{v_1} & \mathfrak{d}^{v_3} & \|\,\mathfrak{od}_7^{w_1..w_3} & =\mathfrak{O}^{\binom{u_1}{v_{1:2}}} & \mathfrak{d}^{u_{12}} & \mathfrak{d}^{u_3} \\
\mathfrak{ed}_8^{w_1..w_3} & =\mathfrak{E}^{\binom{u_3}{v_3}} & \mathfrak{d}^{v_1} & \mathfrak{d}^{v_2} & \|\,\mathfrak{od}_8^{w_1..w_3} & =\mathfrak{O}^{\binom{u_3}{v_{3:2}}} & \mathfrak{d}^{u_1} & \mathfrak{d}^{u_{23}} \\
\mathfrak{ed}_9^{w_1..w_3} & =\mathfrak{E}^{\binom{u_{123}}{v_1}} & \mathfrak{d}^{v_{2:1}} & \mathfrak{d}^{v_{3:1}} & \|\,\mathfrak{od}_9^{w_1..w_3} & =\mathfrak{O}^{\binom{u_3}{v_3}} & \mathfrak{d}^{u_1} & \mathfrak{d}^{u_2}
\end{array}
$$

$$
\begin{array}{lllll|lllll}
\mathfrak{ed}_1^{w_1..w_4} & =\mathfrak{E}^{\binom{u_1}{v_{1:2}}} & \mathfrak{E}^{\binom{u_{12}}{v_{2:3}}} & \mathfrak{E}^{\binom{u_{123}}{v_{3:4}}} & \mathfrak{E}^{\binom{u_{1234}}{v_4}} & \|\,\mathfrak{od}_1^{w_1..w_4} & =\mathfrak{O}^{\binom{u_1}{v_{1:2}}} & \mathfrak{O}^{\binom{u_{12}}{v_{2:3}}} & \mathfrak{O}^{\binom{u_{123}}{v_{3:4}}} & \mathfrak{O}^{\binom{u_{1234}}{v_4}} \\
\ldots\ldots & = \ldots\ldots & \ldots\ldots & \ldots\ldots & \ldots\ldots & \|\,\ldots\ldots & = \ldots\ldots & \ldots\ldots & \ldots\ldots & \ldots\ldots \\
\mathfrak{ed}_{14}^{w_1..w_4} & =\mathfrak{E}^{\binom{u_{1234}}{v_1}} & \mathfrak{E}^{\binom{u_{234}}{v_{2:1}}} & \mathfrak{E}^{\binom{u_{34}}{v_{3:2}}} & \mathfrak{E}^{\binom{u_4}{v_{4:3}}} & \|\,\mathfrak{od}_{14}^{w_1..w_4} & =\mathfrak{O}^{\binom{u_{1234}}{v_1}} & \mathfrak{O}^{\binom{u_{234}}{v_{2:1}}} & \mathfrak{O}^{\binom{u_{34}}{v_{3:2}}} & \mathfrak{O}^{\binom{u_4}{v_{4:3}}} \\
\mathfrak{ed}_{15}^{w_1..w_4} & =\mathfrak{E}^{\binom{u_1}{v_1}} & \mathfrak{E}^{\binom{u_3}{v_3}} & \mathfrak{d}^{v_2} & \mathfrak{d}^{v_4} & \|\,\mathfrak{od}_{15}^{w_1..w_4} & =\mathfrak{O}^{\binom{u_1}{v_1}} & \mathfrak{O}^{\binom{u_3}{v_{3:2}}} & \mathfrak{d}^{u_{23}} & \mathfrak{d}^{u_4} \\
\mathfrak{ed}_{16}^{w_1..w_4} & =\mathfrak{E}^{\binom{u_1}{v_1}} & \mathfrak{E}^{\binom{u_4}{v_4}} & \mathfrak{d}^{v_2} & \mathfrak{d}^{v_3} & \|\,\mathfrak{od}_{16}^{w_1..w_4} & =\mathfrak{O}^{\binom{u_1}{v_1}} & \mathfrak{O}^{\binom{u_4}{v_{4:3}}} & \mathfrak{d}^{u_{34}} & \mathfrak{d}^{u_2} \\
\mathfrak{ed}_{17}^{w_1..w_4} & =\mathfrak{E}^{\binom{u_2}{v_2}} & \mathfrak{E}^{\binom{u_4}{v_4}} & \mathfrak{d}^{v_1} & \mathfrak{d}^{v_3} & \|\,\mathfrak{od}_{17}^{w_1..w_4} & =\mathfrak{O}^{\binom{u_2}{v_{2:3}}} & \mathfrak{O}^{\binom{u_4}{v_4}} & \mathfrak{d}^{u_{23}} & \mathfrak{d}^{u_1} \\
\mathfrak{ed}_{18}^{w_1..w_4} & =\mathfrak{E}^{\binom{u_{12}}{v_2}} & \mathfrak{E}^{\binom{u_1}{v_{1:2}}} & \mathfrak{d}^{v_3} & \mathfrak{d}^{v_4} & \|\,\mathfrak{od}_{18}^{w_1..w_4} & =\mathfrak{O}^{\binom{u_1}{v_{1:2}}} & \mathfrak{O}^{\binom{u_4}{v_4}} & \mathfrak{d}^{u_{12}} & \mathfrak{d}^{u_3} \\
\mathfrak{ed}_{19}^{w_1..w_4} & =\mathfrak{E}^{\binom{u_{12}}{v_1}} & \mathfrak{E}^{\binom{u_2}{v_{2:1}}} & \mathfrak{d}^{v_3} & \mathfrak{d}^{v_4} & \|\,\mathfrak{od}_{19}^{w_1..w_4} & =\mathfrak{O}^{\binom{u_2}{v_{2:1}}} & \mathfrak{O}^{\binom{u_{12}}{v_1}} & \mathfrak{d}^{u_3} & \mathfrak{d}^{u_4} \\
\mathfrak{ed}_{20}^{w_1..w_4} & =\mathfrak{E}^{\binom{u_{23}}{v_3}} & \mathfrak{E}^{\binom{u_2}{v_{2:3}}} & \mathfrak{d}^{v_1} & \mathfrak{d}^{v_4} & \|\,\mathfrak{od}_{20}^{w_1..w_4} & =\mathfrak{O}^{\binom{u_1}{v_{1:2}}} & \mathfrak{O}^{\binom{u_{12}}{v_2}} & \mathfrak{d}^{u_3} & \mathfrak{d}^{u_4} \\
\mathfrak{ed}_{21}^{w_1..w_4} & =\mathfrak{E}^{\binom{u_{23}}{v_2}} & \mathfrak{E}^{\binom{u_3}{v_{3:2}}} & \mathfrak{d}^{v_1} & \mathfrak{d}^{v_4} & \|\,\mathfrak{od}_{21}^{w_1..w_4} & =\mathfrak{O}^{\binom{u_4}{v_{4:3}}} & \mathfrak{O}^{\binom{u_{34}}{v_3}} & \mathfrak{d}^{u_1} & \mathfrak{d}^{u_2} \\
\mathfrak{ed}_{22}^{w_1..w_4} & =\mathfrak{E}^{\binom{u_{34}}{v_4}} & \mathfrak{E}^{\binom{u_3}{v_{3:4}}} & \mathfrak{d}^{v_1} & \mathfrak{d}^{v_2} & \|\,\mathfrak{od}_{22}^{w_1..w_4} & =\mathfrak{O}^{\binom{u_3}{v_{3:4}}} & \mathfrak{O}^{\binom{u_{34}}{v_4}} & \mathfrak{d}^{u_1} & \mathfrak{d}^{u_2} \\
\mathfrak{ed}_{23}^{w_1..w_4} & =\mathfrak{E}^{\binom{u_{34}}{v_3}} & \mathfrak{E}^{\binom{u_4}{v_{4:3}}} & \mathfrak{d}^{v_1} & \mathfrak{d}^{v_2} & \|\,\mathfrak{od}_{23}^{w_1..w_4} & =\mathfrak{O}^{\binom{u_1}{v_{1:2}}} & \mathfrak{O}^{\binom{u_3}{v_{3:4}}} & \mathfrak{d}^{u_{12}} & \mathfrak{d}^{u_{34}} \\
\mathfrak{ed}_{24}^{w_1..w_4} & =\mathfrak{E}^{\binom{u_{1234}}{v_4}} & \mathfrak{E}^{\binom{u_1}{v_{1:4}}} & \mathfrak{w}^{v_{2:4}} & \mathfrak{d}^{v_{3:4}} & \|\,\mathfrak{od}_{24}^{w_1..w_4} & =\mathfrak{O}^{\binom{u_1}{v_{1:4}}} & \mathfrak{O}^{\binom{u_{1234}}{v_4}} & \mathfrak{d}^{u_2} & \mathfrak{d}^{u_3} \\
\mathfrak{ed}_{25}^{w_1..w_4} & =\mathfrak{E}^{\binom{u_{1234}}{v_3}} & \mathfrak{E}^{\binom{u_2}{v_{2:3}}} & \mathfrak{d}^{v_{1:3}} & \mathfrak{d}^{v_{4:3}} & \|\,\mathfrak{od}_{25}^{w_1..w_4} & =\mathfrak{O}^{\binom{u_4}{v_{4:1}}} & \mathfrak{O}^{\binom{u_{1234}}{v_1}} & \mathfrak{d}^{u_2} & \mathfrak{d}^{u_3} \\
\mathfrak{ed}_{26}^{w_1..w_4} & =\mathfrak{E}^{\binom{u_{1234}}{v_2}} & \mathfrak{E}^{\binom{u_3}{v_{3:2}}} & \mathfrak{d}^{v_{4:2}} & \mathfrak{d}^{v_{1:2}} & \|\,\mathfrak{od}_{26}^{w_1..w_4} & =\mathfrak{O}^{\binom{u_2}{v_{2:1}}} & \mathfrak{O}^{\binom{u_{12}}{v_{1:3}}} & \mathfrak{d}^{u_{123}} & \mathfrak{d}^{u_4} \\
\mathfrak{ed}_{27}^{w_1..w_4} & =\mathfrak{E}^{\binom{u_{1234}}{v_1}} & \mathfrak{E}^{\binom{u_4}{v_{4:1}}} & \mathfrak{d}^{v_{3:1}} & \mathfrak{d}^{v_{2:1}} & \|\,\mathfrak{od}_{27}^{w_1..w_4} & =\mathfrak{O}^{\binom{u_1}{v_{1:2}}} & \mathfrak{O}^{\binom{u_{12}}{v_{2:3}}} & \mathfrak{d}^{u_{123}} & \mathfrak{d}^{u_4} \\
\mathfrak{ed}_{28}^{w_1..w_4} & =\mathfrak{E}^{\binom{u_{1234}}{v_4}} & \mathfrak{E}^{\binom{u_{123}}{v_{1:4}}} & \mathfrak{d}^{v_{2:1}} & \mathfrak{d}^{v_{3:1}} & \|\,\mathfrak{od}_{28}^{w_1..w_4} & =\mathfrak{O}^{\binom{u_{34}}{v_{4:2}}} & \mathfrak{O}^{\binom{u_3}{v_{3:4}}} & \mathfrak{d}^{u_{234}} & \mathfrak{d}^{u_1} \\
\mathfrak{ed}_{29}^{w_1..w_4} & =\mathfrak{E}^{\binom{u_{1234}}{v_1}} & \mathfrak{E}^{\binom{u_{234}}{v_{4:1}}} & \mathfrak{d}^{v_{3:4}} & \mathfrak{d}^{v_{2:4}} & \|\,\mathfrak{od}_{29}^{w_1..w_4} & =\mathfrak{O}^{\binom{u_{34}}{v_{3:2}}} & \mathfrak{O}^{\binom{u_4}{v_{4:3}}} & \mathfrak{d}^{u_{234}} & \mathfrak{d}^{u_1} \\
\mathfrak{od}_{30}^{w_1..w_4} & =\mathfrak{d}^{v_1} & \mathfrak{d}^{v_2} & \mathfrak{d}^{v_3} & \mathfrak{d}^{v_4} & \|\,\mathfrak{od}_{30}^{w_1..w_4} & =\mathfrak{d}^{u_1} & \mathfrak{d}^{u_2} & \mathfrak{d}^{u_3} & \mathfrak{d}^{u_4}.
\end{array}
$$

12.2 Table 2: basis for *Flexin*($\mathfrak{E}$)

In Section 4.1 we introduced three series of bimoulds $\{\mathfrak{me}_r^\bullet\}$, $\{\mathfrak{ne}_r^\bullet\}$, $\{\mathfrak{re}_r^\bullet\}$, each of which, under *mu*-multiplication, produces a linear basis for *Flexin*($\mathfrak{E}$). For the first two series, the inductive definitions $\mathfrak{me}_r^\bullet := amit(\mathfrak{me}_{r-1}^\bullet).\mathfrak{E}^\bullet$ and $\mathfrak{ne}_r^\bullet := anit(\mathfrak{ne}_{r-1}^\bullet).\mathfrak{E}^\bullet$ straightaway generate atoms

$$\mathfrak{me}_r^{\binom{u_1\ ,\dots,\ u_r}{v_1\ ,\dots,\ v_r}} = \mathfrak{E}^{\binom{u_1}{v_{1:2}}}\,\mathfrak{E}^{\binom{u_{1,2}}{v_{2:3}}}\dots\,\mathfrak{E}^{\binom{u_{1,\dots,r}}{v_r}} \tag{12.1}$$

$$\mathfrak{ne}_r^{\binom{u_1\ ,\dots,\ u_r}{v_1\ ,\dots,\ v_r}} = \mathfrak{E}^{\binom{u_{1,\dots,r}}{v_1}}\mathfrak{E}^{\binom{u_{2,\dots,r}}{v_{2:1}}}\dots\mathfrak{E}^{\binom{u_r}{v_{r:r-1}}} \tag{12.2}$$

in *all cases*, *i.e.* whether $\mathfrak{E}$ is a flexion unit or not. Not so with the more important – because *alternal* – third series. Here, the inductive rule $\mathfrak{re}_r^\bullet := arit(\mathfrak{re}_{r-1}^\bullet).\mathfrak{E}^\bullet$ produces 2^{r-1} summands. If $\mathfrak{E}$ is a flexion unit, this far exceeds the minimal number of atoms required, which is always r. Moreover, in the polar realisation $\mathfrak{E} = Pi$, the mechanical application of the induction rule produces illusory poles. To remedy these drawbacks, we may use any one of these three alternative expressions:

$$\mathfrak{re}_r^\bullet = \sum_{r_1+r_2=r}^{0\le r_1} (-1)^{r_1}\, r_2\ \mathrm{mu}(\mathfrak{ne}_{r_1}^\bullet, \mathfrak{me}_{r_2}^\bullet) \tag{12.3}$$

$$\mathfrak{re}_r^\bullet = \sum_{r_1+r_2=r}^{0\le r_2} -(-1)^{r_1}\, r_1\ \mathrm{mu}(\mathfrak{ne}_{r_1}^\bullet, \mathfrak{me}_{r_2}^\bullet) \tag{12.4}$$

$$\mathfrak{re}_r^\bullet = \sum_{r_1+r_2=r}^{0\le r_1, r_2} (-1)^{r_1}\, \frac{r_2 - r_1}{2}\ \mathrm{mu}(\mathfrak{ne}_{r_1}^\bullet, \mathfrak{me}_{r_2}^\bullet) \tag{12.5}$$

with the convention $\mathfrak{me}_0^\bullet = \mathfrak{ne}_0^\bullet := 1$. These sums produce indeed the minimal number of atoms[178] and do away with illusory poles, but they are of course valid only if $\mathfrak{E}$ is a flexion unit. Only the last expression is left-right symmetric, and renders the alternality of the $\mathfrak{re}_r^\bullet$ 'visually' obvious.

12.3 Table 3: basis for *Flexinn*($\mathfrak{E}$)

To produce an explicit basis, we must first express the iterated *preari* products $\mathfrak{Re}_r^\bullet$ of the basic alternal bimoulds $\mathfrak{re}_r^\bullet$, calculated as usual from left to right:

$$\mathfrak{Re}_{r_1}^\bullet := \mathfrak{re}_{r_1}^\bullet \quad \textit{and} \quad \mathfrak{Re}_{r_1,\dots,r_s}^\bullet := \mathrm{preari}(\mathfrak{Re}_{r_1,\dots,r_{s-1}}^\bullet, \mathfrak{re}_{r_s}^\bullet).$$

---

178 Strictly speaking, this applies only to the first two sums. For r odd, the last sum produces a supererogatory atom.

Technically, however, it is more convenient to consider the swappees $\mathfrak{Rö}^\bullet_{\boldsymbol{r}} := swap.\mathfrak{Re}^\bullet_{\boldsymbol{r}}$. The formula for expressing them as *minimal* sums of *inflected atoms* may seem forbiddingly complex, but it is still very useful, and in some contexts even indispensable. It reads:

$$\mathfrak{Rö}^\bullet_{r_1,\ldots,r_s} := \sum \mathfrak{Po}^\bullet_{\binom{n_1}{\boldsymbol{r}^1},\ldots,\binom{n_t}{\boldsymbol{r}^t}} \; H^{\boldsymbol{r}^1}_{\overline{n}_1,\underline{n}_1} \ldots H^{\boldsymbol{r}^t}_{\overline{n}_t,\underline{n}_t} \tag{12.6}$$

(i) with a sum extending to all partitions of $\boldsymbol{r} = (r_1,\ldots,r_s)$ into any number of partial sequences $\boldsymbol{r}^i$, and to all choices of integers n_i, subject only to the following constraints: the internal order of each $\boldsymbol{r}^i$ must be compatible with that of $\boldsymbol{r}$, whereas the various $\boldsymbol{r}^i$ may be positioned in *any* order; and the integers n_i need only verify $r^*_{i-1} < n_i \leq r^*_i$ with $r^*_i := ||\boldsymbol{r}^1|| + \ldots + ||\boldsymbol{r}^i||$;
(ii) with half-integers $\underline{n}_i$ and integers $\overline{n}_i$ defined by

$$\underline{n}_i := n_i - r^*_i - \tfrac{1}{2} \quad \text{with } r^*_i := ||\boldsymbol{r}^1|| + \ldots + ||\boldsymbol{r}^i||$$
$$\overline{n}_i := 1 + r^* - n_i \quad \text{with } r^* := ||\boldsymbol{r}^1|| + \ldots + ||\boldsymbol{r}^t|| = r_1 + \ldots + r_s;$$

(iii) with inflected atoms of type:[179]

$$\mathfrak{Po}^\bullet_{\binom{n_1}{\boldsymbol{r}^1},\ldots,\binom{n_t}{\boldsymbol{r}^t}} := \prod_{1\leq i\leq t} \left(\mho^{\binom{u_1+\ldots+u_{r^*_i}}{v_{n_i}-v_{n_{i+1}}}} \prod_{\binom{n^*_{i-1}<n\leq n^*_i}{n\neq n_i}} \mho^{\binom{u_n}{v_n - v_{n_i}}} \right); \tag{12.7}$$

(iv) with coefficients $H^{\boldsymbol{r}}_{\overline{n},\underline{n}}$ given by the sums

$$H^{\boldsymbol{r}}_{\overline{n},\underline{n}} := \sum_{\boldsymbol{r}^+\cup\boldsymbol{r}^-=\boldsymbol{r}} \mathrm{sign}(||\boldsymbol{r}^+||-\underline{n})\; F_{\boldsymbol{r}^+}(\overline{n})\; F_{\boldsymbol{r}^-}(\overline{n}) \tag{12.8}$$

ranging over all partitions $\boldsymbol{r}^+ \cup \boldsymbol{r}^-$ of $\boldsymbol{r}$.

If $\boldsymbol{r}^+ = (r^+_1,\ldots,r^+_p)$ and $\boldsymbol{r}^- = (r^-_1,\ldots,r^-_q)$, the two factors $F_{\boldsymbol{r}^\pm}$ are defined as follows:

$$F_{\boldsymbol{r}^+}(\overline{n}) := (\overline{n})\,(r^+_2 + r^+_3 + .. + r^+_p - \overline{n})(r^+_3 + .. + r^+_p - \overline{n})\ldots(r^+_p - \overline{n})$$
$$F_{\boldsymbol{r}^-}(\overline{n}) := (\overline{n})\,(r^-_2 + r^-_3 + .. + r^-_q + \overline{n})(r^-_3 + .. + r^-_q + \overline{n})\ldots(r^-_p + \overline{n}).$$

If p (respectively q) is 1, then $F_{\boldsymbol{r}^+}(\overline{n})$ (resp $F_{\boldsymbol{r}^-}(\overline{n})$) reduces to $\overline{n}$.

179 For extreme values of the index i, we must of course set $n^*_0 := 0$ and $v_{n_{t+1}} := 0$.

Lastly, for the extreme partitions $(\boldsymbol{r}^+, \boldsymbol{r}^-) = (\boldsymbol{r}, \boldsymbol{\emptyset})$ or $(\boldsymbol{\emptyset}, \boldsymbol{r})$, we must replace the product $F_{\boldsymbol{r}^+}(\overline{n})\, F_{\boldsymbol{r}^-}(\overline{n})$ respectively by

$$F_{\boldsymbol{r},\boldsymbol{\emptyset}}(\overline{n}) := +\overline{n}\,(r_2 + r_3 + .. + r_p - \overline{n})(r_3 + .. + r_p - \overline{n}) \ldots (r_p - \overline{n})$$

$$F_{\boldsymbol{\emptyset},\boldsymbol{r}}(\overline{n}) := -\overline{n}\,(r_2 + r_3 + .. + r_q + \overline{n})(r_3 + .. + r_q + \overline{n}) \ldots (r_q + \overline{n}).$$

Remark 1: massive pole cancellations.
From formula (12.6) and the shape (12.7) of the atoms involved, we immediately infer a huge difference between the specialisations $(\mathfrak{E}, \mathfrak{O}) = (Pa, Pi)$ and $(\mathfrak{E}, \mathfrak{O}) = (Pi, Pa)$. In the first case, the $\mathfrak{Re}^\bullet_{\boldsymbol{r}}$ and $\mathfrak{R}\ddot{\mathfrak{o}}^\bullet_{\boldsymbol{r}}$ are saddled with a maximal number of poles, namely $r\,(r+1)/2$. In the second case, they possess far fewer – as little as $2r - 1$. This results from massive and rather extraordinary compensations that occur during the iteration of the *preari* product *when applied to the* $\mathfrak{re}_{r_i}$. Were we, however, to subject the *separate components* of the $\mathfrak{re}_{r_i}$ (as given for instance by (12.5)) to *preari*-iteration, no such compensations would take place.

Remark 2: bases of ***Flexinn***$(\mathfrak{E})$**.**
On their own, the $\mathfrak{Re}^\bullet_{\boldsymbol{r}}$ span, not *Flexinn*$(\mathfrak{E})$, but the larger *Flexin*$(\mathfrak{E})$. If however we restrict ourselves to combinations of the form[180]

$$^{\Gamma}\mathfrak{Re}^\bullet_{\{r_1,\ldots,r_s\}} := \sum_{\{\boldsymbol{r}'\}=\{\boldsymbol{r}\}} \Gamma^{r'_1,\ldots,r'_s}\ \mathfrak{Re}^\bullet_{r'_1,\ldots,r'_s} \qquad (\Gamma \textit{ symmetral}) \qquad (12.9)$$

then the new $^{\Gamma}\mathfrak{Re}^\bullet_{\{\boldsymbol{r}\}}$ do constitute a basis of *Flexinn*$(\mathfrak{E})$, and that too *irrespective of the choice of the symmetral mould* Γ, provided Γ^{r_1} be $\neq 0$ for all indices r_1. Three choices stand out:

$$\Gamma_1^{r_1,\ldots,r_s} := 1/s! \qquad (12.10)$$

$$\Gamma_2^{r_1,\ldots,r_s} := \prod_{1\le i\le s} \frac{1}{r_1 + \ldots + r_i} \qquad (12.11)$$

$$\Gamma_3^{r_1,\ldots,r_s} := (-1)^s \prod_{1\le i\le s} \frac{1}{r_i + \ldots + r_s} \qquad (12.12)$$

180 With a sum ranging over all permutations $\boldsymbol{r}'$ of the sequence $\boldsymbol{r}$.)

(i) The basis ${}^{\Gamma_1}\mathfrak{Re}^{\bullet}_{\{r\}}$ permits the expression of the elements $\mathfrak{Se}^{\bullet}_{f}$ of $GARI_{<\mathfrak{se}>}$ in terms of the coefficients ϵ_r of the infinitesimal generator f_* of f.
(ii) The basis ${}^{\Gamma_2}\mathfrak{Re}^{\bullet}_{\{r\}}$ permits the expression of the elements $\mathfrak{Se}^{\bullet}_{f}$ of $GARI_{<\mathfrak{se}>}$ in terms of the coefficients γ_r of the (direct) infinitesimal dilator $f_\#$ of f.
(iii) The basis ${}^{\Gamma_3}\mathfrak{Re}^{\bullet}_{\{r\}}$ permits the expression of the elements $\mathfrak{Se}^{\bullet}_{f}$ of $GARI_{<\mathfrak{se}>}$ in terms of the coefficients δ_r of the inverse infinitesimal dilator $(f^{-1})_\#$.

12.4 Table 4: the universal bimould $\mathfrak{ess}^\bullet$

$$
\begin{aligned}
\mathfrak{ess}^{w_1} &= -\tfrac{1}{2}\ \mathfrak{E}^{\binom{u_1}{v_1}} \\
\mathfrak{ess}^{w_1,w_2} &= +\tfrac{1}{12}\ \mathfrak{E}^{\binom{u_1}{v_{1:2}}}\ \mathfrak{E}^{\binom{u_{12}}{v_2}} \\
&\quad +\tfrac{1}{12}\ \mathfrak{E}^{\binom{u_1}{v_1}}\ \mathfrak{E}^{\binom{u_2}{v_2}} \\
\mathfrak{ess}^{w_1,w_2,w_3} &= -\tfrac{1}{24}\ \mathfrak{E}^{\binom{u_1}{v_{1:2}}}\ \mathfrak{E}^{\binom{u_{12}}{v_2}}\ \mathfrak{E}^{\binom{u_3}{v_3}} \\
\mathfrak{ess}^{w_1,w_2,w_3,w_4} &= -\tfrac{1}{720}\ \mathfrak{E}^{\binom{u_1}{v_{1:2}}}\ \mathfrak{E}^{\binom{u_{12}}{v_{2:3}}}\ \mathfrak{E}^{\binom{u_{123}}{v_{3:4}}}\ \mathfrak{E}^{\binom{u_{1234}}{v_4}} \\
&\quad -\tfrac{1}{240}\ \mathfrak{E}^{\binom{u_1}{v_1}}\ \mathfrak{E}^{\binom{u_2}{v_{2:3}}}\ \mathfrak{E}^{\binom{u_{23}}{v_{3:4}}}\ \mathfrak{E}^{\binom{u_{234}}{v_4}} \\
&\quad -\tfrac{1}{240}\ \mathfrak{E}^{\binom{u_{12}}{v_1}}\ \mathfrak{E}^{\binom{u_2}{v_{2:1}}}\ \mathfrak{E}^{\binom{u_3}{v_{3:4}}}\ \mathfrak{E}^{\binom{u_{34}}{v_4}} \\
&\quad +\tfrac{1}{180}\ \mathfrak{E}^{\binom{u_1}{v_{1:2}}}\ \mathfrak{E}^{\binom{u_{12}}{v_{2:3}}}\ \mathfrak{E}^{\binom{u_{123}}{v_3}}\ \mathfrak{E}^{\binom{u_4}{v_4}} \\
&\quad +\tfrac{1}{120}\ \mathfrak{E}^{\binom{u_1}{v_1}}\ \mathfrak{E}^{\binom{u_2}{v_{2:3}}}\ \mathfrak{E}^{\binom{u_{23}}{v_3}}\ \mathfrak{E}^{\binom{u_4}{v_4}} \\
&\quad -\tfrac{1}{720}\ \mathfrak{E}^{\binom{u_{12}}{v_1}}\ \mathfrak{E}^{\binom{u_2}{v_{2:1}}}\ \mathfrak{E}^{\binom{u_3}{v_3}}\ \mathfrak{E}^{\binom{u_4}{v_4}} \\
\mathfrak{ess}^{w_1,w_2,w_3,w_4,w_5} &= -\tfrac{1}{240}\ \mathfrak{E}^{\binom{u_1}{v_{1:2}}}\ \mathfrak{E}^{\binom{u_{12}}{v_2}}\ \mathfrak{E}^{\binom{u_3}{v_{3:4}}}\ \mathfrak{E}^{\binom{u_{34}}{v_4}}\ \mathfrak{E}^{\binom{u_5}{v_5}} \\
&\quad +\tfrac{1}{480}\ \mathfrak{E}^{\binom{u_1}{v_{1:2}}}\ \mathfrak{E}^{\binom{u_{12}}{v_2}}\ \mathfrak{E}^{\binom{u_3}{v_{3:4}}}\ \mathfrak{E}^{\binom{u_{34}}{v_{4:5}}}\ \mathfrak{E}^{\binom{u_{345}}{v_5}} \\
&\quad +\tfrac{1}{480}\ \mathfrak{E}^{\binom{u_1}{v_{1:2}}}\ \mathfrak{E}^{\binom{u_{123}}{v_2}}\ \mathfrak{E}^{\binom{u_3}{v_{3:2}}}\ \mathfrak{E}^{\binom{u_4}{v_{4:5}}}\ \mathfrak{E}^{\binom{u_{45}}{v_5}} \\
&\quad +\tfrac{1}{1440}\ \mathfrak{E}^{\binom{u_1}{v_{1:2}}}\ \mathfrak{E}^{\binom{u_{12}}{v_{2:3}}}\ \mathfrak{E}^{\binom{u_{123}}{v_{3:4}}}\ \mathfrak{E}^{\binom{u_{1234}}{v_4}}\ \mathfrak{E}^{\binom{u_5}{v_5}} \\
&\quad +\tfrac{1}{1440}\ \mathfrak{E}^{\binom{u_1}{v_{1:2}}}\ \mathfrak{E}^{\binom{u_{123}}{v_2}}\ \mathfrak{E}^{\binom{u_3}{v_{3:2}}}\ \mathfrak{E}^{\binom{u_4}{v_4}}\ \mathfrak{E}^{\binom{u_5}{v_5}}.
\end{aligned}
$$

For $r = 6$ or larger, the number of summands increases dramatically. However, one gets markedly simpler expressions when expanding $\mathfrak{ess}^\bullet$ in the bases $\{\mathfrak{me}^{\bullet}_{r_1,\ldots,r_s}\}$, $\{\mathfrak{ne}^{\bullet}_{r_1,\ldots,r_s}\}$, $\{\mathfrak{re}^{\bullet}_{r_1,\ldots,r_s}\}$ of $Flexin(\mathfrak{E}) \subset Flex(\mathfrak{E})$: see Section 4.1.

12.5 Table 5: the universal bimould $\mathfrak{esz}_\sigma^\bullet$

$$\mathfrak{esz}_\sigma^{w_1} =$$

$$+\sigma \qquad \times\ \mathfrak{E}^{\left(\begin{smallmatrix}u_1\\ v_1\end{smallmatrix}\right)}$$

$$\mathfrak{esz}_\sigma^{w_1,w_2} =$$

$$+\tfrac{1}{3}\,\sigma\,(1+2\,\sigma) \qquad \times\ \mathfrak{E}^{\left(\begin{smallmatrix}u_{12}\\ v_2\end{smallmatrix}\right)}\ \mathfrak{E}^{\left(\begin{smallmatrix}u_1\\ v_{1:2}\end{smallmatrix}\right)}$$

$$-\tfrac{1}{3}\,\sigma\,(1-\sigma) \qquad \times\ \mathfrak{E}^{\left(\begin{smallmatrix}u_{12}\\ v_1\end{smallmatrix}\right)}\ \mathfrak{E}^{\left(\begin{smallmatrix}u_2\\ v_{2:1}\end{smallmatrix}\right)}$$

$$\mathfrak{esz}_\sigma^{w_1,w_2,w_3} =$$

$$+\tfrac{1}{6}\,\sigma\,(1+2\,\sigma)\,(1+\sigma) \qquad \times\ \mathfrak{E}^{\left(\begin{smallmatrix}u_{123}\\ v_3\end{smallmatrix}\right)}\ \mathfrak{E}^{\left(\begin{smallmatrix}u_{12}\\ v_{2:3}\end{smallmatrix}\right)}\ \mathfrak{E}^{\left(\begin{smallmatrix}u_1\\ v_{1:2}\end{smallmatrix}\right)}$$

$$-\tfrac{1}{6}\,\sigma\,(1-\sigma) \qquad \times\ \mathfrak{E}^{\left(\begin{smallmatrix}u_{123}\\ v_3\end{smallmatrix}\right)}\ \mathfrak{E}^{\left(\begin{smallmatrix}u_{12}\\ v_{1:3}\end{smallmatrix}\right)}\ \mathfrak{E}^{\left(\begin{smallmatrix}u_2\\ v_{2:1}\end{smallmatrix}\right)}$$

$$-\tfrac{1}{3}\,(1-\sigma)\,\sigma^2 \qquad \times\ \mathfrak{E}^{\left(\begin{smallmatrix}u_{123}\\ v_2\end{smallmatrix}\right)}\ \mathfrak{E}^{\left(\begin{smallmatrix}u_1\\ v_{1:2}\end{smallmatrix}\right)}\ \mathfrak{E}^{\left(\begin{smallmatrix}u_3\\ v_{3:2}\end{smallmatrix}\right)}$$

$$-\tfrac{1}{6}\,\sigma\,(1-\sigma) \qquad \times\ \mathfrak{E}^{\left(\begin{smallmatrix}u_{123}\\ v_1\end{smallmatrix}\right)}\ \mathfrak{E}^{\left(\begin{smallmatrix}u_{23}\\ v_{3:1}\end{smallmatrix}\right)}\ \mathfrak{E}^{\left(\begin{smallmatrix}u_2\\ v_{2:3}\end{smallmatrix}\right)}$$

$$+\tfrac{1}{6}\,\sigma\,(1-\sigma) \qquad \times\ \mathfrak{E}^{\left(\begin{smallmatrix}u_{123}\\ v_1\end{smallmatrix}\right)}\ \mathfrak{E}^{\left(\begin{smallmatrix}u_{23}\\ v_{2:1}\end{smallmatrix}\right)}\ \mathfrak{E}^{\left(\begin{smallmatrix}u_3\\ v_{3:2}\end{smallmatrix}\right)}$$

$$\mathfrak{esz}_\sigma^{w_1,w_2,w_3,w_4} =$$

$$+\tfrac{1}{30}\,\sigma\,(1+2\,\sigma)\,(1+\sigma)\,(3+2\,\sigma) \qquad \times\ \mathfrak{E}^{\left(\begin{smallmatrix}u_{1234}\\ v_4\end{smallmatrix}\right)}\ \mathfrak{E}^{\left(\begin{smallmatrix}u_{123}\\ v_{3:4}\end{smallmatrix}\right)}\ \mathfrak{E}^{\left(\begin{smallmatrix}u_{12}\\ v_{2:3}\end{smallmatrix}\right)}\ \mathfrak{E}^{\left(\begin{smallmatrix}u_1\\ v_{1:2}\end{smallmatrix}\right)}$$

$$-\tfrac{1}{90}\,\sigma\,(1-\sigma)\,(9+2\,\sigma-2\,\sigma^2) \qquad \times\ \mathfrak{E}^{\left(\begin{smallmatrix}u_{1234}\\ v_4\end{smallmatrix}\right)}\ \mathfrak{E}^{\left(\begin{smallmatrix}u_{123}\\ v_{3:4}\end{smallmatrix}\right)}\ \mathfrak{E}^{\left(\begin{smallmatrix}u_{12}\\ v_{1:3}\end{smallmatrix}\right)}\ \mathfrak{E}^{\left(\begin{smallmatrix}u_2\\ v_{2:1}\end{smallmatrix}\right)}$$

$$-\tfrac{1}{6}\,(1-\sigma)\,\sigma^2 \qquad \times\ \mathfrak{E}^{\left(\begin{smallmatrix}u_{1234}\\ v_4\end{smallmatrix}\right)}\ \mathfrak{E}^{\left(\begin{smallmatrix}u_{123}\\ v_{2:4}\end{smallmatrix}\right)}\ \mathfrak{E}^{\left(\begin{smallmatrix}u_1\\ v_{1:2}\end{smallmatrix}\right)}\ \mathfrak{E}^{\left(\begin{smallmatrix}u_3\\ v_{3:2}\end{smallmatrix}\right)}$$

$$-\tfrac{1}{90}\,\sigma\,(1-\sigma)\,(9+2\,\sigma-2\,\sigma^2) \qquad \times\ \mathfrak{E}^{\left(\begin{smallmatrix}u_{1234}\\ v_4\end{smallmatrix}\right)}\ \mathfrak{E}^{\left(\begin{smallmatrix}u_{123}\\ v_{1:4}\end{smallmatrix}\right)}\ \mathfrak{E}^{\left(\begin{smallmatrix}u_{23}\\ v_{3:1}\end{smallmatrix}\right)}\ \mathfrak{E}^{\left(\begin{smallmatrix}u_2\\ v_{2:3}\end{smallmatrix}\right)}$$

$$+\tfrac{1}{90}\,\sigma\,(1-\sigma)\,(9-8\,\sigma+8\,\sigma^2) \qquad \times\ \mathfrak{E}^{\left(\begin{smallmatrix}u_{1234}\\ v_4\end{smallmatrix}\right)}\ \mathfrak{E}^{\left(\begin{smallmatrix}u_{123}\\ v_{1:4}\end{smallmatrix}\right)}\ \mathfrak{E}^{\left(\begin{smallmatrix}u_{23}\\ v_{2:1}\end{smallmatrix}\right)}\ \mathfrak{E}^{\left(\begin{smallmatrix}u_3\\ v_{3:2}\end{smallmatrix}\right)}$$

$$-\tfrac{1}{9}\,(1-\sigma)\,(1+2\,\sigma)\,\sigma^2 \qquad \times\ \mathfrak{E}^{\left(\begin{smallmatrix}u_{1234}\\ v_3\end{smallmatrix}\right)}\ \mathfrak{E}^{\left(\begin{smallmatrix}u_{12}\\ v_{2:3}\end{smallmatrix}\right)}\ \mathfrak{E}^{\left(\begin{smallmatrix}u_1\\ v_{1:2}\end{smallmatrix}\right)}\ \mathfrak{E}^{\left(\begin{smallmatrix}u_4\\ v_{4:3}\end{smallmatrix}\right)}$$

$$+\tfrac{1}{9}\,\sigma^2\,(1-\sigma)^2 \qquad \times\ \mathfrak{E}^{\left(\begin{smallmatrix}u_{1234}\\ v_3\end{smallmatrix}\right)}\ \mathfrak{E}^{\left(\begin{smallmatrix}u_{12}\\ v_{1:3}\end{smallmatrix}\right)}\ \mathfrak{E}^{\left(\begin{smallmatrix}u_2\\ v_{2:1}\end{smallmatrix}\right)}\ \mathfrak{E}^{\left(\begin{smallmatrix}u_4\\ v_{4:3}\end{smallmatrix}\right)}$$

$$-\tfrac{1}{6}\,(1-\sigma)\,\sigma^2 \qquad \times\ \mathfrak{E}^{\left(\begin{smallmatrix}u_{1234}\\ v_2\end{smallmatrix}\right)}\ \mathfrak{E}^{\left(\begin{smallmatrix}u_1\\ v_{1:2}\end{smallmatrix}\right)}\ \mathfrak{E}^{\left(\begin{smallmatrix}u_{34}\\ v_{4:2}\end{smallmatrix}\right)}\ \mathfrak{E}^{\left(\begin{smallmatrix}u_3\\ v_{3:4}\end{smallmatrix}\right)}$$

$$+\tfrac{1}{6}\,(1-\sigma)\,\sigma^2 \qquad \times\ \mathfrak{E}^{\left(\begin{smallmatrix}u_{1234}\\ v_2\end{smallmatrix}\right)}\ \mathfrak{E}^{\left(\begin{smallmatrix}u_1\\ v_{1:2}\end{smallmatrix}\right)}\ \mathfrak{E}^{\left(\begin{smallmatrix}u_{34}\\ v_{3:2}\end{smallmatrix}\right)}\ \mathfrak{E}^{\left(\begin{smallmatrix}u_4\\ v_{4:3}\end{smallmatrix}\right)}$$

$$-\tfrac{1}{90}\,\sigma\,(1-\sigma)\,(9+2\,\sigma-2\,\sigma^2) \qquad \times\ \mathfrak{E}^{\left(\begin{smallmatrix}u_{1234}\\ v_1\end{smallmatrix}\right)}\ \mathfrak{E}^{\left(\begin{smallmatrix}u_{234}\\ v_{4:1}\end{smallmatrix}\right)}\ \mathfrak{E}^{\left(\begin{smallmatrix}u_{23}\\ v_{3:4}\end{smallmatrix}\right)}\ \mathfrak{E}^{\left(\begin{smallmatrix}u_2\\ v_{2:3}\end{smallmatrix}\right)}$$

$$+\tfrac{1}{90}\,\sigma\,(1-\sigma)\,(9-8\,\sigma+8\,\sigma^2) \qquad \times\ \mathfrak{E}^{\left(\begin{smallmatrix}u_{1234}\\ v_1\end{smallmatrix}\right)}\ \mathfrak{E}^{\left(\begin{smallmatrix}u_{234}\\ v_{4:1}\end{smallmatrix}\right)}\ \mathfrak{E}^{\left(\begin{smallmatrix}u_{23}\\ v_{2:4}\end{smallmatrix}\right)}\ \mathfrak{E}^{\left(\begin{smallmatrix}u_3\\ v_{3:2}\end{smallmatrix}\right)}$$

$$+\tfrac{1}{9}\,\sigma^2\,(1-\sigma)^2 \qquad \times\ \mathfrak{E}^{\left(\begin{smallmatrix}u_{1234}\\ v_1\end{smallmatrix}\right)}\ \mathfrak{E}^{\left(\begin{smallmatrix}u_{234}\\ v_{3:1}\end{smallmatrix}\right)}\ \mathfrak{E}^{\left(\begin{smallmatrix}u_2\\ v_{2:3}\end{smallmatrix}\right)}\ \mathfrak{E}^{\left(\begin{smallmatrix}u_4\\ v_{4:3}\end{smallmatrix}\right)}$$

$$+\tfrac{1}{90}\,\sigma\,(1-\sigma)\,(9-8\,\sigma+8\,\sigma^2) \qquad \times\ \mathfrak{E}^{\left(\begin{smallmatrix}u_{1234}\\ v_1\end{smallmatrix}\right)}\ \mathfrak{E}^{\left(\begin{smallmatrix}u_{234}\\ v_{2:1}\end{smallmatrix}\right)}\ \mathfrak{E}^{\left(\begin{smallmatrix}u_{34}\\ v_{4:2}\end{smallmatrix}\right)}\ \mathfrak{E}^{\left(\begin{smallmatrix}u_3\\ v_{3:4}\end{smallmatrix}\right)}$$

$$-\tfrac{1}{90}\,\sigma\,(1-\sigma)\,(9+2\,\sigma-2\,\sigma^2) \qquad \times\ \mathfrak{E}^{\left(\begin{smallmatrix}u_{1234}\\ v_1\end{smallmatrix}\right)}\ \mathfrak{E}^{\left(\begin{smallmatrix}u_{234}\\ v_{2:1}\end{smallmatrix}\right)}\ \mathfrak{E}^{\left(\begin{smallmatrix}u_{34}\\ v_{3:2}\end{smallmatrix}\right)}\ \mathfrak{E}^{\left(\begin{smallmatrix}u_4\\ v_{4:3}\end{smallmatrix}\right).}$$

12.6 Table 6: the bitrigonometric bimould $taal^\bullet / tiil^\bullet$

For simplicity, we drop the c in Qaa_c and Qii_c.

$$\mathrm{taal}^{w_1} = -\tfrac{1}{2}\,\mathrm{Qaa}^{\binom{u_1}{v_1}}$$

$$\begin{aligned}
\mathrm{taal}^{w_1,w_2} = {}&\\
&+\tfrac{1}{12}\,\mathrm{Qaa}^{\binom{u_{12}}{v_1}}\,\mathrm{Qaa}^{\binom{u_2}{v_{2:1}}}\\
&+\tfrac{1}{6}\,\mathrm{Qaa}^{\binom{u_{12}}{v_2}}\,\mathrm{Qaa}^{\binom{u_1}{v_{1:2}}}\\
&+\tfrac{1}{8}\,c^2\,\delta^{v_1}\,\delta^{v_2}
\end{aligned}$$

$$\begin{aligned}
\mathrm{taal}^{w_1,w_2,w_3} = {}&\\
&-\tfrac{1}{24}\,\mathrm{Qaa}^{\binom{u_1}{v_{1:2}}}\,\mathrm{Qaa}^{\binom{u_{12}}{v_2}}\,\mathrm{Qaa}^{\binom{u_3}{v_3}}\\
&-\tfrac{1}{48}\,c^2\,\mathrm{Qaa}^{\binom{u_1}{v_1}}\,\delta^{v_2}\delta^{v_3}\\
&+\tfrac{1}{24}\,c^2\,\mathrm{Qaa}^{\binom{u_2}{v_2}}\,\delta^{v_1}\,\delta^{v_3}\\
&-\tfrac{1}{24}\,c^2\,\mathrm{Qaa}^{\binom{u_3}{v_3}}\,\delta^{v_1}\,\delta^{v_2}
\end{aligned}$$

$$\begin{aligned}
\mathrm{taal}^{w_1,w_2,w_3,w_4} = {}&\\
&-\tfrac{1}{720}\,\mathrm{Qaa}^{\binom{u_1}{v_1}}\mathrm{Qaa}^{\binom{u_2}{v_2}}\mathrm{Qaa}^{\binom{u_3}{v_3}}\mathrm{Qaa}^{\binom{u_4}{v_4}}\\
&-\tfrac{1}{240}\,\mathrm{Qaa}^{\binom{u_1}{v_{1:4}}}\mathrm{Qaa}^{\binom{u_2}{v_{2:4}}}\mathrm{Qaa}^{\binom{u_3}{v_{3:4}}}\mathrm{Qaa}^{\binom{u_{1234}}{v_4}}\\
&+\tfrac{1}{240}\,\mathrm{Qaa}^{\binom{u_1}{v_{1:3}}}\mathrm{Qaa}^{\binom{u_2}{v_{2:3}}}\mathrm{Qaa}^{\binom{u_4}{v_{4:3}}}\mathrm{Qaa}^{\binom{u_{1234}}{v_3}}\\
&+\tfrac{1}{180}\,\mathrm{Qaa}^{\binom{u_1}{v_{1:2}}}\mathrm{Qaa}^{\binom{u_{12}}{v_2}}\mathrm{Qaa}^{\binom{u_3}{v_3}}\mathrm{Qaa}^{\binom{u_4}{v_4}}\\
&+\tfrac{1}{120}\,\mathrm{Qaa}^{\binom{u_1}{v_{1:2}}}\mathrm{Qaa}^{\binom{u_{12}}{v_{2:4}}}\mathrm{Qaa}^{\binom{u_3}{v_{3:4}}}\mathrm{Qaa}^{\binom{u_{1234}}{v_4}}\\
&+\tfrac{1}{720}\,\mathrm{Qaa}^{\binom{u_1}{v_{1:2}}}\mathrm{Qaa}^{\binom{u_{12}}{v_{2:3}}}\mathrm{Qaa}^{\binom{u_4}{v_{4:3}}}\mathrm{Qaa}^{\binom{u_{1234}}{v_3}}\\
&+\tfrac{7}{720}\,c^2\,\mathrm{Qaa}^{\binom{u_1}{v_1}}\mathrm{Qaa}^{\binom{u_2}{v_2}}\delta^{v_3}\delta^{v_4}\\
&+\tfrac{7}{1440}\,c^2\,\mathrm{Qaa}^{\binom{u_1}{v_{1:2}}}\,\mathrm{Qaa}^{\binom{u_{12}}{v_2}}\,\delta^{v_3}\,\delta^{v_4}\\
&-\tfrac{5}{288}\,c^2\,\mathrm{Qaa}^{\binom{u_1}{v_1}}\,\mathrm{Qaa}^{\binom{u_3}{v_3}}\,\delta^{v_2}\,\delta^{v_4}\\
&+\tfrac{19}{1440}\,c^2\,\mathrm{Qaa}^{\binom{u_1}{v_1}}\,\mathrm{Qaa}^{\binom{u_4}{v_4}}\,\delta^{v_2}\,\delta^{v_3}\\
&-\tfrac{1}{480}\,c^2\,\mathrm{Qaa}^{\binom{u_2}{v_2}}\,\mathrm{Qaa}^{\binom{u_3}{v_3}}\,\delta^{v_1}\,\delta^{v_4}\\
&+\tfrac{1}{1440}\,c^2\,\mathrm{Qaa}^{\binom{u_2}{v_2}}\,\mathrm{Qaa}^{\binom{u_4}{v_4}}\,\delta^{v_1}\,\delta^{v_3}\\
&+\tfrac{1}{288}\,c^2\,\mathrm{Qaa}^{\binom{u_3}{v_3}}\,\mathrm{Qaa}^{\binom{u_4}{v_4}}\,\delta^{v_1}\,\delta^{v_2}\\
&-\tfrac{1}{480}\,c^2\,\mathrm{Qaa}^{\binom{u_{1234}}{v_2}}\,\mathrm{Qaa}^{\binom{u_1}{v_{1:2}}}\,\delta^{v_{2:4}}\,\delta^{v_{2:3}}\\
&-\tfrac{1}{288}\,c^2\,\mathrm{Qaa}^{\binom{u_{1234}}{v_1}}\,\mathrm{Qaa}^{\binom{u_2}{v_{2:1}}}\,\delta^{v_{1:4}}\,\delta^{v_{1:3}}\\
&+\tfrac{11}{1440}\,c^2\,\mathrm{Qaa}^{\binom{u_{1234}}{v_1}}\,\mathrm{Qaa}^{\binom{u_3}{v_{3:1}}}\,\delta^{v_{1:4}}\,\delta^{v_{1:2}}\\
&-\tfrac{1}{480}\,c^2\,\mathrm{Qaa}^{\binom{u_{1234}}{v_1}}\,\mathrm{Qaa}^{\binom{u_4}{v_{4:1}}}\,\delta^{v_{1:3}}\,\delta^{v_{1:2}}\\
&+\tfrac{7}{5760}\,c^4\,\delta^{v_1}\,\delta^{v_2}\,\delta^{v_3}\,\delta^{v_4}
\end{aligned}$$

$$\mathrm{tiil}^{w_1} = -\tfrac{1}{2}\,\mathrm{Qii}^{\binom{u_1}{v_1}}$$

$$\begin{aligned}
\mathrm{tiil}^{w_1,w_2} = {}&\\
&+\tfrac{1}{6}\,\mathrm{Qii}^{\binom{u_{12}}{v_2}}\,\mathrm{Qii}^{\binom{u_1}{v_{1:2}}}\\
&+\tfrac{1}{12}\,\mathrm{Qii}^{\binom{u_{12}}{v_1}}\,\mathrm{Qii}^{\binom{u_2}{v_{2:1}}}\\
&-\tfrac{1}{8}\,c^2\,\delta^{u_1}\,\delta^{u_2}
\end{aligned}$$

$$\begin{aligned}
\mathrm{tiil}^{w_1,w_2,w_3} = {}&\\
&-\tfrac{1}{24}\,\mathrm{Qii}^{\binom{u_1}{v_{1:2}}}\,\mathrm{Qii}^{\binom{u_{12}}{v_2}}\,\mathrm{Qii}^{\binom{u_3}{v_3}}\\
&-\tfrac{1}{24}\,c^2\,\mathrm{Qii}^{\binom{u_1}{v_{1:2}}}\,\delta^{u_{12}}\,\delta^{u_3}\\
&+\tfrac{1}{24}\,c^2\,\mathrm{Qii}^{\binom{u_2}{v_{2:3}}}\,\delta^{u_1}\delta^{u_{23}}\\
&+\tfrac{1}{48}\,c^2\,\mathrm{Qii}^{\binom{u_3}{v_3}}\,\delta^{u_1}\,\delta^{u_2}
\end{aligned}$$

$$\begin{aligned}
\mathrm{tiil}^{w_1,w_2,w_3,w_4} = {}&\\
&-\tfrac{1}{720}\,\mathrm{Qii}^{\binom{u_1}{v_{1:2}}}\mathrm{Qii}^{\binom{u_{12}}{v_{2:3}}}\mathrm{Qii}^{\binom{u_{123}}{v_{3:4}}}\mathrm{Qii}^{\binom{u_{1234}}{v_4}}\\
&-\tfrac{1}{240}\,\mathrm{Qii}^{\binom{u_1}{v_1}}\mathrm{Qii}^{\binom{u_2}{v_{2:3}}}\mathrm{Qii}^{\binom{u_{23}}{v_{3:4}}}\mathrm{Qii}^{\binom{u_{234}}{v_4}}\\
&-\tfrac{1}{240}\,\mathrm{Qii}^{\binom{u_{12}}{v_1}}\mathrm{Qii}^{\binom{u_2}{v_{2:1}}}\mathrm{Qii}^{\binom{u_3}{v_{3:4}}}\mathrm{Qii}^{\binom{u_{34}}{v_4}}\\
&+\tfrac{1}{180}\,\mathrm{Qii}^{\binom{u_1}{v_{1:2}}}\mathrm{Qii}^{\binom{u_{12}}{v_{2:3}}}\mathrm{Qii}^{\binom{u_{123}}{v_3}}\mathrm{Qii}^{\binom{u_4}{v_4}}\\
&+\tfrac{1}{120}\,\mathrm{Qii}^{\binom{u_1}{v_1}}\mathrm{Qii}^{\binom{u_2}{v_{2:3}}}\mathrm{Qii}^{\binom{u_{23}}{v_3}}\mathrm{Qii}^{\binom{u_4}{v_4}}\\
&-\tfrac{1}{720}\,\mathrm{Qii}^{\binom{u_2}{v_{2:1}}}\mathrm{Qii}^{\binom{u_{12}}{v_1}}\mathrm{Qii}^{\binom{u_3}{v_3}}\mathrm{Qii}^{\binom{u_4}{v_4}}\\
&-\tfrac{1}{480}\,\mathrm{Qii}^{\binom{u_1}{v_1}}\,\mathrm{Qii}^{\binom{u_4}{v_4}}\,\delta^{u_2}\,\delta^{u_3}\\
&-\tfrac{1}{480}\,\mathrm{Qii}^{\binom{u_3}{v_3}}\,\mathrm{Qii}^{\binom{u_4}{v_4}}\,\delta^{u_1}\,\delta^{u_2}\\
&-\tfrac{5}{288}\,c^2\,\mathrm{Qii}^{\binom{u_2}{v_{2:3}}}\,\mathrm{Qii}^{\binom{u_4}{v_4}}\,\delta^{u_1}\,\delta^{u_{23}}\\
&+\tfrac{1}{360}\,c^2\,\mathrm{Qii}^{\binom{u_3}{v_{3:4}}}\,\mathrm{Qii}^{\binom{u_{34}}{v_4}}\,\delta^{u_1}\,\delta^{u_2}\\
&+\tfrac{19}{1440}\,c^2\,\mathrm{Qii}^{\binom{u_1}{v_{1:2}}}\,\mathrm{Qii}^{\binom{u_4}{v_4}}\,\delta^{u_{12}}\,\delta^{u_3}\\
&-\tfrac{1}{288}\,c^2\,\mathrm{Qii}^{\binom{u_1}{v_1}}\,\mathrm{Qii}^{\binom{u_3}{v_{3:4}}}\,\delta^{u_2}\,\delta^{u_{34}}\\
&-\tfrac{11}{1440}\,c^2\,\mathrm{Qii}^{\binom{u_1}{v_1}}\,\mathrm{Qii}^{\binom{u_3}{v_{3:2}}}\,\delta^{u_{23}}\,\delta^{u_4}\\
&+\tfrac{1}{480}\,c^2\,\mathrm{Qii}^{\binom{u_{12}}{v_1}}\,\mathrm{Qii}^{\binom{u_2}{v_{2:1}}}\,\delta^{u_3}\,\delta^{u_4}\\
&+\tfrac{1}{1440}\,c^2\,\mathrm{Qii}^{\binom{u_1}{v_{1:2}}}\,\mathrm{Qii}^{\binom{u_3}{v_{3:4}}}\,\delta^{u_{12}}\,\delta^{u_{34}}\\
&+\tfrac{1}{288}\,c^2\,\mathrm{Qii}^{\binom{u_1}{v_{1:2}}}\,\mathrm{Qii}^{\binom{u_{12}}{v_{2:3}}}\,\delta^{u_{123}}\,\delta^{u_4}\\
&-\tfrac{1}{480}\,c^2\,\mathrm{Qii}^{\binom{u_2}{v_{2:3}}}\,\mathrm{Qii}^{\binom{u_{23}}{v_{3:4}}}\,\delta^{u_1}\,\delta^{u_{234}}\\
&-\tfrac{1}{640}\,c^4\,\delta^{u_1}\,\delta^{u_2}\,\delta^{u_3}\,\delta^{u_4}
\end{aligned}$$

12.7 Index of terms and notations

Slight liberties have been taken with the alphabetical order, so as to regroup similar objects or notions.

ALAL: Section 2.4, Section 5.7, Section 7, Section 8.4.
ASAS: Section 2.8.
al/al, $\underline{al}/\underline{al}$: Section 2.7.

as/as, $\underline{as}/\underline{as}$: Section 2.8.
ALIL: Section 4.7, Section 5.7.
ASIS: Section 4.7.
ALIIL, ASIIS: Section 4.7
al/il, $\underline{al}/\underline{il}$: Section 5.7.
as/is, $\underline{as}/\underline{is}$: Section 4.7.
alternal: Section 2.4, (2.72).
alternil: Section 3.4.
anti: Section 2.1, (2.6).
ami, amit, ani, anit, ari, arit: Section 2.2.
axi, axit: Section 2.1.
approximate flexion unit: Section 3.2 (towards the end).
bialternal: Section 2.7, Section 7, Section 8.
bisymmetral: Section 2.8, Section 9.1.
carma$^\bullet$/carmi$^\bullet$, corma$^\bullet$/cormi$^\bullet$, curma$^\bullet$/curmi$^\bullet$: Sections 7.3, 7.7.
conjugate flexion units: Section 3.2.
dilator (infinitesimal): Sections 4.1, 11.8, 11.10, 12.3.
dimorphy, *dimorphic*: Section 1.1, Section 2, Section 10.1.
doma$^\bullet$/domi$^\bullet$: Section 7.2.
ekma$^\bullet$/ekmi$^\bullet$: Section 7.3.
$\mathfrak{E}^\bullet$: Section 3.1, Section 3.2.
$\mathfrak{E}^\bullet$-alternal: Section 3.4.
$\mathfrak{E}^\bullet$-symmetral: Section 3.4.
$\mathfrak{E}^\bullet$-mantar: Section 3.4, (3.46).
$\mathfrak{E}^\bullet$-gantar: Section 3.4, (3.49).
$\mathfrak{E}^\bullet$-push: Section 3.4, (3.53), (3.54).
$\mathfrak{E}^\bullet$-gush: Section 3.4, (3.60).
$\mathfrak{E}^\bullet$-neg: Section 3.4, (3.52).
$\mathfrak{E}^\bullet$-geg: Section 3.4, (3.59).
$\mathfrak{es}^\bullet$, $\mathfrak{ez}^\bullet$: Section 4.3, (4.70), (4.71).
$\mathfrak{ess}^\bullet$, $\mathfrak{esz}^\bullet$: Section 4.2, (4.35), (4.36), Sections 11.9, 11.10, 12.4, 12.5.
expari: Section 2.2, (2.50).
Exter(Qi_c): Section 11.5.
flexion: Section 2.1.
flexion unit: Section 3.2.
flexion structure: Section 2.
gami, gamit, gani, ganit, gari, garit: Section 2.2.
gantar, gantir: Section 2.3, (2.74), (2.75), Section 3.4.
gepar: Section 4.1, (4.10), Section 11.8.
hepar: Section 11.8, (4.10).
gegu, gegi: Section 3.5, (3.65).
gus: Section 2.4, (2.74), (2.75).

gusi, gusu: Section 3.4.
gush: Section 2.4, (2.76).
gushi, gushu: Section 3.4.
invmu: Section 2.1, (2.2).
invgami, invgani, invgari: Section 2.2, (2.58).
Inter(Qi_c): Section 11.5.
lama$^\bullet$/lami$^\bullet$: Section 6.5.
loma$^\bullet$/lomi$^\bullet$: Section 6.6.
luma$^\bullet$/lumi$^\bullet$: Section 6.7.
$\mathfrak{O}^\bullet$: Section 3.2.
$\mathfrak{me}^\bullet_r$: Section 4.1, Section 12.2.
$\mathfrak{ne}^\bullet_r$: Section 4.1, Section 12.2.
mantar, mantir: Section 2.1, (2.7), Section 3.4.
minu: Section 2.1, (2.4).
neg: Section 2.1, (2.8).
negi, negu: Section 3.4, (3.61).
pari: Section 2.1, (2.5).
P: $P(t) := 1/t$.
pac$^\bullet$/pic$^\bullet$, paj$^\bullet$/pij$^\bullet$: Section 4.3.
pal$^\bullet$/pil$^\bullet$, par$^\bullet$/pir$^\bullet$: Section 4.2 (last but one para).
perinomal: Section 9.4, Section 9.5, Section 9.6.
preami, preani, preari: Section 2.2.
predoma: Section 7.5.
precarma: Section 7.6.
pus: Section 2.1, (2.10).
pusi, pusu: Section 3.4.
push: Section 2.1, (2.11), (2.12).
pushi, pushu: Section 3.4, (3.62), (3.63).
Q, Q_c: $Q(t) := 1/\tan(t)$, $Q_c(t) := c/\tan(c\,t)$.
$\mathfrak{re}^\bullet_r$: Section 4.1, Section 12.2.
$\mathfrak{Re}^\bullet_f$, $\mathfrak{R\ddot{o}}^\bullet_f$: Section 12.3.
sap, swap, syap: Section 2.2, (2.9), Section 3.3, (4.37), (4.38), (4.70), (4.71).
separ: Section 10.9.
$\mathfrak{se}^\bullet_r$: Section 4.1.
$\mathfrak{sse}^\bullet_{12}$: Section 4.2.
$\mathfrak{Se}^\bullet_f$, $\mathfrak{S\ddot{o}}^\bullet_f$: Section 4.1, Section 11.8.
slank, srank, sang: Section 5.4, Section 5.5.
sen: Section 5.1.
senk, seng: Section 5.3.
singulator, singuland, singulate etc: Section 5.
symmetral: Section 2.4 (2.72).

symmetril: Section 3.5.
symmetry types (straight): Section 2.4.
symmetry types (twisted): Section 3.5.
subsymmetries (simple or double, straight): Section 2.4.
subsymmetries (simple or double, twisted): Section 3.5.
$\mathrm{tac}^\bullet/\mathrm{tic}^\bullet$, $\mathrm{taj}^\bullet/\mathrm{tij}^\bullet$: Section 4.2.
$\mathrm{tal}^\bullet/\mathrm{til}^\bullet$, $\mathrm{taal}^\bullet/\mathrm{tiil}^\bullet$: Section 12.6.
tripartite relation: Section 3.2, (3.9).
wandering bialternals: Section 6.9, Section 9.1.
$\mathrm{Wa}^\bullet$: Section 1.1.
$\mathrm{Za}^\bullet$: Section 1.2 (after (1.8)).
$\mathrm{Ze}^\bullet$: Section 1.1.
$\mathrm{Zag}^\bullet/\mathrm{Zig}^\bullet$: Section 1.2, Section 9.

References

[1] D. J. BROADHURST, "Conjectured Enumeration of irreducible Multiple Zeta Values, from Knots and Feynman diagrams", preprint Physics Dept., Open Univ. Milton Keynes, MK76AA, 1996.

[2] C. BREMBILLA, "Elliptic flexion units", to appear as part of an Orsay PhD Thesis, 2012.

[3] J. ECALLE, *A Tale of Three Structures: the Arithmetics of Multizetas, the Analysis of Singularities, the Lie algebra ARI* , In: "Diff. Equ. and the Stokes Phenomenon", B. L. Braaksma, G. K.Immink, M.v.d. Put, J. Top (eds.), World Scient. Publ., vol. 17, 2002, 89–146.

[4] J. ECALLE, *ARI/GARI, La dimorphie et l'arithmétique des multizêtas: un premier bilan*, Journal de Théorie des Nombres de Bordeaux **15** (2003), 411–478.

[5] J. ECALLE, *Multizetas, perinomal numbers, arithmetical dimorphy*, Ann. Fac. Toulouse **4** (2004), 683–708.

[6] J. ECALLE, *Multizeta Cleansing: the formula for eliminating all unit indices from harmonic sums*, forthcoming.

[7] J. ECALLE, *Remark on monogenous flexion algebras*, forthcoming.

[8] J. ECALLE, *Weighted products and parametric resurgence*, Travaux en Cours, 47, Hermann (ed.), 1994,7–49.

[9] J. ECALLE, *Singular perturbation and co-equational resurgence*, forthcoming.

[10] V. G. DRINFEL'D, *On quasi-triangular quasi-Hopf algebras and some groups related to* $Gal(\overline{\mathbb{Q}}/\mathbb{Q})$, Lening. Math. J. **2** (1991), 829–860.

N.B.: There exists of course a vast literature on multizetas and related lore: polylogarithms, associators, knots, Feynman diagrams, etc. Ample references are readily available at the end of papers dealing with any of these topics. The present article, however, is not primarily about multizetas, but about the *flexion structure*, which happens to be a new subject. Hence the paucity of our bibliographical references.

On the parametric resurgence for a certain singularly perturbed linear differential equation of second order

Augustin Fruchard and Reinhard Schäfke

Abstract. We consider the 1-dimensional stationary Schrödinger equation

$$\varepsilon^2 \frac{d^2 y}{dx^2} = P(x)y, \tag{0.1}$$

where x is a complex variable, ε is a small complex parameter and P is a polynomial, for the example $P(x) = x^2(x-1)^2$. Based on a global study of its actual solutions, especially their Stokes phenomena, we present an analytic proof of the resurgence of its formal solutions. Many details are given, but the complete proof will be presented in a future article.

1 Introduction

The 1-dimensional stationary Schrödinger equation (0.1) has been extensively studied from many viewpoints, also from the viewpoint of formal series and associated sums. We mention here only Aoki, Kawai, Takei [2], Voros [21], Delabaere, Dillinger, Pham [5]. We will not give any details concerning previous works in the present article, but only present the objects necessary for our approach; they are all classical. Our work is based on ideas of Sibuya [19, 20] and Ramis [17, 18] and can be seen as an example for the general theory of Ecalle [7–9].

The well known Liouville-Green (also called WKB)-approximation

$$y(x,\varepsilon) \approx P(x)^{-1/4} \exp\left(\pm\frac{1}{\varepsilon}\int^x \sqrt{P(x)}\,dx\right)$$

provides a good approximation for a fundamental system of solutions of (0.1)

- for fixed ε and large x,
- for a small neighborhood of a fixed x satisfying $P(x) \neq 0$ and small ε.

This can be found, for example, in the books of Olver [16] and Wasow [22].

The Liouville-Green approximation is not adequate near zeros of P. For simple zeros, $\sqrt{P(x)}$ is not even an analytic function. The zeros of P are the so-called *turning points* of (0.1). They also have been extensively studied, see for example the book of Wasow [22].

Except near the turning points, the Liouville-Green approximation can be refined to a formal series solution that we will call the WKB solution

$$\widehat{y}(x,\varepsilon)=e^{F(x)/\varepsilon}P(x)^{-1/4}\widehat{v}(x,\varepsilon), \text{ where } \widehat{v}(x,\varepsilon)=1+\sum_{n=1}^{\infty} y_n(x)\varepsilon^n, \qquad (1.1)$$

$F(x) = \int_0^x P(\xi)^{1/2}\, d\xi$ and $y_n(x)$ are locally analytic. As we will see, WKB solutions are not unique unless they are normalized appropriately. In order to define the normalizations most convenient for the subsequent sections, we use the following notation.

Definition 1.1. A *path of steepest descent for F, x and ψ* is a continuous function $\gamma : [0, +\infty[\to \mathbb{C}$ such that

$$\gamma(0) = x, \quad F(\gamma(\tau)) = F(x) - \tau e^{\psi i}.$$

The *Stokes lines for F and ψ* are the sets of x such that a path of steepest descent contains one of the turning points.

We call a WKB solution *normalized for F, x_0 and ψ* if x_0 is not on a Stokes line for F and ψ and for the corresponding path $\gamma : [0, +\infty[\to \mathbb{C}$ of steepest descent and for all $n \in \mathbb{N}$, we have $y_n(\gamma(\tau)) \to 0$ as $\tau \to \infty$.

Such a normalization is always possible, as we will see. Of course, any path homotopic to a path of steepest descent yields the same normalization.

The Liouville-Green approximation can also be refined to a formal solution

$$\widehat{y}(x, \varepsilon) = e^{F(x)/\varepsilon} P(x)^{-1/4} \left(1 + \sum_{n=1}^{\infty} b_n(\varepsilon) x^{-n}\right),$$

where the $b_n(\varepsilon)$ are polynomials. Equation (0.1) is a *doubly* singular equation at $x = \infty$, $\varepsilon = 0$ and the theory of *monomial summability* [3] can be applied. Details will not be given here.

The point of view of the theory of exact WKB expansions is that *an appropriately normalized WKB solution encodes all important information for the equation*:

- the series are 1-summable (except at turning points), the 1-sums are solutions,
- information about their analytic continuation can be derived from the fact – accepted without a complete published proof – that its *Borel transform* is *resurgent.*

The notion of "resurgent functions" introduced by Ecalle [7] can have different meanings in different contexts. In this article we restrict our study to a simple non-trivial example and prove the following result.

Theorem 1.2. *Consider equation* (0.1) *with* $P(x) = x^2(x-1)^2$, *i.e.*

$$\varepsilon^2 y'' = x^2(x-1)^2 y \tag{1.2}$$

$x_0 \in \mathbb{C}$, $x \neq 0, 1$ *and a* WKB *solution normalized for* F, x_0 *and* ψ. *Then, for* x *in some neighborhood of* x_0, *the Borel transform* $\widetilde{v}(x,t) = 1 + \sum_{n=1}^{\infty} y_n(x)\frac{t^n}{n!}$ *of* $\varepsilon\widehat{v}(x,\varepsilon)$ *converges in a neighborhood of* $t = 0$ *and the resulting function can be continued analytically along every path*[1] *in the complex plane avoiding the set* $\mathcal{S} = \big(2F(x) + \frac{1}{3}\mathbb{Z}\big) \cup \frac{1}{3}\mathbb{Z}$. *For any* $\zeta \in \mathcal{S}$ *and any such analytic continuation to a neighborhood of* ζ, *there exists* $m \in \mathbb{N}$ *such that* $\widetilde{v}(x,t) = \mathcal{O}\big(\log(t-\zeta)^m\big)$ *as* $t \to \zeta$ *in any sector with vertex* ζ. *Moreover, the analytic continuation along an infinite path that is eventually linear and not horizontal has at most exponential growth.*

Other normalizations will be discussed in Section 9.

The example we chose is simple because $F(x) = \int_0^x \xi(1-\xi)\,d\xi = \frac{1}{2}x^2 - \frac{1}{3}x^3$ is a polynomial — in general, F would be a hyperelliptic integral. The example is also non-trivial because there are two double turning points at $x = 0$ and $x = 1$ and the WKB solution contains logarithms as we will see. We are convinced that our method applies to prove the resurgence of formal solutions of any equation of the form (0.1).

The article is organized as follows. After introducing the solutions of (1.2) we work with (Section 2), we state their Stokes relations and begin our investigation of the wronskians – the functions of ε determining the global behavior of the solutions (Section 3). The behavior of the

[1] We only consider paths of class C^1.

solutions near the turning points implies asymptotic expansions for the most interesting wronskians (Section 4), but these contain logarithms. In the principal part of the article, we establish Stokes relations for wronskians and functions related to their asymptotic expansions, in particular the residue, which is essentially the factor of the logarithm (Sections 5, 6 and 7, in particular Theorem 7.1). As a consequence of the form of the Stokes relations we derive the resurgence of the formal series associated to the wronskians (Section 8) and are finally able to deduce the resurgence of the WKB series (Section 9). We finish with some remarks and perspectives.

2 The distinguished solutions

In the rest of the article, we focus on equation (1.2). If $x_0 \neq 0, 1$ then (1.2) admits WKB solutions

$$\widehat{y}^{\pm}(x,\varepsilon) = e^{\pm F(x)/\varepsilon} x^{-1/2}(x-1)^{-1/2}\widehat{v}^{\pm}(x,\varepsilon), \tag{2.1}$$

with $\widehat{v}^{\pm}(x,\varepsilon) = 1 + \sum_{n=1}^{\infty} y_n(x)(\pm\varepsilon)^n$, where $y_n(x)$ are analytic near x_0. These WKB solutions are not unique because multiplication by any formal series $c(\varepsilon) = 1 + \sum_{n=1}^{\infty} c_n \varepsilon^n$ independent of x leads again to a WKB solution.

We will mostly study $\widehat{y}^{+}$ because results for $\widehat{y}^{-}$ are obtained by the symmetry $\varepsilon \mapsto -\varepsilon$. In our example, also $x \mapsto 1-x$ is a symmetry that we can apply to carry over the results from the turning point 0 to the turning point 1. The latter symmetry will simplify the presentation in our example and reduce the number of functions to consider. It is, however, not essential for our approach.

Here and later on, it will be useful to study a Riccati equation associated to (1.2); this is a classical strategy. Putting $u = \varepsilon y'/y$ and $u = x(1-x) + z$, we obtain

$$\varepsilon z' = -2x(1-x)z - z^2 - \varepsilon(1-2x). \tag{2.2}$$

The WKB series corresponds to its uniquely determined formal solution

$$\widehat{z}^{+}(x,\varepsilon) = \sum_{n=1}^{\infty} z_n(x)\varepsilon^n \tag{2.3}$$

which has rational coefficients $z_n(x)$ having poles only at 0, 1. By using the recursion formula

$$2x(1-x)z_n(x) = -z_{n-1}'(x) - \sum_{k=1}^{n-1} z_k(x) z_{n-k}(x),\ n \geq 2,$$

with initial term $z_1(x) = \frac{2x-1}{2x(1-x)}$, we also see that $z_n(x) = \mathcal{O}(x^{-3n+2})$ as $x \to \infty$.

As

$$\widehat{y}^+(x,\varepsilon) = e^{F(x)/\varepsilon} x^{-1/2}(x-1)^{-1/2} \exp\left(\frac{1}{\varepsilon}\int^x (\widehat{z}^+(\xi,\varepsilon) - \varepsilon z_1(x))d\xi\right), \quad (2.4)$$

this implies that the coefficients $y_n(x)$ of the WKB solutions are combinations of $\log x$, $\log(x-1)$ and rational functions, more precisely elements of $\mathbb{C}(x)[\log x, \log(x-1)]$. It also shows that we can normalize the WKB solution $\widehat{y}^+(x,\varepsilon)$ by taking the integral from infinity to x along some path avoiding 0 and 1. As stated in the introduction, normalizations along paths of steepest descent are most convenient for us.

The first important property of the WKB solution is the following result.

Theorem 2.1. *Suppose that x_0 is not on a Stokes line for F and ψ. If the* WKB *solution $\widehat{y}^+(x,\varepsilon)$ is normalized for F, x_0 and ψ according to Definition* 1.1 *then it is* 1*-summable in the direction* $\arg\varepsilon = \psi$ *uniformly for x near x_0, i.e. $\widetilde{v}(x,t) = 1 + \sum_{n=1}^{\infty} y_n(x)\frac{t^n}{n!}$ converges for small t, $\widetilde{v}(x,t)$ can be continued analytically along* $\arg t = \psi$,

$$y(x,\varepsilon) = e^{F(x)/\varepsilon} x^{-1/2}(x-1)^{-1/2}\frac{1}{\varepsilon}\int_{\arg t=\psi} e^{-t/\varepsilon}\,\widetilde{v}(x,t)\,dt$$

converges and yields a solution of (1.2) *for ε small,* $|\arg\varepsilon - \psi| < \frac{1}{2}\pi$.

One proof of this result consists in a reduction of (1.2) to the framework of [6] by using the change of variables $\zeta = F(x)$ in a neighborhood of the path of steepest descent. This yields an equation $\varepsilon^2\frac{d^2y}{d\zeta^2} = \big(1 + \varepsilon^2\psi(\zeta)\big)y$ that satisfies the conditions of [6] for $\zeta_0 = F(x_0)$. An independent proof can be found in Costin *et al.* [4].

We would like to present two other ideas of proof of this result. For the first one, we consider again the corresponding Riccati equation (2.2). For small positive δ, let $M(x_0,\psi,\delta)$ denote the domain of all x for which there exist $\widetilde{x}_0$, $\widetilde{\psi}$ with $|\widetilde{x}_0 - x_0| < \delta$, $\left|\widetilde{\psi} - \psi\right| < \delta$, such that x is on the path of steepest descent for F, $\widetilde{x}_0$ and $\widetilde{\psi}$. Let $S(\psi,\theta)$ denote the sector of all ε with $0 < |\varepsilon| < \delta$, $|\arg\varepsilon - \psi| < \theta$. On the set of functions holomorphic and bounded on $M(x_0,\psi,\delta) \times S(\psi, \frac{1}{2}\pi - 2\delta)$, it turns out that the Banach fixed point theorem applies to the equation

$$z = \mathcal{T}z,\ (\mathcal{T}w)(x,\varepsilon) := -\frac{1}{\varepsilon}\int_{\infty}^{x} e^{(-2F(x)+2F(t))/\varepsilon}\left(w(t,\varepsilon)^2 + \varepsilon(1-2t)\right)dt.$$

Its solution z satisfies (2.2) and is bounded. By classical arguments of Gevrey theory (see *e.g.* [12] Section 6), $z(x, \varepsilon)$ has an asymptotic expansion of Gevrey order 1 as $S(\psi, \frac{1}{2}\pi - 2\delta) \ni \varepsilon \to 0$, uniformly on $M(x_0, \psi, \delta)$. By the uniqueness of the formal solution (2.3) of (2.2), we have

$$z(x, \varepsilon) \sim \widehat{z}(x, \varepsilon) = \sum_{n\geq 1} z_n(x)\varepsilon^n \text{ as } S\left(\psi, \tfrac{\pi}{2} - 2\delta\right) \ni \varepsilon \to 0. \tag{2.5}$$

If δ is small enough, it can be seen that, for all $\xi \in M(x_0, \psi, \delta)$ and all $|\mu| < 3\delta$, the solutions of the analogous fixed point equations on $M(\xi, \psi + \mu, \delta) \times S(\psi + \mu, \frac{1}{2}\pi - 2\delta)$ exist and are analytic continuations of the above solution. Combining them yields a function $z : M(x_0, \psi, \delta) \times S(\psi, \frac{1}{2}\pi + \delta) \to \mathbb{C}$ having a Gevrey-1 asymptotic expansion as $S(\psi, \frac{1}{2}\pi + \delta) \ni \varepsilon \to 0$, uniformly on $M(x_0, \psi, \delta)$. This is one of the classical definitions of 1-summability (*c.f.* [18]). The result for y now follows by solving $\varepsilon y'/y = z + x(1 - x)$, *i.e.* using

$$y(x, \varepsilon) = e^{F(x)/\varepsilon} x^{-1/2}(x-1)^{-1/2} \exp\left(\frac{1}{\varepsilon}\int_\infty^x \big(z(t, \varepsilon) - \varepsilon z_1(t)\big)\, dt\right). \tag{2.6}$$

Here the fact that $z(x, \varepsilon) - \varepsilon z_1(x) = \mathcal{O}(x^{-4}\varepsilon^2)$ is used and the determinations of $\arg x$ and $\arg(x - 1)$ are chosen continuously such that $x^{-1/2}(x - 1)^{-1/2} \sim x^{-1}$ as $M(x_0, \psi, \delta) \ni x \to \infty$.

The third proof is only valid for sufficiently large $|x_0|$. Here the theory of monomial summability [3] can be applied to (2.2) and yields monomial summability of the formal double series $\sum_{n,m} z_{nm}x^{-m}\varepsilon^n$ obtained by replacing the coefficients $z_n(x)$ of (2.3) by their Taylor expansions at $x = \infty$. For large $|x_0|$, this implies summability with respect to ε except for some sectors the angular opening of which tends to 0 as $|x_0|$ tends to ∞. This is detailed in [3]. Again, the result is carried over to the second order equation by using (2.6).

Observe that (2.6) also implies that the solution y of the theorem is uniquely determined by the asymptotic approximation

$$y(x, \varepsilon) = e^{F(x)/\varepsilon} x^{-1}\big(1 + \mathcal{O}(x^{-1})\big) \tag{2.7}$$

as $x \to \infty$ in $M(x_0, \psi, \delta)$ for $\varepsilon \in S(\psi, \frac{1}{2}\pi - 2\delta)$. This characterization by the asymptotic behavior as $x \to \infty$ allows to define $y(x, \varepsilon)$ for all values of $\varepsilon \in S(\psi, \frac{1}{2}\pi - 2\delta)$, $0 < |\varepsilon| < \infty$.

As $F(x) \approx -\frac{1}{3}x^3$ for large x, there are three possible "valleys" at $x = \infty$ where the paths of steepest descent might lead. This yields three recessive solutions (with asymptotic expansion $\widehat{y}^+$) defined by summability.

An animated gif available on the web[2] shows how the three summability domains change with $\arg\varepsilon = \psi$. Of course, these three domains are disjoint as a series cannot have two 1-sums. The summability domains can also be seen in the right column of Figure 2.1.

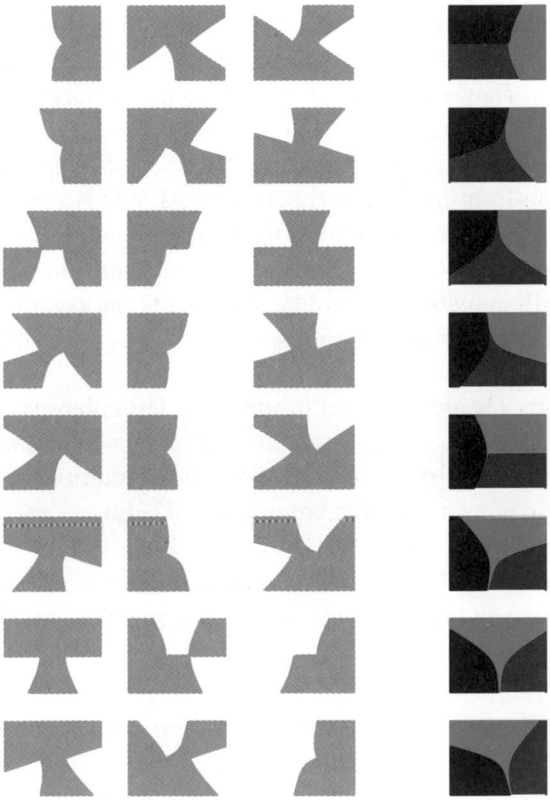

Figure 2.1. The domains $\mathcal{D}_\phi$, $\mathcal{D}_{\phi+\frac{2}{3}\pi}$ and $\mathcal{D}_{\phi+\frac{4}{3}\pi}$ and the corresponding three summability domains for $\phi = \ell\frac{\pi}{12} + \delta$, where $\ell = 0, \dots, 7$ and $\delta > 0$ is small.

These solutions defined as sums can be continued analytically[3] without change in their asymptotic behavior along any path that is *ascending* for F and ψ, *i.e.* along which $\mathrm{Re}\,(F(\gamma(\tau))e^{-\psi i})$ is increasing. This can be seen for (2.2) by using a fixed point argument or by applying Lemma 7 of [13]; it is carried over to the linear equation by means of (2.6). Again,

[2] `http://www-irma.u-strasbg.fr/~schaefke/summability.gif`

[3] Of course, as solutions of a linear differential equation, these functions are entire with respect to x, but they only have the asymptotic behavior specified below in the indicated domain.

the turning points are special as $F(x) \approx \frac{1}{2}x^2$ near 0 and $F(x) \approx -\frac{1}{2}(x-1)^2$ near 1: they are saddle points for $x \mapsto \mathrm{Re}\,(F(x)\,e^{-\psi i})$. In their neighborhood there are thus *two* "valleys" and *two* "mountains" for F and ψ.

Consider now, for a given value of ψ, the maximal domains into which our three solutions can be analytically continued with the same asymptotic expansion. These domains are denoted by $\mathcal{D}_\phi$, $\mathcal{D}_{\phi+2\pi/3}$ and $\mathcal{D}_{\phi+4\pi/3}$ with $\phi = \frac{1}{3}\psi$. They are bounded by certain paths starting at the turning points along which $\arg(F(x) - F(1))$ or $\arg F(x)$ is constant equal to $\psi + \frac{1}{2}\pi$ or $\psi - \frac{1}{2}\pi$. To be more precise, the above asymptotic expansions are not only uniform for x on compact subsets but also on infinite closed subdomains $\mathcal{D}_\phi^\delta$ of $\mathcal{D}_\phi$ not containing 0 and 1, with $\delta > 0$. These are constructed similarly to [11], Section 2.1, and their union is $\mathcal{D}_\phi$. In the sequel, we will sometimes simplify statements on asymptotic behavior and not mention uniformity with respect to the other variable as we have done here.

The domains $\mathcal{D}_\phi$ are sketched in the three left columns of Figure 2.1. The dependence of $\mathcal{D}_\phi$ upon ϕ can also be seen in an animated gif.[4] For simplicity, let us identify $\mathcal{D}_\varphi$ and $\mathcal{D}_\phi$ if $\varphi - \phi$ is a multiple of 2π. Observe that the turning point 1 is on the boundary of $\mathcal{D}_\phi$ for $-\frac{5}{6}\pi < \phi < \frac{5}{6}\pi$ and that $x = 0$ is on its boundary for $\frac{1}{6}\pi < \phi < \frac{11}{6}\pi$.

Let us summarize the situation as follows: there are domains $\mathcal{D}_\phi$ containing x large with $|\arg x - \phi| < \frac{1}{2}\pi$ [5], such that for $x \in \mathcal{D}_\phi$ away from the turning points and $\arg\varepsilon = 3\phi$

$$y_\phi^+(x,\varepsilon) = e^{F(x)/\varepsilon} x^{-1/2}(x-1)^{-1/2} v_\phi^+(x,\varepsilon), \text{ where}$$
$$v_\phi^+(x,\varepsilon) \sim \widehat{v}^+(x,\varepsilon) = 1 + \sum_{n=1}^{\infty} y_n(x)\varepsilon^n \text{ as } \varepsilon \to 0.$$

Here, $\arg x$ and $\arg(x-1)$ are considered as continuous (and hence analytic) functions on $\mathcal{D}_\phi$ that are determined by their values $\arg(x-1) \approx \arg x$ for large $x \in \mathcal{D}_\phi$ with $|\arg x - \phi| < \frac{1}{2}\pi$. For all n we have $y_n(x) \to 0$ as $x \to \infty$ in $|\arg x - \phi| < \frac{1}{2}\pi$. Moreover, y_ϕ^+ is uniquely determined by

$$y_\phi^+(x,\varepsilon) = e^{F(x)/\varepsilon} x^{-1}\big(1 + \mathcal{O}(x^{-1})\big) \tag{2.8}$$

[4] http://www-irma.u-strasbg.fr/~schaefke/domains.gif

[5] More precisely, for all $\delta > 0$ the exists $K > 0$ such that $\mathcal{D}_\phi$ contains all x with $|x| > K$, $|\arg\varepsilon - \phi| < \frac{1}{2}\pi - \delta$.

as $x \to \infty$, $|\arg x - \phi| < \frac{1}{2}\pi$. Therefore y_ϕ^+ and y_φ^+ coincide if $\phi - \varphi$ is a multiple of 2π.

The functions $y_\phi^+(x,\varepsilon)$, $y_{\phi+2\pi/3}^+(x,\varepsilon)$, and $y_{\phi+4\pi/3}^+(x,\varepsilon)$ coexist for the same value of $\arg\varepsilon = 3\phi \mod 2\pi$ and they are the three functions we discussed above.

The dependence of $y_\phi^+(x,\varepsilon)$ upon ε is also analytic: all the functions $y_\phi^+(x,\varepsilon)$ come from **one** function $y^+(x,\varepsilon)$ holomorphic with respect to x and ε — with the indicated asymptotic behavior — in the following domain. We consider ε on a threefold covering of the punctured disk $0 < |\varepsilon| < \infty$ by identifying points on the Riemann surface of the logarithm $\log\varepsilon$ whose arguments differ by 6π. Then the domain of y^+ is the set of all (x,ε) such that $x \in \mathcal{D}_\phi$, where $\phi = \frac{1}{3}\arg\varepsilon$. In the sequel we prefer to use only the functions y_ϕ^+, however, and only consider $\arg\varepsilon \mod 2\pi$. Moreover, it would be natural to consider y_ϕ as a function of ε in a small sector bissected by 3ϕ, but for simplicity we will consider y_ϕ only on the ray $\arg\varepsilon = 3\phi$.

The functions y_ϕ^- and v_ϕ^- corresponding to the WKB solutions $\widehat{y}^-$ are defined by symmetry. Precisely

$$y_\phi^-(x,\varepsilon) = y_{\phi+\pi}^+(x,-\varepsilon)\,, \qquad v_\phi^-(x,\varepsilon) = v_{\phi+\pi}^+(x,-\varepsilon) \tag{2.9}$$

for ε with $\arg\varepsilon = 3\phi$ and $x \in \mathcal{D}_\phi^- := \mathcal{D}_{\phi+\pi}$.

The following symmetry relation for $x \mapsto 1-x$ will simplify the presentation for our example.

$$y_\phi^+(1-x,\varepsilon) = -e^{1/6\varepsilon}\, y_\phi^-(x,\varepsilon) \text{ for } x \in \mathcal{D}_\phi^- \tag{2.10}$$

where $\arg\varepsilon = 3\phi$. Here the formula $F(1-x) = \frac{1}{6} - F(x)$ is used.

3 Stokes relations for the functions

In order to find out the asymptotic expansions of the functions $y_\phi^\pm$ in other regions, *Stokes relations* between the solutions are useful and necessary. Consider, for example, a point x_0 in the left half-plane to the left of the curve $F(x) \in -\mathbb{R}_+$ containing 0. Then there is a unique $\psi \in]\pi, 3\pi[$ such that the turning point 0 is on the path of steepest descent for F, x_0 and ψ. For small $\delta > 0$, the path of steepest descent for $\psi + 3\delta$ has $\arg x = \phi + \delta$, $\phi = \frac{1}{3}\psi$, as asymptote as $x \to \infty$; the analogous path for $\psi - 3\delta$ has $\arg x = \phi - \delta + \frac{2}{3}\pi$ as asymptote. Hence, $x_0 \in \mathcal{D}_\phi \cap \mathcal{D}_{\phi+2\pi/3}$ and the 1-sum of the formal solution of Theorem 2.1 for x_0 and $\psi + 3\delta$ is $y_{\phi+\delta}^+(x_0,\varepsilon)$, whereas for $\psi - 3\delta$, it is $y_{\phi-\delta+2\pi/3}^+(x_0,\varepsilon)$.

Together with the two functions y_ϕ^+, $y_{\phi+2\pi/3}^+$, we consider

$$y_{\phi-2\pi/3}^-(x,\varepsilon) = y_{\phi+\pi/3}(x,-\varepsilon),$$

because the paths of steepest descent for F, x_0 and $\psi + \pi \pm \delta$ are both close to $\arg x = \phi + \frac{1}{3}\pi$ as $x \to \infty$ and pose no problem. Then we have a Stokes relation

$$y^+_{\phi+2\pi/3}(x,\varepsilon) = \alpha(\varepsilon) y^+_\phi(x,\varepsilon) + \beta(\varepsilon) y^-_{\phi-2\pi/3}(x,\varepsilon), \tag{3.1}$$

where $\alpha(\varepsilon)$,$\beta(\varepsilon)$ are independent of x. First of all, this relation is interesting in the neighborhood of x_0, but as the solutions are entire functions with respect to x, they are valid throughout $\mathbb{C}$. The asymptotic behavior of the three functions as $x \to \infty$, $\arg x = \phi + \pi/3$ yields easily $\alpha(\varepsilon) = 1$. Hence this Stokes relation contains one interesting Stokes multiplier $\beta(\varepsilon)$.

As we have coupled $\arg \varepsilon = 3\phi$, (3.1) defines $\beta(\varepsilon)$ for one argument of ε only. We have seen, however, that $\{y^\pm_\phi\}_{\phi\in[0,2\pi]}$ come from two analytic functions of (x,ε) and hence $\beta(\varepsilon)$ is analytic[6] in $0 < \left|\varepsilon^{1/3}\right| < \infty$. Formula (3.1) is also a Stokes relation for the functions v_ϕ.

$$v^+_{\phi+2\pi/3}(x,\varepsilon) = v^+_\phi(x,\varepsilon) + \beta(\varepsilon) e^{-2F(x)/\varepsilon}\, v^-_{\phi-2\pi/3}(x,\varepsilon)\,. \tag{3.2}$$

As we have seen, these relations give information about the 1-sums concerning analytic continuation with respect to ε as $\arg\varepsilon$ changes, when x is fixed, but also concerning analytic continuation with respect to x across Stokes lines, when ε is fixed. *So it would be very useful to know more about the coefficient function $\beta(\varepsilon)$.*

For that purpose we use *wronskians*: for two solutions y_1, y_2 of (1.2) put

$$[y_1, y_2](\varepsilon) = \tfrac{\varepsilon}{2}(y_1 y_2' - y_2 y_1')(x,\varepsilon);$$

this is independent of x. Our definition contains the unusual factor $\frac{\varepsilon}{2}$ for a reason that will become apparent in the sequel. The wronskians are bilinear and antisymmetric and they satisfy the relations

$$\begin{aligned} &[y_1, y_2]y_3 + [y_2, y_3]y_1 + [y_3, y_1]y_2 = 0,\\ &[y_1, y_2][y_3, y_4] - [y_1, y_3][y_2, y_4] + [y_1, y_4][y_2, y_3] = 0 \end{aligned} \tag{3.3}$$

for any three respectively four solutions of (1.2). From their asymptotic expansions as $x \to \infty$ one finds

$$[y^+_\phi, y^-_{\phi-2\pi/3}] = [y^+_{\phi+2\pi/3}, y^-_{\phi-2\pi/3}] = 1. \tag{3.4}$$

---

[6] Actually, a change of variable $x = \varepsilon^{1/3}X$ shows that the limit $\lim_{\varepsilon\to\infty}\beta(\varepsilon)$ exists and hence β is an entire function of $\varepsilon^{-1/3}$ as is mentioned in [9].

Indeed, since $y_\phi^+(x,\varepsilon) = e^{F(x)/\varepsilon}x^{-1}\big(1+\mathcal{O}(x^{-1})\big)$ and $y_{\phi-2\pi/3}^-(x,\varepsilon) = e^{-F(x)/\varepsilon}x^{-1}\big(1+\mathcal{O}(x^{-1})\big)$ as $x \to \infty$, $\arg x \in]\phi - \frac{\pi}{6}, \phi + \frac{1}{2}\pi[$, a little calculation yields the first result. The proof of the second is analogous. The factors in the definition of the wronskians have been chosen in order to obtain 1 in (3.4).

Using the wronskians and (3.1), we find that $\beta(\varepsilon) = [y_\phi^+, y_{\phi+2\pi/3}^+]$. Thus the wronskians $[y_\phi^+, y_{\phi+2\pi/3}^+]$, $[y_\phi^-, y_{\phi+2\pi/3}^-]$ and, for Stokes relations like

$$y_\phi^+(x,\varepsilon) = A(\varepsilon)y_{\phi+2\pi/3}^+(x,\varepsilon) + B(\varepsilon)y_\phi^-(x,\varepsilon), \tag{3.5}$$

also $[y_\phi^+, y_\phi^-]$ have to be studied. The latter Stokes relation occurs if the considerations of the beginning of this section are carried out for x_0 between the lines $F(x) \in -\mathbb{R}_+$ and $F(x) - F(1) \in \mathbb{R}_+$.

Observe that using the asymptotic behavior of $y_\phi^\pm$ as $x \to \infty$ only results in $[y_{\phi+2\pi/3}^\pm, y_\phi^\pm] = \mathcal{O}\left(e^{\pm 2F(x)/\varepsilon}x^{-n}\right)$ for all n because both functions have the *same* asymptotic expansion. As we will see, it is possible to determine their asymptotic behavior as $\varepsilon \to 0$ from the behavior of the solutions at the turning points.

4 Behavior near the turning points

The behavior near the turning points is again first studied for the solutions of the associated Riccati equation (2.2). The formula $z_\phi^+ = \varepsilon {y_\phi^+}'/y_\phi^+ - x(1-x)$ provides a solution of (2.2) on the domain $\mathcal{D}_\phi$ having the formal solution $\widehat{z}^+$ of (2.3) as asymptotic expansion away from the turning points.

Our recent work [14] yields a *composite asymptotic expansion* near the turning points. By symmetry, is is sufficient to state this for the turning point 0. For those values of ϕ where z_ϕ^+ is defined near 0, *i.e.* $\frac{1}{6}\pi < \phi < \frac{11}{6}\pi$, by the results of [14], Chapter 6, there are functions a_n and g_n such that

$$z_\phi^+(x,\varepsilon) \sim g_1(T)\,\varepsilon^{1/2} + \sum_{n=2}^{\infty}\big(a_n(x) + g_n(T)\big)\,\varepsilon^{n/2}. \tag{4.1}$$

with $T = x\varepsilon^{-1/2}$, $\arg\varepsilon = 3\phi$ and $\varepsilon \to 0$; the expansion is uniform with respect to x in a domain containing the turning point 0 on its boundary. Precisely, for any $\delta > 0$ there exists $\mu > 0$ such that the expansion is uniform for $x \in \mathcal{D}_\phi$ such that $|x| \le \mu$ and $\frac{3}{2}\phi - \frac{5}{4}\pi + \delta \le \arg x \le \frac{3}{2}\phi + \frac{1}{4}\pi - \delta$.[7] Here the functions

- $a_{2n}(x) = z_n(x) - p_n(x)$, where $z_n(x)$ is a coefficient of the formal solution (2.3) and $p_n(x)$ is its principle part at $x = 0$. Hence they are polynomials in $(x-1)^{-1}$, $a_{2n+1} = 0$,
- $g_n(T)$ are analytic in an infinite sector $\arg T \in]-\frac{5}{4}\pi, \frac{1}{4}\pi[$
- and have 2-summable asymptotic expansions $g_n(T) \sim \sum_{m=1}^{\infty} g_{nm} T^{-m}$ as $T \to \infty$, $g_{nm} = 0$ if $n + m$ is odd.

As explained in the memoir [14], such a composite expansion yields the *outer* expansion (2.3) for $x \in \mathcal{D}_\phi, 0 < \widetilde{\delta} \le |x| \le \delta$ by using the asymptotics of the g_n and an *inner* expansion valid for x of order $\varepsilon^{1/2}$ using the Taylor series of the functions a_n. The right hand side of (4.1) (and also those of the inner and outer expansion) is a formal solution of (2.2); for the elementary operations with composite formal series, see [14]. Especially, the function g_1 satisfies the *inner equation*

$$g_1' = -2Tg_1 - g_1^2 - 1$$

associated to (2.2). This inner equation is obtained by replacing x by $\varepsilon^{1/2}T$, z by $\varepsilon^{1/2}g$, dividing by ε and then replacing ε by 0.

The inner equation is associated to the equation $h' = T^2 - h^2$ by $h = T + g_1$ and this is the Riccati equation associated to $w'' = T^2 w$ – the inner equation corresponding to (1.2). It can be shown that $g_1(T) = \frac{w_0'}{w_0}(T) - T$, where w_0 is the unique (recessive) solution of $w'' = T^2 w$ equivalent to $e^{T^2/2}T^{-1/2}$ as $T \to \infty$, $\arg T \in]-\frac{5}{4}\pi, \frac{1}{4}\pi[$. Hence $w_0(T) = 2^{1/4}e^{\pi i/4}U(0, i\sqrt{2}T)$, where $U(a,x)$ denotes the Whittaker function, see *e.g.* [1], Chapter 19. Numerically, it turns out that the function w_0 has no zeros in the closed sector $\arg T \in [-\frac{5}{4}\pi, \frac{1}{4}\pi]$, hence g_1 can be analytically continued to a neighborhood of this sector. By Corollary 6.17 of [14], this implies that z_ϕ can be analytically continued to a neighborhood of the closure of $\mathcal{D}_\phi$, especially to a neighborhood of order $\varepsilon^{1/2}$ of $x = 0$, and that the composite asymptotic expansion remains valid there.

Using (2.6) and methods — such as integration of composite asymptotic expansions — developed in [14] for analogous linear second order equations, it is possible to obtain a complete composite expansion for y_ϕ^+

---

[7] The Stokes lines limiting $\mathcal{D}_\phi$ near $x = 0$ are tangent to the rays $\arg x = \frac{3}{2}\phi - \frac{5}{4}\pi$ and $\arg x = \frac{3}{2}\phi + \frac{1}{4}\pi$.

near 0. Here, however, we only give an asymptotic formula for $y_\phi^+(0,\varepsilon)$, $\frac{1}{6}\pi < \phi < \frac{11}{6}\pi$:

$$y_\phi^+(0,\varepsilon) \sim -\varepsilon^{-1/4} \exp\left(\tfrac{1}{2}\widehat{r}(\varepsilon)\log\varepsilon\right) \sum_{n=0}^{\infty} \beta_n \varepsilon^{n/2} \qquad (4.2)$$
$$\text{as } \arg\varepsilon = 3\phi,\ \varepsilon \to 0,$$

where $\widehat{r}(\varepsilon) = \sum_{n=1}^{\infty} g_{2n+1,1}\varepsilon^n$ is the series of the coefficients of $1/T$ in the expansions of the $g_{2n+1}(T)$ and $\beta_0 = w_0(0) = e^{\pi i/4}\sqrt{\pi}\frac{1}{\Gamma}\left(\frac{3}{4}\right)$. Of course, such a formula could also be found by other methods. The series $\widehat{r}(\varepsilon)$, which appears frequently, will be called the "residue" in the sequel.

It can be shown for a general polynomial P that $\widehat{r}$ is odd, *i.e.* $\widehat{r}(-\varepsilon) = -\widehat{r}(\varepsilon)$. In our example, this follows from three facts: first, $g_{2n-1,1}$ is the residue at $x = 0$ of $z_n(x)$. Secondly $z_n(1-x) = (-1)^n z_n(x)$ for all n and hence $\operatorname{res}(z_n, 0) = (-1)^{n+1}\operatorname{res}(z_n, 1)$. Finally the sum of the residues of z_n at 0 and 1 is zero for $n \geq 2$ because the residue at $x = \infty$ vanishes.

The expansion for $y_\phi^{+\prime}(0,\varepsilon)$ is similar. It is obtained by the formula $y_\phi^{+\prime}(0,\varepsilon) = \frac{1}{\varepsilon} y_\phi^+(0,\varepsilon) z_\phi^+(0,\varepsilon)$ using (4.1). It can be concluded that the wronskian has an expansion

$$[y_\phi^+, y_{\phi+2\pi/3}^+](\varepsilon) \sim \exp\left(\widehat{r}(\varepsilon)\log\varepsilon\right)\widehat{\gamma}^+(\varepsilon),\ \widehat{\gamma}^+(\varepsilon) = \sum_{n=0}^{\infty} \gamma_n\, \varepsilon^{n/2} \qquad (4.3)$$

with certain γ_n, in the case where both $\mathcal{D}_\phi$ and $\mathcal{D}_{\phi+2\pi/3}$ contain 0 on their boundary, *i.e.* $\frac{1}{6}\pi < \phi < \frac{7}{6}\pi$; here we still have $\arg\varepsilon = 3\phi$. As $w_0(T) = 2^{1/4}e^{\pi i/4}U(0, i\sqrt{2}\,T)$, the formulas of [1], Chapter 19, yield $\gamma_0 = i\,w_0(0)w_0'(0) = i\sqrt{2}$.

We now prove that $\widehat{\gamma}^+(\varepsilon)$ is actually a formal series in ε. Indeed, if $\frac{1}{6}\pi < \phi < \frac{1}{2}\pi$ then on the one hand (4.2) is valid with $\phi + \frac{4}{3}\pi$ instead of ϕ; then $\widehat{r}(\varepsilon)$ and $\varepsilon^{n/2}$ remain unchanged but $\log\varepsilon$ has to be changed into $\log\varepsilon + 4\pi i$ and $\varepsilon^{-1/4}$ into $-\varepsilon^{-1/4}$, giving

$$y_{\phi+4\pi/3}^+(0,\varepsilon) \sim e^{2\pi i\widehat{r}(\varepsilon)}\varepsilon^{-1/4} \exp\left(\tfrac{1}{2}\widehat{r}(\varepsilon)\log\varepsilon\right) \sum_{n=0}^{\infty} \beta_n \varepsilon^{n/2}$$
$$\text{as } \arg\varepsilon = 3\phi,\ \varepsilon \to 0.$$

In other terms, the quotient $y_{\phi+4\pi/3}^+(0,\varepsilon)/y_\phi^+(0,\varepsilon)$ is asymptotic to $-e^{2\pi i\widehat{r}(\varepsilon)}$; the same is true for the quotient $(y_{\phi+4\pi/3}^+)'(0,\varepsilon)/(y_\phi^+)'(0,\varepsilon)$.

On the other hand, if $\frac{1}{6}\pi < \phi < \frac{1}{2}\pi$ then (4.3) is also valid with $\phi+\frac{2}{3}\pi$ instead of ϕ; then $\widehat{r}(\varepsilon)$ is unchanged, $\log\varepsilon$ is changed into $\log\varepsilon + 2\pi i$ and $\widehat{\gamma}^+(\varepsilon)$ into $\widehat{\gamma}^+\big(\varepsilon e^{2\pi i}\big) = \sum_{n=0}^{\infty}(-1)^n\gamma_n\,\varepsilon^{n/2}$. This gives

$$[y^+_{\phi+4\pi/3}, y^+_{\phi+2\pi/3}](\varepsilon) \sim -\exp\big(\widehat{r}(\varepsilon)(\log\varepsilon + 2\pi i)\big)\widehat{\gamma}^+(\varepsilon e^{2\pi i}).$$

By comparison, we obtain $\sum_{n=0}^{\infty}\gamma_n\,\varepsilon^{n/2} = \sum_{n=0}^{\infty}(-1)^n\gamma_n\,\varepsilon^{n/2}$, *i.e.* $\gamma_n = 0$ for odd n. In the sequel, γ_{2n} is denoted by c_n^+, *i.e.*

$$\widehat{\gamma}^+(\varepsilon) = \sum_{n=0}^{\infty} c_n^+\,\varepsilon^n, \quad c_0^+ = i\sqrt{2}.$$

By the definition (2.9) of y_ϕ^-, we find an expansion for the second wronskian

$$[y_\phi^-, y^-_{\phi+2\pi/3}](\varepsilon) \sim \exp\big(-\widehat{r}(\varepsilon)\log\varepsilon\big)\widehat{\gamma}^-(\varepsilon), \tag{4.4}$$

if $-\frac{5}{6}\pi < \phi < \frac{1}{6}\pi$, $\arg\varepsilon = 3\phi$, where $\widehat{\gamma}^-(\varepsilon) = -e^{-3\pi i\widehat{r}(\varepsilon)}\sum_{n=0}^{\infty}c_n^+(-\varepsilon)^n$.

Observe that the wronskians have a different asymptotic behavior in the complementary ϕ-sectors, because the values of both functions near $x = 1$ are needed and thus (2.10) has to be used. For example, we have

$$\begin{aligned}[y_\phi^+, y^+_{\phi+2\pi/3}](\varepsilon) &= -e^{1/3\varepsilon}[y_\phi^-, y^-_{\phi+2\pi/3}](\varepsilon)\\ &\sim -e^{1/3\varepsilon}\exp\big(-\widehat{r}(\varepsilon)\log\varepsilon\big)\widehat{\gamma}^-(\varepsilon),\end{aligned} \tag{4.5}$$

for $-\frac{5}{6}\pi < \phi < \frac{1}{6}\pi$. Thus the wronskians are entire functions of $\varepsilon^{-1/3}$ with different asymptotic behavior as $\varepsilon \to 0$ in different sectors.

5 The residue

We construct here two functions r_1 and r_2 on sectors S_1, S_2 forming a good covering of the origin which are asymptotic to the residue $\widehat{r}$ appearing in (4.2). It will turn out that r_1 and r_2 are also 1-sums of $\widehat{r}$ on appropriate sectors. These functions are also related to the third kind of wronskians we need to study. For $\phi \in]\frac{1}{6}\pi, \frac{5}{6}\pi[$, $\arg\varepsilon = 3\phi$, x can tend to infinity in $\mathcal{D}_\phi$ in the direction *opposite* to the one used for normalization. By (2.6), we obtain

$$y_\phi^+(x,\varepsilon) = e^{F(x)/\varepsilon}x^{-1}\big(-\exp(-2\pi i r_1(\varepsilon)) + \mathcal{O}(x^{-1})\big)$$

as $\mathcal{D}_\phi \ni x \to \infty$, $\arg x \approx \phi + \pi$, where

$$r_1(\varepsilon) = -\frac{1}{2\pi i\varepsilon}\int_{\infty e^{i\phi}}^{\infty e^{i(\phi+\pi)}}\big(z_\phi(x,\varepsilon) - \varepsilon z_1(x)\big)\,dx. \tag{5.1}$$

This defines a function analytic in the sector

$$S_1 = \{\varepsilon \in \mathbb{C}^* \,;\ \arg\varepsilon \in \left]\tfrac{1}{2}\pi, \tfrac{5}{2}\pi\right[\mod 2\pi\}.$$

The minus sign in front of $\exp\big(-2\pi i r_1(\varepsilon)\big)$ is due to the fact that, on the path from $\arg x \approx \phi$ to $\arg x \approx \phi + \pi$ considered here, the argument of $x-1$ changes from $\arg(x-1) \approx \phi$ to $\arg(x-1) \approx \phi - \pi$, therefore $x^{-1/2}(x-1)^{-1/2} \sim -x^{-1}$ as $\mathcal{D}_\phi \ni x \to \infty$, $\arg x \approx \phi + \pi$. As before, the asymptotic behavior $y_\phi^-(x,\varepsilon) = y_{\phi+\pi}^+(x,-\varepsilon) = e^{-F(x)/\varepsilon}x^{-1}\big(1 + \mathcal{O}(x^{-1})\big)$ as $x \to \infty$, $\arg x \approx \phi+\pi$ yields $[y_\phi^-, y_\phi^+](\varepsilon) = \exp\big(-2\pi i\, r_1(\varepsilon)\big)$. By (2.5), we have $r_1(\varepsilon) \sim -\frac{1}{2\pi i}\sum_{n=1}^{\infty} \varepsilon^n \int_{\infty e^{i\phi}}^{\infty e^{i(\phi+\pi)}} z_{n+1}(x)\,dx$ and a careful study shows that this is exactly $\widehat{r}(\varepsilon)$ by the residue theorem of complex analysis and the correspondance between the composite asymptotic expansion (4.1) and the outer expansion (2.5), see [14] for details. We will prove later (see Section 7) that the expansion $r_1 \sim \widehat{r}$ is Gevrey of order 1 on S_1, hence r_1 is the 1-sum of $\widehat{r}$ in any direction in $]\pi, 2\pi[$.

In a similar way, we find that the function $r_2(\varepsilon) = \mathcal{O}(\varepsilon)$ satisfying $[y_{\phi+4\pi/3}^-, y_{\phi+4\pi/3}^+](\varepsilon) = \exp\big(+2\pi i\, r_2(\varepsilon)\big)$ for $\arg\varepsilon \in]-\frac{\pi}{2}, \frac{3}{2}\pi[$, $\phi = \frac{1}{3}\arg\varepsilon$, also has the residue $\widehat{r}(\varepsilon)$ as its asymptotic expansion. This function r_2 is analytic and asymptotic to $\widehat{r}$ in the sector

$$S_2 = \{\varepsilon \in \mathbb{C}^* \,;\ \arg\varepsilon \in \left]-\tfrac{1}{2}\pi, \tfrac{3}{2}\pi\right[\mod 2\pi\},$$

and turns out to be its 1-sum for directions in $]0,\pi[$. We emphasize that S_1 and S_2 are subsets of the complex plane. We need later that their union is the punctured plane in order to apply the Ramis-Sibuya theorem. Nevertheless, we can fix a branch of the logarithm on S_1 or S_2 by specifying the range of the argument. This will and can be done differently in different contexts.

For later convenience, we introduce functions $w_j^0 : S_j \to \mathbb{C}$ whose values are wronskians by $w_j^0(\varepsilon) = \exp\big((-1)^j 2\pi i r_j(\varepsilon)\big)$. Then we have

$$\begin{aligned} &w_1^0(\varepsilon) = [y_\phi^-, y_\phi^+](\varepsilon)\\ &\qquad \text{if } \phi = \tfrac{1}{3}\arg\varepsilon \text{ and } \arg\varepsilon \in \left]\tfrac{1}{2}\pi, \tfrac{5}{2}\pi\right[\text{ is used for } \varepsilon \in S_1\\ &w_2^0(\varepsilon) = [y_\phi^-, y_\phi^+](\varepsilon)\\ &\qquad \text{if } \phi = \tfrac{1}{3}\arg\varepsilon \text{ and } \arg\varepsilon \in \left]\tfrac{7}{2}\pi, \tfrac{11}{2}\pi\right[\text{ is used for } \varepsilon \in S_2. \end{aligned} \tag{5.2}$$

We will use later that $r_1(-\varepsilon) = -r_2(\varepsilon)$ and $w_1^0(-\varepsilon) = w_2^0(\varepsilon)$; this is a consequence of (2.9).

Observe that not for all values of ϕ, the wronskian $[y_\phi^+, y_\phi^-]$ can be expressed in terms of r_1, r_2. This is only the case for $\phi \in]\frac{1}{6}\pi, \frac{5}{6}\pi[\cup]\frac{7}{6}\pi, \frac{11}{6}\pi[$ modulo 2π.

6 Stokes relations for the wronskians

In order to prove summability and eventually resurgence of the series $\widehat{r}(\varepsilon)$ and $\widehat{\gamma}^{\pm}(\varepsilon)$, we first establish Stokes relations for the wronskians. In the next section, we carry them over to the functions $r_j(\varepsilon)$ and new functions $\gamma_j^{\pm}(\varepsilon)$ the definitions of which are suggested by (4.3) and (4.4).

These Stokes relations for the wronskians contain again other wronskians. As a consequence, the Stokes relations of the $r_j(\varepsilon)$ and $\gamma_j^{\pm}(\varepsilon)$ have a special structure that will lead to their resurgence.

Here the symmetry $x \mapsto 1 - x$ reduces the number of functions to consider.

Consider again the sectors S_j introduced in the previous section. In view of (4.3), put

$$\begin{aligned} & w_1^+(\varepsilon) = [y_\phi^+, y_{\phi+2\pi/3}^+](\varepsilon) \\ & \text{where } \varepsilon \in S_1 \text{ and } \arg\varepsilon \in \left]\tfrac{1}{2}\pi, \tfrac{5}{2}\pi\right[, \phi = \tfrac{1}{3}\arg\varepsilon \text{ is used,} \\ & w_2^+(\varepsilon) = [y_\phi^+, y_{\phi+2\pi/3}^+](\varepsilon) \\ & \text{where } \varepsilon \in S_2 \text{ and } \arg\varepsilon \in \left]\tfrac{3}{2}\pi, \tfrac{7}{2}\pi\right[, \phi = \tfrac{1}{3}\arg\varepsilon \text{ is used.} \end{aligned} \tag{6.1}$$

Observe that $w_1^+(\varepsilon) = w_2^+(\varepsilon)$ if $\operatorname{Re}\varepsilon > 0$. Because of the definition (2.9) of y_ϕ^-, put $w_1^-(\varepsilon) = -w_2^+(-\varepsilon)$ for $\varepsilon \in S_1$ and $w_2^-(\varepsilon) = -w_1^+(-\varepsilon)$ for $\varepsilon \in S_2$. Then we have

$$\begin{aligned} & w_1^-(\varepsilon) = [y_{\phi-2\pi/3}^-, y_\phi^-](\varepsilon) \\ & \text{where } \varepsilon \in S_1 \text{ and } \arg\varepsilon \in \left]\tfrac{1}{2}\pi, \tfrac{5}{2}\pi\right[, \phi = \tfrac{1}{3}\arg\varepsilon \text{ is used,} \\ & w_2^-(\varepsilon) = [y_{\phi-2\pi/3}^-, y_\phi^-](\varepsilon) \\ & \text{where } \varepsilon \in S_2 \text{ and } \arg\varepsilon \in \left]-\tfrac{1}{2}\pi, \tfrac{3}{2}\pi\right[, \phi = \tfrac{1}{3}\arg\varepsilon \text{ is used.} \end{aligned} \tag{6.2}$$

Now we apply (3.3) to y_ϕ^+, $y_{\phi+2\pi/3}^+$, $y_{\phi+4\pi/3}^+$ and y_ϕ^- for $\arg\varepsilon = 3\phi \approx \pi$ together with (3.4), (4.5), (5.2) and (6.1) and obtain

$$w_2^+(\varepsilon) = \big(w_1^+(\varepsilon) - e^{1/3\varepsilon} w_1^-(\varepsilon)\big)/w_1^0(\varepsilon) \text{ for } \operatorname{Re}\varepsilon < 0. \tag{6.3}$$

By definition, we have $w_2^-(\varepsilon) = w_1^-(\varepsilon)$ for these values of ε. Using (3.3) for y_ϕ^+, $y_{\phi-2\pi/3}^+$, y_ϕ^- and $y_{\phi-2\pi/3}^-$ and $\arg\varepsilon = 3\phi \approx \pi$ yields similarly

$$w_2^0(\varepsilon) = \big(1 - e^{1/3\varepsilon} w_1^-(\varepsilon)^2\big)/w_1^0(\varepsilon) \text{ for } \operatorname{Re}\varepsilon < 0. \tag{6.4}$$

Replacing ε by $-\varepsilon$ now yields, besides $w_1^+(\varepsilon) = w_2^+(\varepsilon)$

$$\begin{aligned} w_1^-(\varepsilon) &= \big(w_2^-(\varepsilon) - e^{-1/3\varepsilon} w_2^+(\varepsilon)\big)/w_2^0(\varepsilon) \\ w_1^0(\varepsilon) &= \big(1 - e^{-1/3\varepsilon} w_2^+(\varepsilon)^2\big)/w_2^0(\varepsilon) \end{aligned} \tag{6.5}$$

if $\operatorname{Re}\varepsilon > 0$. The relations (6.3), (6.4) and (6.5) are a complete set of Stokes relations for the wronskians only containing other wronskians. They share this property with the relations (5.33) of [9] which are essentially due to Voros.

As $w_j^{+/0/-}(\varepsilon)$ are all bounded as $\varepsilon \to 0$, we immediately conclude that $w_1^0(\varepsilon)$ and $1/w_2^0(\varepsilon)$ are exponentially close on both components of $S_1 \cap S_2$. By the Ramis-Sibuya Theorem, this yields that both have asymptotic expansions of Gevrey order 1 and hence they are the 1-sums of their asymptotic expansions in any direction $\theta \in]\pi, 2\pi[$ respectively $\theta \in]0, \pi[$.

In our example, there is another relation between the functions w_j^+, w_j^- and w_j^0. Using (3.3) for $y_\phi^+, y_{\phi+2\pi/3}, y_{\phi-2\pi/3}^-$ and y_ϕ^- with $\arg\varepsilon \in]\frac{1}{2}\pi, \frac{5}{2}\pi[$, $\phi = \frac{1}{3}\arg\varepsilon$ yields together with (6.1) and (6.2)

$$w_1^+(\varepsilon) w_1^-(\varepsilon) = w_1^0(\varepsilon) + 1 \text{ for } \varepsilon \in S_1. \tag{6.6}$$

Replacing ε by $-\varepsilon$ shows that the analogous relation on S_2 is also true.

This relation could be used to eliminate the functions w_j^- altogether, but this is not convenient for the next section, because the resulting formulas contain $1/w_j^+$. It should be noted that, with this relation, the Stokes formulas (6.3), (6.4) and (6.5) imply that after analytic continuation of the functions $w_j^{+/0/-}$ three times around $\varepsilon = 0$ we obtain the original functions. This can be verified by a straightforward calculation and can be seen as a necessary condition for the correctness of the Stokes formulas. The relation also implies a relation for $\widehat{\gamma}^+$. Actually, (4.3) and (4.4) yield that

$$-\widehat{\gamma}^+(\varepsilon)\widehat{\gamma}^+(-\varepsilon) = e^{\pi i \widehat{r}(\varepsilon)} + e^{-\pi i \widehat{r}(\varepsilon)}. \tag{6.7}$$

The value $c_0^+ = i\sqrt{2}$ we obtained above is compatible with this formula. Observe that this relation does not determine $\widehat{\gamma}^+(\varepsilon)$ uniquely because the exponential $\widehat{f}(\varepsilon)$ of any odd formal series satisfies $\widehat{f}(\varepsilon)\widehat{f}(-\varepsilon) = 1$. The even part of $\log\left(\gamma^+(\varepsilon)/c_0^+\right)$, however, can be calculated in terms of $\widehat{r}$.

We end this section by considering the wronskian $[y_\phi^-, y_\phi^+]$ for $\phi \approx 0$. Using (3.3) for $y_\phi^-, y_\phi^+, y_{\phi+2\pi/3}^-$ and $y_{\phi+2\pi/3}^+$ yields

$$[y_\phi^-, y_\phi^+](\varepsilon) = \left(1 - e^{1/3\varepsilon} w_1^-(\varepsilon)^2\right)/w_1^0(\varepsilon). \tag{6.8}$$

The same result is obtained by continuing (6.4) analytically below 0 to the right halfplane and using (5.2) to show that the left hand side becomes the wanted wronskian.

7 Stokes relations for the factors of the wronskians

In this section, we establish Stokes relations for the functions r_j and the functions $\gamma_j^\pm$ suggested by (4.3) and (4.4).

First, because of $w_j^0(\varepsilon) = \exp\big((-1)^j 2\pi i r_j(\varepsilon)\big)$, we have the following Stokes relations

$$\begin{aligned} r_2(\varepsilon) &= r_1(\varepsilon) + \frac{1}{2\pi i} \log\big(1 - e^{1/3\varepsilon} w_1^-(\varepsilon)^2\big) \text{ for } \operatorname{Re}\varepsilon < 0, \\ r_1(\varepsilon) &= r_2(\varepsilon) - \frac{1}{2\pi i} \log\big(1 - e^{-1/3\varepsilon} w_2^+(\varepsilon)^2\big) \text{ for } \operatorname{Re}\varepsilon > 0 \end{aligned} \tag{7.1}$$

if $|\varepsilon|$ is sufficiently small. They imply that $r_1(\varepsilon)$ and $r_2(\varepsilon)$ are exponentially close on $S_1 \cap S_2$ and hence they are the sums of their asymptotic expansion $\widehat{r}(\varepsilon)$, as is mentioned in Section 5.

For the Stokes relations involving $w_j^{\pm}$, the factors $1/w_j^0(\varepsilon)$ complicate the discussion. This is natural, as we can only expect that the factors $\widehat{\gamma}^{\pm}(\varepsilon)$ of (4.3) and (4.4) are summable. Hence we are lead to define

$$\gamma_j^+(\varepsilon) = w_j^+(\varepsilon) \exp\big(-r_j(\varepsilon)\log\varepsilon\big) \tag{7.2}$$

for $\varepsilon \in S_j$, $j = 1, 2$, where $\arg\varepsilon \in]\frac{1}{2}\pi, \frac{5}{2}\pi[$ in the case $j = 1$ and $\arg\varepsilon \in]\frac{3}{2}\pi, \frac{7}{2}\pi[$ in the case $j = 2$. Formula (4.3) shows that indeed $\gamma_j^+(\varepsilon) \sim \widehat{\gamma}^+(\varepsilon)$ as $\varepsilon \to 0$ in the two sectors.

Using (7.2), the Stokes relations for w_j^+ and r_j imply Stokes relations for γ_j^+ which turn out to be rather involved

$$\begin{aligned} \gamma_2^+(\varepsilon) &= \big(\gamma_1^+(\varepsilon) - e^{1/3\varepsilon}\exp(-r_1(\varepsilon)\log\varepsilon) w_1^-(\varepsilon)\big) \\ &\quad \times \exp\left(-\frac{1}{2\pi i}\log\big(1 - e^{1/3\varepsilon} w_1^-(\varepsilon)^2\big)(\log\varepsilon + 2\pi i)\right), \\ \gamma_1^+(\varepsilon) &= \gamma_2^+(\varepsilon)\exp\left(\frac{1}{2\pi i}\log\big(1 - e^{-1/3\varepsilon} w_2^+(\varepsilon)^2\big)(\log\varepsilon + 2\pi i)\right). \end{aligned} \tag{7.3}$$

Here $\operatorname{Re}\varepsilon < 0$ and $\arg\varepsilon \in]\frac{1}{2}\pi, \frac{3}{2}\pi[$ is used in the first line, $\operatorname{Re}\varepsilon > 0$ and $\arg\varepsilon \in] -\frac{1}{2}\pi, \frac{1}{2}\pi[$ in the second line; again, $|\varepsilon|$ has to be sufficiently small. In particular $\gamma_1^+(\varepsilon)$ and $\gamma_2^+(\varepsilon)$ are exponentially close on the two components of $S_1 \cap S_2$ and hence are the 1-sums of $\widehat{\gamma}^+(\varepsilon)$ for the directions $\theta \in]\pi, 2\pi[$ respectively $\theta \in]0, \pi[$.

The series $\widehat{\gamma}^-(\varepsilon)$ are treated similarly. We introduce the functions

$$\gamma_j^-(\varepsilon) = w_j^-(\varepsilon)\exp(r_j(\varepsilon)\log\varepsilon) \tag{7.4}$$

for $\varepsilon \in S_j$, $j = 1, 2$, where $\arg\varepsilon \in] -\frac{3}{2}\pi, \frac{1}{2}\pi[$ is used in the case $j = 1$ and $\arg\varepsilon \in] -\frac{5}{2}\pi, -\frac{1}{2}\pi[$ in the case $j = 2$. Again, we have $\gamma_j^-(\varepsilon) \sim \widehat{\gamma}^-(\varepsilon)$ as $\varepsilon \to 0$ in both sectors. The Stokes relations for γ_j^-

are

$$\begin{aligned}\gamma_2^-(\varepsilon)&=\gamma_1^-(\varepsilon)\exp\left(\frac{1}{2\pi i}\log\left(1-e^{1/3\varepsilon}w_1^-(\varepsilon)^2\right)(\log\varepsilon-2\pi i)\right),\\ \gamma_1^-(\varepsilon)&=\left(\gamma_2^-(\varepsilon)-e^{-1/3\varepsilon}\exp\left(r_2(\varepsilon)(\log\varepsilon-2\pi i)\right)w_2^+(\varepsilon)\right)\\ &\quad\times\exp\left(-\frac{1}{2\pi i}\log\left(1-e^{-1/3\varepsilon}w_2^+(\varepsilon)^2\right)\log\varepsilon\right),\end{aligned}\tag{7.5}$$

where again $\mathrm{Re}\,\varepsilon<0$ and $\arg\varepsilon\in\,]\frac{1}{2}\pi,\frac{3}{2}\pi[$ on the first line and $\mathrm{Re}\,\varepsilon>0$, $\arg\varepsilon\in\,]-\frac{1}{2}\pi,\frac{1}{2}\pi[$ on the second line. As before, we obtain the 1-summability of $\widehat{\gamma}^-(\varepsilon)$ in all directions except the positive and negative real axes.

Now we replace w_1^- and w_2^+ using (7.4) and (7.2) in the Stokes relations (7.1), (7.3) and (7.5) in order to obtain relations only for r_j and $\gamma_j^\pm$. Then let us introduce $e_j^\pm(\varepsilon)=\exp(\pm 2\pi i r_j(\varepsilon))$. These functions $e_j^\pm$ are closely related to $(w_j^0)^{\pm 1}$ and satisfy Stokes relations of the same form as $r_j,\gamma_j^\pm$. Finally, we expand the right-hand sides of these relations into convergent series (often called *transseries*) using Taylor expansions for exp and log. We find that each of the functions $r_j,\gamma_j^\pm,e_j^\pm$ has a convergent expansion of the form

$$\sum_{m=0}^{\infty}e^{\pm m/3\varepsilon}\exp(\mp 2m\,r_k(\varepsilon)\log\varepsilon)u_m(\varepsilon)$$

where u_m are polynomials of the functions $\log\varepsilon, r_k, \gamma_k^\pm$ and $e_k^\pm$, $k=3-j$. Clearly, sums and products of two functions having such an expansion also do. Thus we obain the main result of this section

Theorem 7.1. *For each $j\in\{1,2\}$, let $\mathcal{A}_j$ denote the algebra of functions generated by the functions r_j, $\gamma_j^\pm$ and $e_j^\pm$. Then, for any $l\in\{1,2\}$ and $f_l\in\mathcal{A}_l$, there exists another function $f_{3-l}\in\mathcal{A}_{3-l}$ such that the following Stokes relations hold for $\{f_1,f_2\}$.*

For $j\in\{1,2\}$ there exist functions $h_{j,m}\in\mathcal{A}_j[\log\varepsilon]$ such that

$$f_{3-j}(\varepsilon)=f_j(\varepsilon)+\sum_{m=1}^{\infty}\exp\left(\pm\frac{m}{3\varepsilon}\right)\exp\left(\mp 2m\,r_j(\varepsilon)\log\varepsilon\right)h_{jm}(\varepsilon)$$
$$\text{if } (-1)^j\mathrm{Re}\,\varepsilon>0$$

and $|\varepsilon|$ is sufficiently small, where the upper sign is used for $j=1$, the lower sign for $j=2$.

Observe that the theorem is valid regardless of the choice of $\arg\varepsilon$. If, for example, $\arg\varepsilon$ in increased by 2π in the right half-plane ($j = 2$), then the change in the second factor can be corrected by multiplying h_{2m} by $(e_2^-)^{2m}$. For convenience, we choose here and later $\arg\varepsilon \in]\frac{1}{2}\pi, \frac{3}{2}\pi[$ if $\operatorname{Re}\varepsilon < 0$ and $\arg\varepsilon \in] -\frac{1}{2}\pi, \frac{1}{2}\pi[$ if $\operatorname{Re}\varepsilon > 0$.

8 Resurgence of the series $\widehat{r}(\varepsilon)$ and $\widehat{\gamma}^{\pm}$

We have already seen in the last section that $\widehat{r}(\varepsilon)$ and $\widehat{\gamma}^{\pm}$ are 1-summable in all directions except 0 and π. As the difference $f_1 - f_2$ of any two corresponding functions in the algebras $\mathcal{A}_j$ of Theorem 7.1 are exponentially small in both components of $S_1 \cap S_2$, we obtain in the same way that their common asymptotic expansion $\widehat{f}$ is 1-summable except in the directions $\theta = 0, \pi$. As the difference is $\mathcal{O}\left(e^{-\mu/|\varepsilon|}\right)$ for all $\mu < \frac{1}{3}$, the formal Borel transform $\widetilde{f}(t)$ of $\varepsilon\widehat{f}(\varepsilon)$ can be analytically continued to the domain $M = \mathbb{C} \setminus \left( ] -\infty, -\frac{1}{3}] \cup [\frac{1}{3}, \infty[\right)$; we designate it by the same symbol and simply call it the Borel transform of $\varepsilon f_j(\varepsilon)$, $j = 1, 2$. In this section, we prove

Theorem 8.1. *The above Borel transforms $\widetilde{f}(t)$ can be continued analytically along any path in the complex plane avoiding $\frac{1}{3}\mathbb{Z}$. For $\zeta \in \frac{1}{3}\mathbb{Z}$ and any path from 0 to a neighborhood of ζ avoiding $\frac{1}{3}\mathbb{Z}$ there is $m \in \mathbb{N}$ such that for any sector with vertex ζ, the analytic continuation is $\mathcal{O}\big(\log(t-\zeta)^m\big)$ as $t \to \zeta$.*

For the proof, we first need a theorem about the transfer from the ε-plane to the Borel plane – this is the opposite of what is usually done.

Given $\delta, C > 0$, let $\Delta = \Delta(-\delta, \delta, C)$ denote the triangle

$$\Delta = \{t \in \mathbb{C}^* \,;\, |\arg t| < \delta,\ \operatorname{Re} t < C\}$$

with vertices 0 and $T_0^{\pm} = C(1 \pm i \tan\delta)$. Consider $m \in \mathbb{N}$, $0 < a_1 < a_2 < \ldots < a_m < C$ and $T_1, \ldots, T_m \in \mathbb{C}$ with real part C as in Figure 8.1, *i.e.* $T_k = a_k + (C - a_k)(1 + i \tan\delta)$. For each $k = 1, \ldots, m$ and $\mu > 0$ small enough, let S_k denote the part of the sector with vertex a_k between the rays $\arg(t - a_k) = \delta \pm \mu$ inside the triangle Δ; μ is chosen so small that the S_k do not intersect. Then we use the subdomain $\mathcal{M}$ of the universal covering of $\mathbb{C} \setminus \{a_1, \ldots, a_m\}$ constructed in the following way. It is the union of $\widetilde{\Delta} = \Delta \backslash \bigcup_{1\le k\le m} S_k$ and *two* copies $S_k^{\pm}$ of each S_k, $k = 1, \ldots, m$. We glue $\widetilde{\Delta}$ and S_k^+ on the ray $\arg(t - a_k) = \delta + \mu$ and glue $\widetilde{\Delta}$ and S_k^- on the ray $\arg(t - a_k) = \delta - \mu$. For $t \in S_k$, let $t_{\pm}$ denote the points in $S_k^{\pm}$ with projection t. The proof of Theorem 8.1 is based on

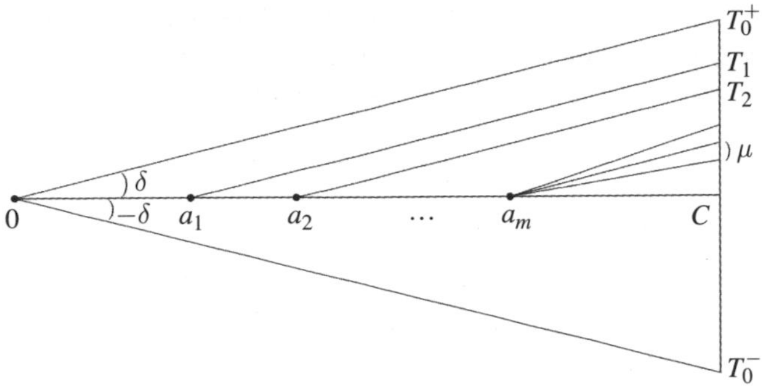

Figure 8.1. Sketch for the construction of the domain $\mathcal{M}$.

Theorem 8.2. *Consider the above notation. Suppose that g is holomorphic in a neigborhood of the segments $]0, T_0^-]$ and $]0, T_0^+]$ and in a sector $S(-\delta, \delta, t_0)$ for some small $t_0 > 0$. Suppose that g_k are holomorphic on the sets $S_k - a_k = \{t \in \mathbb{C} \; ; \; t + a_k \in S_k\}$, $k = 1, \ldots m$. Assume that there exist $K, L > 0$ such that $|g(t)|$ and all $|g_k(t)|$ are bounded by $K \, |\log t|^L$. Let*

$$f^{\pm}(\varepsilon) = \int_0^{T_0^{\pm}} e^{-t/\varepsilon} g(t)\, dt, \;\; f_k^{+}(\varepsilon) = \int_0^{T_k - a_k} e^{-t/\varepsilon} g_k(t)\, dt$$

denote truncated Laplace transforms of g and g_k, $k = 1, \ldots m$.

We suppose the following Stokes relation

$$f^-(\varepsilon) = f^+(\varepsilon) - \sum_{k=1}^{m} e^{-a_k/\varepsilon} f_k^+(\varepsilon) + \mathcal{O}\left(e^{-C/\varepsilon}\right)$$

for $|\arg \varepsilon| < \frac{1}{2}\pi - \delta$.

Then g can be continued analytically to the domain $\mathcal{M}$ to a function called g again. For $k \in \{1, \ldots, m\}$, we have $g(t) = \mathcal{O}\left(\log(t - a_k)^{L+1}\right)$ as $t \to a_k$ and the differences of the restrictions of g to $S_k^{\pm}$ are essentially g_k; more precisely

$$g\mid_{S_k^+}(t_+) - g\mid_{S_k^-}(t_-) = g_k(t - a_k) \textit{ for } t \in S_k.$$

This theorem seems natural and the correspondance between ε-plane and Borel plane is widely used, but we have found no proof of it in the literature we consulted. A concise proof will be given at the end of this section.

The form of the Stokes relations in Theorem 7.1 makes it necessary to consider sets of functions larger than $\mathcal{A}_1, \mathcal{A}_2$. For $j = 1, 2$ let $\mathcal{A}_j^{\text{exp}}$ denote the algebra generated by $\mathcal{A}_j$, $\log \varepsilon$ and

$$E_j^{\pm}(\varepsilon) = \exp\big(\pm 2\pi i r_j(\varepsilon) \log \varepsilon\big).$$

At first, we will only use the Stokes relations of Theorem 7.1 for $\operatorname{Re} \varepsilon > 0$. Hence we choose here $\arg \varepsilon \in]-\frac{3}{2}\pi, \frac{1}{2}\pi[$ on S_1 and $\arg \varepsilon \in]-\frac{1}{2}\pi, \frac{3}{2}\pi[$ on S_2, so that the arguments of ε agree in both sets for $\operatorname{Re} \varepsilon > 0$. Because of the Stokes relation (7.1) for $r_j(\varepsilon)$, Theorem 7.1 can be generalized for the algebras $\mathcal{A}_j^{\text{exp}}$ in the right half-plane.

Corollary 8.3. *For any $l \in \{1, 2\}$ and $f_l \in \mathcal{A}_l^{\text{exp}}$, there exists $f_{3-l} \in \mathcal{A}_{3-l}^{\text{exp}}$ such that the following Stokes relation holds. There exist $h_m \in \mathcal{A}_2^{\text{exp}}$ such that*

$$f_1(\varepsilon) = f_2(\varepsilon) + \sum_{m=1}^{\infty} \exp\Big(-\frac{m}{3\varepsilon}\Big) h_m(\varepsilon) \tag{8.1}$$

if $\operatorname{Re} \varepsilon > 0$ and $|\varepsilon|$ is sufficiently small.

Observe that the Stokes relation in $\operatorname{Re} \varepsilon < 0$ would be much more complicated because of the above choice of the branch of the logarithm. The present choice is convenient for the analytic continuation across segments in $\mathbb{R}_+$; for the negative x-axis we will choose $\arg \varepsilon$ differently.

Next, we show that functions in $\varepsilon \mathcal{A}_j^{\text{exp}}$ are Laplace transforms. As $\varepsilon (\log \varepsilon)^n$ is a Laplace transform of some polynomial of $\log t$ of degree n, also $r_j(\varepsilon) \log \varepsilon$ are Laplace transforms of some function, say $\widetilde{R}(t)$, holomorphic in $M_0 := M \setminus [-\frac{1}{3}, 0] = \mathbb{C} \setminus \big(]-\infty, 0] \cup [\frac{1}{3}, \infty[\big)$. More precisely,

$$r_1(\varepsilon) \log \varepsilon = \int_0^{\infty e^{-i\theta}} e^{-t/\varepsilon} \widetilde{R}(t) dt, \; r_2(\varepsilon) \log \varepsilon = \int_0^{\infty e^{i\theta}} e^{-t/\varepsilon} \widetilde{R}(t) dt, \tag{8.2}$$

where $\theta \in]0, \pi[$ and $\varepsilon \in S_1$, $|\arg \varepsilon + \theta| < \frac{1}{2}\pi$ for the first formula, $\varepsilon \in S_2$, $|\arg \varepsilon - \theta| < \frac{1}{2}\pi$ for the second formula. Here and in the sequel, we abbreviate the above statements by "$\widetilde{R}(t)$ is the Borel transform of $r_j(\varepsilon) \log \varepsilon$, $j = 1, 2$".

It follows that $E_j^{\pm}$ have Borel transforms $\widetilde{E}^{\pm}(t) = \exp^*\big(\pm 2\pi i \widetilde{R}(t)\big) := D + \sum_{k=1}^{\infty} \frac{(\pm 2\pi i)^k}{k!} \widetilde{R}^{*k}(t)$, where D denotes the Dirac function and $\widetilde{R}^{*k}$ the convolution of k factors $\widetilde{R}$. It can be shown that $\widetilde{E}^{\pm}(t)$ are again analytic in M_0. Finally, as the Borel transform of a product is the convolution of the Borel transforms, we obtain that all $\varepsilon f_j \in \varepsilon \mathcal{A}_j^{\text{exp}}$ have Borel transforms.

Consider now two corresponding functions f_j in $\mathcal{A}_j^{\exp}$ and the Stokes relation (8.1) for them in $\operatorname{Re}\varepsilon > 0$. Let $\widetilde{h}_m(t)$ denote the Borel transforms of $\varepsilon h_m(\varepsilon)$, and $\widetilde{f}(t)$ denote that of $\varepsilon f_j(\varepsilon)$, $j = 1, 2$. They are all analytic in M_0 and for every m, there is a positive integer L_m, such that for all sectors $S \subset M_0$ with vertex at $t = 0$, we have $\widetilde{h}_m(t) = \mathcal{O}\big((\log t)^{L_m}\big)$ as $S \ni t \to 0$.

At this point, Theorem 8.2 becomes applicable for any positive C. It shows that $\widetilde{f}$ can be continued analytically from the lower half-plane to the upper half-plane across any of the intervals $]\frac{m}{3}, \frac{m+1}{3}[$, $m \in \mathbb{N}$; moreover the resulting functions $\widetilde{f}_m$ satisfy $\widetilde{f}_m(t) = \widetilde{f}_{m-1}(t) + \widetilde{h}_m(t - \frac{m}{3})$, where $\widetilde{f}_0 = \widetilde{f}$. Thus we have $\widetilde{f}_m \in \mathcal{B}_m^+$, where $\mathcal{B}_m^+$ denotes the set of all sums $\widetilde{g}_0(t) + \sum_{k=1}^m \widetilde{g}_k(t - \frac{k}{3})$, where $\widetilde{g}_k$ are Borel transforms of functions in $\varepsilon\mathcal{A}_j^{\exp}$, $j = 1, 2$, considered as functions analytic in the upper half-plane. Since $\widetilde{f}_m$ is analytic in the lower half-plane and in a neighborhood of $]\frac{m}{3}, \frac{m+1}{3}[$, this shows that $\widetilde{f}_m$ is analytic in $M_m = M_0 + \frac{m}{3} = \mathbb{C} \setminus \big(]-\infty, \frac{m}{3}] \cup [\frac{m+1}{3}, \infty[\big)$. We also obtain from Theorem 8.2, that for every m, there exists a positive integer L_m, such that for all sectors S with vertex at $t = m$, we have $\widetilde{f}_m(t) = \mathcal{O}\big(\log(t - m)^{L_m}\big)$ as $S \ni t \to m$. An analogous statement holds at $t = m + 1$.

Now for $f_j \in \mathcal{A}_j$, $j = 1, 2$, the relations of Theorem 7.1 also imply relations of the form

$$f_2(\varepsilon) = f_1(\varepsilon) + \sum_{m=1}^{k} \exp\left(-\frac{m}{3\varepsilon}\right) \exp\left(2m\, r_1 \log\varepsilon\right) g_{1m}(\varepsilon)$$
$$+ \mathcal{O}\left(e^{-\left(\frac{k}{3}+\frac{1}{6}\right)/\varepsilon}\right) \quad \text{if } \operatorname{Re}\varepsilon > 0$$

with certain $g_{1m} \in \mathcal{A}_1[\log\varepsilon]$. Consequently, for $f_j \in \mathcal{A}_j^{\exp}$ and any $k \in \mathbb{N}$, there are $g_{1m} \in \mathcal{A}_1^{\exp}$ such that

$$f_2(\varepsilon) = f_1(\varepsilon) + \sum_{m=1}^{k} \exp\left(-\frac{m}{3\varepsilon}\right) g_{1m}(\varepsilon) + \mathcal{O}\left(e^{-\left(\frac{k}{3}+\frac{1}{6}\right)/\varepsilon}\right) \quad \text{if } \operatorname{Re}\varepsilon > 0.$$

Using the statement analogous to Theorem 8.2 for $\arg(T_k - a_k) = -\delta$ or going over to the conjugate functions, this implies that crossing the segments $]\frac{m}{3}, \frac{m+1}{3}[$ from the upper to the lower half-plane also yields a function in $\mathcal{B}_m^+$. Combining the two statements proves that we can continue functions in $\mathcal{B}_0^+$ analytically along any path starting near 0 (on the positive real axis) avoiding $\frac{1}{3}\mathbb{N}$ along which the real part of t is increasing and that the resulting function is in some $\mathcal{B}_m^+$.

Now we show that the Borel transforms $\widetilde{f}(t)$ of functions εf_j in $\varepsilon\mathcal{A}_j^{\exp}$ can be continued analytically across the segments $]-\frac{m+1}{3},-\frac{m}{3}[, m \in \mathbb{N}$. Here the cases of crossing $]-\frac{1}{3},0[$ from the upper to the lower half-plane and crossing from the lower to the upper half-plane have to be treated individually, as $\widetilde{f}$ is only a single valued function if $f_j \in \mathcal{A}_j$, $j = 1, 2$.

First consider the same expressions $f_j \in \mathcal{A}_j^{\exp}$, $j = 1, 2$, as before but now fix the logarithms by choosing $\arg\varepsilon \in]\frac{1}{2}\pi, \frac{5}{2}\pi[$ for $\varepsilon \in S_1$ and $\arg\varepsilon \in]-\frac{1}{2}\pi, \frac{3}{2}\pi[$ for $\varepsilon \in S_2$. Then the arguments agree on the left half-plane. These functions are Laplace transforms of a common function $\widetilde{f}^+$ analytic on $M \setminus [0, \frac{1}{3}]$ in a way similar to (8.2); only the paths of integration are close to the negative real axis. The functions $\widetilde{f}^+$ and the previous Borel transform $\widetilde{f}$ coincide in the upper half-plane. Therefore $\widetilde{f}^+$ is the analytic continuation of $\widetilde{f}$ across the segment $]-\frac{1}{3},0[$ from the upper to the lower half-plane. We can now proceed as before and obtain that $\widetilde{f}^+$ can be analytically continued across any segment $]-\frac{m+1}{3},-\frac{m}{3}[$, $m \in \mathbb{N}$ and that the resulting functions are sums of $\widetilde{f}^+(t)$ and of shifted functions $t \mapsto \widetilde{h}_k\big(t+\frac{k}{3}\big)$, where the Laplace transforms of $\widetilde{h}_k$ (along paths near the negative real axis) are also functions in $\varepsilon\mathcal{A}_j^{\exp}$.

The case of crossing $]-\frac{1}{3},0[$ from the lower to the upper haf-plane is completely analogous, only $\arg\varepsilon$ is chosen in $]-\frac{3}{2}\pi, \frac{1}{2}\pi[$ for $\varepsilon \in S_1$ and in $]-\frac{5}{2}\pi, -\frac{1}{2}\pi[$ for $\varepsilon \in S_2$.

Now we combine the above facts and obtain that the Borel transforms $\widetilde{f}(t)$ of functions εf_j in $\varepsilon\mathcal{A}_j^{\exp}$ can be continued analytically along any path in the complex plane avoiding $\frac{1}{3}\mathbb{Z}$. As the resulting functions are sums of shifted Borel transforms of functions in $\varepsilon\mathcal{A}_j^{\exp}$, there is, for every $\zeta \in \frac{1}{3}\mathbb{Z}$, a positive L such that the resulting function is $\mathcal{O}\big(\log(t-\zeta)^L\big)$ as $t \to \zeta$ (restricted to any sector with vertex ζ). This proves Theorem 8.1.

Concise proof of Theorem 8.2 First, construct a function h analytic in $\mathcal{M}$ with the property we want to prove for g, *i.e.* $h\,|_{S_k^+}\,(t_+) - h\,|_{S_k^-}\,(t_-) = g_k(t-a_k)$ for $t \in S_k$. This can be done using Cauchy integrals

$$h(t) = \frac{1}{2\pi i}\sum_{k=1}^{m}\int_{a_k}^{T_k}\frac{g_k(\tau-a_k)}{\tau-t}\,d\tau;$$

the analytic continuation to S_k^- (respectively S_k^+) is done by "pushing" the path of integration to the upper (lower) boundary of S_k.

Then define the truncated Laplace transforms

$$H^{\pm}(\varepsilon) = \int_0^{T_0^{\pm}} e^{-t/\varepsilon}h(t)\,dt.$$

By deforming the paths of integration in a classical way we obtain that

$$H^-(\varepsilon) = H^+(\varepsilon) - \sum_{k=1}^{m} e^{-a_k/\varepsilon} f_k^+(\varepsilon) + \mathcal{O}\left(e^{-C/\varepsilon}\right)$$

for small ε, $|\arg \varepsilon| < \frac{1}{2}\pi - \delta$, a relation that we have supposed for $f^{\pm}$. Therefore the differences $D^{\pm} = f^{\pm} - H^{\pm}$ satisfy $D^+(\varepsilon) - D^-(\varepsilon) = \mathcal{O}\left(e^{-C/\varepsilon}\right)$. The following extension of our result [10], applied to $F^{\pm}(\varepsilon) = \varepsilon^2 D^{\pm}(\varepsilon)$, then implies that the difference $f(t) - h(t)$ can be analytically continued to Δ. We also use the injectivity of the Laplace transform and that integration correponds to multiplication by ε via the Laplace transform.

Lemma 8.4. *With certain positive* δ, C, r_0 *assume that*

$$F^+ : S\left(0, -\delta, \tfrac{1}{2}\pi + \delta, r_0\right) \to \mathbb{C}, \; F^- : S\left(0, -\tfrac{1}{2}\pi - \delta, \delta, r_0\right) \to \mathbb{C}$$

are holomorphic, bounded and $\mathcal{O}(\varepsilon^2)$ *as* $\varepsilon \to 0$. *We suppose that*

$$F^+(\varepsilon) - F^-(\varepsilon) = \mathcal{O}\left(e^{-C/\varepsilon}\right) \; \textit{if} 0 < \varepsilon < r_0.$$

Then there exists $G : \Delta = \Delta(-\delta, \delta, C) \to \mathbb{C}$ *holomorphic and bounded such that*

$$F^{\pm}(\varepsilon) - \int_0^{T_0^{\pm}} e^{-t/\varepsilon} G(t)\, dt = \mathcal{O}\left(e^{-T_0^{\pm}/\varepsilon}\right) \; \textit{if} \arg \varepsilon \in \left]-\tfrac{1}{2}\pi, \tfrac{1}{2}\pi\right[$$

where $T_0^{\pm} = C(1 \pm i \tan \delta)$.

The lemma can be proved similarly to [10]. The starting point is a formula for the Borel transform

$$\begin{aligned}
2\pi i\, G(t) = & \int_0^{P_-} e^{t/s} F^-(s) \frac{ds}{s^2} + \int_{P_-}^{P_0} e^{t/s} F^-(s) \frac{ds}{s^2} \\
& + \int_0^{P_0} e^{t/s} \left(F^+(s) - F^-(s)\right) \frac{ds}{s^2} + \\
& + \int_{P_0}^{P_+} e^{t/s} F^+(s) \frac{ds}{s^2} - \int_0^{P_+} e^{t/s} F^+(s) \frac{ds}{s^2},
\end{aligned}$$

where $P_{\pm} = Ce^{\pm\left(\frac{1}{2}\pi + \delta\right)i}$, $P_0 = C$ and the integrals are taken over segments containing 0 or circular arcs.

9 Resurgence of the WKB solution

Finally, we can come back to the formal series

$$\widehat{v}^{\pm}(x,\varepsilon) = 1 + \sum_{n=1}^{\infty} y_n(x)(\pm\varepsilon)^n$$

in our WKB solution

$$\widehat{y}^{\pm}(x,\varepsilon) = e^{\pm F(x)/\varepsilon} x^{-1/2}(x-1)^{-1/2}\widehat{v}^{\pm}(x,\varepsilon).$$

As in the beginning of Section 3, consider x_0 to the left of the line $F(x) \in -\mathbb{R}_+$ passing through 0, $\psi = \arg F(x_0) \in]\pi, 3\pi[$ and $\phi = \frac{1}{3}\psi \in]\frac{1}{3}\pi, \pi[$. Then the paths of steepest descent for F, x_0 and all other directions are homotopic and we assume that $\widehat{y}^{\pm}$ is normalized for all of them. We discussed there that the only singular direction of $\widehat{v}^{\pm}(x_0,\varepsilon)$ with respect to ε is ψ modulo 2π — in all other directions, it is 1-summable.

Recall the corresponding Stokes relation (3.2)

$$v^{+}_{\phi+2\pi/3}(x,\varepsilon) = v^{+}_{\phi}(x,\varepsilon) + [y^{+}_{\phi}, y^{+}_{\phi+2\pi/3}](\varepsilon)e^{-2F(x)/\varepsilon}\, v^{-}_{\phi-2\pi/3}(x,\varepsilon);$$

here we only need it for $x = x_0$. Because of the range of ϕ, the wronskian is equal to $w_1^+(\varepsilon)$ or $w_2^+(\varepsilon)$ if $\phi \in]\frac{1}{3}\pi, \frac{5}{6}\pi[$ resp. $\phi \in]\frac{1}{2}\pi, \pi[$ (the two functions agree on the intersection). For simplicity, let us consider the first range only.

As the asymptotic expansion of all $v^-_\mu(x_0,\varepsilon)$, $\mu \in \mathbb{R}$ is $\widehat{v}^+(x_0,-\varepsilon)$, we also need to consider the latter formal series. Its only singular direction is $\psi - \pi$ modulo 2π and the corresponding Stokes relation is obtained from the above one by replacing ε by $-\varepsilon$ and using (2.9) and (6.2). We find

$$v^{-}_{\phi-\pi/3}(x_0,\varepsilon) = v^{-}_{\phi-\pi}(x_0,\varepsilon) - w_2^-(\varepsilon)e^{2F(x_0)/\varepsilon}\, v^{+}_{\phi+\pi/3}(x_0,\varepsilon). \qquad (9.1)$$

By Theorem 8.2 and a convenient rotation in the ε- and t-planes, we see that the only singularity of the Borel transform $\widetilde{v}(x_0,t)$ of $\varepsilon\widehat{v}^+(x_0,\varepsilon)$ visible for analytic continuation along straight lines is $2F(x_0)$ and that the function can be continued analytically across the ray $\arg(t-2F(x_0)) = \psi$. The resulting function equals $\widetilde{v}(x_0,t) + \widetilde{V}_1\big(x_0, t-2F(x_0)\big)$, where $\widetilde{V}_1(x_0,t)$ denotes the Borel transform of the formal series $\varepsilon\widehat{\gamma}^+(\varepsilon)\exp\big(-\widehat{r}(\varepsilon)\log\varepsilon\big)\widehat{v}^+(x_0,-\varepsilon)$ that is the asymptotic expansion of

$$\varepsilon w_1^+(\varepsilon) v^-_{\phi-2\pi/3}(x_0,\varepsilon).$$

An analogous statement follows for the Borel transform $\widetilde{v}(x_0, -t)$ of $-\varepsilon\widehat{v}^+(x_0, -\varepsilon)$.

As we have to consider the analytic continuation of $\widetilde{V}_1(x_0, t)$, we have to consider the Stokes relations for the product $w_1^+(\varepsilon)v_{\phi-2\pi/3}^-(x_0, \varepsilon)$ and other products. This is the reason why we first studied the Stokes relations and the resurgence of the wronskians and their factors.

Thus we are lead to consider the following sets. First, the $\mathbb{C}$-algebra $\widehat{\mathcal{A}}$ of formal series generated by $\widehat{r}(\varepsilon)$, $\widehat{\gamma}^{\pm}(\varepsilon)$ and the series $\widehat{e}^{\pm}(\varepsilon) = \exp\big(\pm 2\pi i\,\widehat{r}(\varepsilon)\big)$ – these are the asymptotic expansions of the functions in $\mathcal{A}_j$ of Theorem 7.1. Then let $\widehat{\mathcal{A}}^{\exp}$ denote the set of all $\exp\big(\pm \ell\widehat{r}(\varepsilon)\log\varepsilon\big)\widehat{h}(\varepsilon)$, $\ell \in \mathbb{N}$, $\widehat{h}(\varepsilon) \in \widehat{\mathcal{A}}[\log\varepsilon]$ and $\mathcal{B}_{\pm}^{\exp}$ the set of all Borel transforms of expressions in $\varepsilon\widehat{\mathcal{A}}^{\exp}\cdot\widehat{v}^+(x_0, \pm\varepsilon)$ in the sense of the last section.

In the sequel let τ_α, $\alpha \in \mathbb{C}$ denote the shift operator defined by

$$(\tau_\alpha f)(\varepsilon) = f(\varepsilon - \alpha).$$

Using the above statements we can show in a way similar to the previous section that

- the functions of $\mathcal{B}_+^{\exp}$ can be analytically continued along any path in $\mathbb{C}\setminus\big(\frac{1}{3}\mathbb{Z}\cup(2F(x_0)+\frac{1}{3}\mathbb{Z})\big)$ closed in some point $t_0 > 0$ close to 0 and that the functions obtained are finite sums

$$\sum_{k=-m}^{m} \tau_{m/3} f_k^+ + \sum_{k=-m}^{m} \tau_{2F(x_0)+m/3} f_k^-,$$

  where $f_k^{\pm} \in \mathcal{B}_{\pm}^{\exp}$,
- the functions of $\mathcal{B}_-^{\exp}$ can be analytically continued along any path in $\mathbb{C}\setminus\big(\frac{1}{3}\mathbb{Z}\cup(-2F(x_0)+\frac{1}{3}\mathbb{Z})\big)$ closed in some point $t_0 > 0$ close to 0 and that the functions obtained are finite sums

$$\sum_{k=-m}^{m} \tau_{m/3} f_k^- + \sum_{k=-m}^{m} \tau_{-2F(x_0)+m/3} f_k^+,$$

  where $f_k^{\pm} \in \mathcal{B}_{\pm}^{\exp}$.

Also the statements about the behavior near the singularities and the growth as $t \to \infty$ follow similarly to the previous section. Thus we have proved our main Theorem 1.2 in the case where x_0 is on the left of the line $F(x) \in -\mathbb{R}_+$ containing 0.

Consider now any path $\mathbf{c}$ from x_0 to ∞ that avoids 0, 1 and is eventually linear. Formula (2.4) shows that we can normalize the WKB solution such that the analytic continuation of the coefficients $y_n(x)$ from a neighborhood of x_0 to infinity along **c** tends to 0. If we compare the corresponding integral in (2.4) with the integral taken along a path of steepest descent, then the difference is an integral along a closed path avoiding 0 and 1. By the residue theorem, we obtain a multiple of

$$2\pi i \operatorname{res}\big(\widehat{z}(x,\varepsilon) - \varepsilon z_1(x), x = 0\big)$$
$$= -2\pi i \operatorname{res}\big(\widehat{z}(x,\varepsilon) - \varepsilon z_1(x), x = 1\big) = 2\pi i \varepsilon \widehat{r}(\varepsilon).$$

Hence this WKB solution differs from the one normalized for a path of steepest descent by a factor $\exp\big(2\pi i m \widehat{r}(\varepsilon)\big)$, $m \in \mathbb{Z}$, that we have discussed in the previous sections. Hence we obtain the resurgence of the WKB solution also for the normalization correponding to **c**.

The cases that x_0 is on the right of the line $F(x) - F(1) \in \mathbb{R}_+$ containing 1 and that x_0 is between these two lines are analogous. In the former case, the exponential $e^{-2F(x_0)/\varepsilon}$ in the Stokes relation is replaced by $e^{-2(F(x_0)-F(1))/\varepsilon}$. In the latter case, there are two Stokes relations, one containing each of the above exponentials, and two homotopy classes for the paths of steepest descent have to be distinguished. The resulting statements about analytic continuation of the Borel transform are not affected. Thus Theorem 1.2 is finally proved in all cases.

10 Remarks and Perspectives

1. The theory presented in this work remains valid for all polynomials of the form $P(x,\varepsilon) = x^2(x-1)^2 + \varepsilon a(\varepsilon)x(1-x) + \varepsilon^2 b(\varepsilon)$, where a and b are bounded functions holomorphic near 0. Only the norm of ε cannot tend to ∞ here.

With some changes, the theory can be carried over to $P(x,\varepsilon,\lambda) = x^2(x-1)^2 + \varepsilon\lambda$ and yields wronskians analytic with respect to λ. This is interesting because $\varepsilon\lambda_n(\varepsilon)$ with zeros $\lambda = \lambda_n(\varepsilon)$ of the wronskian $[y_\phi^+, y_\phi^-](\varepsilon,\lambda)$, $\phi \approx 0$ are exactly the eigenvalues of the operator $y \mapsto \varepsilon^2 y'' - x^2(x-1)^2 y$. This will be detailed in a future article.

2. The resurgence of the factors of the wronskians and thus of the WKB series is based on the Stokes relations (6.3), (6.4) and (6.5) which only contain other wronskians. The only additional information needed is the boundedness of $w_j^{\pm}(\varepsilon)$ and $(w_j^0(\varepsilon))^{\pm 1}$ as $\varepsilon \to 0$. André Voros called this *analytic bootstrap*. The Ramis-Sibuya theorem yields the existence of an asymptotic expansion. Thus, the asymptotic approximations in Section 4 are not needed to prove the resurgence, but they are useful, because they allow to characterize the coefficients of the asymptotic expansions.

3. The behavior of the analytic continuation of the Borel transforms at their singuarities is rather involved – it is related to Borel transforms of products containing $\exp(\widehat{r}(\varepsilon)\log\varepsilon)$. This indicates that it might be difficult to prove this behavior using the partial differential equation

$$\frac{\partial^2 w}{\partial x^2} + x(1-x)\frac{\partial^2 w}{\partial x\partial t} + (1-2x)\frac{\partial w}{\partial t} = 0$$

satisfied by $w(x,t) = x^{-1/2}(x-1)^{-1/2}\widetilde{v}(x,t)$, where $\widetilde{v}(x,t)$ denotes the Borel transform of $\varepsilon\widehat{v}^+(x,\varepsilon)$, *c.f.* (2.1).

4. It is an interesting question, whether the Stokes relations (6.3), (6.4), (6.5), the relation (6.6), the fact that the functions are entire functions of $\varepsilon^{-1/3}$, their values at $\varepsilon = \infty$ and their limits as $\varepsilon \to 0$ in some sectors uniquely characterize the wronskians $w_j^{\pm}$ and w_j^0.

5. With additional effort, the resurgence of the formal series $\sum_{n=0}^{\infty}\beta_n\varepsilon^{n/2}$ of (4.2) can also be shown.

6. Our theory also implies the resurgence of the formal solution (2.1) normalized to $y_n(x_0) = 0$ for all $n \in \mathbb{N}$ at some point $x_0 \neq 0, 1$. As it is the quotient of *our* formal solution normalized at infinity divided by the series $1 + \sum_n y_n(x_0)\varepsilon^n$, the singularities in the Borel plane at some point x are not only $\frac{1}{3}\mathbb{Z}$ and $2F(x) + \frac{1}{3}\mathbb{Z}$, but also $2F(x_0) + \frac{1}{3}\mathbb{Z}$. This indicates the advantage of a normalization at infinity.

7. The theory presented here implies the parametric resurgence of the Riccati equation associated to the WKB equation – we do not give details of the proof using $z = \varepsilon y'/y - x(1-x)$. The set of singularities is larger here: it is the lattice $\frac{1}{3}\mathbb{Z} + 2F(x)\mathbb{Z}$. This indicates that it might be possible to carry over the method presented here to other first order nonlinear equations with turning points. This is the topic of work in progress with Eric Matzinger concerning the forced Van der Pol equation.

References

[1] M. ABRAMOWITZ and I. STEGUN, "Handbook of mathematical functions with formulas, graphs, and mathematical tables", National Bureau of Standards Applied Mathematics Series, U.S. Government Printing Office, Washington, D.C. 1964.

[2] T. AOKI, T. KAWAI and Y. TAKEI, *Algebraic analysis of singular perturbations—exact WKB analysis*, (Japanese) Sugaku **45** n. 4 (1993), 299–315, translation: *Sugaku Expositions* **8** n. 2 (1995), 217–240.

[3] M. CANALIS-DURAND, J. MOZO-FERNANDEZ and R. SCHÄFKE, *Monomial summability and doubly singular differential equations*, J. Differential Equations **233** (2007), 485–511.

[4] O. COSTIN, L. DUPAIGNE and M. D. KRUSKAL, *Borel summation of adiabatic invariants*, Nonlinearity **17** (2004), 1509–1519.

[5] E. DELABAERE, H. DILLINGER and F. PHAM, *Résurgence de Voros et périodes des courbes hyperelliptiques*, Ann. Inst. Fourier, Grenoble **43** n. 1 (1993), 163–199.

[6] T. M. DUNSTER, D. A. LUTZ and R. SCHÄFKE, *Convergent Liouville-Green expansions for second-order linear differential equations, with an application to Bessel functions*, Proc. Roy. Soc. London **440** Ser. A (1993), 37–54.

[7] J. ECALLE, "Les fonctions résurgentes", vol. I, II et III, Publ. Math. Orsay, 1981–85.

[8] J. ECALLE, "Singularités irrégulières et résurgence multiple, Cinq Applications des Fonctions Résurgentes, Preprint n. 84T62, Dept. Math., Univ. Paris-Sud, Orsay, 1984.

[9] J. ECALLE, "Weighted products and parametric resurgence", Analyse algébrique des perturbations singulières, I, Marseille-Luminy, 1991, Travaux en Cours, 47, 7–49.

[10] A. FRUCHARD and R. SCHÄFKE, *On the Borel transform*, C. R. Acad. Sci. Paris Sér. I Math. **323** n. 9 (1996), 999–1004.

[11] A. FRUCHARD and R. SCHÄFKE, *Exceptional complex solutions of the forced van der Pol equation*, Funkcialaj Ekvacioj **42** n. 2 (1999), 201–223.

[12] A. FRUCHARD and R. SCHÄFKE, *Analytic solutions of difference equations with small step size*, In memory of W. A. Harris, Jr., J. Differ. Equations Appl. **7** n. 5 (2001), 651–684.

[13] A. FRUCHARD and R. SCHÄFKE, *Overstability and resonance*, Ann. Inst. Fourier, Grenoble **53** n. 1 (2003), 227–264.

[14] A. FRUCHARD and R. SCHÄFKE, "Développements asymptotiques combinés et points tournants", arXiv: 1004.5254.

[15] T. KAWAI and Y. TAKEI, "Algebraic analysis of singular perturbation theory", translated from the 1998 Japanese original by Goro Kato, Translations of Mathematical Monographs, 227. AMS Providence, RI, 2005.

[16] F. W. J. OLVER, "Asymptotics and special functions", Academic Press, New York-London, 1974.

[17] J.-P. RAMIS, *Dévissage Gevrey*, Astérisque **59-60** (1978), 173–204.

[18] J.-P. RAMIS, *Les séries k-sommables et leurs applications*, In: "Complex Analysis, Microlocal Calcul and Relativistic Quantum Theory", Lect. Notes Physics **126** (1980), 178–199.

[19] Y. SIBUYA, "Linear differential equations in the complex domain, problems of analytic continuation", American Mathematical Society, Providence (RI), 1990.
[20] Y. SIBUYA, *Uniform simplification in a full neighborhood of a transition point*, Memoirs of the American Mathematical Society, **149** (1974).
[21] A. VOROS, *The return of the quartic oscillator: the complex WKB method*, Ann. Inst. H. Poincaré, Sect. A (N.S.) **39** (1983), 211–338.
[22] W. WASOW, "Linear Turning Point Theory", Springer-Verlag, New York, 1985.

On a Schrödinger equation with a merging pair of a simple pole and a simple turning point — Alien calculus of WKB solutions through microlocal analysis

Shingo Kamimoto, Takahiro Kawai, Tatsuya Koike
and Yoshitsugu Takei

Abstract. This report shows how effective microlocal analysis is in alien calculus of WKB solutions. Concretely speaking, we describe how to analyze the structure of fixed singularities of Borel transformed WKB solutions of an MPPT (= a merging pair of a simple pole and a simple turning point) Schrödinger equation by finding out its microlocal canonical form. The microlocal canonical form of the Borel transformed MPPT equation is the Borel transformed Whittaker equation (Theorem 3 and Theorem 5), whose Borel transformed Voros coefficient can be concretely given (Theorem 2). These results give a tangible description of the alien derivative of WKB solutions of an MPPT equation (Theorem 6). The details are given in [5] and [11].

The purpose of this report is to present the core results of [5] and [11] with emphasis on their background. The object studied in these papers is, in somewhat rough description, a Schrödinger equation

$$\left(\frac{d^2}{dx^2} - \eta^2 Q(x,\eta)\right)\psi(x,\eta) = 0 \quad (\eta : \text{a large parameter}) \tag{1}$$

with one simple turning point and with a simple pole in the potential Q. Now that satisfactory results have been obtained by [3] concerning the WKB theoretic structure of a Schrödinger equation with two simple turning points, it is high time for us to study the above equation in view of the fact that a simple pole in the potential gives the Borel transformed WKB solutions of (1) essentially the same effect as a simple turning point does ([8,9]).

In studying this problem we have to analyse two (or more) singularities of the Borel transformed WKB solutions whose relative location is fixed (the so-called "fixed singularities" (*cf.* [4]; see also [7,14])). This means

The research of the authors has been supported in part by JSPS grants-in-aid No.20340028, No.21740098, No.21340029 and No. S-19104002.

that the usual technique (*cf.* [2,7]) of relating Borel transformed WKB solutions through integral operators determined by some microdifferential operators (*cf.* [1, 6, 13]) requires the domain of definition of the relevant operators to be sufficiently large. To circumvent this problem, following the idea in [3], we introduce an auxiliary parameter a to the potential Q so that the turning point and the pole in question merge as the parameter a tends to 0. Interestingly enough, we then naturally encounter the so-called ghost equation (*cf.* [5, 10]) at $a = 0$, the top degree part $Q_0(x)$ of whose potential contains neither zeros nor poles. The transformation of a ghost equation to its canonical form is known ([10]; see also [5, Section 1]), and by perturbing the transformation with respect to the parameter a we can find the WKB-theoretic canonical operator of an appropriately defined (Definition 1 below) class of Schrödinger operators with a simple turning point and a simple pole (Theorem 1 below).

A mathematical formulation of the intuitive picture of such an "appropriate" class is given by the following

Definition 1. The Schrödinger equation (1) is called an equation with a merging pair of a simple pole and a simple turning point, or, for short, an MPPT equation if its potential Q depends also on an auxiliary complex parameter a and has the following form:

$$Q = \frac{Q_0(x,a)}{x} + \eta^{-1}\frac{Q_1(x,a)}{x} + \eta^{-2}\frac{Q_2(x,a)}{x^2}, \tag{2}$$

where $Q_j(x,a)$ $(j = 0, 1, 2)$ are holomorphic near $(x,a) = (0,0)$ and $Q_0(x,a)$ satisfies the following conditions (3) and (4):

$$\left(\frac{\partial Q_0}{\partial a}\right)(0,0) \neq 0, \tag{3}$$

$Q_0(x,0) = c_0^{(0)}x + O(x^2)$ holds with $c_0^{(0)}$ being a constant different from 0. (4)

Remark 1. In [5] a slightly weaker condition

$$Q_0(0,a) \neq 0 \ \text{ if } \ a \neq 0 \tag{3'}$$

is imposed instead of (3).

It follows from the above definition that there exists a unique holomorphic function $x(a)$ near $a = 0$ that satisfies

$$Q_0(x(a),a) = 0, \tag{5}$$

$$x(a) \neq 0 \ \text{ if } \ a \neq 0. \tag{6}$$

Then the assumption (4) guarantees that $x = x(a)$ ($a \neq 0$, $|a| \ll 1$) is a simple turning point. Thus the above assumptions visualize our intuitive picture of the equation. The following Theorem 1 guarantees the appropriateness of the above definition. For the clarity of description we put $\tilde{}$ to quantities relevant to a general MPPT equation to distinguish them from those of the canonical equation (16).

Theorem 1. *Let*

$$\tilde{L}\tilde{\psi} = \left(\frac{d^2}{d\tilde{x}^2} - \eta^2 \tilde{Q}(\tilde{x}, a, \eta)\right) \tilde{\psi}(\tilde{x}, a, \eta) = 0 \tag{7}$$

be an MPPT *equation in the sense of Definition* 1, *that is, the potential* $\tilde{Q}(\tilde{x}, a, \eta)$ *is of the form* (2) *and the conditions* (3) *and* (4) *are satisfied. Then there exist an open neighborhood* U *of* $\tilde{x} = 0$, *holomorphic functions* $x_k^{(j)}(\tilde{x})$ *defined on* U *and constants* $\alpha_k^{(j)}$ $(j, k \geq 0)$ *for which the following conditions* (8) $\sim$ (12) *are satisfied:*

$$\frac{dx_0^{(0)}}{d\tilde{x}}(0) \neq 0, \tag{8}$$

$$x_k^{(j)}(0) = 0 \quad \textit{for every } j \textit{ and } k, \tag{9}$$

$$\alpha_0^{(0)} = 0, \tag{10}$$

$$\sup_{\tilde{x} \in U} |x_k^{(j)}(\tilde{x})|, \ |\alpha_k^{(j)}| \leq A C_1^j C_2^k k! \tag{11}$$

with some positive constants A, C_1 *and* C_2,

$$\begin{aligned} &\tilde{Q}(\tilde{x}, a, \eta) \\ &= \left(\frac{\partial x(\tilde{x}, a, \eta)}{\partial \tilde{x}}\right)^2 \left(\frac{1}{4} + \frac{\alpha(a, \eta)}{x(\tilde{x}, a, \eta)} + \eta^{-2} \frac{\tilde{Q}_2(0, a)}{x(\tilde{x}, a, \eta)^2}\right) \\ &\quad - \frac{1}{2}\eta^{-2}\{x; \tilde{x}\}, \end{aligned} \tag{12}$$

where

$$x(\tilde{x}, a, \eta) = \sum_{k \geq 0} \sum_{j \geq 0} x_k^{(j)}(\tilde{x}) a^j \eta^{-k}, \tag{13}$$

$$\alpha(a, \eta) = \sum_{k \geq 0} \sum_{j \geq 0} \alpha_k^{(j)} a^j \eta^{-k} \tag{14}$$

and $\{x; \tilde{x}\}$ *denotes the Schwarzian derivative*

$$\frac{d^3x/d\tilde{x}^3}{dx/d\tilde{x}} - \frac{3}{2}\left(\frac{d^2x/d\tilde{x}^2}{dx/d\tilde{x}}\right)^2. \tag{15}$$

This theorem combined with the general WKB theory (*cf.* [7]) asserts that the WKB theoretically canonical equation of an MPPT equation $\tilde{L}\tilde{\psi} = 0$ is given by the following

$$M\psi = \left(\frac{d^2}{dx^2} - \eta^2\left(\frac{1}{4} + \frac{\alpha(a,\eta)}{x} + \eta^{-2}\frac{\tilde{Q}_2(0,a)}{x^2}\right)\right)\psi = 0. \tag{16}$$

In parallel with the usage of the name "∞-Weber equation" in [3], we call the equation $M\psi = 0$ an ∞-Whittaker equation.

An important point is that in the double series $x(\tilde{x}, a, \eta)$ and $\alpha(a, \eta)$ in Theorem 1 the growth order property of $|x_k^{(j)}|$ and $|\alpha_k^{(j)}|$ as j tends to ∞ and that as k tends to ∞ are substantially different despite the fact that their construction is done in a symmetric way with respect to indexes j and k (*cf.* [5, Remark 2.1]). In particular,

$$x_k(\tilde{x}, a) = \sum_{j\geq 0} x_k^{(j)}(\tilde{x})a^j \tag{17}$$

and

$$\alpha_k(a) = \sum_{j\geq 0} \alpha_k^{(j)} a^j \tag{18}$$

are holomorphic respectively on $U \times V$ and on V for some open neighborhood V of $a = 0$, while $x(\tilde{x}, a, \eta)$ and $\alpha(a, \eta)$ are only Borel transformable series in the sense of [7]. Although the problem is of singular perturbative character, it seems that it is of regular perturbative character in the variable a. Actually our reasoning indicates that the singular perturbative character originates from the part $\eta^{-2}(d^3x_k^{(j)}/d\tilde{x}^3)\big/(dx_k^{(j)}/d\tilde{x})$ in the defining equation of $x_k^{(j)}$, which does not affect much the behavior of $x_k^{(j)}$ as j tends to infinity. (See [5, (B.64)].)

Although an ∞-Whittaker equation contains an infinite series $\alpha(a,\eta) = \sum_{k\geq 0}\alpha_k(a)\eta^{-k}$ in its coefficients, this series has a simple structure; it is a translation of a complex number $\alpha_0(a)$ by $\alpha_+(a, \eta) = \sum_{k\geq 1}\alpha_k(a)\eta^{-k}$. At first sight, this translation might look terrible because of the (possible) divergence of $\alpha_+(a, \eta)$. But, as we know ([2, 7]), such a translation by a Borel transformable series defines a microdifferential operator $\mathcal{A}$ in Theorem 5, which acts on the Borel transforms. Since our eventual target is a Borel transformed WKB solution, we may say, by using an intuitive expression to be formulated precisely later in Theorem 5, that the canonical equation $M\psi = 0$ is further reduced to the following Whittaker equation with a large parameter:

$$M_0\chi = \left(\frac{d^2}{dx^2} - \eta^2\left(\frac{1}{4} + \frac{\alpha_0}{x} + \eta^{-2}\frac{\gamma(\gamma+1)}{x^2}\right)\right)\chi = 0, \qquad (19)$$

where α_0 and γ are complex numbers. Concerning the Whittaker equation with a large parameter for $\alpha_0 \neq 0$ we know ([11]) the following Theorem 2: Let $\chi_\pm(x,\eta)$ be WKB solutions of the Whittaker equation normalized as

$$\chi_\pm(x,\eta) = \frac{1}{\sqrt{S_{\text{odd}}}}\exp\left(\pm\int_{-4\alpha_0}^{x} S_{\text{odd}}dx\right), \qquad (20)$$

where S_{odd} is the odd part of the formal power series solution

$$S = \eta S_{-1}(x) + S_0(x) + \eta^{-1}S_1(x) + \cdots$$

of the associated Riccati equation (*cf.* [5]). Then the following holds.

Theorem 2. *Suppose $\alpha_0 \neq 0$. Then the Borel transform $\chi_{+,B}(x,y)$ of χ_+ has fixed singularities at $y = -y_+(x) + 2m\pi i\alpha_0$ $(m = \pm1, \pm2, \cdots)$, where*

$$y_+(x) = \int_{-4\alpha_0}^{x} S_{-1}dx = \int_{-4\alpha_0}^{x}\sqrt{\frac{x+4\alpha_0}{4x}}dx \qquad (21)$$

and its alien derivative is explicitly given by

$$\begin{aligned}&\left(\Delta_{y=-y_+(x)+2m\pi i\alpha_0}\chi_+\right)_B(x,y)\\ &= \frac{\exp(2m\pi i\gamma) + \exp(-2m\pi i\gamma)}{2m}\chi_{+,B}(x, y - 2m\pi i\alpha_0).\end{aligned} \qquad (22)$$

Note that the relative location between two singular points $-y_+(x) + 2m\pi i\alpha_0$ and $-y_+(x) + 2m'\pi i\alpha_0$ does not vary, that is, their difference $2(m-m')\pi i\alpha_0$ is a constant independent of x. The proof of Theorem 2 can be done by using the following expression of the Borel transform of the Voros coefficient ϕ:

$$\begin{aligned}&\phi_B(\alpha_0,\gamma;y)\\ &= \frac{1}{2y}\left(\frac{\exp(y/\alpha_0)+1}{\exp(y/\alpha_0)-1}\right)\cosh\left(\frac{\gamma y}{\alpha_0}\right) - \frac{\alpha_0}{y^2} + \frac{1}{2y}\sinh\left(\frac{\gamma y}{\alpha_0}\right),\end{aligned} \qquad (23)$$

where the Voros coefficient of the Whittaker equation (19) is defined by

$$\phi(\alpha_0,\gamma;\eta) = \int_{-4\alpha_0}^{\infty}(S_{\text{odd}} - \eta S_{-1})dx. \qquad (24)$$

See [11] for the details. Since the concrete computation in alien calculus is normally performed on the Borel plane (*cf.* [4, 12]), we have to study the Borel transformed version of Theorem 1. To employ Theorem 2, we assume $a \neq 0$ in what follows. Thanks to the estimate (11), the Borel transformed version of Theorem 1 is endowed with a clear analytic meaning in terms of microdifferential operators, as is done in the case mentioned earlier concerning the reduction of an ∞-Whittaker equation $M\psi = 0$ to $M_0\chi = 0$. To state Theorem 3 and Theorem 4 below we make the following notational preparations: Let $g(x, a)$ be the inverse function of $x_0(\tilde{x}, a)$, *i.e.*, a holomorphic function that satisfies

$$x = x_0\big(g(x, a), a\big), \quad \tilde{x} = g\big(x_0(\tilde{x}, a), a\big) \tag{25}$$

on a neighborhood of $(x, a) = (0, 0)$. Then we consider the Borel transform of $\tilde{L}$ in (x, y, a)-variable:

$$\begin{aligned} \mathcal{L} &\overset{\text{def}}{=} \left(\frac{\partial g}{\partial x}\right)^2 \times \left(\text{Borel transform of } \tilde{L}\right)\Big|_{\tilde{x}=g(x,a)} \\ &= \frac{\partial^2}{\partial x^2} - \left(\frac{\partial^2 g/\partial x^2}{\partial g/\partial x}\right)\frac{\partial}{\partial x} - \left(\frac{\partial g}{\partial x}\right)^2 \tilde{Q}\left(g(x, a), a, \frac{\partial}{\partial y}\right). \end{aligned} \tag{26}$$

Similarly let $\mathcal{M}$ (respectively $\mathcal{M}_0$) be the Borel transform of M (respectively M_0):

$$\mathcal{M} = \frac{\partial^2}{\partial x^2} - \left(\frac{1}{4} + \frac{\alpha(a, \partial/\partial y)}{x}\right)\frac{\partial^2}{\partial y^2} - \frac{\tilde{Q}_2(0, a)}{x^2}, \tag{27}$$

$$\mathcal{M}_0 = \frac{\partial^2}{\partial x^2} - \left(\frac{1}{4} + \frac{\alpha_0}{x}\right)\frac{\partial^2}{\partial y^2} - \frac{\gamma(\gamma+1)}{x^2}. \tag{28}$$

Theorem 3. *Suppose $a \neq 0$. Let ω_0 be a sufficiently small open neighborhood of $x = 0$, and set*

$$\Omega_0 = \{(x, y; \xi, \eta) \in T^*\mathbb{C}^2_{(x,y)};\, x \in \omega_0, \eta \neq 0\}. \tag{29}$$

Then there exist microdifferential operators $\mathcal{X}$ and $\mathcal{Y}$ defined on Ω_0 that satisfy

$$\mathcal{L}\mathcal{X} = \mathcal{Y}\mathcal{M} \tag{30}$$

for $x \neq 0$. The concrete form of operators $\mathcal{X}$ and $\mathcal{Y}$ is as follows:

$$\mathcal{X} = :\left(\frac{\partial g}{\partial x}\right)^{1/2}\left(1 + \frac{\partial r}{\partial x}\right)^{-1/2} \exp\big(r(x, a, \eta)\xi\big):, \tag{31}$$

$$\mathcal{Y} = :\left(\frac{\partial g}{\partial x}\right)^{1/2}\left(1 + \frac{\partial r}{\partial x}\right)^{3/2} \exp\big(r(x, a, \eta)\xi\big):, \tag{32}$$

where

$$r(x, a, \eta) = \sum_{k \geq 1} x_k\bigl(g(x, a), a\bigr)\eta^{-k} \tag{33}$$

and : : *designates the normal ordered product* (*cf.* [1]).

Theorem 3 implies that the operators $\mathcal{L}$ and $\mathcal{M}$ are microlocally equivalent. This fact indicates that the singularities of $\tilde{\psi}_B(g(x, a), y)$ that satisfies $\mathcal{L}\tilde{\psi}_B = 0$ and those of $\psi_B(x, y)$ that satisfies $\mathcal{M}\psi_B = 0$ are the same. This is really visualized by the following Theorem 4:

Theorem 4. *The action of the microdifferential operator $\mathcal{X}$ upon the Borel transformed WKB solution $\psi_{+,B}$ of the ∞-Whittaker equation is expressed as an integro-differential operator of the following form*:

$$\mathcal{X}\psi_{+,B} = \int_{y_0}^{y} K(x, a, y - y', \partial/\partial x)\psi_{+,B}(x, a, y')dy', \tag{34}$$

where $K(x, a, y, \partial/\partial x)$ is a differential operator of infinite order that is defined on $\{(x, a, y) \in \mathbb{C}^3;\ (x, a) \in \omega$ for an open neighborhood ω of the origin and $|y| < C$ for some positive constant $C\}$, and y_0 is a constant that fixes the action of $(\partial/\partial y)^{-1}$ as an integral operator.

Since a differential operator of infinite order acts on the sheaf of holomorphic functions as a sheaf homomorphism, we can immediately locate the singularities of $\mathcal{X}\psi_{+,B}$ through the integral representation (34). Another important point in the integral representation (34) is that its domain of definition enjoys the uniformity with respect to the parameter a, that is, the open neighborhood ω is taken to be of the form

$$\{x \in \mathbb{C};\ |x| < \delta_1\} \times \{a \in \mathbb{C};\ |a| < \delta_2\} \tag{35}$$

for some positive constants δ_1 and δ_2. Note that since $\alpha_0(a)$ tends to 0 as a tends to 0 by (10), (δ_1, δ_2) can be chosen so that $\{|x| < \delta_1\}$ contains $x = -4\alpha_0(a)$ for every a in $\{|a| < \delta_2\}$. This is the precise meaning of saying "To circumvent the problem (of the existence of a large domain of definition of relevant integral operators)" at the beginning of this report.

In parallel with Theorem 3, we can show that $\mathcal{M}$ and $\mathcal{M}_0$ are also microlocally equivalent. For simplicity we employ $\alpha_0(a)$ as an independent variable in substitution for a (this substitution of variable is guaranteed by (3)). Thanks to the estimate (11) we obtain the following

Theorem 5. *Let $\mathcal{A}$ be a microdifferential operator on*

$$\{(\alpha_0, y; \theta, \eta) \in T^*\mathbb{C}^2;\ |\alpha_0| < \delta_0, \eta \neq 0\} \tag{36}$$

for some positive constant δ_0 defined by

$$\mathcal{A} = :\exp\left((\alpha_1(\alpha_0)\eta^{-1} + \alpha_2(\alpha_0)\eta^{-2} + \cdots)\theta\right): . \tag{37}$$

Here θ and η are respectively identified with the symbol $\sigma(\partial/\partial\alpha_0)$ and the symbol $\sigma(\partial/\partial y)$. Then the following holds:

$$\mathcal{M}\mathcal{A} = \left(\mathcal{A}\mathcal{M}_0\right)\big|_{\gamma(\gamma+1)=\tilde{Q}_2(0,a)} \tag{38}$$

for $x \neq 0$.

Although the target variable is α_0, not x, as is the case for the microdifferential operator $\mathcal{X}$, the operator $\mathcal{A}$ also has a concrete expression as an integro-differential operator stated in Theorem 4. On the other hand, as is indicated in Theorem 2, a fixed singular point of $\psi_{+,B}(x, y)$ ("fixed" with respect to $y = -y_+(x)$) is located at $y = -y_+(x) + 2m\pi i\alpha$. Thus, by the same reasoning for the case of $\mathcal{X}$, each individual fixed singular point of $\tilde{\psi}_{+,B}(x, y)$ is contained, for sufficiently small a, in the domain of definition of the integro-differential operator $\mathcal{A}$.

Summing up all these results, we finally obtain

Theorem 6. *Suppose $a \neq 0$ and let $\tilde{\psi}_+(\tilde{x}, a, \eta)$ be a WKB solution of an MPPT equation normalized at its turning point $\tilde{x}_0(a)$ as follows:*

$$\tilde{\psi}_+(x, a, \eta) = \frac{1}{\sqrt{\tilde{S}_{\mathrm{odd}}}} \exp\left(\int_{\tilde{x}_0(a)}^{x} \tilde{S}_{\mathrm{odd}}dx\right), \tag{39}$$

where $\tilde{S}_{\mathrm{odd}}$ is the odd part of the formal power series solution $\tilde{S}$ of the associated Riccati equation. Then for each positive integer m the following relation (40) holds for sufficiently small a:

$$\begin{aligned}&\left(\Delta_{y=-y_+(\tilde{x},a)+2m\pi i\alpha_0(a)}\tilde{\psi}_+\right)_B(\tilde{x}, a, y)\\ &= \frac{\exp(2m\pi i\gamma(a)) + \exp(-2m\pi i\gamma(a))}{2m}\\ &\quad\times:\exp\left(-2m\pi i(\alpha_1(a) + \alpha_2(a)\eta^{-1} + \cdots)\right):\tilde{\psi}_{+,B}(\tilde{x}, a, y - 2m\pi i\alpha_0(a)),\end{aligned} \tag{40}$$

where

$$y_+(\tilde{x}, a) = \int_{\tilde{x}_0(a)}^{\tilde{x}} \sqrt{\frac{\tilde{Q}_0(\tilde{x}, a)}{\tilde{x}}}\, d\tilde{x}, \tag{41}$$

$$\gamma(a)^2 + \gamma(a) = \tilde{Q}_2(0, a) \tag{42}$$

and

$$\alpha_j(a) = \frac{1}{2\pi i} \oint_{\tilde{\Gamma}(a)} \tilde{S}_{\mathrm{odd},j-1}(\tilde{x}, a) d\tilde{x} \tag{43}$$

with $\tilde{\Gamma}(a)$ being a closed curve encircling $\tilde{x}_0(a)$ and the origin as in Figure 1 and with $\tilde{S}_{\mathrm{odd},k}$ designating the degree k part of $\tilde{S}_{\mathrm{odd}}$.

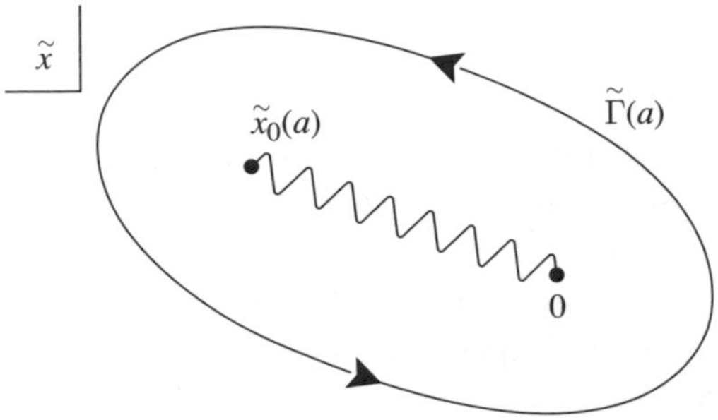

Figure 1.

References

[1] T. AOKI, *Symbols and formal symbols of pseudodifferential operators*, Advanced Studies in Pure Mathematics, Kinokuniya **4** (1984), 181–208.

[2] T. AOKI, T. KAWAI and T. TAKEI, "The Bender-Wu analysis and the Voros theory", Special Functions, Springer-Verlag, 1991, 1–29.

[3] T. AOKI, T. KAWAI and T. TAKEI, *The Bender-Wu analysis and the Voros theory*, II, Advanced Studies in Pure Mathematics, Math. Soc. Japan **54** (2009), 19–94.

[4] E. DELABAERE and F. PHAM, *Resurgent methods in semi-classical asymptotics*, Ann. Inst. Henri Poincaré **71** (1999), 1–94.

[5] S. KAMIMOTO, T. KAWAI, T. KOIKE and Y. TAKEI, *On the WKB theoretic structure of a Schrödinger operator with a merging pair of a simple pole and a simple turning point*, Kyoto Journal of Mathematics **50** (2010), 101–164.

[6] M. KASHIWARA, T. KAWAI and T. KIMURA, "Foundations of Algebraic Analysis", Princeton University Press, Princeton, 1986.

[7] T. KAWAI and Y. TAKEI, "Algebraic Analysis of Singular Perturbation Theory", Amer. Math. Soc., 2005.

[8] T. KOIKE, "On a Regular Singular Point in the Exact WKB Analysis", Toward the Exact WKB Analysis of Differential Equations, Linear or Non-Linear, Kyoto Univ. Press, 2000, 39–54.

[9] T. KOIKE, *On the exact WKB analysis of second order linear ordinary differential equations with simple poles*, Publ. RIMS, Kyoto Univ. **36** (2000), 297–319.

[10] T. KOIKE, *On "new" turning points associated with regular singular points in the exact WKB analysis*, RIMS Kôkyûroku, **1159** RIMS (2000), 100–110.

[11] T. KOIKE and Y. TAKEI, *On the Voros coefficient for the Whittaker equation with a large parameter – Some progress around Sato's conjecture in exact WKB analysis*, Publ. RIMS, Kyoto Univ. **47** (2011), 375–395.

[12] F. PHAM, "Resurgence, quantized canonical transformations, and multi-instanton expansion", Algebraic Analysis, Vol. II, Academic Press, 1988, 699–726.

[13] M. SATO, T. KAWAI and M. KASHIWARA, "Microfunctions and pseudo-differential equations", Lect. Notes in Math., **287**, Springer, 1973, 265–529.

[14] A. VOROS, *The return of the quartic oscillator – The complex WKB method*, Ann. Inst. Henri Poincaré **39** (1983), 211–338.

On the turning point problem for instanton-type solutions of Painlevé equations

Yoshitsugu Takei

Abstract. The turning point problems for instanton-type solutions of Painlevé equations with a large parameter are discussed. Generalizing the main result of [4] near a simple turning point, we report in this paper that Painlevé equations can be transformed to the second Painlevé equation and the most degenerate third Painlevé equation near a double turning point and near a simple pole, respectively. An outline of the proof based on the theory of isomonodromic deformations of associated linear differential equations is also explained.

1 Background and main results

The purpose of this report is to discuss the turning point problem for instanton-type solutions of Painlevé equations from the viewpoint of exact WKB analysis.

In our series of papers ([1,3,4]) we develop the exact WKB analysis of Painlevé equations (P_J) with a large parameter η (> 0):

$$\frac{d^2\lambda}{dt^2} = G_J\left(\lambda, \frac{d\lambda}{dt}, t\right) + \eta^2 F_J(\lambda, t). \qquad (P_J)$$

Here J runs over the following set of indices

$$\mathcal{I} = \{\mathrm{I},\ \mathrm{II},\ \mathrm{III}',\ \mathrm{III}'(\mathrm{D7}),\ \mathrm{III}'(\mathrm{D8}),\ \mathrm{IV},\ \mathrm{V},\ \mathrm{VI}\}, \qquad (1.1)$$

and $F_J(\lambda, t)$ and $G_J(\lambda, \mu, t)$ are some rational functions of (λ, t) and (λ, μ, t), respectively. For the concrete form of $F_J(\lambda, t)$ and $G_J(\lambda, \mu, t)$ see Table 1 below. Note that instead of the usual third Painlevé equation (P_{III}) we use $(P_{\mathrm{III}'})$, which is equivalent to (P_{III}), for the sake of convenience in this paper. Note also that it is now considered to be natural to distinguish the degenerate third Painlevé equations of type (D7) and (D8) from the generic third Painlevé equation since the type of their affine Weyl group symmetries is different from that of the generic third

Painlevé equation. In this paper, being conformed to this convention, we have listed up ($P_{\mathrm{III}'(D7)}$) and ($P_{\mathrm{III}'(D7)}$) as well in Table 1. These Painlevé equations (P_J) are related to one another according to the so-called coalescence diagram described in Table 2.

Table 1. Painlevé equations with a large parameter η. Here λ denotes an unknown function, t is an independent variable and c, c_0 etc. are complex parameters.

$$(P_{\mathrm{I}}) \quad \frac{d^2\lambda}{dt^2} = \eta^2(6\lambda^2 + t).$$

$$(P_{\mathrm{II}}) \quad \frac{d^2\lambda}{dt^2} = \eta^2(2\lambda^3 + t\lambda + c).$$

$$(P_{\mathrm{III}'}) \quad \frac{d^2\lambda}{dt^2} = \frac{1}{\lambda}\left(\frac{d\lambda}{dt}\right)^2 - \frac{1}{t}\frac{d\lambda}{dt} + \eta^2\left[\frac{c_\infty\lambda^3}{t^2} - \frac{c'_\infty\lambda^2}{t^2} + \frac{c'_0}{t} - \frac{c_0}{\lambda}\right].$$

$$(P_{\mathrm{III}'(D7)}) \quad \frac{d^2\lambda}{dt^2} = \frac{1}{\lambda}\left(\frac{d\lambda}{dt}\right)^2 - \frac{1}{t}\frac{d\lambda}{dt} - \eta^2\left[\frac{2\lambda^2}{t^2} + \frac{c}{t} + \frac{1}{\lambda}\right].$$

$$(P_{\mathrm{III}'(D8)}) \quad \frac{d^2\lambda}{dt^2} = \frac{1}{\lambda}\left(\frac{d\lambda}{dt}\right)^2 - \frac{1}{t}\frac{d\lambda}{dt} + \eta^2\left[\frac{\lambda^2}{t^2} - \frac{1}{t}\right].$$

$$(P_{\mathrm{IV}}) \quad \frac{d^2\lambda}{dt^2} = \frac{1}{2\lambda}\left(\frac{d\lambda}{dt}\right)^2 - \frac{2}{\lambda} + \eta^2\left[\frac{3}{2}\lambda^3 + 4t\lambda^2 + \left(2t^2 + c_1\right)\lambda - \frac{c_0}{\lambda}\right].$$

$$(P_{\mathrm{V}}) \quad \frac{d^2\lambda}{dt^2} = \left(\frac{1}{2\lambda} + \frac{1}{\lambda - 1}\right)\left(\frac{d\lambda}{dt}\right)^2 - \frac{1}{t}\frac{d\lambda}{dt} + \frac{(\lambda-1)^2}{t^2}\left(2\lambda - \frac{1}{2\lambda}\right)$$
$$+ \eta^2\frac{2\lambda(\lambda-1)^2}{t^2}\left[c_\infty - \frac{c_0}{\lambda^2} - \frac{c_2 t}{(\lambda-1)^2} - \frac{c_1 t^2(\lambda+1)}{(\lambda-1)^3}\right].$$

$$(P_{\mathrm{VI}}) \quad \frac{d^2\lambda}{dt^2} = \frac{1}{2}\left(\frac{1}{\lambda} + \frac{1}{\lambda-1} + \frac{1}{\lambda-t}\right)\left(\frac{d\lambda}{dt}\right)^2 - \left(\frac{1}{t} + \frac{1}{t-1} + \frac{1}{\lambda-t}\right)\frac{d\lambda}{dt}$$
$$+ \frac{2\lambda(\lambda-1)(\lambda-t)}{t^2(t-1)^2}\left[1 - \frac{\lambda^2 - 2t\lambda + t}{4\lambda^2(\lambda-1)^2}\right.$$
$$\left. + \eta^2\left\{c_\infty - \frac{c_0 t}{\lambda^2} + \frac{c_1(t-1)}{(\lambda-1)^2} - \frac{c_t t(t-1)}{(\lambda-t)^2}\right\}\right].$$

Table 2. Coalescence diagram of Painlevé equations.

$$\begin{array}{ccccccccc} (P_{\mathrm{VI}}) & \longrightarrow & (P_{\mathrm{V}}) & \longrightarrow & (P_{\mathrm{III}'}) & \longrightarrow & (P_{\mathrm{III}'(D7)}) & \longrightarrow & (P_{\mathrm{III}'(D8)}) \\ & & & \searrow & & \searrow & & & \\ & & & & (P_{\mathrm{IV}}) & \longrightarrow & (P_{\mathrm{II}}) & \longrightarrow & (P_{\mathrm{I}}) \end{array}$$

As can be readily confirmed, every Painlevé equation (P_J) $(J \in \mathcal{I})$ admits the following formal power series solution (in η^{-1}) called a "0-parameter solution":

$$\lambda_J^{(0)}(t, \eta) = \lambda_0(t) + \eta^{-2}\lambda_2(t) + \eta^{-4}\lambda_4(t) + \cdots , \tag{1.2}$$

where the top term $\lambda_0(t)$ satisfies an algebraic equation

$$F_J(\lambda_0(t), t) = 0 \tag{1.3}$$

and the other terms $\lambda_{2j}(t)$ $(j \geq 1)$ are recursively determined once $\lambda_0(t)$ is fixed. Furthermore, by using the multiple-scale method, we have constructed in [1] the following formal solution of (P_J), called a "2-parameter solution" or an "instanton-type solution", containing 2 free complex parameters (α, β):

$$\lambda_J(t, \eta; \alpha, \beta) = \lambda_0(t) + \eta^{-1/2}\lambda_{1/2}(t, \eta) + \eta^{-1}\lambda_1(t, \eta) + \cdots . \tag{1.4}$$

Here the leading term $\lambda_0(t)$ is the same as that of a 0-parameter solution and the other terms $\lambda_{j/2}(t, \eta)$ $(j \geq 1)$ are of the form

$$\lambda_{j/2}(t, \eta) = \sum_{k=0}^{j} b_{j-2k}^{(j/2)}(t) \exp\left((j-2k)\Phi_J\right), \tag{1.5}$$

where $\Phi_J = \Phi_J(t, \eta)$, sometimes called an "instanton", is defined by

$$\Phi_J(t, \eta) = \eta \int^t \sqrt{\frac{\partial F_J}{\partial \lambda}(\lambda_0(s), s)}ds + \alpha\beta \log\left(\eta^2\theta_J(t)\right) \tag{1.6}$$

with an appropriately defined function $\theta_J(t)$ and $b_l^{(j/2)}(t)$ $(l = j, j-2, \ldots, -j)$ are functions of t depending also on α and β but not on η, that is, the η-dependence of $\lambda_{j/2}(t, \eta)$ comes only from the l-instanton terms $\exp(l\Phi_J)$. In particular, the subleading term $\lambda_{1/2}(t, \eta)$ is of the form

$$\lambda_{1/2}(t, \eta) = \mu_J(t)\left(\alpha \exp(\Phi_J) + \beta \exp(-\Phi_J)\right). \tag{1.7}$$

For the explicit forms of $\theta_J(t)$ and $\mu_J(t)$ we refer the reader to [4, Section 1].

The subject of our series of papers ([1, 3, 4]) is the analysis of the structure of $\lambda_J^{(0)}(t,\eta)$ and $\lambda_J(t,\eta;\alpha,\beta)$ near a simple turning point of (P_J). Here a (simple) turning point of (P_J) is defined as follows:

Definition 1.1. Let

$$\frac{d^2}{dt^2}\Delta\lambda = \eta^2\frac{\partial F_J}{\partial\lambda}(\lambda_J^{(0)},t)\Delta\lambda + \frac{\partial G_J}{\partial\lambda}\left(\lambda_J^{(0)},\frac{d\lambda_J^{(0)}}{dt},t\right)\Delta\lambda + \frac{\partial G_J}{\partial\mu}\left(\lambda_J^{(0)},\frac{d\lambda_J^{(0)}}{dt},t\right)\frac{d}{dt}\Delta\lambda \tag{ΔP_J}$$

be the linearized equation (or the Frechét derivative) of (P_J) at its 0-parameter solution $\lambda_J^{(0)}$. Then a turning point of (P_J) is, by definition, a turning point of (ΔP_J). That is, a turning point of (P_J) is a zero of $(\partial F_J/\partial\lambda)(\lambda_0(t),t)$. In particular, a point t satisfying

$$\frac{\partial F_J}{\partial\lambda}(\lambda_0(t),t) = 0 \quad\text{and}\quad \frac{\partial^2 F_J}{\partial\lambda^2}(\lambda_0(t),t) \neq 0 \tag{1.8}$$

is called a simple turning point. Similarly, a Stokes curve of (P_J) is defined as a Stokes curve of (ΔP_J), that is, a Stokes curve of (P_J) is an integral curve of the direction field $\operatorname{Im}\sqrt{(\partial F_J/\partial\lambda)(\lambda_0(t),t)}\,dt = 0$ emanating from a turning point.

The main result of [4] is then described in the following Theorem 1.2 (where we put $\tilde{}$ to the variables relevant to (P_J) to distinguish them from those relevant to the first Painlevé equation (P_{I})).

Theorem 1.2. *Let $\tilde{t} = \tilde{t}_*$ be a simple turning point of (P_J) and $\tilde{\sigma}$ a point on a Stokes curve emanating from $\tilde{t}_*$. Then there exists a neighborhood $\tilde{V}$ of $\tilde{\sigma}$ so that every 2-parameter instanton-type solution $\tilde{\lambda}_J(\tilde{t},\eta;\tilde{\alpha},\tilde{\beta})$ of (P_J) is formally transformed to a 2-parameter instanton-type solution $\lambda_{\mathrm{I}}(t,\eta;\alpha,\beta)$ of (P_{I}) in $\tilde{V}$. To be more specific, there exist a formal transformation $t(\tilde{t},\eta)$ of an independent variable and a formal transformation $x(\tilde{x},\tilde{t},\eta)$ of an unknown function of the form*

$$t(\tilde{t},\eta) = \sum_{j\geq 0} t_{j/2}(\tilde{t},\eta)\eta^{-j/2}, \tag{1.9}$$

$$x(\tilde{x},\tilde{t},\eta) = \sum_{j\geq 0} x_{j/2}(\tilde{x},\tilde{t},\eta)\eta^{-j/2}, \tag{1.10}$$

where $t_{j/2}$ and $x_{j/2}$ are holomorphic in both $\tilde{x}$ and $\tilde{t}$, that satisfy the following relation:

$$x(\tilde{\lambda}_J(\tilde{t},\eta;\tilde{\alpha},\tilde{\beta}),\tilde{t},\eta) = \lambda_{\mathrm{I}}(t(\tilde{t},\eta),\eta;\alpha,\beta). \tag{1.11}$$

Theorem 1.2 implies that the first Painlevé equation (P_{I}) can be thought of as a canonical equation (or a normal form) near a simple turning point of Painlevé equations (P_J). For instanton-type solutions of (P_{I}) we have the following connection formula on its Stokes curve, say, on $\{\arg t = \pi\}$ (*cf.* [9]):

$$\frac{\beta' \, 2^{2\alpha'\beta'}}{\Gamma(2\alpha'\beta' + 1)} = \frac{\beta \, 2^{2\alpha\beta}}{\Gamma(2\alpha\beta + 1)}, \tag{1.12}$$

$$e^{2i\pi\alpha'\beta'} \frac{\alpha' \, 2^{-2\alpha'\beta'}}{\Gamma(-2\alpha'\beta' + 1)} = e^{2i\pi\alpha\beta} \frac{\alpha \, 2^{-2\alpha\beta}}{\Gamma(-2\alpha\beta + 1)} - ie^{4i\pi\alpha\beta}, \tag{1.13}$$

where $\lambda_{\mathrm{I}}(t, \eta; \alpha, \beta)$ (respectively, $\lambda_{\mathrm{I}}(t, \eta; \alpha', \beta')$) is an instanton-type solution of (P_{I}) in $\{\arg t < \pi\}$ (respectively, $\{\arg t > \pi\}$). In particular, the analytic continuation across the Stokes curve $\{\arg t = \pi\}$ of a 0-parameter solution $\lambda_{\mathrm{I}}^{(0)}(t, \eta) = \lambda_{\mathrm{I}}(t, \eta; 0, 0)$ in $\{\arg t < \pi\}$ is given by $\lambda_{\mathrm{I}}(t, \eta; -i/(2\sqrt{\pi}), 0)$ in $\{\arg t > \pi\}$. In view of Theorem 1.2 it is expected that the same connection formula as (1.12) and (1.13) should hold also for an instanton-type solution of (P_J) on its Stokes curve emanating from a simple turning point.

The aim of this report is to discuss some generalizations of Theorem 1.2. Now, **what kind of generalizations of Theorem 1.2 is possible?** To consider possible generalizations of Theorem 1.2, we first briefly review a simpler case, that is, the case of second order linear ordinary differential equations

$$\left(-\frac{d^2}{dx^2} + \eta^2 Q(x)\right)\psi = 0. \tag{1.14}$$

It is well-known that at a simple turning point Equation (1.14) can be transformed into the Airy equation (*i.e.*, Equation (1.14) with $Q(x) = x$). In fact, such a transformation is constructed in the framework of exact WKB analysis as well (*cf.* [5, Chapter 2]) and Theorem 1.2 can be regarded as a nonlinear analogue of this result. For linear equations (1.14) several generalizations of this result are also known. For example, at a double turning point (*i.e.*, a double zero of $Q(x)$) (1.14) can be transformed into the Weber equation (*i.e.*, Equation (1.14) with $Q(x) = x^2 + \eta^{-1}E$ with some constant E). Furthermore at a simple pole of $Q(x)$ (1.14) is transformed into the Whittaker equation (*i.e.*, Equation (1.14) with $Q(x) = 1/x + \eta^{-2}\gamma/x^2$ with some constant γ). This fact means that a simple pole of $Q(x)$ also plays a role of turning points for Equation (1.14) and in the framework of exact WKB analysis this fact is verified by Koike in [6].

In parallel to the case of linear equations (1.14) we are then able to consider some generalizations of Theorem 1.2 for Painlevé equations, that is, generalizations to a transformation near a double turning point and that near a simple pole. First, near a double turning point, we can prove the following

Theorem 1.3. *Near a double turning point every 2-parameter instanton-type solution of* (P_J) *is formally transformed to that of the following second Painlevé equation* $(P_{\mathrm{II,deg}})$ *(in the same sense as in Theorem* 1.2):

$$\frac{d^2\lambda}{dt^2} = \eta^2(2\lambda^3 + t\lambda + \eta^{-1}c). \qquad (P_{\mathrm{II,deg}})$$

Note that $(P_{\mathrm{II,deg}})$ is different from the ordinary second Painlevé equation (P_{II}) in that the parameter c is multiplied by η^1 in $(P_{\mathrm{II,deg}})$ (while it is multiplied by η^2 in (P_{II})). Next, near a simple pole, we have

Theorem 1.4. *Near a simple pole every 2-parameter instanton-type solution of* (P_J) *is formally transformed to that of the most degenerate third Painlevé equation* $(P_{\mathrm{III}'(\mathrm{D8})})$ *(in the same sense as in Theorem* 1.2).

Theorem 1.3 and Theorem 1.4 are the main results of this report. Their precise statements will be given below in Theorem 2.5 and Theorem 3.1, respectively. Theorem 1.4 has been announced also in [10].

The plan of this report is as follows: In Section 2, after discussing the exact WKB theoretic structure of $(P_{\mathrm{II,deg}})$, we give the definition of a double turning point of (P_J) and explain an outline of the proof of Theorem 1.3. A key idea is to use the relationship between Painlevé equations and the theory of isomonodromic deformations of the associated linear differential equations. Then in Section 3 we review the discussion of [10], that is, we consider the transformation near a simple pole in a way parallel to Section 2. The details will be discussed in our forthcoming paper(s).

ACKNOWLEDGEMENTS. The author is deeply grateful to Professor Takashi Aoki for his great assistance in completing the proof of Proposition 2.4 and to Professor Tatsuya Koike for his kind help in drawing Figure 2.1. He expresses his sincere gratitude also to Professor Takahiro Kawai for the stimulating discussions with him.

2 Transformation near a double turning point

2.1 Exact WKB theoretic structure of $(P_{\mathrm{II,deg}})$

In this section we consider transformation near a double turning point. We first investigate the exact WKB theoretic structure of the canonical equation

$$\frac{d^2\lambda}{dt^2} = \eta^2(2\lambda^3 + t\lambda + \eta^{-1}c). \qquad (P_{\mathrm{II,deg}})$$

As was explained in Section 1, the top term $\lambda_0 = \lambda_0(t)$ of the 0-parameter solution

$$\lambda^{(0)}_{\mathrm{II,deg}}(t,\eta) = \lambda_0(t) + \eta^{-1}\lambda_1(t) + \eta^{-2}\lambda_2(t) + \cdots \qquad (2.1)$$

of $(P_{\mathrm{II,deg}})$ is determined by an algebraic equation

$$F_{\mathrm{II,deg}}(\lambda_0, t) = 2\lambda_0^3 + t\lambda_0 = \lambda_0(2\lambda_0^2 + t) = 0. \qquad (2.2)$$

Among the solutions of (2.2) we pick up a solution of $2\lambda_0^2 + t = 0$, *i.e.*,

$$\lambda_0(t) = \sqrt{-\frac{t}{2}} \quad \left(\text{or} \quad -\sqrt{-\frac{t}{2}}\right), \qquad (2.3)$$

when we consider a double turning point. Note that, as $(P_{\mathrm{II,deg}})$ contains an odd order term ηc (with respect to η), $\lambda^{(0)}_{\mathrm{II,deg}}(t,\eta)$ also contains odd order terms $\lambda_1(t), \lambda_3(t), \ldots$. Then, starting with this top term $\lambda_0(t)$ given by (2.3), we can construct a 2-parameter instanton-type solution of $(P_{\mathrm{II,deg}})$ of the form

$$\begin{aligned} &\lambda_{\mathrm{II,deg}}(t,\eta;\alpha,\beta) \qquad (2.4)\\ &= \lambda_0(t) + \eta^{-1/2}(6\lambda_0^2 + t)^{-1/4}\big(\alpha\exp(\Phi_{\mathrm{II,deg}}) + \beta\exp(-\Phi_{\mathrm{II,deg}})\big) + \cdots \end{aligned}$$

with

$$\Phi_{\mathrm{II,deg}}(t,\eta) = \eta\int_0^t \sqrt{6\lambda_0^2 + s}\,ds + (2\alpha\beta + c/2)\log\left(\eta^2(6\lambda_0^2 + t)^3\right) \qquad (2.5)$$

by employing the multiple-scale method. (Since $(P_{\mathrm{II,deg}})$ contains an odd order term ηc, the form of the instanton $\Phi_{\mathrm{II,deg}}$ is slightly different from the general form (1.6) of Φ_J.)

The linearized equation of $(P_{\mathrm{II,deg}})$ at a 0-parameter solution (2.1) is given by

$$\frac{d^2}{dt^2}\Delta\lambda = \eta^2\left(6(\lambda^{(0)}_{\mathrm{II,deg}})^2 + t\right)\Delta\lambda = \eta^2\left(-2t + O(\eta^{-1})\right)\Delta\lambda. \qquad (\Delta P_{\mathrm{II,deg}})$$

Hence ($P_{\mathrm{II,deg}}$) has a unique turning point at $t = 0$. Note that this turning point $t = 0$ is also an algebraic branch point of the Riemann surface of $\lambda_0(t)$. The Stokes curves of ($P_{\mathrm{II,deg}}$), *i.e.*, integral curves of the direction field $\mathrm{Im}\sqrt{6\lambda_0^2 + t}\,dt = \mathrm{Im}\sqrt{-2t}\,dt = 0$ emanating from the turning point $t = 0$, thus consist of the following three lines:

$$\{t \in \mathbb{C} \mid \arg t = \pi + 2n\pi/3 \ (n \in \mathbb{Z})\}. \tag{2.6}$$

It is expected that a Stokes phenomenon should be observed on each Stokes curve for instanton-type solutions $\lambda_{\mathrm{II,deg}}(t, \eta; \alpha, \beta)$. To analyze the Stokes phenomenon, we make use of the well-known relationship between the Painlevé equation and the theory of isomonodromic deformations of the associated linear differential equation (*cf.* [2, 8]). In the case of ($P_{\mathrm{II,deg}}$) the relationship is formulated as follows: Let ($SL_{\mathrm{II,deg}}$) and ($D_{\mathrm{II,deg}}$) be the following linear differential equations, respectively.

$$\left(-\frac{\partial^2}{\partial x^2} + \eta^2 Q_{\mathrm{II,deg}}\right)\psi = 0, \tag{$SL_{\mathrm{II,deg}}$}$$

$$\frac{\partial \psi}{\partial t} = A_{\mathrm{II,deg}}\frac{\partial \psi}{\partial x} - \frac{1}{2}\frac{\partial A_{\mathrm{II,deg}}}{\partial x}\psi, \tag{$D_{\mathrm{II,deg}}$}$$

where

$$Q_{\mathrm{II,deg}} = x^4 + tx^2 + 2\eta^{-1}cx + 2K_{\mathrm{II,deg}} - \eta^{-1}\frac{\nu}{x-\lambda} + \eta^{-2}\frac{3}{4(x-\lambda)^2}, \tag{2.7}$$

$$A_{\mathrm{II,deg}} = \frac{1}{2(x-\lambda)}, \tag{2.8}$$

with

$$K_{\mathrm{II,deg}} = \frac{1}{2}\left[\nu^2 - (\lambda^4 + t\lambda^2 + 2\eta^{-1}c\lambda)\right]. \tag{2.9}$$

Then the compatibility condition of ($SL_{\mathrm{II,deg}}$) and ($D_{\mathrm{II,deg}}$) is represented by the Hamiltonian system

$$\frac{d\lambda}{dt} = \eta\frac{\partial K_{\mathrm{II,deg}}}{\partial \nu}, \quad \frac{d\nu}{dt} = -\eta\frac{\partial K_{\mathrm{II,deg}}}{\partial \lambda}, \tag{$H_{\mathrm{II,deg}}$}$$

which is equivalent to the second order differential equation ($P_{\mathrm{II,deg}}$) for λ. As its consequence, we find that the monodromy data of ($SL_{\mathrm{II,deg}}$) should be independent of the deformation parameter t if a solution of ($H_{\mathrm{II,deg}}$) or ($P_{\mathrm{II,deg}}$) is substituted into the coefficients of ($SL_{\mathrm{II,deg}}$).

To determine the connection formula which describes the Stokes phenomenon for $\lambda_{\mathrm{II,deg}}(t, \eta; \alpha, \beta)$ on a Stokes curve of ($P_{\mathrm{II,deg}}$), we then substitute $\lambda_{\mathrm{II,deg}}(t, \eta; \alpha, \beta)$ into the coefficients of ($SL_{\mathrm{II,deg}}$) and compute its monodromy data by employing the exact WKB analysis. The following is a key proposition in executing the computation of the monodromy data.

Proposition 2.1. *If an instanton-type solution of* $(H_{\mathrm{II,deg}})$ *or* $(P_{\mathrm{II,deg}})$ *is substituted into the coefficients of* $(SL_{\mathrm{II,deg}})$*, the following hold*:

(i) *The top term (with respect to* η^{-1}*)* Q_0 *of the potential* $Q_{\mathrm{II,deg}}$ *of* $(SL_{\mathrm{II,deg}})$ *is factorized as*

$$Q_0 = (x - \lambda_0(t))^2 (x + \lambda_0(t))^2. \tag{2.10}$$

That is, $(SL_{\mathrm{II,deg}})$ *has two double turning points* $x = \lambda_0(t)$ *and* $x = -\lambda_0(t)$.

(ii) *When* t *lies on a Stokes curve* (2.6) *of* $(P_{\mathrm{II,deg}})$*, there exists a Stokes curve of* $(SL_{\mathrm{II,deg}})$ *that connects the two double turning points* $x = \pm\lambda_0(t)$ *of* $(SL_{\mathrm{II,deg}})$. (*Cf.* Figure 2.1, (ii), *where the configuration of Stokes curves is shown when* t *lies on a Stokes curve* $\arg t = \pi$.)

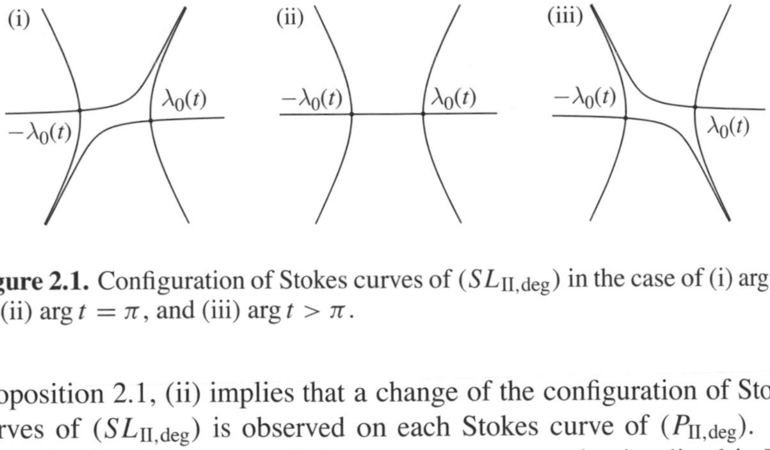

Figure 2.1. Configuration of Stokes curves of $(SL_{\mathrm{II,deg}})$ in the case of (i) $\arg t < \pi$, (ii) $\arg t = \pi$, and (iii) $\arg t > \pi$.

Proposition 2.1, (ii) implies that a change of the configuration of Stokes curves of $(SL_{\mathrm{II,deg}})$ is observed on each Stokes curve of $(P_{\mathrm{II,deg}})$. For example, the change on a Stokes curve $\arg t = \pi$ is visualized in Figure 2.1. This change of the configuration causes a Stokes phenomenon for $\lambda_{\mathrm{II,deg}}(t, \eta; \alpha, \beta)$ to occur on a Stokes curve of $(P_{\mathrm{II,deg}})$. As a matter of fact, by substituting an instanton-type solution $\lambda_{\mathrm{II,deg}}(t, \eta; \alpha, \beta)$ into the coefficients of $(SL_{\mathrm{II,deg}})$ and employing the exact WKB analysis for linear equations, we obtain the following

Proposition 2.2. *Suppose that an instanton-type solution of* $(P_{\mathrm{II,deg}})$ *is substituted into the coefficients of* $(SL_{\mathrm{II,deg}})$*. Let* $m_1^{(\pm)}$ *and* $m_2^{(\pm)}$ *be two independent monodromy data (i.e., Stokes multipliers around* $x = \infty$ *in this case) of* $(SL_{\mathrm{II,deg}})$ *when* t *belongs to the region* $\Omega_{\pm} = \{t \mid \pm(\arg t -$

$\pi) > 0\}$, *respectively. Then* $(m_1^{(\pm)}, m_2^{(\pm)})$ *can be explicitly computed as follows*:

(*When* $t \in \Omega_+$)

$$m_1^{(+)} = -2\sqrt{\pi}\frac{i\beta\, 2^{2\alpha\beta}}{\Gamma(2\alpha\beta+1)}, \tag{2.11}$$

$$m_2^{(+)} = -2\sqrt{\pi}\, e^{2i\pi\alpha\beta}\frac{\alpha\, 2^{-2\alpha\beta}}{\Gamma(-2\alpha\beta+1)}. \tag{2.12}$$

(*When* $t \in \Omega_-$)

$$m_1^{(-)} = -2\sqrt{\pi}\frac{i\beta\, 2^{2\alpha\beta}}{\Gamma(2\alpha\beta+1)}, \tag{2.13}$$

$$m_2^{(-)} = -2\sqrt{\pi}\left(e^{2i\pi\alpha\beta}\frac{\alpha\, 2^{-2\alpha\beta}}{\Gamma(-2\alpha\beta+1)} - e^{4i\pi\alpha\beta}\frac{i\, 2^{c+2\alpha\beta-1/2}}{\Gamma(c+2\alpha\beta-1/2)}\right). \tag{2.14}$$

Since the computation of monodromy data through the exact WKB analysis heavily depends on the configuration of Stokes curves, the concrete expression of $(m_1^{(+)}, m_2^{(+)})$ becomes different from that of $(m_1^{(-)}, m_2^{(-)})$, as one can readily see in Proposition 2.2, due to the difference of two configurations of Stokes curves shown in Figure 2.1. On the other hand, thanks to the isomonodromic property, the monodromy data should coincide if two instanton-type solutions in the two regions $\Omega_\pm$ correspond to the same analytic solution. Thus, if an instanton-type solution $\lambda_{\mathrm{II,deg}}(t, \eta; \alpha, \beta)$ in Ω_- corresponds to the same analytic solution with $\lambda_{\mathrm{II,deg}}(t, \eta; \alpha', \beta')$ in Ω_+, we obtain the following relation in view of Proposition 2.2:

$$\frac{\beta'\, 2^{2\alpha'\beta'}}{\Gamma(2\alpha'\beta'+1)} = \frac{\beta\, 2^{2\alpha\beta}}{\Gamma(2\alpha\beta+1)}, \tag{2.15}$$

$$\begin{aligned} &e^{2i\pi\alpha'\beta'}\frac{\alpha'\, 2^{-2\alpha'\beta'}}{\Gamma(-2\alpha'\beta'+1)} \\ &= e^{2i\pi\alpha\beta}\frac{\alpha\, 2^{-2\alpha\beta}}{\Gamma(-2\alpha\beta+1)} - e^{4i\pi\alpha\beta}\frac{i\, 2^{c+2\alpha\beta-1/2}}{\Gamma(c+2\alpha\beta-1/2)}. \end{aligned} \tag{2.16}$$

In particular, the 0-parameter solution $\lambda_{\mathrm{II,deg}}^{(0)}(t, \eta) = \lambda_{\mathrm{II,deg}}(t, \eta; 0, 0)$ in the region Ω_- should be analytically continued to

$$\lambda_{\mathrm{II,deg}}(t, \eta; -i2^{c-1/2}/\Gamma(c+1/2), 0)$$

in Ω_+ across the Stokes curve $\{\arg t = \pi\}$. This is the mechanism for a Stokes phenomenon to occur for instanton-type solutions of $(P_{\mathrm{II,deg}})$ on its Stokes curve. The above relations (2.15) and (2.16) describe the connection formula on $\{\arg t = \pi\}$.

2.2 Transformation theory to $(P_{\mathrm{II,deg}})$ near a double turning point

As we have observed in Subsection 2.1, in the case of $(P_{\mathrm{II,deg}})$ its linearized equation $(\Delta P_{\mathrm{II,deg}})$ has a unique turning point at $t = 0$ and three Stokes curves emanate from there. We call this kind of turning points a double turning point in general. To be more specific, we define a double turning point of (P_J) as follows:

Definition 2.3. A point $t = \tau_d$ is said to be a double turning point of (P_J) if the following two conditions are satisfied.

(i) $t = \tau_d$ is an algebraic branch point of the Riemann surface of $\lambda_0(t)$.
(ii) Near $t = \tau_d$, $(\partial F_J/\partial\lambda)(\lambda_0(t), t)$ has a simple zero, that is,

$$\frac{\partial F_J}{\partial \lambda}(\lambda_0(t), t) = c(t - \tau_d) + \cdots \tag{2.17}$$

holds with a non-zero constant c.

In particular, from each double turning point of (P_J) three Stokes curves emanate thanks to the condition (2.17). Note that at a simple turning point of (P_J) $(\partial F_J/\partial\lambda)(\lambda_0(t), t)$ has a square-root branch point (and hence the condition (i) is automatically satisfied there). Thus a double turning point is a turning point more degenerate than a simple turning point.

A Painlevé equation (P_J) does not always have a double turning point. For example, the first Painlevé equation (P_{I}) has a unique turning point at the origin $t = 0$ which is simple. Similarly, the degenerate third Painlevé equation $(P_{\mathrm{III}'(\mathrm{D}7)})$ or $(P_{\mathrm{III}'(\mathrm{D}8)})$ does not possess any double turning point. In order that (P_J) may have a double turning point, the parameters contained in (P_J) should satisfy some algebraic condition, the explicit form of which is described in the following

Proposition 2.4.

(i) (P_{I}), $(P_{\mathrm{III}'(\mathrm{D}7)})$ *and* $(P_{\mathrm{III}'(\mathrm{D}8)})$ *have no double turning points.*
(ii) *A double turning point appears for a Painlevé equation* (P_J) $(J = \mathrm{II}, \mathrm{III}', \mathrm{IV}, \mathrm{V}, \mathrm{VI})$ *if and only if the parameters contained in* (P_J)

should satisfy the following relations:

$$c = 0 \qquad \textit{for } J = \mathrm{II}, \tag{2.18}$$

$$c_0(c'_\infty)^2 - c_\infty(c'_0)^2 = 0 \qquad \textit{for } J = \mathrm{III}', \tag{2.19}$$

$$2c_0 - c_1^2 = 0 \qquad \textit{for } J = \mathrm{IV}, \tag{2.20}$$

$$16c_1^2c_\infty^2 - 8c_0c_1c_2^2 - 8c_\infty c_1c_2^2 + c_2^4 = 0 \quad \textit{for } J = \mathrm{V}, \tag{2.21}$$

$$\begin{aligned} &16(c_0^2c_1^2 + c_1^2c_t^2 + c_t^2c_0^2) \\ &-32(c_0^2c_1c_t + c_0c_1^2c_t + c_0c_1c_t^2) \\ &-64c_0c_1c_t\tilde{c}_\infty \\ &-8(c_0c_1 + c_1c_t + c_tc_0)\tilde{c}_\infty^2 + \tilde{c}_\infty^4 = 0 \end{aligned} \qquad \textit{for } J = \mathrm{VI}, \tag{2.22}$$

where $\tilde{c}_\infty = c_\infty - (c_0 + c_1 + c_t)$ *in the case of* $J = \mathrm{VI}$.

Throught this subsection we assume that the conditions (2.18) $\sim$ (2.22) listed in Proposition 2.4, (ii) are satisfied. The problem we want to discuss is to develop transformation theory near a double turning point. Let $t = \tau_d$ be a double turning point of (P_J) ($J = \mathrm{II}, \mathrm{III}', \mathrm{IV}, \mathrm{V}, \mathrm{VI}$). Generalizing the transformation theory (Theorem 1.2) near a simple turning point, we can then prove the following theorem which claims that every 2-parameter instanton-type solution of (P_J) is transformed to that of $(P_{\mathrm{II,deg}})$ near $t = \tau_d$. (In stating Theorem 2.5, we put $\tilde{}$ to the variables relevant to (P_J) to distinguish them from those relevant to $(P_{\mathrm{II,deg}})$.)

Theorem 2.5. *Suppose that the conditions* (2.18) $\sim$ (2.22) *are satisfied. Let* $\tilde{t} = \tilde{\tau}_d$ *be a double turning point of* (P_J) ($J = \mathrm{II}, \mathrm{III}', \mathrm{IV}, \mathrm{V}, \mathrm{VI}$) *and* $\tilde{\sigma}$ *be a point on a Stokes curve emanating from* $\tilde{\tau}_d$. *Then we can find a neighborhood* $\tilde{V}$ *of* $\tilde{\sigma}$ *and a formal power series of* η^{-1} *with constant coefficients*

$$c(\eta) = c_0 + \eta^{-1}c_1 + \eta^{-2}c_2 + \cdots \tag{2.23}$$

such that in $\tilde{V}$ *every* 2*-parameter instanton-type solution* $\tilde{\lambda}_J(\tilde{t}, \eta; \tilde{\alpha}, \tilde{\beta})$ *of* (P_J) *is formally transformed to a* 2*-parameter instanton-type solution* $\lambda_{\mathrm{II,deg}}(t, \eta; \alpha, \beta)$ *of the degenerate second Painlevé equation*

$$\frac{d^2\lambda}{dt^2} = \eta^2(2\lambda^3 + t\lambda + \eta^{-1}c(\eta)) \tag{2.24}$$

with the infinite series $c(\eta)$ *of* (2.23) *being substituted into its coefficient. To be more specific, there exist a formal transformation* $t = t(\tilde{t}, \eta)$ *of an independent variable and a formal transformation* $x = x(\tilde{x}, \tilde{t}, \eta)$ *of an unknown function of the form*

$$t(\tilde{t},\eta) = \sum_{j\geq 0} \eta^{-j/2} t_{j/2}(\tilde{t},\eta), \tag{2.25}$$

$$x(\tilde{x},\tilde{t},\eta) = \sum_{j\geq 0} \eta^{-j/2} x_{j/2}(\tilde{x},\tilde{t},\eta), \tag{2.26}$$

where $t_{j/2}$ *and* $x_{j/2}$ *are holomorphic in both* $\tilde{x}$ *and* $\tilde{t}$, *that satisfy the following relation:*

$$x(\tilde{\lambda}_J(\tilde{t},\eta;\tilde{\alpha},\tilde{\beta}),\tilde{t},\eta) = \lambda_{\text{II,deg}}(t(\tilde{t},\eta),\eta;\alpha,\beta). \tag{2.27}$$

Hence the (degenerate) second Painlevé equation ($P_{\text{II,deg}}$) can be regarded as the canonical equation of Painlevé equations near a double turning point. Theorem 2.5 suggests that the connection formula (2.15) and (2.16) for ($P_{\text{II,deg}}$) described in Section 2.1 should hold also for an instanton-type solution of (P_J) on its Stokes curve emanating from a double turning point.

Let us explain an outline of the construction of the transformations $t(\tilde{t},\eta)$ and $x(\tilde{x},\tilde{t},\eta)$. It is done in a parallel way to the transformation theory near a simple turning point; we again make use of the relationship between Painlevé equations and the theory of isomonodromic deformations of linear differential equations, that is, we use the fact that (P_J) is equivalent to the compatibility condition of a system of linear differential equations

$$\left(-\frac{\partial^2}{\partial x^2} + \eta^2 Q_J\right)\psi = 0, \tag{SL_J}$$

$$\frac{\partial \psi}{\partial t} = A_J \frac{\partial \psi}{\partial x} - \frac{1}{2}\frac{\partial A_J}{\partial x}\psi. \tag{D_J}$$

(See [3] or [5, Chapter 4] for the concrete form of Q_J and A_J.) A key proposition in constructing the transformations is then the following Proposition 2.6, which is a generalization of Proposition 2.1 to (P_J).

Proposition 2.6. *Suppose that the conditions* (2.18)∼(2.22) *are satisfied and let* $t = \tau_d$ *be a double turning point of* (P_J) ($J =$ II, III′, IV, V, VI). *If an instanton-type solution* $\lambda_J(t,\eta;\alpha,\beta)$ *of* (P_J) *is substituted into the coefficients of* (SL_J), *then the following hold*:

(i) *The top term (with respect to* η^{-1}) Q_0 *of the potential* Q_J *of* (SL_J) *has two double zeros, one of which is given by the top term* $\lambda_0(t)$ *of the instanton-type solution* $\lambda_J(t,\eta;\alpha,\beta)$. *In what follows the other double zero is denoted by* $\kappa(t)$. *Hence* (SL_J) *has two double turning points* $x = \lambda_0(t)$ *and* $x = \kappa(t)$.

(ii) *When t lies on a Stokes curve of (P_J) emanating from a double turning point $t = \tau_d$, there exists a Stokes curve of (SL_J) that connects the two double turning points $x = \lambda_0(t)$ and $x = \kappa(t)$ of (SL_J).*

Using this Proposition 2.6 of geometric character, we construct the transformations in the following manner. (In what follows we again adopt the convention of putting $\tilde{}$ to the variables relevant to (P_J) and (SL_J) to distinguish them from those relevant to $(P_{\mathrm{II,deg}})$ and $(SL_{\mathrm{II,deg}})$.) Let $\tilde{t} = \tilde{\sigma}$ be a point on a Stokes curve of (P_J) emanating from a double turning point $\tilde{t} = \tilde{\tau}_d$ and let $\tilde{\gamma}$ denote a Stokes curve of (SL_J) that connects the two double turning points $\tilde{x} = \tilde{\lambda}_0(\tilde{t})$ and $\tilde{x} = \tilde{\kappa}(\tilde{t})$ at $\tilde{t} = \tilde{\sigma}$ (whose existence is guaranteed by Proposition 2.6, (ii)). Then we can construct an invertible formal transformation $(x(\tilde{x}, \tilde{t}, \eta), t(\tilde{t}, \eta))$ which brings the simultaneous equations (SL_J) and (D_J) into $(SL_{\mathrm{II,deg}})$ and $(D_{\mathrm{II,deg}})$ in a neighborhood of $\tilde{\gamma} \times \{\tilde{\sigma}\}$. That is, we have

Theorem 2.7. *Under the above geometric situation there exist a neighborhood $\tilde{U}$ of the Stokes curve $\tilde{\gamma}$, a neighborhood $\tilde{V}$ of $\tilde{\sigma}$, and a formal coordinate transformation*

$$x = x(\tilde{x}, \tilde{t}, \eta) = \sum_{j \geq 0} \eta^{-j/2} x_{j/2}(\tilde{x}, \tilde{t}, \eta), \tag{2.28}$$

$$t = t(\tilde{t}, \eta) = \sum_{j \geq 0} \eta^{-j/2} t_{j/2}(\tilde{t}, \eta) \tag{2.29}$$

with $x_{j/2}(\tilde{x}, \tilde{t}, \eta)$ and $t_{j/2}(\tilde{t}, \eta)$ being holomorphic on $\tilde{U} \times \tilde{V}$ and $\tilde{V}$, respectively, for which the following conditions (i) $\sim$ (v) *are satisfied:*

(i) *The function $x_0(\tilde{x}, \tilde{t}, \eta)$ is independent of η and $\partial x_0/\partial \tilde{x}$ never vanishes on $\tilde{U} \times \tilde{V}$.*

(ii) *The function $t_0(\tilde{t}, \eta)$ is also independent of η and $dt_0/d\tilde{t}$ never vanishes on $\tilde{V}$.*

(iii) *$x_0(\tilde{x}, \tilde{t})$ and $t_0(\tilde{t})$ satisfy*

$$x_0(\tilde{\lambda}_0(\tilde{t}), \tilde{t}, \eta) = \lambda_0(t_0(\tilde{t})) = \sqrt{-\frac{t_0(\tilde{t})}{2}}, \tag{2.30}$$

$$x_0(\tilde{\kappa}_0(\tilde{t}), \tilde{t}, \eta) = -\lambda_0(t_0(\tilde{t})) = -\sqrt{-\frac{t_0(\tilde{t})}{2}}. \tag{2.31}$$

(iv) *$x_{1/2}$ and $t_{1/2}$ identically vanish.*

(v) *If* $\psi(x, t, \eta)$ *is a WKB solution of* $(SL_{\mathrm{II,deg}})$ *that satisfies* $(D_{\mathrm{II,deg}})$ *also, then* $\tilde{\psi}(\tilde{x}, \tilde{t}, \eta)$ *defined by*

$$\tilde{\psi}(\tilde{x}, \tilde{t}, \eta) = \left(\frac{\partial x(\tilde{x}, \tilde{t}, \eta)}{\partial \tilde{x}}\right)^{-1/2} \psi(x(\tilde{x}, \tilde{t}, \eta), t(\tilde{t}, \eta), \eta) \quad (2.32)$$

satisfies both (SL_J) *and* (D_J) *on* $\tilde{U} \times \tilde{V}$.

The transformations (2.25) and (2.26) that provide a local equivalence (2.27) between $\tilde{\lambda}_J(\tilde{t}, \eta; \tilde{\alpha}, \tilde{\beta})$ and $\lambda_{\mathrm{II,deg}}(t, \eta; \alpha, \beta)$ in Theorem 2.5 are given by the semi-global transformation (2.28) and (2.29) constructed in Theorem 2.7. Otherwise stated, by considering a transformation for the underlying system (SL_J) and (D_J) of linear differential equations, we can find a transformation of the Painlevé equation (P_J). This is a sketch of the proof of Theorem 2.5. The details will be discussed in our forthcoming paper.

3 Transformation near a simple pole

As was outlined in [10], the transformation theory near a simple pole, *i.e.*, Theorem 1.4, is proved in a parallel way to the case of the transformation theory near a double turning point discussed in Section 2. In this section we briefly review the discussion of [10] to explain the transformation near a simple pole.

In view of the list of Painlevé equations (Table 1) we readily find that the Painlevé equations (P_J) have the following singular points:

$$\begin{aligned} &(P_{\mathrm{I}}),\ (P_{\mathrm{II}}),\ (P_{\mathrm{IV}}) &&: \{\infty\}, \\ &(P_{\mathrm{III'}}),\ (P_{\mathrm{III'(D7)}}),\ (P_{\mathrm{III'(D8)}}),\ (P_{\mathrm{V}}) &&: \{0, \infty\}, \\ &(P_{\mathrm{VI}}) &&: \{0, 1, \infty\}. \end{aligned} \quad (3.1)$$

Among them a pair of a Painlevé equation and its singular point contained in the following list is of “the first kind” or of “regular singular type”.

$$\begin{aligned} &((P_{\mathrm{III'}}), 0),\ ((P_{\mathrm{III'(D7)}}), 0),\ ((P_{\mathrm{III'(D8)}}), 0),\ ((P_{\mathrm{V}}), 0), \\ &((P_{\mathrm{VI}}), 0),\ ((P_{\mathrm{VI}}), 1),\ ((P_{\mathrm{VI}}), \infty). \end{aligned} \quad (3.2)$$

At a singular point of the first kind, in addition to a double pole type 0-parameter solution, there exists a simple pole type 0-parameter solution, that is, for any pair $((P_J), \tau_s)$ in (3.2), there exists a 0-parameter solution whose top term $\lambda_0(t)$ has a branch point at $t = \tau_s$ and satisfies

$$\frac{\partial F_J}{\partial \lambda}(\lambda_0(t), t) = O((t - \tau_s)^{-3/2}) \quad \text{as} \quad t \to \tau_s, \quad (3.3)$$

where $F_J(\lambda, t)$ denotes the coefficient of η^2 in the expression of (P_J). Note that the condition (3.3) guarantees that the corresponding linearized equation (ΔP_J) of (P_J) at the 0-parameter solution in question has a simple pole type singularity at $t = \tau_s$ after a new independent variable $\tilde{t} = (t - \tau_s)^{1/2}$, which is a local parameter of the Riemann surface of $\lambda_0(t)$ at $t = \tau_s$, is introduced. Consequently, if $((P_J), \tau_s)$ is a simple pole, only one Stokes curve of (P_J) emanates from $t = \tau_s$.

Using the top term $\lambda_0(t)$ of a simple pole type 0-parameter solution, we can also construct a 2-parameter instanton-type solution $\lambda_J(t, \eta; \alpha, \beta)$ of simple pole type for each pair $((P_J), \tau_s)$ listed in (3.3). The problem we want to discuss is then to develop transformation theory for these instanton-type solutions $\lambda_J(t, \eta; \alpha, \beta)$ of simple pole type. The precise formulation of the main result (*i.e.*, Theorem 1.4) in this case is the following theorem (where we again adopt the convention of putting $\tilde{}$ to the variables relevant to (P_J) to distinguish them from those relevant to $(P_{\mathrm{III}'(\mathrm{D8})})$).

Theorem 3.1. *Let $\tilde{\lambda}_J(\tilde{t}, \eta; \tilde{\alpha}, \tilde{\beta})$ be a 2-parameter instanton-type solution of simple pole type for one of the pairs $((P_J), \tilde{\tau}_s)$ of a Painlevé equation and its singular point listed in* (3.2). *Let $\tilde{\sigma}$ be a point on a Stokes curve emanating from $\tilde{\tau}_s$. Then we can find a neighborhood $\tilde{V}$ of $\tilde{\sigma}$ and a 2-parameter instanton-type solution $\lambda_{\mathrm{III}'(\mathrm{D8})}(t, \eta; \alpha, \beta)$ of $(P_{\mathrm{III}'(\mathrm{D8})})$ such that $\tilde{\lambda}_J(\tilde{t}, \eta; \tilde{\alpha}, \tilde{\beta})$ is formally transformed to $\lambda_{\mathrm{III}'(\mathrm{D8})}(t, \eta; \alpha, \beta)$ in $\tilde{V}$. To be more specific, there exist a formal transformation $t = t(\tilde{t}, \eta)$ of an independent variable and a formal transformation $x = x(\tilde{x}, \tilde{t}, \eta)$ of an unknown function of the form*

$$t(\tilde{t}, \eta) = \sum_{j \geq 0} \eta^{-j/2} t_{j/2}(\tilde{t}, \eta), \tag{3.4}$$

$$x(\tilde{x}, \tilde{t}, \eta) = \sum_{j \geq 0} \eta^{-j/2} x_{j/2}(\tilde{x}, \tilde{t}, \eta), \tag{3.5}$$

where $t_{j/2}$ and $x_{j/2}$ are holomorphic in both $\tilde{x}$ and $\tilde{t}$, that satisfy the following relation:

$$x(\tilde{\lambda}_J(\tilde{t}, \eta; \tilde{\alpha}, \tilde{\beta}), \tilde{t}, \eta) = \lambda_{\mathrm{III}'(\mathrm{D8})}(t(\tilde{t}, \eta), \eta; \alpha, \beta). \tag{3.6}$$

Thus $(P_{\mathrm{III}'(\mathrm{D8})})$ can be thought of as a canonical equation of Painlevé equations near a simple pole.

The proof of Theorem 3.1 is done in a parallel way to that of Theorem 2.5. We again make use of the fact that a Painlevé equation (P_J) is equivalent to the compatibility condition of (SL_J) and (D_J) given in Section 2. A key geometric proposition in this case is the following

Proposition 3.2. *Suppose that an instanton-type solution $\lambda_J(t,\eta;\alpha,\beta)$ of simple pole type of (P_J) is substituted into the coefficients of (SL_J). Then the following hold:*

(i) *The top term (with respect to η^{-1}) Q_0 of the potential Q_J of (SL_J) has a double zero at $x=\lambda_0(t)$, that is, (SL_J) has a double turning point at $x=\lambda_0(t)$.*

(ii) *When t lies on a Stokes curve of (P_J) emanating from a simple pole type singular point τ_s in question, there exists a Stokes curve of (SL_J) that starts from $\lambda_0(t)$ and returns to $\lambda_0(t)$ after encircling several singular points and/or turning points of (SL_J).*

For example, in the case of the canonical equation, *i.e.*, the most degenerate third Painlevé equation $(P_{\mathrm{III}'(\mathrm{D8})})$,

$$\lambda^{(0)}_{\mathrm{III}'(\mathrm{D8})}(t,\eta)=\sqrt{t} \tag{3.7}$$

is a 0-parameter solution and the linearized equation of $(P_{\mathrm{III}'(\mathrm{D8})})$ at this 0-parameter solution is given by

$$\frac{d^2}{dt^2}\Delta\lambda=\eta^2\left(\frac{2}{t^{3/2}}-\eta^{-2}\frac{1}{4t^2}\right)\Delta\lambda. \qquad (\Delta P_{\mathrm{III}'(\mathrm{D8})})$$

Hence $t=0$ is a simple pole type singularity (and a unique turning point) of $(P_{\mathrm{III}'(\mathrm{D8})})$ and only one Stokes curve

$$\{t\in\mathbb{C}\mid \arg t=4n\pi\ (n\in\mathbb{Z})\} \tag{3.8}$$

(*i.e.*, the positive real axis) emanates from $t=0$. Since the potential $Q_{\mathrm{III}'(\mathrm{D8})}$ of the associated linear equation $(SL_{\mathrm{III}'(\mathrm{D8})})$ has the form

$$\begin{aligned} Q_{\mathrm{III}'(\mathrm{D8})} &= \frac{t}{2x^3}+\frac{1}{2x}+\frac{\lambda^2}{x^2}\left[\nu^2-\left(\frac{t}{2\lambda^3}+\frac{1}{2\lambda}\right)\right] \\ &\quad -\eta^{-1}\lambda\nu\left[\frac{1}{x^2}+\frac{1}{x(x-\lambda)}\right]+\eta^{-2}\frac{3}{4(x-\lambda)^2} \end{aligned} \tag{3.9}$$

(where $\nu=\eta^{-1}(td\lambda/dt+\lambda)/(2\lambda^2)$), its top term $Q_0(x,t)$ becomes

$$Q_0(x,t)=\frac{(x-\sqrt{t})^2}{2x^3} \tag{3.10}$$

after the substitution of an instanton-type solution of simple pole type of $(P_{\mathrm{III}'(\mathrm{D8})})$ beginning with the leading term $\lambda_0(t):=\sqrt{t}$. Using (3.10), we thus find that when t lies on a Stokes curve of $(P_{\mathrm{III}'(\mathrm{D8})})$, *i.e.*, when

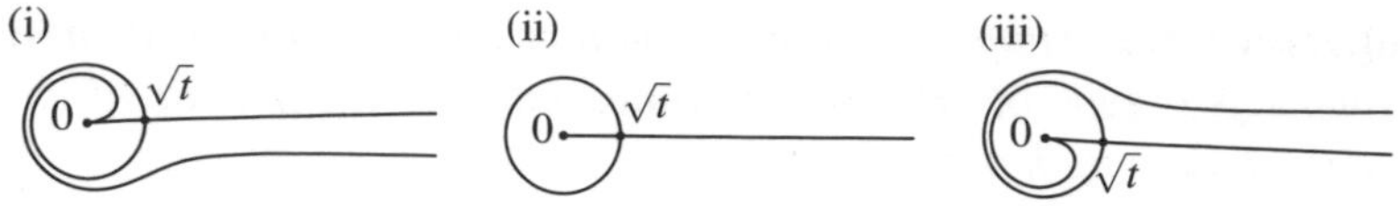

Figure 3.1. Configuration of Stokes curves of $(SL_{\mathrm{III}'(\mathrm{D8})})$ in the case of (i) $\arg t > 0$, (ii) $\arg t = 0$, and (iii) $\arg t < 0$.

$t > 0$, a circle $\{|x| = \sqrt{t}\}$ is a Stokes curve of $(SL_{\mathrm{III}'(\mathrm{D8})})$ that starts from $\sqrt{t}$, encircles the simple pole $t = 0$, and returns to $\sqrt{t}$, as is indicated in Figure 3.1, (ii).

In parallel to Theorem 2.7, near a Stokes curve γ of (SL_J) that starts from a double turning point $\lambda_0(t)$ and returns to $\lambda_0(t)$ at a point $t = \sigma$ on a Stokes curve of (P_J) emanating from a simple pole whose existence is guaranteed by Proposition 3.2, (ii), we can construct an invertible formal transformation which brings (SL_J) and (D_J) into $(SL_{\mathrm{III}'(\mathrm{D8})})$ and $(D_{\mathrm{III}'(\mathrm{D8})})$ in a neighborhood of $\gamma \times \{\sigma\}$. That is, we have

Theorem 3.3. *Let $((P_J), \tilde{\tau}_s)$ be one of the pairs in the list* (3.2) *and $\tilde{\sigma}$ a point on a Stokes curve emanating from $\tilde{\tau}_s$. Suppose that an instanton-type solution $\tilde{\lambda}_J(\tilde{t}, \eta; \tilde{\alpha}, \tilde{\beta})$ of simple pole type of (P_J) is substituted into the coefficients of (SL_J). Then there exist a neighborhood $\tilde{U}$ of a Stokes curve $\tilde{\gamma}$ of (SL_J) that starts from and returns to $\tilde{\lambda}_0(\tilde{t})$ at $\tilde{t} = \tilde{\sigma}$, a neighborhood $\tilde{V}$ of $\tilde{\sigma}$, and a formal coordinate transformation of the form*

$$x = x(\tilde{x}, \tilde{t}, \eta) = \sum_{j \geq 0} \eta^{-j/2} x_{j/2}(\tilde{x}, \tilde{t}, \eta), \tag{3.11}$$

$$t = t(\tilde{t}, \eta) = \sum_{j \geq 0} \eta^{-j/2} t_{j/2}(\tilde{t}, \eta) \tag{3.12}$$

with $x_{j/2}(\tilde{x}, \tilde{t}, \eta)$ and $t_{j/2}(\tilde{t}, \eta)$ being holomorphic on $\tilde{U} \times \tilde{V}$ and $\tilde{V}$, respectively, for which the following conditions (i) $\sim$ (v) *are satisfied:*

(i) *The function $x_0(\tilde{x}, \tilde{t}, \eta)$ is independent of η and $\partial x_0 / \partial \tilde{x}$ never vanishes on $\tilde{U} \times \tilde{V}$.*
(ii) *The function $t_0(\tilde{t}, \eta)$ is also independent of η and $dt_0 / d\tilde{t}$ never vanishes on $\tilde{V}$.*
(iii) *$x_0(\tilde{x}, \tilde{t})$ and $t_0(\tilde{t})$ satisfy*

$$x_0(\tilde{\lambda}_0(\tilde{t}), \tilde{t}, \eta) = \lambda_0(t_0(\tilde{t})) = \sqrt{t_0(\tilde{t})}. \tag{3.13}$$

(iv) *$x_{1/2}$ and $t_{1/2}$ identically vanish.*

(v) *If* $\psi(x, t, \eta)$ *is a WKB solution of* $(SL_{\mathrm{III}'(\mathrm{D}8)})$ *that satisfies* $(D_{\mathrm{III}'(\mathrm{D}8)})$ *also, then* $\tilde{\psi}(\tilde{x}, \tilde{t}, \eta)$ *defined by*

$$\tilde{\psi}(\tilde{x}, \tilde{t}, \eta) = \left(\frac{\partial x(\tilde{x}, \tilde{t}, \eta)}{\partial \tilde{x}} \right)^{-1/2} \psi(x(\tilde{x}, \tilde{t}, \eta), t(\tilde{t}, \eta), \eta) \quad (3.14)$$

satisfies both (SL_J) *and* (D_J) *on* $\tilde{U} \times \tilde{V}$.

The semi-global transformation (3.11) and (3.12) constructed in Theorem 3.3 again provides a local equivalence (3.6) between $\tilde{\lambda}_J(\tilde{t}, \eta; \tilde{\alpha}, \tilde{\beta})$ and $\lambda_{\mathrm{III}'(\mathrm{D}8)}(t, \eta; \alpha, \beta)$ in Theorem 3.1. This is a sketch of the proof of Theorem 3.1. The details will be discussed in our forthcoming paper.

In the case of the canonical equation $(P_{\mathrm{III}'(\mathrm{D}8)})$, we can explicitly compute the monodromy data of the associated linear equation $(SL_{\mathrm{III}'(\mathrm{D}8)})$ by using exact WKB analysis for linear differential equations. Combining this computation with Proposition 3.2, we obtain the following connection formula for instanton-type solutions of $(P_{\mathrm{III}'(\mathrm{D}8)})$ on its Stokes curve $\arg t = 0$: Let $\lambda(t, \eta; \alpha, \beta)$ and $\lambda(t, \eta; \alpha', \beta')$ be instanton-type solutions of $(P_{\mathrm{III}'(\mathrm{D}8)})$ in the region $\Omega_- = \{\arg t < 0\}$ and $\Omega_+ = \{\arg t > 0\}$, respectively. If $\lambda(t, \eta; \alpha', \beta')$ is the analytic continuation of $\lambda(t, \eta; \alpha, \beta)$ across the Stokes curve $\arg t = 0$, then we have

$$\frac{\alpha' \, 2^{-2\alpha'\beta'}}{\Gamma(-2\alpha'\beta' + 1)} = \frac{\alpha \, 2^{-2\alpha\beta}}{\Gamma(-2\alpha\beta + 1)}, \quad (3.15)$$

$$\begin{aligned} &\frac{i\beta' \, 2^{2\alpha'\beta'}}{\Gamma(2\alpha'\beta' + 1)} + e^{2i\pi\alpha'\beta'} \frac{\alpha' \, 2^{-2\alpha'\beta'}}{\Gamma(-2\alpha'\beta' + 1)} \\ &= \frac{i\beta \, 2^{2\alpha\beta}}{\Gamma(2\alpha\beta + 1)} - e^{-2i\pi\alpha\beta} \frac{\alpha \, 2^{-2\alpha\beta}}{\Gamma(-2\alpha\beta + 1)}. \end{aligned} \quad (3.16)$$

See [11, Section 5] for the computation of the monodromy data of $(SL_{\mathrm{III}'(\mathrm{D}8)})$. Theorem 3.1 then suggests that the same connection formula as (3.15) and (3.16) should hold also for instanton-type solutions of simple pole type of (P_J) listed in (3.2) on its Stokes curve emanating from a simple pole.

References

[1] T. Aoki, T. Kawai. and Y. Takei, "WKB Analysis of Painlevé Transcendents with a Large Parameter, II", Structure of Solutions of Differential Equations, World Scientific Publishing, 1996, 1–49.

[2] M. Jimbo, T. Miwa. and K. Ueno, *Monodromy preserving deformation of linear ordinary differential equations with rational coefficients*, I, Phys. D **2** (1981), 306–352.

[3] T. Kawai and Y. Takei, *WKB analysis of Painlevé transcendents with a large parameter*, I, Adv. Math. **118** (1996), 1–33.

[4] T. Kawai and Y. Takei, *WKB analysis of Painlevé transcendents with a large parameter*, III, Adv. Math. **134** (1998), 178–218.

[5] T. Kawai and Y. Takei, *Algebraic Analysis of Singular Perturbation Theory*, AMS, 2005.

[6] T. Koike, *On the exact WKB analysis of second order linear ordinary differential equations with simple poles*, Publ. Res. Inst. Math. Sci. **36** (2000), 297–319.

[7] Y. Ohyama, H. Kawamuko, H. Sakai and K. Okamoto, *Studies on the Painlevé equations*, V, J. Math. Sci. Univ. Tokyo **13** (2006), 145–204.

[8] K. Okamoto, *Isomonodromic deformation and Painlevé equations, and the Garnier systems*, J. Fac. Sci. Univ. Tokyo Sect. IA **33** (1986), 575–618.

[9] Y, Takei, "An Explicit Description of the Connection Formula for the First Painlevé Equation", Toward the Exact WKB Analysis of Differential Equations, Linear or Non-Linear, Kyoto Univ. Press, 2000, 271–296.

[10] Y, Takei, "On the Role of the Degenerate Third Painlevé Equation of Type (D8) in the Exact WKB Analysis", preprint (RIMS-1660), 2009.

[11] Y. Takei and H. Wakako, *Exact WKB analysis for the degenerate third Painlevé equation of type* (D_8), Proc. Japan Acad., Ser. A **83** (2007), 63–68.

CRM Series
Publications by the Ennio De Giorgi Mathematical Research Center Pisa

The Ennio De Giorgi Mathematical Research Center in Pisa, Italy, was established in 2001 and organizes research periods focusing on specific fields of current interest, including pure mathematics as well as applications in the natural and social sciences like physics, biology, finance and economics. The CRM series publishes volumes originating from these research periods, thus advancing particular areas of mathematics and their application to problems in the industrial and technological arena.

Published volumes

1. Matematica, cultura e società 2004 (2005). ISBN 88-7642-158-0
2. Matematica, cultura e società 2005 (2006). ISBN 88-7642-188-2
3. M. GIAQUINTA, D. MUCCI, *Maps into Manifolds and Currents: Area and* $W^{1,2}$-, $W^{1/2}$-, BV*-Energies*, 2006. ISBN 88-7642-200-5
4. U. ZANNIER (editor), *Diophantine Geometry.* Proceedings, 2005 (2007). ISBN 978-88-7642-206-5
5. G. MÉTIVIER, *Para-Differential Calculus and Applications to the Cauchy Problem for Nonlinear Systems*, 2008. ISBN 978-88-7642-329-1
6. F. GUERRA, N. ROBOTTI, *Ettore Majorana. Aspects of his Scientific and Academic Activity*, 2008. ISBN 978-88-7642-331-4
7. Y. CENSOR, M. JIANG, A. K. LOUISR (editors), *Mathematical Methods in Biomedical Imaging and Intensity-Modulated Radiation Therapy (IMRT)*, 2008. ISBN 978-88-7642-314-7
8. M. ERICSSON, S. MONTANGERO (editors), *Quantum Information and Many Body Quantum systems.* Proceedings, 2007 (2008). ISBN 978-88-7642-307-9
9. M. NOVAGA, G. ORLANDI (editors), *Singularities in Nonlinear Evolution Phenomena and Applications.* Proceedings, 2008 (2009). ISBN 978-88-7642-343-7

Matematica, cultura e società 2006 (2009). ISBN 88-7642-315-4

10. H. HOSNI, F. MONTAGNA (editors), *Probability, Uncertainty and Rationality*, 2010. ISBN 978-88-7642-347-5
11. L. AMBROSIO (editor), *Optimal Transportation, Geometry and Functional Inequalities*, 2010. ISBN 978-88-7642-373-4
12. O. COSTIN, F. FAUVET, F. MENOUS, D. SAUZIN (editors), *Asymptotics in Dynamics, Geometry and PDEs; Generalized Borel Summation*, vol. I, 2011. ISBN 978-88-7642-374-1, e-ISBN 978-88-7642-379-6
13. O. COSTIN, F. FAUVET, F. MENOUS, D. SAUZIN (editors), *Asymptotics in Dynamics, Geometry and PDEs; Generalized Borel Summation*, vol. II, 2011. ISBN 978-88-7642-376-2, e-ISBN 978-88-7642-377-2

Volumes published earlier

Dynamical Systems. Proceedings, 2002 (2003)
Part I: *Hamiltonian Systems and Celestial Mechanics.*
ISBN 978-88-7642-259-1
Part II: *Topological, Geometrical and Ergodic Properties of Dynamics.*
ISBN 978-88-7642-260-1

Matematica, cultura e società 2003 (2004). ISBN 88-7642-129-7

Ricordando Franco Conti, 2004. ISBN 88-7642-137-8

N.V. KRYLOV, *Probabilistic Methods of Investigating Interior Smoothness of Harmonic Functions Associated with Degenerate Elliptic Operators,* 2004. ISBN 978-88-7642-261-1

Phase Space Analysis of Partial Differential Equations. Proceedings, vol. I, 2004 (2005). ISBN 978-88-7642-263-1

Phase Space Analysis of Partial Differential Equations. Proceedings, vol. II, 2004 (2005). ISBN 978-88-7642-263-1

Fotocomposizione "CompoMat" Loc. Braccone, 02040 Configni (RI) Italy
Finito di stampare nel mese di luglio 2011
dalla CSR, Via di Pietralata 157, 00158 Roma